Statistical Methods for Quality Improvement

WILEY SERIES IN PROBABILITY AND STATISTICS
APPLIED PROBABILITY AND STATISTICS SECTION

Established by WALTER A. SHEWHART and SAMUEL S. WILKS

A complete list of the titles in this series appears at the end of this volume.

Statistical Methods for Quality Improvement

Second Edition

THOMAS P. RYAN
Department of Statistics
Case Western Reserve University

A Wiley-Interscience Publication
JOHN WILEY & SONS, INC.
New York • Chichester • Weinheim • Brisbane • Singapore • Toronto

This book is printed on acid-free paper. ♾

Published simultaneously in Canada.

Library of Congress Cataloging-in-Publication Data:

Ryan, Thomas P.
Statistical methods for quality improvement / by Thomas P. Ryan.—2nd ed.
p. cm.
"A Wiley-Interscience publication."
Includes bibliographical references and index.
ISBN 0-471-19775-0 (alk. paper)
1. Quality control—Statistical methods. 2. Process control—Statistical methods. I. Title.
TS156.R9 2000
658.5′62—dc21 99-15784

Printed in the United States of America.

10 9 8 7 6 5 4

To my parents

Contents

PART III BEYOND CONTROL CHARTS: GRAPHICAL AND STATISTICAL METHODS

Preface

There have been many developments in statistical process control (SPC) during the past ten years, and many of those developments have been incorporated into this edition.

In particular, major changes were made to the chapters on process capability and multivariate control charts as much material has been added with the result that these chapters are now considerably longer.

Chapter 10 has also been considerably expanded and now includes sections on short-run control charts, pre-control, autocorrelated data, nonparametric control charts, and various other topics that were not covered in the first edition.

Chapter 13 on the design of experiments is noticeably longer, in part because of the addition of material on robust design considerations. Chapter 14 on Taguchi methods and alternatives while retaining the material from the first edition now includes considerable discussion and illustration of combined arrays and product arrays.

Chapter 17 is a new chapter on using SPC tools together as is done in Six Sigma programs. These programs are also discussed in the chapter.

Other significant additions include material on probability-type limits for attribute charts and cause-selecting (regression-type) control charts.

In addition to new material, retained material from the first edition has been extensively reorganized. In particular, cumulative sum (CUSUM) and exponentially weighted moving average (EWMA) methods are now in a separate chapter, and are covered in considerable detail.

The first edition has been used in college courses as well as in short courses. Chapters 4–10 of the second edition could form the basis for a course that covers control charts and process capability. Instructors who wish to cover only basic concepts might cover Chapters 1, 2, as much of 3 as is necessary, 4, 5, and 6, and selectively choose from 7, 8 and 10.

The book might also be used in a course on design of experiments, especially a special topics course. There are some topics in Chapters 13 and 14 that have not been covered in experimental design texts, and evolutionary operation and analysis of means (Chapters 15 and 16, respectively) are not covered to any extent in design texts. So an atypical design course could be put together using Chapters 13–16 as a basis.

I am indebted to the researchers who have made many important contributions during the past 10 years, and I am pleased to present their work, in addition to my own work.

Many people have commented on the strengths and weaknesses of the first edition and their comments were considered for the second edition.

I am also indebted to Bill Woodall and Dennis Lin who made many helpful suggestions when the second edition was in manuscript form, and Rich Charnigo's proofreading assistance is gratefully acknowledged. I also appreciate the support of my editor at Wiley, Steve Quigley, and the work of the production people, especially Rosalyn Farkas.

THOMAS P. RYAN

July 1999

Preface to the First Edition

A moderate number of books have been written on the subject of statistical quality control, which in recent years has also been referred to as *statistical process control* (SPC). These range from books that contain only the basic control charts to books that also contain material on acceptance sampling and selected statistical methods such as regression and analysis of variance.

Statistical Methods for Quality Improvement was written in recognition of the fact that quality improvement requires the use of more than just control charts. In particular, it would be difficult to keep a particular process characteristic "in control" without some knowledge of the factors affecting that characteristic. Consequently, chapters 13–16 were written to provide insight into statistically designed experiments and related topics.

The first two chapters provide an overview of the use of statistics in quality improvement in the United States and Japan. Chapter 3 presents statistical distributions that are needed for the rest of the book, and also reviews basic concepts in probability and statistics. Basic control chart principles are discussed in Chapter 4, and Chapters 5, 6, 8, and 9 contain the material on the various control charts. This material has several unique features. In particular, there is some emphasis on cumulative sum (CUSUM) procedures, and an entire chapter (Chapter 9) is devoted to multivariate charts. Chapter 7 discusses the commonly used process capability indices and compares them. The bibliography of control chart applications at the end of Chapter 10 is another unique feature of the book.

Quality improvement practitioners are beginning to recognize what can be accomplished using statistical design of experiments, but progress has been slow. With this in mind, Chapter 13 was written to show what can be accomplished using experimental design principles.

In recent years there has been much interest and discussion regarding a set of statistical and nonstatistical tools referred to as *Taguchi methods*. These are critically examined in Chapter 14. Evolutionary Operation is presented in Chapter 15; Chapter 16 is an updated treatment of Analysis of Means. The latter is a valuable tool that allows nonstatisticians, in particular, to analyze data from designed experiments.

In general, there has been a conscious attempt to bring the reader up to date in regard to the various topics that are presented in each chapter. There was also a concerted effort to use simple heuristics and intuitive reasoning, rather than relying heavily upon mathematical and statistical formalism and symbolism. The control chart material, in particular, has also been written under the assumption that a sizable percentage of readers will have access to a computer for control charting.

Chapters 4–10 could be used for a one-semester course devoted exclusively to control charts, and Chapters 13–16 could from the core for a course on design of experiments. Short-course instructors will also find ample material from which to pick and choose.

A book of this type is the end product of the combined efforts of many people, even though the book has only one author. The architects of many of the statistical tools presented herein have indirectly contributed greatly to the quality of the book. In particular, Jim Lucas's work on cumulative sum procedures is presented in detail for the first time in a statistics book, and the same can be said for Frank Alt's work on multivariate charts. I have also contributed some new control chart procedures, which hopefully will be viewed as improvements on the standard procedures.

Much of the material in the book has been presented in industrial short courses and college courses; the feedback from some of the participants has been valuable. There are also a number of colleagues who have read parts of the manuscript and have made helpful suggestions. Those deserving particular mention are Johannes Ledolter, Frank Alt, Jon Cryer, and Jim Lucas. The contributions of the editorial reviewers are also appreciated, as is the work of Joy Klammer who typed most of the manuscript. Permission from MINITAB, INC. to use MINITAB for generating certain tables is also gratefully acknowledged, as is permission from SQC SYSTEMS, INC. to use SQCS in producing many of the control charts and CUSUM tabulations that are contained in the book. Permission from various publications to reproduce certain materials is also appreciated, as are the efforts of the editorial and production people at Wiley, especially Isabel Stein and Shirley Thomas. Lastly, I am very much indebted to my editor, Bea Shube, whose patience and steadfast support made writing the book a less arduous task than it could have been, particularly during trying times.

THOMAS P. RYAN

Iowa City, Iowa
October, 1988

Statistical Methods for
Quality Improvement

PART I

Fundamental Quality Improvement and Statistical Concepts

CHAPTER 1

Introduction

This is a book about using statistical methods to improve quality. It is not a book about Total Quality Management (TQM), Total Quality Assurance (TQA), just-in-time (JIT) manufacturing, benchmarking, QS-9000, or the ISO 9000 series. In other words, the scope of the book is essentially restricted to statistical techniques. Although standards such as QS-9000 and ISO 9000 are *potentially* useful, they are oriented toward the *documentation* of quality problems, not the identification or eradication of problems. Furthermore, many people feel that companies tend to believe that all they need to do is acquire ISO 9000 certification, thus satisfying only a minimum requirement.

Statistical techniques, on the other hand, are useful for identifying trouble spots and their causes as well as predicting major problems before they occur. Then it is up to the appropriate personnel to take the proper corrective action.

The emphasis is on quality *improvement*, not quality control. On July 1, 1997, the American Society for Quality Control (ASQC) became simply the American Society for Quality (ASQ). The best choice for a new name is arguable, as some would undoubtedly prefer American Society for Quality Improvement (the choice of the late Bill Hunter, former professor of statistics at the University of Wisconsin). Nevertheless, the name change reflects an appropriate movement away from quality *control*.

What is quality? How do we know when we have it? Can we have too much quality? The "fitness-for-use" criterion is usually given in defining quality. Specifically, a quality product is defined as a product that meets the needs of the marketplace. Those needs are not likely to be static, however, and will certainly be a function of product quality. For example, if automakers build cars that are free from major repairs for 5 years, the marketplace is likely to accept this as a quality standard. However, if another automaker builds its cars in such a way that they will probably be trouble free for 7 years, the quality standard is likely to shift upward. This is what happened in the Western world some years ago as the marketplace discovered that Japanese products, in particular, are of high quality.

A company will know that it is producing high-quality products if those products satisfy the demands of the marketplace.

We could possibly have too much quality. What if we could build a car that would last for 50 years. Would anyone want to drive the same car for 50 years even if he or she lived long enough to do so? Obviously styles and tastes change. This is particularly true for high-technology products that might be obsolete after a year or two. How long should a personal computer be built to last?

In statistical terms, quality is largely determined by the amount of variability in what is being measured. Assume that the target for producing certain invoices is 15 days, with anything less than, say, 10 days being almost physically impossible. If records for a 6-month period showed that all invoices of this type were processed within 17 days, this invoice-processing operation would seem to be of high quality.

In general, the objective should be to reduce variability and to "hit the target" if target values exist for process characteristics. The latter objective has been influenced by Genichi Taguchi (see Chapter 14), who has defined quality as the "cost to society."

1.1 QUALITY AND PRODUCTIVITY

One impediment to achieving high quality has been the misconception of some managers that there is an inverse relationship between productivity and quality. Specifically, it has been believed (by some) that steps taken to improve quality will simultaneously cause a reduction in productivity.

This issue has been addressed by a number of authors, including Fuller (1986), who related that managers at Hewlett-Packard began to realize many years ago that productivity rose measurably when nonconformities (i.e., product defects) were reduced. This increase was partly attributable to a reduction in rework that resulted from the reduction of nonconformities. Other significant gains resulted from the elimination of problems such as the late delivery of materials. These various problems contribute to what the author terms "complexity" in the workplace, and he discusses ways to eliminate complexity so as to free the worker for productive tasks. Other examples of increased productivity resulting from improved quality can be found in Chapter 1 of Deming (1982).

1.2 QUALITY COSTS (OR DOES IT?)

It is often stated that "quality doesn't cost, it pays." Although Crosby (1979) said that quality is free (the title of his book) and reiterated this in Crosby (1996), companies such as Motorola and General Electric, which have launched massive training programs, would undoubtedly disagree. The large amount of money that GE has committed to a particular training program, Six Sigma, is discussed in, for example, the January 13, 1997 issue of the *Wall Street Journal*. Wall Street is beginning to recognize Six Sigma companies as companies that, for example, operate efficiently and have greater customer satisfaction. Six Sigma is discussed in detail in Chapter 17.

What is the real cost of a quality improvement program? That cost is impossible to determine precisely, since it would depend in part on the quality costs for a given time period without such a program as well as the costs of the program for the same time period. Obviously we cannot both have a program and not have a program at the same point in time, so the quality costs that would be present if the program were not in effect would have to be estimated from past data.

Such a comparison would not give the complete picture, however. Any view of quality costs that does not include the effect that a quality improvement program will have on sales and customers' perceptions is a myopic view of the subject. Should a supplier consider the cost of a statistical quality control program before deciding whether or not to institute such a program? The supplier may not have much choice if it is to remain a supplier. As a less extreme example, consider an industry that consists of 10 companies. If two of these companies implement a statistical quality improvement program and, as a result, the public soon perceives their products to be of higher quality than their competitors' products, should their competitors consider the cost of such a program before following suit? Definitely not, unless they can adequately predict the amount of lost sales and weigh that against the cost of the program.

1.3 THE NEED FOR STATISTICAL METHODS

Generally, statistical techniques are needed to determine if abnormal variation has occurred in whatever is being monitored, to determine changes in the values of process parameters, and to identify factors that are influencing process characteristics. Methods for achieving each of these objectives are discussed in subsequent chapters. Statistics is generally comparable to medicine in the sense that there are many subareas in statistics, just as there are many medical specialties. Quality "illnesses" generally can be cured and quality optimized only through the sagacious use of combinations of statistical techniques, as discussed in Chapter 17.

1.4 EARLY USE OF STATISTICAL METHODS FOR IMPROVING QUALITY

Although statistical methods have been underutilized and underappreciated in quality control/improvement programs for decades, such methods are extremely important. Occasionally their importance may even be overstated. In discussing the potential impact of statistical methods, Hoerl (1994) points out that Ishikawa (1985, pp. 14–15) stated the following: "One might even speculate that the second world war was won by quality control and by the utilization of modern statistics. Certain statistical methods researched and utilized by the Allied powers were so effective that they were classified as military secrets until the surrender of Nazi Germany." Although such a conclusion is clearly arguable, statistical methods did clearly play a role in World War II.

Shortly after the war, the American Society for Quality Control was formed in 1946; it published the journal *Industrial Quality Control*, the first issue of which had appeared in July 1944. In 1969 the journal was essentially split into two publications—the *Journal of Quality Technology* and *Quality Progress*. The former contains technical articles whereas the latter contains less technical articles and also has news items. The early issues of *Industrial Quality Control* contained many interesting articles on how statistical procedures were being used in firms in various industries, whereas articles in the *Journal of Quality Technology* are oriented more toward the proper use of existing procedures as well as the introduction of new procedures. Publication of *Quality Engineering* began in 1988, with case studies featured in addition to statistical methodology. The Annual Quality Congress has been held every year since the inception of the ASQC, and the proceedings of the meeting are published as the *ASQ Annual Quality Transactions*.

Other excellent sources of information include the Fall Technical Conference, which is jointly sponsored by ASQ and the American Statistical Association (ASA), the annual Quality and Productivity Research Conference, and the annual meetings of ASA, which are referred to as the Joint Statistical Meetings (JSM).

There are also various "applied" statistics journals that contain important articles relevant to industry, including *Technometrics*, published jointly by ASQ and ASA, *Quality and Reliability Engineering International, IIE Transactions, Applied Statistics (Journal of The Royal Statistical Society, Series C)*, and *The Statistician (Journal of the Royal Statistical Society, Series D)*. The latter two are British publications.

Readers interested in the historical development of statistical quality control in Great Britain are referred to Pearson (1935, 1973). An enlightening look at the early days of quality control practices in the United States, as seen through the eyes of Joseph M. Juran, can be found in Juran (1997). See also Montgomery (1996, pp. 10–11) for a chronology of some important events in the history of quality improvement.

1.5 INFLUENTIAL QUALITY EXPERTS

Walter A. Shewhart (1891–1967) came first. As discussed more fully in Chapter 2, he invented the idea of a control chart, with certain standard charts now commonly referred to as "Shewhart charts." Shewhart (1931) is still cited by many writers as an authoritative source on process control. The book was reprinted in 1980 by the ASQC. Shewhart (1939) was Shewhart's other well-known book.

W. Edwards Deming (1900–1993) was such a prominent statistician and quality and productivity consultant that his passing was noted on the front page of leading newspapers. His "14 points for management" for achieving quality have been frequently cited (and also changed somewhat over the years). It has been claimed that there are as many as eight versions. One version is as follows:

1. Create a constancy of purpose.
2. Adopt a new philosophy.
3. Cease dependence on inspection.
4. Work constantly to improve the system.
5. Break down barriers between departments.
6. Do not award business to suppliers solely on the basis of price.
7. Drive out fear.
8. Eliminate numerical goals, targets, and slogans.
9. Eliminate work standards and substitute leadership.
10. Institute a program of training and education for all employees.
11. Institute modern training methods.
12. Remove the barriers that make it difficult for employees to do their jobs.
13. Institute and practice modern methods of supervision.
14. Create a management climate that will facilitate the attainment of these objectives.

Although these 14 points are typically applied in industrial settings, they can be slightly modified and applied in other settings. For an application that is certainly far removed from manufacturing, Guenther (1997) gives a closely related list of 14 points for parenting.

There is one point of clarification that should be made. When Deming argued against target values, he was arguing against targets for production quotas, not target values for process characteristics. The use of target values for process characteristics is advocated and illustrated in Chapter 14.

Deming was constantly berating American management, believing that about 90% of quality problems were caused by management. Deming's views on the shortcomings of American management can be found in many places, including Chapter 2 of Deming (1986). In general, Deming claimed that management (1) emphasizes short-term thinking and quarterly profits rather than long-term strategies, (2) is inadequately trained and does not possess an in-depth knowledge of the company, and (3) is looking for quick results.

Deming has also been given credit for the PDCA (plan–do–check–act) cycle, although in his later years his preference was that it be called the PDSA cycle, with 'study' replacing 'check.' This has been termed *Deming's wheel*, but Deming referred to it as Shewhart's cycle. The cycle consists of planning a study, performing the study, checking or studying the results, and acting in accordance with what was learned from the study. See, for example, Cryer and Miller (1994) for additional information on the PDCA cycle.

Several books have been written about Deming; one of the best-known books was written by Mary Walton, a journalist (Walton, 1986). See also Walton (1990), which is a book of case studies, and Voehl (1995). The latter is an edited volume that contains chapters written by some prominent people in the field of quality improvement.

Joseph M. Juran (1904–) is another prominent quality figure. He is mentioned only briefly here, however, because his contributions have been to quality management rather than to the use of statistical methods for achieving quality improvement. His quality control handbook, which appropriately enough was renamed *Juran's Quality Control Handbook* when the fourth edition came out in 1988, does contain a few chapters on statistical techniques, however. The first edition was published in 1951 and has been used as a reference book by countless quality practitioners.

Eugene L. Grant (1897–1996) has not been accorded the status of other quality pioneers, but nevertheless deserves to be mentioned with the others in this section. In Struebing (1996), Juran is quoted as saying, "His contribution to statistical methodology was much greater than (W. Edwards) Deming's. Even though his impact on quality was profound and he was much more instrumental in advancing quality than Deming, the media—which overstated Deming's contribution—didn't publicize Grant's contributions." Grant has been described as a quiet worker who did not seek to extol his accomplishments. He was an academic who spent over 30 years on the faculty of Stanford University. In the field of quality improvement he was best known for his classic book *Statistical Quality Control*, first published in 1946. Recent editions of the book have been co-authored by Richard S. Leavenworth. The seventh edition was published in 1996. A very large number of copies of the book were sold through the various editions, but some observers felt that his teaching of statistical quality control during World War II contributed at least as much to the increase in the use of quality techniques as has his well-known book.

George E. P. Box (1919–) is not generally listed as a quality leader or "guru," but his contributions to statistical methods for improving quality are well known. His recent book, Box and Luceño (1997), extols the authors' ideas and suggested approaches for improving quality. The primary message of the book is that control charts and engineering process control should be used in tandem. This idea is discussed in Chapter 17. He is the author of several other books, the best known of which is Box, Hunter, and Hunter (1978). Box also had a column entitled *George's Corner* during the early years of the journal *Quality Engineering*. He was named an Honorary Member of the ASQ by the ASQ Board of Directors in 1997 in recognition of his contributions to quality improvement.

There are, of course, many other quality leaders, but they won't be listed here for fear of leaving someone out.

1.6 SUMMARY

Statistical methods should be used to identify unusual variation and to pinpoint the causes of such variation, whether it be for a manufacturing process or for general business. The use of statistical methods should produce improvements in quality, which, in turn, should result in increased productivity. The tools for accomplishing this are presented in Parts II and III.

REFERENCES

Box, G. E. P. and A. Luceño (1997). *Statistical Control by Monitoring and Feedback Adjustment.* New York: Wiley.

Box, G. E. P., W. G. Hunter, and J. S. Hunter (1978). *Statistics for Experimenters.* New York: Wiley.

Crosby, P. (1979). *Quality Is Free: The Art of Making Quality Certain.* New York: McGraw-Hill.

Crosby, P. (1996). *Quality Is Still Free: Making Quality Certain in Uncertain Times.* New York: McGraw-Hill.

Cryer, J. D. and R. B. Miller (1994). *Statistics for Business: Data Analysis and Modeling*, 2nd ed. Belmont, CA: Duxbury.

Deming, W. E. (1982). *Quality, Productivity, and Competitive Position.* Cambridge, MA: Massachusetts Institute of Technology, Center for Advanced Engineering Study.

Deming, W. E. (1986). *Out of the Crisis.* Cambridge, MA: Massachusetts Institute of Technology, Center for Advanced Engineering Study.

Fuller, F. T. (1986). Eliminating complexity from work: Improving productivity by enhancing quality. Report No. 17, Center for Quality and Productivity Improvement, University of Wisconsin — Madison.

Guenther, M. (1997). Letter to the Editor. *Quality Progress 30*(10): 12–14.

Hoerl, R. (1994). Enhancing the bottom line impact of statistical methods. W. J. Youden Memorial Address given at the 38th Annual Fall Technical Conference. *Chemical and Process Industries Division Newsletter, American Society for Quality Control*, Winter, pp. 1–9.

Ishikawa, K. (1985). *What Is Total Quality Control? The Japanese Way.* Englewood Cliffs, NJ: Prentice-Hall.

Juran, J. M. (1997). Early SQC: A historical supplement. *Quality Progress 30*(9): 73–81.

Juran, J. M., editor-in-chief, and F. M. Gryna, associate editor (1988). *Juran's Quality Control Handbook*, 4th ed. New York: McGraw-Hill.

Montgomery, D. C. (1996). *Introduction to Statistical Quality Control*, 3rd ed. New York: Wiley.

Pearson, E. S. (1935). *The Application of Statistical Methods to Industrial Standardisation and Quality Control.* London: British Standards Association.

Pearson, E. S. (1973). Some historical reflections on the introduction of statistical methods in industry. *The Statistician 22*(3): 165–179.

Shewhart, W. A. (1931). *Economic Control of Quality of Manufactured Product*. New York: Van Nostrand. (Reprinted in 1980 by the American Society for Quality Control.)

Shewhart, W. A. (1939). *Statistical Method from the Viewpoint of Quality Control*. Washington, DC: Graduate School, Department of Agriculture (editorial assistance by W. Edwards Deming).

Struebing, L. (1996). Eugene L. Grant: 1897–1996. *Quality Progress 29*(11): 81–83.

Voehl, F., ed. (1995). *Deming: The Way We Knew Him*. Delray Beach, FL: St. Lucie.

Walton, M. (1986). *The Deming Management Method*. New York: Dodd and Mead.

Walton, M. (1990). *Deming Management at Work*. New York: Putnam.

CHAPTER 2

Basic Tools for Improving Quality

Various statistical and nonstatistical tools have been used extensively in quality improvement work. In particular, there are seven simple tools that have often been referred to as "the seven basic tools," with the late Kaoru Ishikawa generally associated with the term. In particular, see Ishikawa (1976). The tools are:

1. Histogram
2. Pareto chart
3. Scatter plot
4. Control chart
5. Check sheet
6. Cause-and-effect diagram
7. Defect concentration diagram

The first four of these are statistical/graphical techniques. They are introduced here and some are covered in greater detail in subsequent chapters. The last three tools are discussed only in this chapter. It is important to realize that although many gains have been made using just these seven tools, there are other tools, such as experimental designs (see Chapter 13), that should additionally be used. See also the "seven new tools," which are discussed in Section 2.8.

2.1 HISTOGRAM

A *histogram* is a bar chart that shows the relative frequencies of observations in each of several classes. For example, Figure 2.1 is a histogram that might represent quality control data that have been grouped into seven classes, such as values of a process characteristic that have been obtained over time. A histogram is thus a pictorial display of the way the data are distributed over the various classes. As such, it can indicate, in particular, whether the data are distributed symmetrically or asymmetrically over the classes. This can be very useful information, as many control charts are based on the implicit assumption of a particular symmetric distribution, a normal distribution, for whatever is being charted. (The normal distribution is covered, along with other distributions, in Chapter 3.)

If we have a set of, say, 100 numerical values that were all obtained at the same time, we should address the question of determining a meaningful way to portray the data graphically so as to provide some insight into the process that generated the numbers. Assume that the 100 numbers are those given in Table 2.1. Such a table, by itself, tells us very little. By looking at Table 2.1, we can determine the largest value and the smallest value, and that is about all.

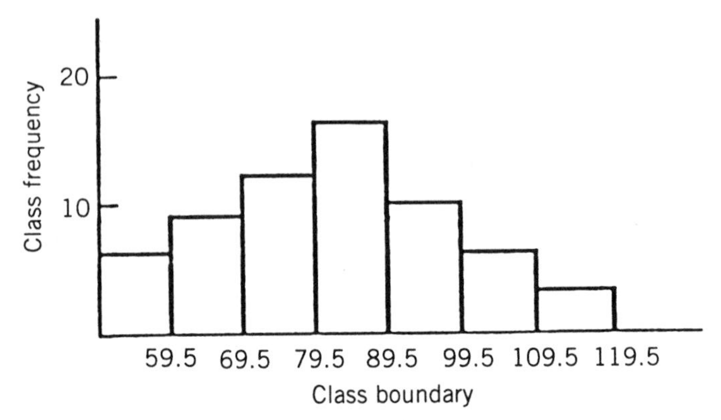

FIGURE 2.1 Histogram.

TABLE 2.1 100 Data Values

24	45	36	59	48
31	70	85	62	87
81	57	68	60	78
27	25	37	56	65
42	50	53	39	57
51	51	40	34	63
58	66	54	46	43
82	55	55	75	66
21	32	49	69	79
54	23	50	68	64
53	64	74	30	65
60	58	52	61	44
32	52	40	59	49
83	84	35	76	67
55	56	41	59	47
64	52	28	76	71
33	33	56	51	69
51	43	72	73	45
41	45	61	42	46
58	58	63	52	62

One or more good graphical displays of the data will tell us much more, however. A commonly used starting point in summarizing data is to put the data into classes and then to construct a histogram from the data that have been thus grouped. This is what is generally covered in the first week or two in an introductory statistics course. We will construct a histogram for the data in Table 2.1, but our choice of a histogram as the first graphical tool to illustrate should not be interpreted as an indication that a histogram is superior to other graphical tools. It is not. There are alternative displays, some of which are presented in Chapter 11, that have clear advantages over the histogram, particularly for small data sets. It is illustrated first simply because it is an often-used display that is well understood by both statisticians and nonstatisticians.

As indicated, a histogram is produced from grouped data. Before data can be grouped, however, there is an obvious need to determine the number of classes that is to be used. From Table 2.1 we can see that the smallest number is 21 and the largest number is 87, so it might seem reasonable to use the following set of classes: 20–29, 30–39, 40–49, . . ., 80–89. This selection of classes produces the *frequency distribution* given in Table 2.2, from which a histogram is then constructed and displayed as in Figure 2.2.

It can be observed that the histogram is simply a bar chart in which the height of each of the seven rectangles corresponds to the frequency of the class that the rectangle represents. Notice that the values along the horizontal axis of the histogram do not correspond to the values of the class intervals in Table 2.2. That is because these are *class boundaries*, which are defined as the average of adjacent class limits (e.g., 29.5 is the average of 29 and 30). To illustrate their use, we might think of the data in Table 2.1 as being rounded to the nearest integer so that values between 29.0 and 29.5 would be rounded down to 29 and thus appear in the first class, whereas values above 29.5 and less than 39.5 would be put in the second class. Also, if the class limits had been used to construct the histogram, there would have been gaps between the rectangles because there is a one-unit gap between 29 and 30, 39 and 40, and so on. If the classes are of equal width, which is generally desirable, the rectangles will then be of equal width.

TABLE 2.2 Frequency Distribution for the Data in Table 2.1

Class	Frequency
20–29	6
30–39	11
40–49	18
50–59	29
60–69	20
70–79	10
80–89	6

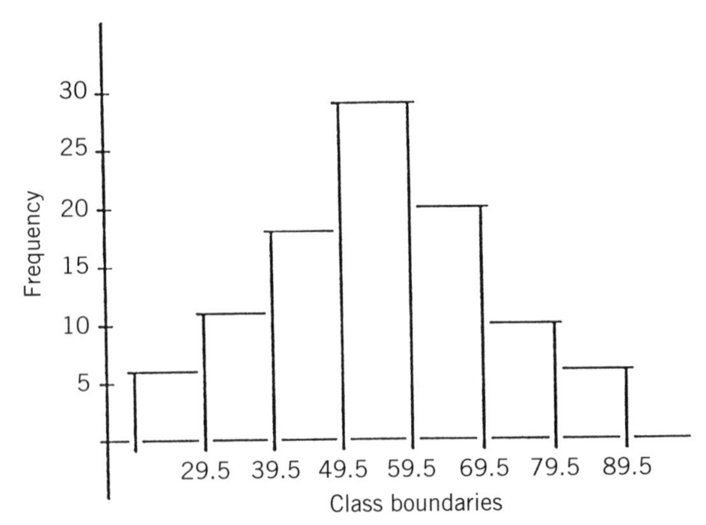

FIGURE 2.2 Histogram of the data in Table 2.1.

In this example the number of classes was implicitly determined from the selection of what seemed to be logical class intervals. The use of the latter is desirable whenever possible, but it is not always possible. What should we have done if there had been only 30 values rather than 100 but the largest and smallest values were still 21 and 87, respectively? If we tried to spread 30 values over seven classes, we might have some empty classes and/or the shape of the histogram could be rather flat. We should keep in mind that one of the main reasons for constructing a histogram is to provide some insight into the shape of the distribution of population values from which the sample values were obtained. We will have a distorted view of that shape if we use either too many or not enough classes.

Therefore, we need a rule for determining the number of classes that is based on the number of observations. One rule that generally works well is the *power-of-2 rule:* for n observations we would use a classes, where $2^{a-1} < n \leq 2^a$. Thus, for $n = 100$, we have $2^6 < 100 < 2^7$, so that seven classes would be used, which is the number that was actually used in Figure 2.1. Another rule of thumb that has been advanced is to let the number of classes equal $\sqrt{n}$, but this will produce quite a few classes when n is well in excess of 100. The first rule seems to be better suited for giving us a good view of the distribution of values.

It should be remembered that these are just rough rules-of-thumb, however. As indicated by, for example, Scott (1979), we need to know the shape of the true distribution for the type of data that we are using in order to determine the number of classes to use so as to provide a good picture of that distribution. (Note that classes are frequently referred to as "bins" in the statistics literature.) Of course, that distribution is generally unknown, so we cannot expect to be able to routinely determine the best number of classes to use.

Histograms can also be constructed using frequencies of individual values. Velleman and Hoaglin (1981) provide a histogram of the chest measurements of 5738 Scottish militiamen; the measurements were recorded to the nearest inch and ranged from 33 inches to 48 inches. With only 16 different values (33–48) there is certainly no need to group them into classes, and, in fact, the *power-of-2 rule* would specify 13 classes anyway. We would expect such anthropometric measurements to be roughly normally distributed, and the histogram did have that general shape.

When used in process capability studies, specification limits can be displayed on a histogram to show what portion of the data exceeds the specifications. Ishikawa (1976) displays these as dotted vertical lines. Process capability is discussed in Chapter 7.

2.2 PARETO CHARTS

A Pareto chart is a graph and it is somewhat similar to a *histogram*. The latter is also in the form of a bar chart, with the heights of the bars (rectangles) representing frequencies, although the frequencies are not arranged in descending order. A Pareto chart draws its name from an Italian economist, but J. M. Juran is credited with being the first to apply it to industrial problems. The idea is quite simple. The causes of whatever is being investigated (e.g., nonconforming items) are listed and percentages assigned to each one so that the total is 100%. The percentages are then used to construct the diagram that is essentially a bar chart. An example is given in Figure 2.3.

In this illustration the percentage of nonconforming condensers of a certain type is 6.39. Since this is quite high, in general, it would obviously be desirable to determine the causes of the nonconforming condensers and to display the percentage (or frequency) for each cause. In essence, we can think of a Pareto chart as an extension of a cause-and-effect diagram (see Section 2.6) in that the causes are not only identified but also listed in order of their frequency of

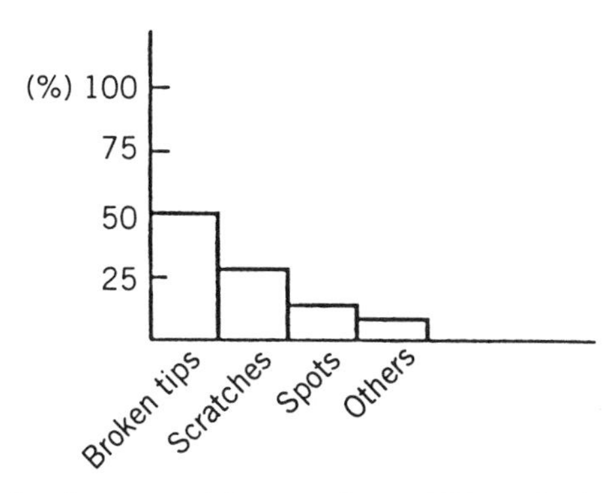

FIGURE 2.3 Pareto chart (percentages). Item, condenser AG1; number inspected, 15,000; number of nonconforming units, 958.

occurrence. It is generally found that there are a "vital few" causes and a "trivial many," as was first claimed many years ago by J. M. Juran.

When Pareto charts are to be shown to management, it is desirable to use "money lost" (or something similar) as the label for the vertical axis, assuming that the losses can be determined, or at least estimated. Thus, the data might be collected and arranged as in Table 2.3.

When the data are collected and tabulated in this manner, it is easy to construct either type of chart. The companion chart to Figure 2.3 that shows the monetary losses is given in Figure 2.4.

Although the order of nonconformities is the same in both diagrams, that will not always be the case. A unit that is considered nonconforming (i.e., unfit for distribution) because of one type of nonconformity could perhaps be reworked, whereas a unit with another type of nonconformity might have to be scrapped. The severity of the nonconformity would also be a factor: how deep is the scratch, how big is the spot, and so on. Although the order of nonconformities is the same, it is apparent that "broken tips" is even more of a problem from a monetary standpoint than from a percentage standpoint but "spots" is less of a problem monetarily.

See Pitt (1974) for more discussion of Pareto charts with monetary values, including an example in which the order of nonconformities is different between the two types of diagrams.

TABLE 2.3 Nonconformities and Associated Monetary Losses[a] ($)

Lot	Date		Scratches		Broken Tips		Spots		Other	
Number	(March)	NI	Number	ML	Number	ML	Number	ML	Number	ML
2014	1	1,000	22	$86	36	$160	6	$20	3	$6
2026	2	1,000	23	88	39	170	3	10	2	3
2013	3	1,000	30	100	41	178	8	24	4	7
2032	4	1,000	18	79	37	164	14	35	5	9
2030	5	1,000	20	81	28	146	15	38	3	6
2028	6	1,000	21	83	39	170	10	28	6	10
2040	7	1,000	19	80	33	152	9	25	2	3
2011	8	1,000	12	66	29	150	5	18	7	12
2010	9	1,000	14	69	31	149	8	24	6	10
2015	10	1,000	16	74	30	148	7	22	9	16
2022	11	1,000	12	66	22	136	4	16	5	9
2021	12	1,000	13	68	27	145	11	27	2	3
2024	13	1,000	21	83	35	158	13	31	1	1
2023	14	1,000	22	86	29	150	10	26	6	10
2018	15	1,000	19	80	23	138	6	20	7	12
		15,000	282	1189	479	2314	129	364	68	117

[a]NI, number inspected; ML, money lost.

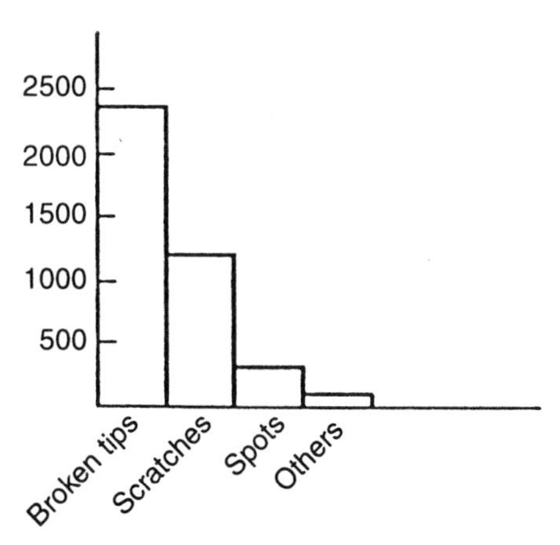

FIGURE 2.4 Pareto chart (monetary losses). Item, condenser AG1; number inspected, 15,000; total dollar loss, $3984.

There are obviously other modifications that could be used. The frequencies could have been used in Figure 2.3 instead of the percentages, but the configuration would be exactly the same. Similarly, percentages of the total dollar loss could have been displayed in Figure 2.4 instead of the individual dollar losses, but, again, the configuration would be the same.

We should realize that the usefulness of a Pareto chart is not limited to data on nonconformities and nonconforming units. It can be used to summarize all types of data. To illustrate, assume that the management of a company wishes to investigate the considerable amount of missing data in surveys that have been conducted to assess customers' views of the quality of their products. The results of the study are given in Figure 2.5.

In this example there is no evidence of a "vital few" and "trivial many," nor would we expect there to be from the nature of the categories A, B, C, D, and E. Nevertheless, such a diagram could be of considerable value to management.

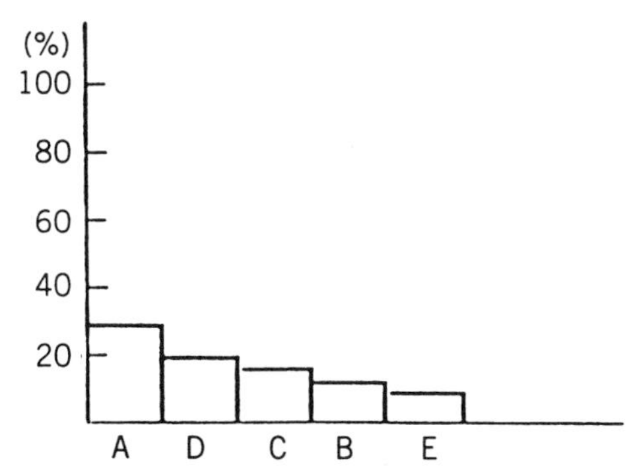

FIGURE 2.5 Pareto chart for survey data: (A) data lost in transcribing; (B) data not turned in by surveyor; (C) unable to contact; (D) customer refuses to answer certain questions; (E) other.

Often the vital few and trivial many that one hopes to see in a Pareto chart does not occur when the data are graphed in one way but does occur when the data are graphed in a different way. This was illustrated in Joiner Associates, Inc. (1996a), which is now Oriel, Inc. The chart of injuries by department in a particular company was flat, but when the same data were used to construct two additional Pareto charts, the additional charts did have the hoped-for shape. Specifically, when the injuries were categorized by body part, finger injuries stood out, and when a chart was constructed showing lost time due to injuries by department, the baking department stood out. In general, it is useful to construct multiple Pareto charts, if possible, for a given data set.

See Chapter 5 and pp. 162–174 of Ishikawa (1976) for further reading on Pareto diagrams, and see Joiner Associates (1995b) for various examples of Pareto charts.

2.3 SCATTER PLOTS

A *scatter plot* is another simple graphical device. The simplest type is a bivariate scatter plot, in which two quantities are plotted. For example, we might want to plot the number of nonconforming units of a particular product against the total production for each month, so as to see the relationship between the two. Figure 2.6 indicates that the relationship is curvilinear.

As a second example, consider the hypothetical data in Table 2.4, which are assumed to be for a large company that has recently initiated a quality improvement program.

A scatter plot of cost of training versus number of employees trained might be expected to show points that virtually form a straight line, so points that deviate from such a pattern might require further investigation. The scatter plot is shown in Figure 2.7.

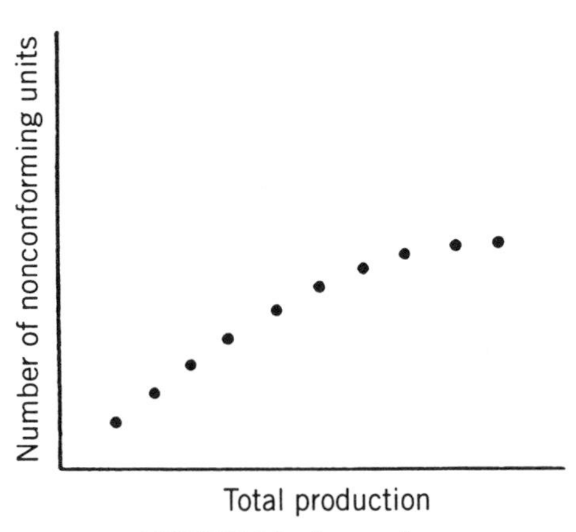

FIGURE 2.6 Scatter plot.

TABLE 2.4 Data for Quality Improvement Program

Month	Number of Employees Trained (000)	Cost of Training (000)
January	12	23
February	10	19
March	10	27
April	11	20
May	9	15
June	6	10
July	8	14
August	5	8
September	6	9
October	3	5
November	2	3
December	2	4

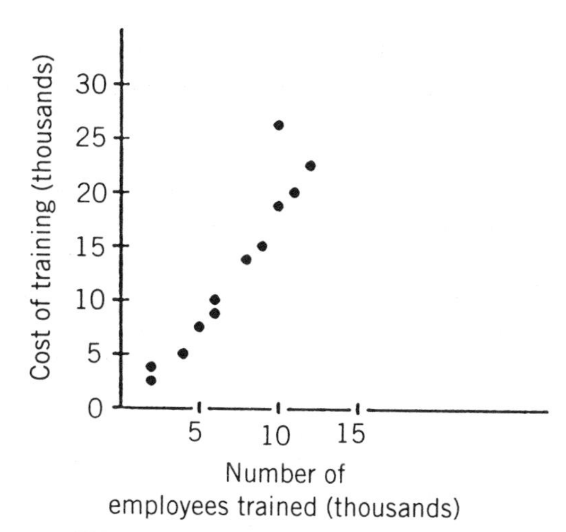

FIGURE 2.7 Scatter plot of Table 2.4 data.

(Note that the horizontal axis has a wider scale than the vertical axis. This is done to produce horizontal separation of the points. A wider vertical scale would have produced more vertical separation.) From the plot we can see rather clearly that there is one point that does not fit in with the others, namely, the point that corresponds to March. (One way to illustrate that it represents the figures for March will be explained shortly.) Perhaps during that month there was a deviation form the usual training program in that outside consultants were used. In any event, the point is highlighted by the scatter plot.

The label for each axis is often determined by whether or not one of the variables could be logically classified as the "dependent" variable. For example, in Figure 2.7 "cost of training" is dependent upon the number of employees trained. Traditionally the dependent variable is placed on the vertical axis.

The scaling of the two axes is somewhat arbitrary, although we should use the same scaling for each axis when the data for each variable are of the same order of magnitude. When this is not the case, we should keep in mind that two variables that are not strongly related can often be depicted as having a linear relationship just by increasing the scale of the axes, which will tend to squeeze the data together. [See Cleveland, Diaconis, and McGill (1982).] Therefore, some thought needs to be given to the choice of scaling.

A *time sequence plot* is a type of scatter plot in that data on one variable are plotted against a second variable, time, where time could be in hours, days, months, etc. Thus, if either of the two variables used in Figure 2.7 was to be graphed against "month" (which would be on the horizontal axis), the result would be a time sequence plot. A control chart (see Section 2.4) can also be thought of as a time sequence plot since sample number or time is generally used for the horizontal axis label. In general, time sequence plots should be constructed in such a way as to be 2–3 inches tall, have 4–10 data points per horizontal inch, and to cover 150% of the range of data values (Joiner Associates, 1996b). The use of time sequence plots with data from designed experiments is illustrated in Chapter 13.

A time sequence plot will often reveal peculiarities in a data set. It is an important graphical tool that should be routinely used whenever data have been collected over time, and the time order has been preserved. A convincing argument of the importance of this type of plot can be found in Ott (1975, pp. 34–36). Specifically, a student completed a course assignment by recording the amount of time for sand to run through a 3-minute egg timer. A time sequence plot of the times exhibited a perfect sawtooth pattern, with hardly any point being close to the median time. This should suggest that the two halves of the egg timer differ noticeably, a difference that might not be easily detected when the egg timer was used in the intended manner. Since the two halves must differ more than slightly, this means that at least one of the two halves is not truly a 3-minute egg timer—a discovery that could be of considerable interest when the timer is applied to eggs instead of sand!

A *probability plot* is another type of scatter plot. There are different ways of constructing a probability plot, depending on what is graphed. One approach is to plot the expected values for a specific probability distribution on the vertical axis, with the observations plotted on the horizontal axis. If a set of data might reasonably have come from a particular distribution, the plot should form approximately a straight line. In particular, a normal probability plot is used for determining if sample data might have come from a population that could be well represented by a normal distribution. Probability plots are best constructed by a computer; they are illustrated in Chapter 3 and used in subsequent chapters, and their construction is described in detail in Section 11.4.

2.3.1 Variations of Scatter Plots

Although it was not a problem in Figure 2.7, when a sizable number of points are plotted in a scatter plot, some of the points will likely have the same value for the two variables. There are several ways to handle this situation. One approach is to use a number instead of the dot, where the number indicates how many values are at that point. When multiple values occur at different points in the same area, however, there will generally be a need to show multiple values using other symbolism. One such approach is to use lines through the dot to designate additional values at that point. For example, ⍿ might represent two values at one point and * might represent five values at one point. Such symbols are referred to as "sunflowers" in Chambers et al. (1983, p. 107), and the reader is referred to that text for more details.

Although a scatter plot is generally in the form of either a "half box" (two sides) as in Figure 2.7, or a full box where the points are enclosed in a box, Tufte (1983) claims that a range frame is more informative. The latter is similar to a half box but differs in that the two lines cover only the range of the data and thus do not meet. A range frame for the data in Table 2.4 is given in Figure 2.8.

Such a display is obviously more informative than the regular scatter plot in that the reader can easily observe that the cost of training ranged from $3000 to $27,000, and the number of employees trained ranged from 2000 to 12,000. Such information is not available from the conventional scatter plot in Figure 2.7. Additional information can be provided by using staggered line segments for the axes so as to show the quartiles and the median for each variable (Tufte, 1983).

2.4 CONTROL CHART

A control chart is a time sequence plot with "decision lines" added. These decision lines are used to try to determine whether or not a process is in control.

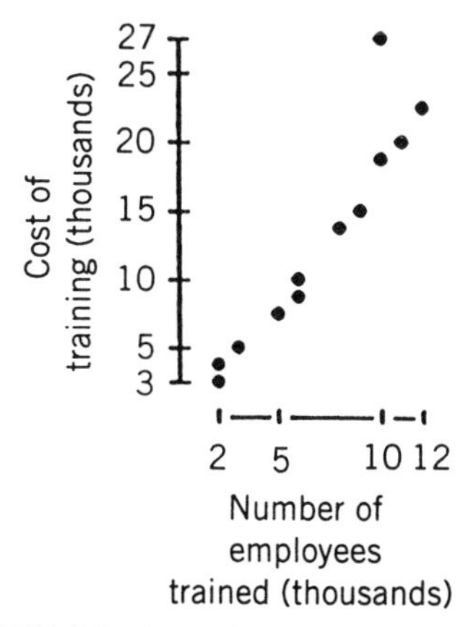

FIGURE 2.8 Range frame of Table 2.4 data.

Control charts, which are a deterrent to process tampering when a process is in control and which can also prevent underreaction when processes are out of control, are discussed extensively in Chapters 4–10. The general idea of a control chart was sketched out in a memorandum that Walter Shewhart of Bell Labs wrote on May 16, 1924. See Juran (1997) for an interesting treatise on the early use of control charts and other quality control techniques at AT&T and Bell Labs.

The construction of a control chart is based on statistical principles. Specifically, the charts are based upon some of the statistical distributions presented in Chapter 3. When used in conjunction with a manufacturing process (or a nonmanufacturing process), a control chart can indicate when a process is "out of control." Ideally, we would want to detect such a situation as soon as possible after its occurrence. Conversely, we would like to have as few "false alarms" as possible. The use of statistics allows us to strike a balance between the two. Basic control chart principles are illustrated in Figure 2.9. The center line of Figure 2.9 could represent an estimate of the process mean, or process standard deviation, or a number of other statistics to be illustrated in the following chapters. The curve to the left of the vertical axis should be viewed relative to the upper and lower control limits. The important detail to notice is that there is very little area under the curve below the lower control limit (LCL) and above the upper control limit (UCL). This is desirable since, as was mentioned in Chapter 3, areas under a curve for a continuous distribution represent probabilities. Since we look for a process that is out of statistical control when we obtain a value that is outside the control limits, we want the probability of conducting such a search to be quite small when the process is, in fact, in control.

A process that is not in a state of statistical control is one for which the variable being measured does not have a stable distribution. For example, a plot

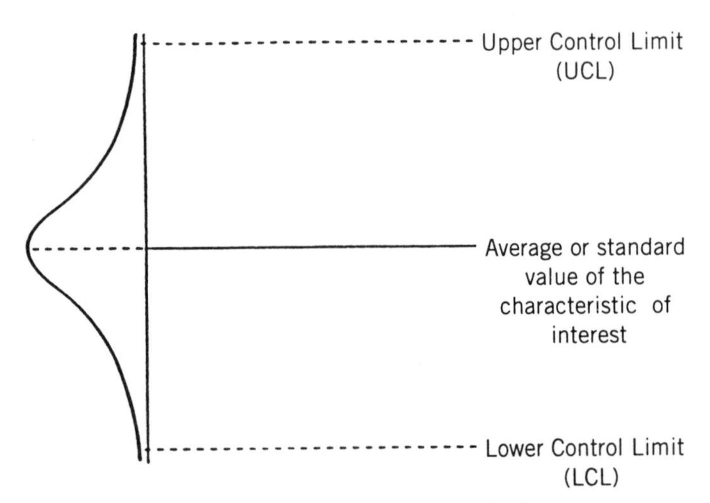

FIGURE 2.9 Basic form of a control chart.

of individual observations against time may suggest that the process mean is fluctuating considerably over time.

2.5 CHECK SHEET

A check sheet is a means of recording historical data on causes of nonconformities or nonconforming units. Thus, it can be used as a source of the data for a Pareto diagram, for example. Although there is no uniform design of check sheets, the general idea is to record all pertinent information relative to nonconformities and nonconforming units, so that the sheets can facilitate process improvement. Such information might include notes on, for example, raw materials, machine performance, and operator changes.

See Ishikawa (1976) for an extensive discussion of check sheets, which are also discussed and illustrated in Montgomery (1996) and Leitnaker, Sanders, and Hild (1996).

2.6 CAUSE-AND-EFFECT DIAGRAM

The cause-and-effect diagram was introduced in Japan in 1943 by Professor Kaoru Ishikawa of the University of Tokyo. For that reason it is sometimes called an Ishikawa diagram; it has also been called a fishbone chart. The reason for the latter name should be apparent from Figure 2.10.

In Figure 2.10, vibration is the effect that is to be reduced, and there are four possible causes that are to be investigated. This is an example of a *dispersion analysis type* of cause-and-effect diagram. The other two major types are production process classification type and cause enumeration type. The production process classification type is set up in accordance with the flow of the production process, whereas the cause enumeration type is simply a listing of all of the possible causes without trying to establish any structure relative to the process.

The diagram in Figure 2.10 is quite simplified and is meant to serve only as an illustration. Useful diagrams in practice will generally have more branches. There will always be a single "effect" (i.e., a quality characteristic) that we

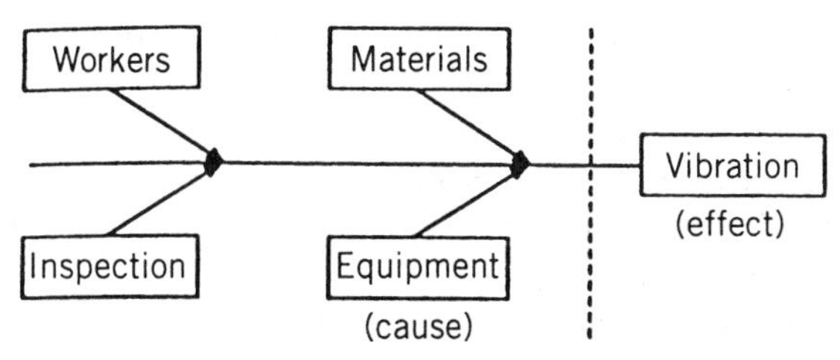

FIGURE 2.10 Cause-and-effect diagram.

wish to improve, control, or eliminate. (In this example we would probably want to be able to eliminate vibration, but perhaps the best we can do is to control it at a reasonable level.) We should list as many possible or probable causes as we can think of without making the diagram too cluttered, maintaining the correct relationship between the causes. When a relationship between the quality characteristic and a cause can be shown quantitatively using numerical information, the cause should be enclosed in a box (as in Figure 2.10). When it is known that a relationship between a cause and an effect does exist, but the relationship cannot be supported with data, the cause should be underlined. Thus, in a typical diagram there will be some causes that will be enclosed in a box and some that will be underlined.

Tribus (1998) makes an interesting point regarding the use of cause-and-effect diagrams. Specifically, he reports that it is much easier to obtain feedback that will allow what he terms a "negative Ishikawa diagram" to be constructed than when the desired effect is positive. He uses such a diagram for redesigning a process and asks workers who are part of a process to describe how to make it the worst possible process. This request elicits many responses, whereas asking workers how to perfect the process results in virtually no responses. The new process is then constructed by ensuring that the components of the negative diagram do not occur. Similarly, Tolman (1998) reports great success in using the negative Ishikawa diagram in a hospital application. Specifically, the redesign of an adolescent services unit was approached by first constructing a negative Ishikawa diagram and then using the negative diagram to construct a positive diagram that is in juxtaposition to the positive one.

In general, cause-and-effect diagrams that are carefully and thoughtfully constructed should enable the cause of a quality problem to be quickly detected, so that corrective action can be taken. Lore (1998) indicates that these diagrams can be used more effectively by asking why certain causes exist. Doing so may lead to additional branches on the tree and hopefully to clear conclusions.

See Chapter 3 of Ishikawa (1976) for a detailed discussion of cause-and-effect diagrams. Some examples of these diagrams can also be found in Joiner Associates (1995a).

2.7 DEFECT CONCENTRATION DIAGRAM

This is simply a schematic diagram that shows the various sides of a unit of production with the positions where nonconformities occur pinpointed. Ishikawa (1976) viewed this as a type of check sheet, whereas others (e.g, Montgomery, 1996) have viewed it as a separate tool. The latter view seems preferable since it is a graph and thus differs in form from a check sheet. Several examples of a defect concentration diagram were given by Ishikawa (1976), who termed the tool a *defect location check sheet*.

The diagram would be analyzed to determine if the locations of the nonconformities in the unit convey any useful information about the potential causes of the nonconformities. In one example given by Ishikawa (1976), use of the diagram revealed that bubbles in laminated automobile windshield glass occurred primarily on the right side. A subsequent investigation revealed that the pressure was out of balance, with the right side receiving less pressure. When the machine was adjusted, the bubbles were almost completely eliminated.

2.8 THE SEVEN NEW TOOLS

In his Youden Memorial Address at the 1993 Fall Technical Conference, Lynne Hare (1993, p. 5) stated: "Teamwork is enhanced by the seven new tools for management (Brassard, 1989) which, in my opinion, deserve more attention from statistically minded people." These tools are affinity diagram, interrelationship digraph, tree diagram, prioritization matrix, matrix diagram, process decision program chart, and activitv network diagram. Brassard (1989) emphasizes that although the tools can be used individually, it is best when they are used together.

2.8.1 Affinity Diagram

An *affinity diagram* is a set of ideas about a particular topic that are grouped into clusters. The diagram is the end product of brainstorming that is performed in a prescribed manner. Specifically, a team of at most five or six people is formed, with the team preferably comprised of people who have worked together previously. A question/issue to be addressed is then selected, and relevant ideas are listed. The ideas are written on cards and then randomly scattered. The ideas are then put into groups, with the finished product—the affinity diagram—being a collection of groups (with appropriate group headings) and a list of the ideas that form each group. An affinity diagram has been recommended for use with large or complex issues and problems and when new approaches are needed. Alloway (1997) illustrates the use of an affinity diagram for minor injuries.

In using "affinity" as short for affinity diagram, Brassard (1989) states: "We have yet to find an issue for which an affinity has not proven helpful."

2.8.2 Interrelationship Digraph

An *interrelationship digraph* is used for identifying and exploring causal relationships between related ideas. This is a step beyond an affinity diagram, as an interrelationship digraph is a figure with arrows indicating relationships between ideas. Therefore, it would be reasonable to construct the digraph after an affinity diagram has been constructed.

2.8.3 Tree Diagram

A *tree diagram* is somewhat similar to a cause-and-effect diagram in that a desired effect (e.g., reducing delivery delays) can be shown pictorially as related to the factors that can lead to the effect. A tree diagram will generally more closely resemble a company organizational chart in appearance than a cause-and-effect diagram, however. A tree diagram is a more structured display than either an affinity diagram or an interrelationship digraph.

2.8.4 Prioritization Matrix

A *prioritization matrix* is, as the name suggests, a relative ranking of issues jobs, objectives, products, etc. The ranking is accomplished by comparing the components pairwise so that a logical and consistent ranking results. The steps used in constructing a prioritization matrix can vary considerably. Brassard (1989) gives three alternative methods: (1) the full analytical criteria method, (2) the consensus criteria method, and (3) the combination interrelationship digraph/matrix method.

With the first approach, the prioritization matrix consists of numbers that represent each pairwise comparison. The relative rankings are then obtained by summing the scores for the various components. This is the most rigorous and analytical of the three approaches.

Pairwise comparisons are not made when the consensus criteria method is used. Rather, as the name implies, the relative ranking results from the composite score from the team members that each component receives.

The end product when the last approach is used is a matrix that contains arrows and other symbols that show the relationships between the components.

2.8.5 Matrix Diagram

A matrix diagram is used for showing relationships between two or more sets of ideas, projects, etc. The matrix can have one of several different forms. As discussed by Brassard (1989, p. 138), at least five forms have been used: C-shaped, L-shaped, T-shaped, X-shaped, and Y-shaped, with L-shaped matrices being the most frequently used. Alloway (1997) illustrates the use of a matrix diagram for determining the components of a first-aid diagram.

2.8.6 Process Decision Program Chart

A process decision program chart is a listing of undesirable events and corresponding contingency actions relative to planned actions. It is used when there is considerable concern about the possibility of negative unanticipated outcomes.

2.8.7 Activity Network Diagram

This is essentially a combination of two well-known techniques: PERT (Program Evaluation and Review Technique) and CPM (Critical Path Method).

These seven tools are discussed in detail in the popular book by Brassard (1989). The latter contains many examples, training suggestions, etc. Software for constructing these diagrams, charts, and matrices is available from GOAL/QPC, the publishers of Brassard (1989).

2.9 SUMMARY

Significant improvements in quality have resulted from the use of the seven basic tools presented in this chapter, and the seven new tools have also been extensively used. It should be understood, however, that the use of only these tools will not be sufficient. These tools should be augmented with the statistical tools that are presented in Part III. Similarly, there are more advanced graphical tools, presented in Chapter 11, that can also be used advantageously.

REFERENCES

Alloway, J. A., Jr. (1997). Be prepared with a affinity diagram. *Quality Progress 30*(7): 75–77.

Brassard, M. (1989). *The Memory Jogger Plus +*. Methuen, MA: Goal/QPC.

Chambers, J. M., W. S. Cleveland, B. Kleiner, and P. A. Tukey (1983). *Graphical Methods for Data Analysis*. Boston: Duxbury.

Cleveland, W. S., P. Diaconis, and R. McGill (1982). Variables on scatterplots look more highly correlated when the scales are increased. *Science 216*(4550): 1138–1141.

Hare, L. B. (1993). Youden Memorial Address. *ASQC Statistics Division Newsletter 13*(4): 4–8.

Ishikawa, K. (1976). *Guide to Quality Control*. White Plains, NY: UNIPUB.

Joiner Associates, Inc. (1995a). Cause-and-effect diagrams: Everything you need to know but are too busy to ask. Madison, WI: Joiner Associates (now Oriel, Inc.)

Joiner Associates, Inc. (1995b). Pareto charts: Everything you need to know about Pareto charts but are too busy to ask. Madison, WI: Joiner Associates, (now Oriel)

Joiner Associates, Inc. (1996a). One-minute lesson: An extra twist on Pareto charts. *Managing for Quality*, Issue 14. Madison, WI: Joiner Associates (now Oriel)

Joiner Associates, Inc. (1996b). One minute lesson: A sense of proportion. *Managing for Quality*, Issue 15. Madison, WI: Joiner Associates (now Oriel)

Juran, J. M. (1997). Early SQC: A historical supplement. *Quality Progress 30*(9), 73–81.

Leitnaker, M. G., R. D. Sanders, and C. Hild (1996). *The Power of Statistical Thinking: Improving Industrial Processes*. Reading, MA: Addison-Wesley.

Lore, J. (1998). A new slant on fishbones. *Quality Progress 31*(9), 128.

Montgomery, D. C. (1996). *Introduction to Statistical Quality Control*, 3rd ed. New York: Wiley.

Ott, E. R. (1975). *Process Quality Control*. New York: McGraw-Hill.

Pitt, H. (1974). Pareto revisited. *Quality Progress 7*(3): 29–30.

Scott, D. W. (1979). On optimal and data-based histograms. *Biometrika 66*, 605–610.

Tolman, A. O. (1998). Contribution to the Deming Electronic Network, February.

Tribus, M. (1998). Contribution to the Deming Electronic Network, February 6.

Tufte, E. R. (1983). *The Visual Display of Quantitative Information*. Cheshire, CT: Gıaphics Press.

Velleman, P. V. and D. C. Hoaglin (1981). *ABC of EDA*. Boston: Duxbury.

CHAPTER 3

Basic Concepts in Statistics and Probability

In this chapter we present the tools that form the foundation for the control charts that are covered in Part II and the other statistical procedures that are presented in Part III.

Much of the material in this chapter is typically found in introductory statistics texts. Some of the material is moderately advanced, so the reader may wish to use this chapter as reference material and read certain sections as needed.

3.1 PROBABILITY

During a typical day many of us will hear statements that are probability-type statements, although the word "probability" might not be used. One example (statement 1) is a weather forecast in which a meteorologist states, "There is a 10% chance of rain tomorrow." What does such a statement actually mean? For one thing, it means that, in the opinion of the meteorologist, it is very unlikely that rain will fall tomorrow. Contrast that with statement 2: "If a balanced coin is tossed, there is a 50% chance that a head will be observed." Disregarding for the moment that the coin could land on its edge (an unlikely possibility), there are two possible outcomes, a head or a tail, and they are equally likely to occur. Finally, consider statement 3: "I found a slightly bent coin in the street. I am going to toss this coin 1000 times and use the results of this experiment to determine an estimate of the likelihood of obtaining a head when the coin is tossed once. Thus, if I observe 522 heads during the 1000 tosses, I estimate there is a 52.2% (=522/1000) chance of observing a head when the coin is tossed once."

There are some important differences between these three statements. Statement 1 has to be at least somewhat subjective since it is not possible to repeat tomorrow 1000 times and observe how many times it rains, nor is it possible to know the exact likelihood of rain, as in the case of the balanced coin. The second and third statements are illustrative examples of the two approaches that will be followed in this book: (i) acting as if the assumption is valid (i.e., the coin is balanced), and (ii) not making any assumption but rather collecting data and then drawing some conclusion from the analysis of that data (e.g., from the 1000 tosses of the coin). The latter approach is obviously preferable if there is any question as to whether or not the assumption is valid and if the consequences of making a false assumption are considerable.

The word probability has not as yet been used relative to the examples in the section; instead *percent chance* has been used. The two terms can be thought of as being virtually synonymous, however. The first statement that was given (the weather forecast) is essentially a subjective probability statement. Statement 2 could be expressed concisely as

$$P(\text{head}) = \tfrac{1}{2}$$

which is read as "the probability of a head equals one-half" on a single toss of the balanced coin. Thus, a 50% chance is equivalent to a probability of $\frac{1}{2}$.

Just as percentages must range from 0 to 100, the probability of some arbitrary "event" (such as observing a head) must be between 0 and 1. An "impossible event" (such as rolling a 7 on a single die) would be assigned a probability of zero. The converse is not true, however: if an event is assigned a probability of zero, it does not mean that the event is an impossible event. As mentioned previously, a coin *could* land on its edge (and most assuredly will if it is tossed enough times), but we customarily assign a probability of zero to that possible event.

With statement 3, no probability was assumed; instead it was "estimated." In practice, this is customary since practical applications of probability go far beyond tossing a balanced coin or rolling a single die. In this instance the probability of observing a head on a single toss of the misshapen coin was estimated by tossing the coin a large number of times and counting the number of heads that was observed. Is 1000 tosses adequate? That depends upon the degree of accuracy that is required in estimating the true probability. The idea of determining the number of trials in an experiment from a stated error of estimation will not be pursued here. In general, however

$$\frac{x}{n} \to p \quad \text{as } n \to \infty$$

where x denotes the number of times that the event in question occurs, n denotes the number of trials, and p is the true probability that the particular event will occur on a single trial. The symbol $\to$ should be read as "approaches" and $n \to \infty$ designates the number of trials becoming large without bound.

How can these concepts be applied in a manufacturing environment? Assume that a particular plant has just opened and we want to estimate the percentage of nonconforming units of a particular product that the process is producing. How many items should we inspect? We would certainly hope that the percentage of nonconforming units is quite small. If it is, and if we were to inspect only a very small number of units, we might not observe any nonconforming units. In that case our estimate of the percentage of nonconforming units produced would be zero (i.e., $x/n = 0$), which could certainly be very misleading. At the other extreme, we could inspect every unit that is produced for a particular week. This would be rather impractical, however, if the production item happened to be steel balls and thousands of them were produced every week. Consequently, a compromise would have to be struck so that a practical number of items would be inspected. (See Section 6.1.3 for a further discussion of this problem in the context of attribute control charts.) For a reasonable number of items to be inspected, the percentage of nonconforming items that the process is producing could then be *estimated* by dividing the number of nonconforming units observed by the total number of items that are inspected (i.e., x/n).

3.2 SAMPLE VERSUS POPULATION

For the example just given, the "reasonable number" of units that is inspected would constitute a *sample* from some *population.* The word "population" in statistics need not refer to people. A statistical population can consist of virtually anything. For the manufacturing scenario presented in the preceding section, the population could be all of the items of a particular type produced by the manufacturing process—past, present, and future. A sample is simply part of the population. There are various types of samples that can be obtained. One of the most common is a *random sample.* A random sample of size n is one in which every possible sample of size n has the same probability of being selected.

An example will be given to illustrate this concept. Suppose a population is defined to consist of the numbers 1, 2, 3, 4, 5, and 6, and you wish to obtain a random sample of size 2 from this population. How might this be accomplished? What about listing all of the possible samples of size 2 and then randomly selecting one? There are 15 such samples and they are given below:

12	15	24	34	45
13	16	25	35	46
14	23	26	36	56

In practice, a population is apt to be much larger than this, however, so this would be a rather cumbersome procedure, even if carried out on a computer.

Another approach would be to use a *random number table* such as Table A in the Appendix to the book. This table could be used as follows. In general, the elements in the population would have to be numbered in some way. In this example the elements are numbers, and since the numbers are single-digit numbers, only one column of Table A need be used. If we arbitrarily select the first column in the first set of four columns, we could proceed down that column; the first number observed is 1 and the second is 5. Thus, our sample of size 2 would consist of those two numbers.

One fact that should be apparent for each of these procedures is that the elements of the population must be enumerated, and if these elements are not numbers, they must be assigned numbers.

Can this be accomplished if a population is defined to consist of all of the units of a product that will be produced by a particular company? Certainly not. Therefore, in practice it is often impossible to obtain a random sample. Consequently, a *convenience sample* is frequently used instead of a random sample. For example, if samples of five units are obtained from an assembly line every 30 minutes, every item produced will not have the same probability of being included in any one of the samples.

The objective, however, should be to obtain samples that are *representative* of the population from which they are drawn. (We should keep in mind, however, that populations generally change over time, and the change might be considerable relative to what we are trying to estimate. Hence, a sample that is representative today may not be representative 6 months later.) How can we tell whether or not convenience samples are likely to give us a true picture of a particular population? We cannot unless we have some ideas as to whether there are any patterns or trends in regard to the units that are produced. Consider the following example. Assume that every twenty-first unit of a particular product is nonconforming. If samples of size 3 happen to be selected in such a way (perhaps every 15 minutes) that one nonconforming unit is included in each sample, the logical conclusion would be that 1 out of every 3 units produced is nonconforming, instead of 1 out of 21.

Consequently, it is highly desirable to acquire a good understanding of the processes with which you will be working before using any "routine" sampling procedure.

3.3 LOCATION

In statistics, various measures are used in "describing" a set of data. For example, assume that you have obtained a sample that consists of the following numbers:

$$15 \quad 13 \quad 11 \quad 17 \quad 19$$

How might you describe or summarize this set of data relative to what would seem to be a typical value?

One possibility would be to use the *median*, which is the middle value (after the numbers have been arranged in ascending order) if the sample consists of an odd number of observations, as it does in this case. Thus, the median is 15. If the sample had consisted of these five numbers plus the number 21, there would then not be a single middle number, but rather two middle numbers. The median would then be defined as the average of the two middle values, which in that case would be 16. Thus, when there is an even number of observations, the median will always be a number that is not observed unless the two middle numbers happen to be the same. This might seem strange, but it should be remembered that the objective is to *estimate* the middle value of the population from which the sample was drawn.

Another measure that is often used to estimate a typical value in the population is the (sample) *average*, which is simply the sum of the values divided by n. For this example the average is also 15 so the average and the median have the same value. This will usually not be the case, however. In fact, the average might not even be close to the center of the sample when the values are ordered from the smallest to largest. For the sample

28 39 40 50 97

the average (50.8) is between the fourth and fifth numbers so it is not particularly close to the middle value. This is the result of the fact that one observation, 97, is considerably larger than the others. Thus, although the average is often referred to as a *measure of center or measure of central tendency*, it often will not be very close to the middle of the data. Consequently, during the past 20 years, in particular, there has been considerable interest in the statistical community in developing estimators that are insensitive to extreme observations (numerical values that differ considerably from the others in the sample). These *robust estimators* have not been used to any extent in practice, however, and this is undoubtedly due to the fact that they are not particularly well known by practitioners. So are robust estimators really necessary? The answer is "yes," and the reader is referred to a (nonmathematical) paper with that title by Rocke, Downs, and Rocke (1982) and the references contained therein. The paper is oriented toward chemistry, but robust estimators can be used in all fields.

A *trimmed average* is one type of robust estimator. If, for example, 10% of the observations in a sample are trimmed from each end (where the observations are ordered), extreme observations, which could have a considerable effect on the average of all of the observations, would thus be deleted. If there are no extreme observations, such "trimming" should have very little effect on the average. For example, if the smallest and largest values are deleted from the sample.

11 13 15 17 19

the average remains unchanged, but if the same trimming is done for the sample

28 39 40 50 97

the average changes from 50.8 to 43.0.

Extreme observations might very well be values that have been recorded incorrectly. In any event, they are not typical observations. If the trimming is not done haphazardly, but rather some trimming procedure is consistently applied, a better estimate of the center of the corresponding population is apt to result. Nevertheless, trimmed averages will not be used for the control charts and other statistical procedures presented in this book. Instead, the conventional procedure of using all of the observations will be used, assuming that the data do not contain any errors. The need to be watchful for errors should be kept in mind, however, since errors do occur quite often in data collection and tabulation.

3.4 VARIATION

There is "natural" variation in everything. Is your driving time to work precisely the same every day? Of course not. It will depend upon factors such as weather conditions and traffic conditions. Assume that your time varies slightly for a particular week, but at the beginning of the second week an accident on the expressway causes you to be 30 minutes late for work. Your travel time for that day is not due to natural, random variation but rather to an "assignable cause"—the accident.

With statistical procedures in general and control charts in particular, a primary objective is to analyze components of variability so that variability due to assignable causes can be detected. If you are the "unit" that is being "measured" for travel time, you know why you were late for work and can thus explain the cause. A ball bearing, however, cannot explain why its diameter is considerably larger than the diameter of the preceding 500 ball bearings that have rolled off the assembly line. Thus, statistical procedures are needed to spotlight the abnormal variation and, we hope, to enable the contributing factor(s) to be pinpointed.

Before we can speak of normal and abnormal variation, however, we must have one or more objective measures of variation. The simplest such measure is the sample *range*, which is defined to be the largest observation in a sample minus the smallest observation. For the sample

$$11 \quad 13 \quad 15 \quad 17 \quad 19$$

the range is 8. It should be observed that only two of the five values are used in obtaining this number; the other three are essentially "thrown away." Because of its simplicity and ease of calculation by hand, the range has been used extensively in quality control work. Nevertheless, it is wasteful of information and will be inferior to good measures of variability that use all of the observations.

I stated in the preface that this book is written under the assumption that many, if not most, of its readers will be using a computer for constructing and maintaining control charts, in particular, as well as for the other statistical procedures presented herein. Therefore, simple and easy-to-use procedures will not be recommended over more efficient, but involved, procedures when they can both be handled with approximately equal ease on a computer.

If we were to start from scratch and devise a measure of variability that uses all of the sample observations, if would seem logical that what we construct should measure how the data vary from the average.

At this point we need to introduce some symbols. The sample average is a statistic and will henceforth be denoted by $\overline{X}$, which is read "x-bar." Its calculation can be expressed as

$$\overline{X} = \frac{\sum_{i=1}^{n} X_i}{n} \tag{3.1}$$

where the Greek letter $\sum$ is read as (capital) sigma and is used to indicate summation. Specifically, what lies to the right of $\sum$ is to be summed. The letter i in X_i is a subscript, which in this case varies from 1 to n. The number at the bottom of $\sum$ indicates where the summation is to start, and the number (or symbol) at the top indicates where it is to end.

Thus, the average for the sample

11 13 15 17 19

can be expressed as

$$\begin{aligned} \overline{X} &= \frac{\sum_{i=1}^{n} X_i}{n} \\ &= \frac{\sum_{i=1}^{5} X_i}{5} \\ &= \frac{X_1 + X_2 + X_3 + X_4 + X_5}{5} \\ &= \frac{11 + 13 + 15 + 17 + 19}{5} \\ &= 15.0 \end{aligned}$$

If we wanted to construct a measure of variability, we might *attempt* to use $\sum_{i=1}^{n}(X_i - \overline{X})$. However, it can be shown that this sum will equal zero for any sample. This is due to the fact that some of the deviations $(X_i - \overline{X})$ will be positive whereas others will be negative, and the positive and negative values add to zero. For the present sample of five numbers,

$$\begin{aligned} &\sum_{i=1}^{5}(X_i - \overline{X}) \\ &\quad = -4 - 2 + 0 + 2 + 4 \\ &\quad = 0 \end{aligned}$$

One obvious way to eliminate the negative deviations would be to square all of the deviations. The *sample variance*, S^2, is usually defined as

$$S^2 = \frac{\sum_{i=1}^{n}(X_i - \overline{X})^2}{n-1} \tag{3.2}$$

although a few authors have chosen to divide by n instead of $n - 1$. Arguments can be given in support of each choice, but a discussion of the merits of each will be delayed until the S^2 chart is presented in Chapter 4. Definition (3.2) will be used for all values of S^2 that are presented in this book.

It could be shown that S^2 can also be calculated as

$$S^2 = \frac{\sum X^2 - (\sum X)^2/n}{n - 1}$$

For the same sample of five numbers,

$$\begin{aligned} S^2 &= \frac{\sum_{i=1}^{n}(X_i - \overline{X})^2}{n - 1} \\ &= \frac{\sum_{i=1}^{5}(X_i - 15)^2}{4} \\ &= \frac{(-4)^2 + (-2)^2 + (0)^2 + (2)^2 + (4)^2}{4} \\ &= \frac{40}{4} \\ &= 10 \end{aligned}$$

The sample variance is not as intuitive as the sample range, nor is it in the same unit of measurement. If the unit of measurement is inches, the range will be in inches but the variance will be in inches squared. This is not sufficient cause for discarding the variance in favor of the range, however.

Another measure of variability that is in terms of the original units is the *sample standard deviation*, which is simply the square root of the sample variance. Therefore,

$$S = \sqrt{\frac{\sum_{i=1}^{n}(X_i - \overline{X})^2}{n - 1}}$$

is the sample standard deviation. For this example $S = \sqrt{10} = 3.16$. The standard deviation is also not as intuitive as the range, but will generally be of the same order of magnitude as the average deviation between the numbers. Thus, with the (ordered) numbers

11 13 15 17 19

the deviation between each pair of adjacent numbers is 2, so the average deviation is also 2. Therefore, a standard deviation of 3.16 is well within reason; a value of, say, 31.6 or .316 should lead us to check our calculations. This can be helpful as a rough check, regardless of the calculating device that is used to obtain the answer.

Another measure of variation, which is used for two variables such as two process characteristics, is *covariance*. This is a measure of how two random variables "covary." For variables X and Y, the sample covariance is given by

$$S_{xy} = \frac{\sum_{i=1}^{n}(X_i - \overline{X})(Y_i - \overline{Y})}{n - 1}$$

Notice that if Y were replaced by X, this would produce the sample variance, S^2, given previously. If S_{xy} were divided by the square root of the products of the sample variances for X and Y, this would produce the sample *correlation coefficient*—a unit-free statistic whose value must lie between -1 and $+1$, inclusive. (A value of $+1$ would result if all of the points could be connected by a straight line with a positive slope; a value of -1 would occur if all of the points could be connected by a straight line with a negative slope. Neither extreme case could be expected to occur in practice, however.)

A few other measures of variability are occasionally used, but the range, variance, and standard deviation have been used most frequently for a single variable, and these will be used in this book. The correlation coefficient is frequently given when there are two variables, and covariance is used in Section 9.1.

Sections 3.5 and 3.6 contain discussions of many statistical distributions. All of these distributions are needed in later chapters, but the reader may wish to return to this chapter to read about certain distributions when they are needed in subsequent chapters, especially those distributions that are presented briefly in these two sections.

3.5 DISCRETE DISTRIBUTIONS

In addition to the concepts of location and variability, it is important to understand what is meant by the word *distribution* in statistics. The word implies (even in a nonstatistical context) that something is distributed in a certain way. In statistics, that "something" is called a *random variable*. In keeping with my intention to explain concepts in a simple manner whenever possible, the mathematically formal definition of a random variable that is usually found in introductory statistics books will be eschewed in favor of a simpler definition. A random variable is literally "something that varies in a random manner," and a discrete random variable is a random variable that can assume only a finite or countably infinite number of possible values (usually integers).

The following example will be used for illustration. Assume that an experiment is defined to consist of tossing a single coin twice and recording the number of heads that is observed. The experiment is to be performed 16 times. The "something" that varies is the number of heads and this is our random variable. It is customary to have a random variable represented by an alphabetical (capital) letter. Thus, we could define

$$X = \text{ number of heads observed in each experiment}$$

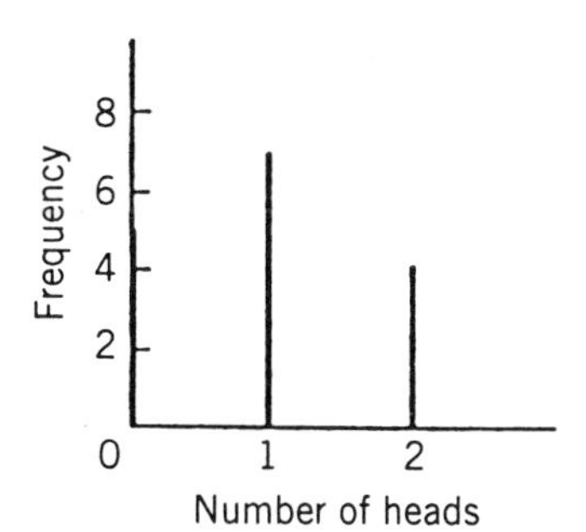

FIGURE 3.1 Empirical distribution.

Assume that the 16 experiments produce the following values of X:

0 2 1 1 2 0 0 1 2 1 1 0 1 1 2 0

There is no apparent pattern in the sequence of numbers, so it can be stated that X *varies* in a *random* manner and is thus a random variable.

A *line diagram* could then be used to portray the *empirical distribution* of X for these 16 experiments. The line diagram is given in Figure 3.1. The height of each line represents the number of occurrences of each of the three numbers 0, 1, and 2. Thus, there were five 0's, seven 1's, and four 2's.

An empirical distribution (such as this one) should not be confused with the theoretical distributions that are presented in the remainder of this chapter. A theoretical distribution is, roughly speaking, what the corresponding empirical distribution could be expected to resemble if there were a very large number of observations (e.g., millions). What should this particular empirical distribution resemble? The answer is the theoretical distribution that is presented in the next section.

3.5.1 Binomial Distribution

The binomial distribution is one of several that will be presented. There is actually a very large number of statistical distributions (hundreds, at least), and the selection of these few distributions for presentation is based upon their wide applicability and also because most of them relate to the control charts and other statistical procedures to be presented later. We should keep in mind that these distributions are used to model reality, with the latter frequently having statistical complexities that cannot be easily represented by commonly used distributions.

The binomial distribution can be used when the following conditions are met:

1. When there are two possible outcomes (e.g., heads and tails). These outcomes are arbitrarily labeled *success* and *failure*, with the outcome that is labeled "success" being the one for which one or more probabilities are to be calculated. There is no intended connotation of good versus bad. For example, if probabilities for various numbers of nonconforming items are to be computed, "nonconforming" is labeled "success".

2. There are n trials (such as n tosses of a coin) and the trials are independent. The trials are called *Bernoulli trials*. The independence assumption means that the outcome for a particular trial is not influenced by the outcome of any preceding trials. Furthermore, the probability of a success on a single trial does not vary from trial to trial.

Clearly the coin-tossing experiment given in the preceding section meets these requirements. The probabilities of observing zero, one, and two heads will now be obtained in a somewhat heuristic manner. It should be apparent that whatever happens on the second toss of the coin is independent of the outcome on the first toss. When two events are independent, the probability of both of them occurring is equal to the product of their separate probabilities of occurrence. Thus, if H_1 represents a head on the first toss and H_2 represents a head on the second toss, then

$$\begin{aligned} P(H_1 \text{ and } H_2) &= P(H_1)P(H_2) \\ &= \left(\tfrac{1}{2}\right)\left(\tfrac{1}{2}\right) \\ &= \tfrac{1}{4} \end{aligned}$$

Similarly, it could be shown that the probability of two tails (zero heads) equals $\frac{1}{4}$. One head could be observed in one of two ways, either on the first toss (followed by a tail) or on the second toss (preceded by a tail). Since the probability for each sequence is $\frac{1}{4}$, the probability of observing one head is equal to the sum of those two probabilities, which is $\frac{1}{2}$.

If we define the random variable X as

$$X = \text{ number of heads observed}$$

we can put together these probabilities to form the probability distribution of X. If we let $P(X)$ represent "the probability of X" (with X assuming the three different values), we then have the following:

X	P(X)
0	$\frac{1}{4}$
1	$\frac{1}{2}$
2	$\frac{1}{4}$

Thus, if this experiment were repeated an extremely large number of times, we would theoretically expect that one head would occur 50% of the time, two heads would occur 25% of the time, and zero heads would also occur 25% of the time.

Compare this theoretical expectation with the results of the 16 experiments depicted in Figure 3.1. The theoretical frequencies are 4, 8, and 4, which are very close to the observed frequencies of 5, 7, and 4. Although the theoretical frequencies of 4, 8, and 4 constitute our "best guess" of what should occur, we should not be surprised if the observed frequencies differ somewhat from the theoretical frequencies. In fact, it may be apparent that the observed frequencies could never equal the theoretical frequencies unless the number of experiments was a multiple of 4. (If the number of experiments is not a multiple of 4, the theoretical frequencies for zero heads and two heads would not be an integer.) The important point is that the difference between the observed and theoretical frequencies should become very small as the number of experiments becomes very large.

The way that the probabilities were found for the coin-tossing experiment would certainly be impractical if the number of tosses, n, was much larger than 2. In virtually any practical application n will be much larger than 2, so there is clearly a need for a general formula that can be used in obtaining binomial probabilities. The following symbols will be used:

p = probability of observing a success on a single trial

$1 - p$ = probability of not observing a success on a single trial (i.e., a failure)

n = number of trials

x = number of successes for which the probability is to be calculated

Regardless of the size of n, it is easy to find $P(x)$ when $x = 0$ or $x = n$. There is only one way to observe either no successes or all successes. Therefore, since the trials are independent, $P(n)$ is simply p multiplied times itself n times, that is, p^n. Similarly, $P(0) = (1 - p)^n$. It is by no means obvious, however, what $P(x)$ equals when x is neither 0 nor n.

If we wanted the probability of x successes followed by $n - x$ failures, that would clearly be

$$p^x(1 - p)^{n-x} \tag{3.3}$$

If, instead, we just wanted the probability of x successes without regard to order, the answer would obviously be larger than what would be produced by Eq. (3.3). For example, if you toss a coin 10 times, there are many different ways that you could observe four heads and six tails, one of which would be four heads followed by six tails.

In general, the number of ways that x successes can be observed in n trials is $\binom{n}{x}$, which is defined as

$$\binom{n}{x} = \frac{n!}{x!(n-x)!} \tag{3.4}$$

where $n! = 1 \cdot 2 \cdot 3 \cdot 4, \ldots, n$. For example, if $n = 5$ and $x = 3$, then

$$\binom{5}{3} = \frac{5!}{3!2!}$$

$$= \frac{120}{(6)(2)}$$

$$= 10$$

If such computation is to be done by hand, it is easier to first simplify the quotient by dividing the larger of the two numbers in the denominator into the numerator. Thus,

$$\frac{5!}{3!} = \frac{1 \cdot 2 \cdot 3 \cdot 4 \cdot 5}{1 \cdot 2 \cdot 3} = 4 \cdot 5 = 20$$

so that

$$\frac{5!}{3!2!} = \frac{20}{2} = 10$$

By putting Eq. (3.4) together with Eq. (3.3), we have the following general expression for the probability of x successes in n trials:

$$P(x) = \binom{n}{x} p^x (1-p)^{n-x} \tag{3.5}$$

Although it was easy to find the probabilities of observing zero, one and two heads without using Eq. (3.5), the direct approach would be to use Eq. (3.5). Thus, we could have found the probability of observing one head as follows:

$$\begin{aligned} P(1) &= \binom{2}{1} (.5)^1 (.5)^1 \\ &= 2(.5)(.5) \\ &= .5 \end{aligned}$$

Notice that if we were to attempt to determine the probability of 2 heads as

$$\begin{aligned} P(2) &= \binom{2}{2} (.5)^2 (.5)^0 \\ &= \frac{2}{2!0!}(.25)(1) \end{aligned}$$

we would have to know what to do with 0!. It is defined to be equal to 1, and it should be apparent that if it were defined in any other way we would obtain an incorrect answer for this example. Specifically, we *know* that $P(2) = (.5)(.5) = .25$; therefore, we must have

$$\begin{aligned} P(2) &= \frac{2!}{2!0!}(.25)(1) \\ &= 1(.25)(1) \\ &= .25 \end{aligned}$$

which will result only if 0! is defined to be equal to 1.

A practical problem might be to determine the probability that a lot of 1000 capacitors contains no more than one nonconforming capacitor, if 1 out of every 100 capacitors produced is nonconforming. There are clearly two possible outcomes (conforming and nonconforming), but are the "trials" independent so that the probability of any particular capacitor being nonconforming does not depend upon whether or not any of the previously produced capacitors were nonconforming? This is the type of question that must be addressed for any manufacturing application and applications of the binomial distribution in general.

If this assumption is not valid, the binomial distribution would be of questionable value as a model for solving this problem. For this example we shall assume that the assumption is valid. The words *no more than one* indicate that we should focus attention upon zero nonconforming units and one nonconforming unit and add the two probabilities together.

Thus,

$$\begin{aligned} P(0) &= \binom{1000}{0}(.01)^0(.99)^{1000} \\ &= (.99)^{1000} \\ &= .000043 \end{aligned}$$

and

$$\begin{aligned} P(1) &= \binom{1000}{1}(.01)^1(.99)^{999} \\ &= .000436 \end{aligned}$$

Therefore, the probability of no more than one nonconforming unit in a lot of 1000 capacitors is $.000043 + .000436 = .000479$.

Both of these individual probabilities are quite small, but that is due primarily to the fact that X (= the number of nonconforming capacitors) could be any one of 1001 numbers (0–1000). Therefore, we should not expect any one probability to be particularly high.

If the binomial distribution is a probability distribution (as it is), all of the probabilities will add to 1. It can be shown with some algebraic manipulation that

$$\sum_{x=0}^{n} \binom{n}{x} p^x (1-p)^{n-x} = 1$$

for any combination of n and p, but this will not be demonstrated here. (You should recall that for the coin-tossing experiment with $n = 2$ and $p = \frac{1}{2}$ the probabilities were $\frac{1}{4}$, $\frac{1}{2}$, and $\frac{1}{4}$, which obviously do add to 1). It should also be noted that the binomial distribution is a *discrete* probability distribution; that is, the random variable X can assume only integer values.

We might ask the following question that relates to both the coin-tossing experiment and the nonconforming capacitor example. What would be our "best guess" for the value of X in a binomial experiment? In other words, how many heads would we expect to observe if we toss a balanced coin twice? How many nonconforming capacitors should we expect to observe in a lot of 1000 capacitors if 1 out of 100 capacitors was nonconforming in the past? We should be able to answer these questions without any knowledge of the binomial distribution. We would certainly expect to observe 1 head in 2 tosses of the coin and 10 nonconforming capacitors in the lot of 1000 capacitors. But what if we wanted to know how many nonconforming capacitors we should expect to find in a lot of 500 capacitors if 1 out of every 75 capacitors has been nonconforming in the past? Now the answer is not quite so obvious. Therefore, we need a way of formally determining the *expected value* of X [which is written $E(X)$] for the binomial distribution.

There are two commonly used ways of obtaining the expected value, one of which is quite easy to use. The Bernoulli trials that were mentioned previously correspond to the Bernoulli distribution. A Bernoulli random variable has two possible outcomes: 0 and 1. If we let 0 represent failure and 1 represent success, then the expected value of the Bernoulli random variable is $(1)(p) + (0)(1 - p) = p$. Since the trials are independent, it follows that for the sum of n independent trials

$$E(X) = np$$

For the coin-tossing experiment $n = 2$ and $p = \frac{1}{2}$, so $np = 1$, and for the first capacitor example $n = 1000$ and $p = .01$, so $np = 10$. Thus, the theoretical results coincide with what common sense would tell us. For the second capacitor example $n = 500$ and $p = .0133$ (1 divided by 75), so $np = 6.67$. This last result was probably not obvious. Notice that $E(X)$ for the binomial distribution will not always be an integer, although X itself must always be an integer.

This should not seem incongruous, however, because $E(X)$ is simply the theoretical "average" value of X and should be very close to the actual average value of X if a binomial experiment is repeated a very large number of times.

The concept of *variance* of a random variable is very important is statistics, in general, and particularly so for the control charts presented in later chapters. You will recall that the sample variance, S^2, was introduced in Section 3.4 of this chapter. It was calculated from the data in a sample and is a *sample statistic*. *If* a person had an entire statistical population, he or she could then compute the population variance, σ^2, which is defined as

$$\sigma^2 = \sum_{i=1}^{N} \frac{(x_i - \mu)^2}{N}$$

where μ represents the average value (mean) of the population and N represents the number of units in the population. However, since we should not expect to ever have an entire population before us, the *sample statistic*, S^2, can be used to estimate the *population parameter*, σ^2. The use of sample statistics to estimate population parameters is one of the primary uses of statistics. Such estimation will be done extensively in constructing control charts in later chapters, as well as in the other statistical procedures to be presented. (See Sections 3.8.2 and 3.8.2.1 for additional information on estimation.)

The variance for a probability distribution is *somewhat* different, however, in that it is not calculated from data. Such a variance will henceforth be denoted by Var(X). If X is a binomial random variable, we may obtain its variance in the following way. For an arbitrary random variable, W, the variance of W is defined as $E(W^2) - [E(W)]^2$. For a discrete random variable, $E(W^2) = \sum w^2 P(W = w)$.

The *covariance* between two random variables, say W and Y, is similarly defined as $E(WY) - E(W)E(Y)$. Covariance measures the manner in which two random variables covary. A positive covariance means that as one variable increases (decreases), the other variable increases (decreases). With a negative covariance, one variable increases (decreases) while the other variable decreases (increases). Note that the sample covariance, S_{xy}, given in Section 3.4 is computed from *data* and is thus different from the covariance between two random variables.

For a Bernoulli random variable it can be easily seen that $E(W^2) = p$, so the variance of a Bernoulli variable is $p(1 - p)$. We can then find the variance of a binomial random variable by using the same general approach as was used for determining the mean. That is, the variance of a binomial random variable is n times the variance of the corresponding Bernoulli variable. Thus, for the binomial random variable we have

$$\text{Var}(X) = np(1 - p)$$

Notice that this depends not on any sample data but on the sample size. It was shown previously that the $E(X)$ for the binomial distribution is quite intuitive, but Var(X) cannot be explained in a similar manner. The square root of Var(X) can, however, be explained in the following manner. If p is close to $\frac{1}{2}$ and n is at least 15 or 20, then

$$P[E(X) - 3\sqrt{\text{Var}(X)} \leq X \leq E(X) + 3\sqrt{\text{Var}(X)}] \doteq .99$$

In words, we would expect the value of x to fall between the two endpoints of the interval almost all of the time. Thus, the square root of the variance (i.e., the standard deviation) can be combined with the mean to determine an interval that should contain x with a high probability. The standard deviation for a probability distribution thus measures the spread of the *possible* values of a random variable, whereas the sample standard deviation, S, is a measure of the spread of the *actual* values of a random variable that are observed in a sample.

The binomial distribution has been used extensively in quality improvement work. For example, it is used in constructing p charts and np charts in Chapter 6.

Binomial tables have been prepared for a number of different combinations of n and p. They are not provided in this book because they are not needed for the methods that are presented. They are needed for acceptance sampling and other applications, however. The reader with a need for such tables is referred to the extensive tables prepared by Weintraub (1963), Harvard Computation Laboratory (1955), and the U.S. Army Materiel Command (1972). Weintraub (1963) is recommended for $p < .01$. Of course, computer software is also quite useful in this regard and is generally preferable to tables.

3.5.2 Beta–Binomial Distribution

The binomial distribution will often be inadequate for modeling binary data because the variation in such data will often be greater than the variation represented by the binomial distribution Consequently, a distribution that captures such variation will be needed in many applications. McCullagh and Nelder (1989, p. 125) indicate that binomial overdispersion frequently occurs and state: "Unless there are good external reasons for relying on the binomial assumption, it seems wise to be cautions and to assume that overdispersion is present to some extent unless and until it is shown to be absent."

The beta–binomial distribution is a mixture of the binomial and beta distributions. (The latter is discussed in Section 3.6.11.) Specifically, whereas p in the binomial distribution is assumed to be constant, with a beta–binomial distribution p is assumed to be a random variable that would be modeled by a beta distribution. A beta distribution can have various shapes depending upon the values of the two shape parameters, so the distribution does provide some flexibility in the modeling of p.

Letting $B(r, s)$ denote the beta distribution with parameters r and s (see Section 3.6.11), for the beta–binomial distribution

$$P(x) = \binom{n}{x} \frac{B(r + x, n + s - x)}{B(r, s)}$$

where n and x have the same representation as for the binomial distribution.

3.5.3 Poisson Distribution

This distribution can be used when dealing with rare events. It has been used in quality improvement work to construct control charts for "nonconformities." (A product can have one or more nonconformities without being proclaimed a "nonconforming unit," so the terms are not synonymous. Specifically, a very minor nonconformity might be considered almost inconsequential.) The Poisson distribution is similar to the binomial distribution in that it is also a discrete distribution. In fact, it can be used to approximate binomial probabilities when n is large and p is small.

Specifically, if we again consider the binomial distribution

$$P(x) = \binom{n}{x} p^x (1 - p)^{n-x}$$

and let $n \to \infty$ and $p \to 0$ in such a way that np remains constant ($= \lambda$, say), it can be shown that

$$\lim_{n\to\infty} P(x) = \frac{e^{-\lambda}\lambda^x}{x!} \qquad x = 0, 1, 2, \ldots \tag{3.6}$$

where "lim" is short for "limit." The letter e represents the nonrepeating and nonterminating mathematical constant 2.71828..., and the right-hand side of Eq. (3.6) is the Poisson distribution, with λ representing the mean of X. Since this distribution can be obtained as a limiting form of the binomial distribution as $n \to \infty$ and $p \to 0$, it stands to reason that it should be possible to use the Poisson distribution to approximate binomial probabilities when n is large and p is small and obtain a reasonably good approximation (if a computer is not available to produce the exact answer).

The Poisson distribution is a valuable distribution in its own right; it is not just for approximating binomial probabilities. The distribution is named for Simeon Poisson (1781–1840), who presented it as a limit of the binomial distribution in 1837. It was stated earlier that the Poisson distribution is used as a basis for constructing control charts for nonconformities. Before the distribution can be applied in a physical setting, however, it must be determined whether or not the assumptions for the distribution are at least approximately met. These assumptions will be illustrated with the following example.

Assume that sheets of steel are being produced in a manufacturing process and the random variable X is the number of surface scratches per square yard. Before it can be claimed that X is a Poisson random variable, the following questions must be addressed. (1) Do the scratches occur randomly and independently of each other? (2) Is the possible number of scratches per square yard quite large (theoretically it should be infinite)? The second question might be answered in the affirmative, but perhaps not the first one. If there are quite a few surface scratches on one section of steel, we might expect to observe a sizable number of scratches on adjacent sections. Or perhaps not. In any event, for this problem and for other practical problems our objective should be to determine whether or not a particular distribution can logically serve as a *model* (and nothing more) for physical phenomena under investigation. We can rarely expect all of the assumptions to be met exactly; we can only hope that they are approximately satisfied. We should be concerned, however, if there is evidence of a radical departure from the assumptions, and our concern should lead us to seek alternative approaches to the problem.

3.5.4 Geometric Distribution

In recent years this distribution has been used as an alternative to the binomial distribution, in particular, and also as an alternative to the Poisson distribution. The need for the geometric distribution is discussed and illustrated in Chapter 4.

The geometric distribution and the negative binomial distribution, a related distribution that is presented in Section 3.5.5, are frequently referred to as "waiting time" distributions. This can be illustrated by contrasting the geometric distribution with the binomial distribution when a single success is considered. With the binomial distribution the single success could occur on the first trial, the nth trial, or any trial in between.

The geometric distribution comes about when the probability is obtained of the single success occurring on the nth trial. In order for the first success to occur on the nth trial, the first $n - 1$ trials must obviously result in failure. The probability of that occurring is $(1 - p)^{n-1}$. Therefore, we multiply this component by the probability of obtaining a success so as to obtain the probability mass function as

$$P(n) = (1 - p)^{n-1} p \qquad n = 1, 2, \ldots$$

where, unlike its designation in Section 3.5.1, here n denotes the number of trials that are needed for the first success to be observed. Thus, n is the random variable for the geometric distribution. The distribution for n depends solely on p. If p is small, the mean of n is large, and conversely. It is demonstrated in the section on ARL calculations in the Appendix to Chapter 4 that the mean is $1/p$. It can also be shown that the variance is $(1 - p)/p^2$.

3.5.5 Negative Binomial Distribution

This distribution is related to the geometric distribution in Section 3.5.4. The difference is that with the negative binomial distribution interest is centered on the probability that the rth success occurs on the nth trial, with r generally greater than 1. Analogous to the way that the probability mass function for the geometric distribution was obtained, it is helpful to view the appropriate function as resulting from the product of two components: the probability of obtaining $r - 1$ successes on the first $n - 1$ trials times the probability of obtaining a success on the nth trial.

The former is a binomial probability, so when that probability is multiplied by p, we obtain

$$P(n) = \binom{n-1}{r-1} p^r (1 - p)^{n-r} \qquad n = r, r + 1, r + 2, \ldots$$

It can be shown that the mean and the variance of n are r/p and $r(1 - p)/p^2$, respectively. Note that these reduce to the geometric mean and variance when $r = 1$.

3.5.6 Hypergeometric Distribution

The last discrete distribution to be presented is the *hypergeometric distribution*. This distribution, like the binomial, has also been used extensively in sampling inspection work. The two distributions are similar in that both assume two possible outcomes. They differ, however, in that the hypergeometric distribution is applicable when sampling is performed without replacement. The following example is used for illustration. Assume it is known that a lot of 1000 condensers contains 12 nonconforming condensers. What is the probability of observing at most one nonconforming condenser when a random sample of 50 is obtained from this lot?

Why is it that the binomial distribution cannot be used to solve this problem using $p = 12/1000 = .012, n = 50$, and $x \leq 1$? Notice a subtle difference between this example and the example with the 1000 capacitors that was used to illustrate the binomial distribution. For the latter there was no sampling from a stated finite population; it was simply a matter of determining the probability of observing at most one nonconforming unit in 1000 when it is known that $p = .01$. For the current problem, however, we cannot say that the probability is .012 that any condenser in the random sample of 50 condensers is nonconforming. This is due to the fact that the probability of any particular condenser being nonconforming depends upon whether or not the previously selected condensers were nonconforming. For example, the probability that the second condenser selected is nonconforming is 12/999 if the first condenser selected is conforming and 11/999 if the first condenser is nonconforming.

Therefore, the binomial distribution cannot be used since p is not constant. We can, however, deduce the answer using the same type of heuristic reasoning as was used for the coin-tossing problem in the section on the binomial distribution. First, how many different samples of 50 are possible out of 1000? Of that number, how many will contain at most one nonconforming unit? The answer to the first question is

$$\binom{1000}{50} = \frac{1000!}{50!\,950!}$$

This is analogous to the earlier example concerning the number of possible ways of obtaining one head in two tosses of a balanced coin:

$$\binom{2}{1} = \frac{2!}{1!1!}$$

Thus, the two tosses were, in essence, partitioned into one for a head and one for a tail. For the current problem the 1000 condensers are partitioned into 50 that will be in the sample and 950 that will not be in the sample.

How many of these $\binom{1000}{50}$ samples will contain at most one nonconforming unit? There are $\binom{12}{0}\binom{988}{50}$ ways of obtaining zero nonconforming units and $\binom{12}{1}\binom{988}{49}$ ways of obtaining exactly one nonconforming unit. Therefore, the probability of having at most one nonconforming condenser in the sample of 50 is

$$\frac{\binom{12}{0}\binom{988}{50} + \binom{12}{1}\binom{988}{49}}{\binom{1000}{50}} = .88254$$

The combinatorics in the numerator [such as $\binom{12}{0}$ and $\binom{988}{50}$] are multiplied together because of a counting rule that states that if one stage of a procedure can be performed in M ways and another stage can be performed in N ways, the two-stage procedure can be performed in MN ways. For this example we can think of the number of ways of obtaining zero nonconforming units out of 12 as constituting one stage and the number of ways of obtaining 50 good items out of 988 as the other stage. Of course, the sample is not collected in two stages, but the sample can be viewed in this manner so as to determine the number of ways that zero nonconforming units can be obtained.

The general form for the hypergeometric distribution is as follows:

$$P(x) = \frac{\binom{D}{x}\binom{N-D}{n-x}}{\binom{N}{n}} \qquad x = 0, 1, 2, \ldots, \ \min\,(n, D)$$

Where N represents the number of items in the finite population of interest, D represents the number of items in the population of the type for which a probability is to be calculated (nonconforming units in the previous example), and $N - D$ represents the number of items of the other type (e.g., conforming units) that are in the population. The sample size is represented by n, of which x are of the type for which the probability is to be calculated and $n - x$ are of the other type. The number D must be at least as large as x; otherwise $\binom{D}{x}$ would be undefined, and, obviously, x cannot exceed n. Thus, x cannot exceed the minimum of n and D [i.e., $\min(n, D)$].

3.6 CONTINUOUS DISTRIBUTIONS

3.6.1 Normal Distribution

It is somewhat unfortunate that any statistical distribution is called "normal" since this could easily create the false impression that this distribution is "typical" and the other hundreds of distributions are "atypical."

The normal distribution (also sometimes called the Gaussian distribution) is the first *continuous* distribution presented in this chapter. A continuous distribution is such that the random variable can assume any value along a continuum. For example, if X is the height of an adult human being, then X can assume any value between, say, 24 inches and 108 inches (Goliath's estimated height). Although a person might actually be 68.1374136 inches tall, that person is not likely to ever have his or her height recorded as such. Thus, random variables that are continuous are, in essence, "discretized" by the use of measuring instruments. Height, for example, is usually not recorded to an accuracy greater than $\frac{1}{4}$ inch. Nevertheless, a normal distribution is often used to approximate the *actual* (unknown) distribution of many random variables. As Geary (1947) pointed out decades ago, there is no such thing as a normal distribution *in practice*.

The equation for the distribution is given by

$$f(x) = \frac{1}{\sigma\sqrt{2\pi}} e^{-(x-\mu)^2/2\sigma^2} \qquad -\infty < x < \infty$$

where μ represents the mean of the distribution, σ represents the standard deviation, π is the mathematical constant 3.14159..., and $-\infty < x < \infty$ indicates that the random variable X can assume any real number. There is actually not just *one* normal distribution since there are different normal distributions for different combinations of μ and σ. The value of σ determines the shape of the distribution, and the value of μ determines the location. This is illustrated in Figures 3.2 and 3.3. [The height of the curve at any point x is given by $f(x)$, but this is usually not of any practical interest.]

In practice, μ and σ are seldom known and must be estimated. This is particularly true when data are to be used in constructing control charts for measurement data.

A normal distribution has an important property that is illustrated in Figure 3.4. The number in each section of the curve denotes the area under the curve in that section. For example, .34134 represents the area under the curve between μ and $\mu + \sigma$, which is also the area between μ and $\mu - \sigma$ since the curve is symmetric with respect to μ. The total area under the curve equals 1, which corresponds to a total probability of 1, with .5 on each side of μ.

The reader should make note of the fact that the area between $\mu - 3\sigma$ and $\mu + 3\sigma$ is .9973 [$= 2(.34134 + .13591 + .02140)$]. Thus, the probability is only $1 - .9973 = .0027$ that a value of the random variable will lie outside this interval.

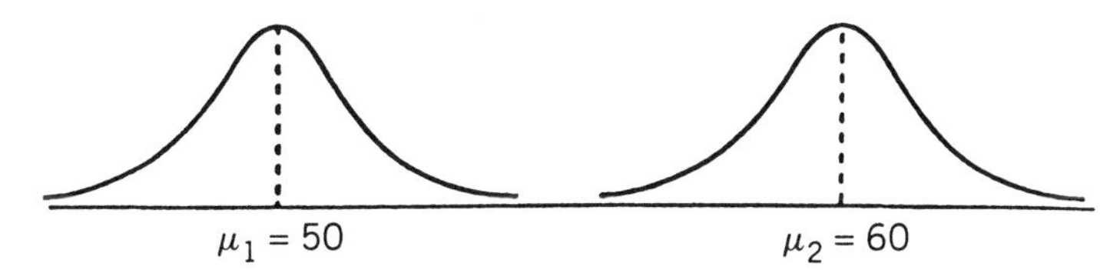

FIGURE 3.2 Two normal distributions with different means but which the same standard deviations ($\sigma_1 = \sigma_2$).

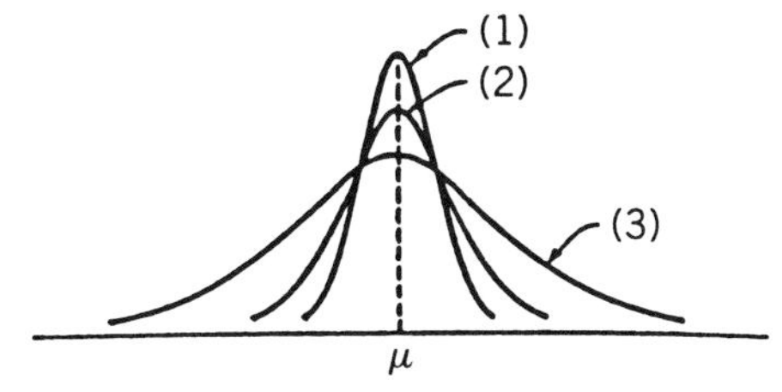

FIGURE 3.3 Three normal distributions with $\sigma_1 < \sigma_2 < \sigma_3$ and with the same mean.

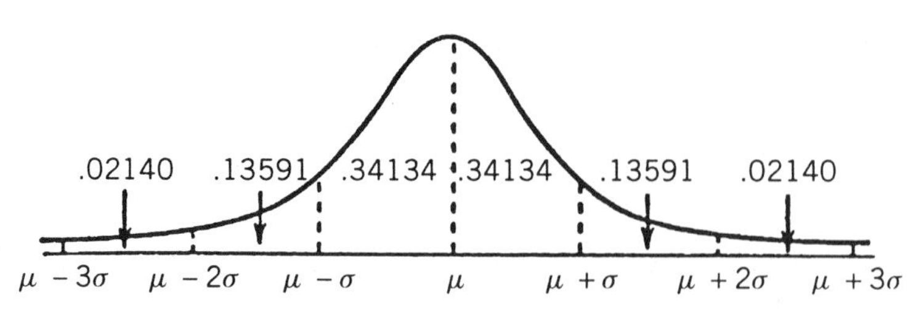

FIGURE 3.4 Areas under a normal curve.

This relates to the "3-sigma limits" on control charts, which are discussed in the following chapters.

The areas given in Figure 3.4 can be determined from Table B in the Appendix to the book. This table is for a *particular* normal distribution, specifically, the distribution with $\mu = 0$ and $\sigma = 1$. The distribution results when the transformation

$$Z = \frac{X - \mu}{\sigma}$$

is used for any normal distribution. This transformation produces the *standard normal distribution* with $\mu = 0$ and $\sigma = 1$, with the value of Z being the number of standard deviations that X is from μ. The transformation must be used before probabilities for any normal distribution can be determined (without the aid of some computing device).

To illustrate, suppose that a shaft diameter has (approximately) a normal distribution with $\mu = .625$ and $\sigma = .01$. If the diameter has an upper specification limit of .65, we might wish to estimate the percentage of shafts whose diameter will exceed .65, in which case a nonconformity will result. Since $.65 = \mu + 2.5\sigma$, we can see from Figure 3.4 that the percentage will be less than 2.275%, since this would be the percentage for the area under the curve beyond $\mu + 2\sigma$. Specifically, we need to determine the shaded area given in Figure 3.5.

By using the transformation

$$Z = \frac{X - \mu}{\sigma}$$

we obtain

$$z = \frac{.65 - .625}{.01} = 2.5$$

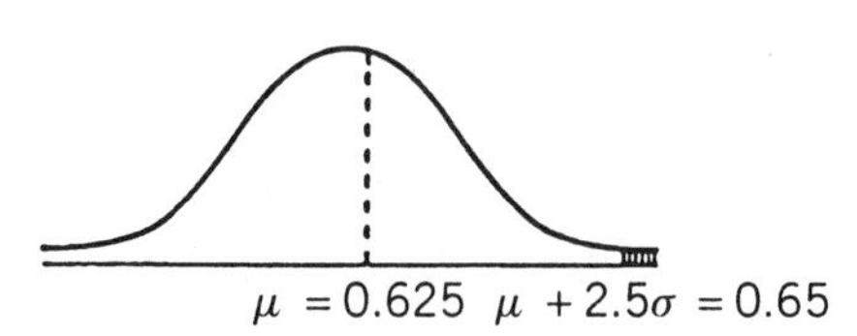

FIGURE 3.5 Distribution of shaft diameter.

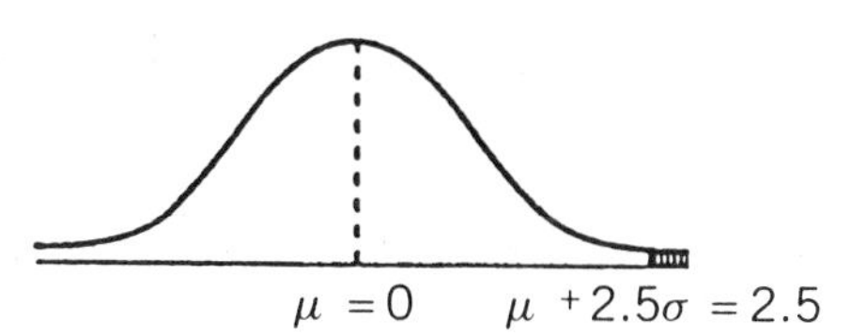

FIGURE 3.6 Standard normal distribution.

Notice that this z-value is the number of standard deviations that the value of x is from μ. This will always be the case.

We can now say that the probability of observing a z-value greater than 2.5 is the same as the probability of observing a shaft diameter in excess of .65. Thus, we need only determine the shaded area in Figure 3.6.

Since this shaded area does not cover $\mu = 0$, we must look up the area for $z = 2.5$ in Table B in the Appendix to the book and subtract it from .5000. We thus obtain $.5000 - .49379 = 0.00621$. Putting all of this together, we obtain

$$P(X > 0.65) = P(Z > 2.5) = .00621$$

Thus, we expect approximately 0.621% of the shafts to not be in conformance with the diameter specification. (The probability is approximate rather than exact since a distribution was assumed that was approximately normal.)

If there had been a lower specification limit, another z-value would have been calculated and the area obtained from that value would be added to .00621 to produce the total percentage of nonconforming shaft diameters. For example, if the lower specification limit were 0.61, this would lead to a z-value of -1.5 and an area of .06681. The total percentage would then be $6.681\% + 0.621\% = 7.302\%$.

In determining areas under the z-curve, it is generally desirable to shade in the appropriate region(s) before going to Table B. This will lessen the chances of making an error such as *subtracting* a number from .5000 for a problem in which the number should instead be *added* to .5000.

The determination of probabilities using the normal distribution should be accomplished in accordance with the following step-by-step procedure:

1. Transform the probability statement on X to the equivalent statement in terms of Z by using the transformation

$$Z = \frac{X - \mu}{\sigma}$$

2. Shade in the appropriate region(s) under the z curve as determined from the probability statement on Z.

3. Find the area(s) in Table B and obtain the answer in accordance with the following.

General Form of the Probability Statement			Action to Be Taken
1.	$P(a < z < b)$	$(a < 0, b > 0)$	Look up the area in the table for $z = -a$ and $z = b$ and add the two areas together to obtain the answer.
2.	$P(a < z < b)$	$(0 < a < b)$	Look up the area for $z = a$ and subtract it from the area for $z = b$.
3.	$P(a < z < b)$	$(a < b < 0)$	Look up the area for $z = -b$ and subtract it from the area for $z = -a$.
4.	$P(z > a)$	$(a > 0)$	Look up the area for $z = a$ and subtract it from 0.5.
5.	$P(z > a)$	$(a < 0)$	Look up the area for $z = -a$ and add it to 0.5.
6.	$P(z < a)$	$(a > 0)$	Look up the area for $z = a$ and add it to 0.5.
7.	$P(z < a)$	$(a < 0)$	Look up the area for $z = -a$ and subtract it from 0.5.
8.	$P(z > a \text{ or } z < b)$	$(a > 0,\ b < 0)$	Look up the area for $z = a$ and $z = -b$, add the two areas together and subtract the sum from 1.0.

It should also be noted that $P(z = a) = 0$ for any value of a. Thus, it is not possible to determine the probability that a shaft will have a diameter of, say, 0.640 feet; a probability can be determined only for an interval. *This is true for any continuous distribution.*

In summary, a normal distribution can often be used to approximate the actual (unknown) distribution of a random variable. We will not, however, encounter a normal distribution *in practice*, so the distribution should be viewed as a model that may or may not be useful in a given situation.

The choice of which distribution to select for the purpose of determining probabilities such as those given in this section depends to a great extent upon how much accuracy is required as well as the availability of tables and/or software. All of the control charts presented in later chapters have an implicit assumption of either a normal distribution (approximately) or the adequacy of the approximation of the normal distribution to another distribution. The consequences of such an assumption when the assumption is untenable will vary from chart to chart. These consequences are discussed when each chart is presented.

3.6.2 *t* Distribution

Although all of the control charts are based upon the assumption of a normal distribution or normal approximation, the statistical techniques presented in Part III utilize other distributions. One such distribution is the t distribution, which is often referred to as *Student's t distribution.* "Student" was the pseudonym used by W. S. Gosset (1876–1937) for the statistical papers that he wrote while employed as a brewer at St. James's Gate Brewery in Dublin, Ireland. The brewery had a rule that prohibited its employees from publishing papers under their own names, hence the need for a pseudonym.

Gosset worked with small samples, often in situations in which it was unreasonable to assume that σ was known. He was also concerned with probability statements on $\overline{X}$ instead of on X. Thus, he was more interested in, say, $P(\overline{X} > a)$ than $P(X > a)$. At the time of his work (c. 1900) it was well known that if X has a normal distribution with mean μ and variance σ^2 [frequently written as $X \sim N(\mu, \sigma^2)$, where $\sim$ is read "is distributed as"], then

$$Z = \frac{\overline{X} - \mu}{\sigma/\sqrt{n}} \tag{3.7}$$

has a normal distribution with $\mu = 0$ and $\sigma = 1$. This stems from the fact that $\overline{X} \sim N(\mu, \sigma^2/n)$ when $X \sim N(\mu, \sigma^2)$. Thus $\overline{X}$ is "standardized" in Eq. (3.7) by first subtracting its mean (which is the same as the mean of X) and then dividing by its standard deviation. This is the same type of standardization that was used for X in obtaining $Z = (X - \mu)/\sigma$. The fact that the mean of $\overline{X}$ is the same as the mean of X is a theoretical result that will not be proven here. It can be easily demonstrated, however, for a finite population using the following example. Assume that a (small) population consists of the numbers 1,2,3, and 4 and we want to find the average of the sample averages in which the averaging is performed over all possible samples of size 2. We would then obtain the results given in Table 3.1.

There are six possible samples of size 2 and the average of the six values for $\overline{X}$ is 2.5. Notice that this is also the average of the numbers 1, 2, 3 and 4. This same result would be obtained for samples of any other size as well as for the populations of any other size.

It should also be observed that there is slightly less variability in the $\overline{X}$ values than in the X values. For infinite populations (or finite populations that are very large)

$$\sigma_{\overline{x}} = \frac{\sigma_x}{\sqrt{n}}$$

where n is the sample size.

TABLE 3.1 Sample Averages

Sample	$\bar{x}$
1, 2	1.5
1, 3	2.0
1, 4	2.5
2, 3	2.5
2, 4	3.0
3, 4	3.5

When n is large (say, $n \geq 30$) the sample standard deviation, s, can be used as a substitute for σ in Eq. (3.7) so as to produce

$$Z = \frac{\overline{X} - \mu}{s/\sqrt{n}} \tag{3.8}$$

and Z will then be approximately normally distributed with $\mu = 0$ and $\sigma^2 \doteq 1$.

Gosset 1908 paper, entitled "The Probable Error of a Mean," led to the t distribution, in which

$$t = \frac{\overline{X} - \mu}{s/\sqrt{n}} \tag{3.9}$$

although his paper did not exactly give Eq. (3.9). The equation for the distribution is

$$f(t) = \frac{1}{\sqrt{(n-1)\pi}} \frac{\Gamma(n/2)}{\Gamma[(n-1)/2]} \left(1 + \frac{t^2}{n-1}\right)^{-n/2} \qquad -\infty < t < \infty$$

although the equation is not generally needed (Γ refers to the gamma function).

Unlike the standard normal distribution and other normal distributions, the shape of the t distribution depends upon the sample size. This is illustrated in Figure 3.7. It can be shown mathematically that the t distribution approaches the standard normal distribution as $n \to \infty$, as is illustrated in Figure 3.7. There is very little difference in the two distributions when $n > 30$, so Eq. (3.8) might be used in place of Eq. (3.9).

Table C in the Appendix to the book is used for the t distribution. Unlike the table for the standard normal distribution (Table B), Table C cannot be used to determine probabilities for the t distribution. Its uses will be illustrated in Part III.

It is worth noting that the t distribution arises not only when averages are used but also for other statistics. In general,

$$t = \frac{\hat{\theta} - \theta}{s_{\hat{\theta}}}$$

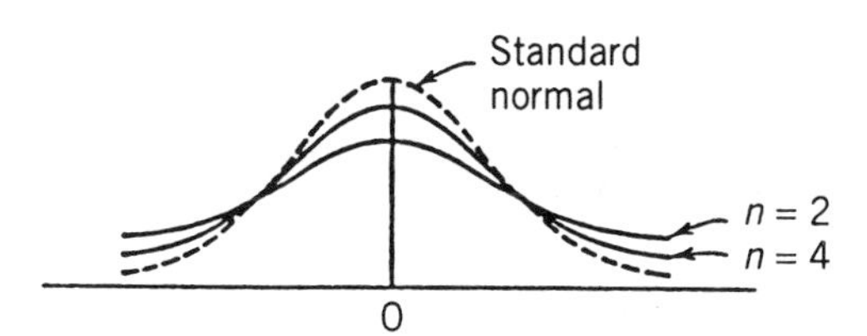

FIGURE 3.7 Student's t distribution for various n.

where θ is the parameter to be estimated, $\hat{\theta}$ is the estimator based upon a sample, and $s_{\hat{\theta}}$ is the sample estimator of $\sigma_{\hat{\theta}}$. The *degrees of freedom* for t are determined by the degrees of freedom for $s_{\hat{\theta}}$. Loosely speaking, a degree of freedom is used when a parameter is estimated. The t statistic in Eq. (3.9) has $n - 1$ degrees of freedom; other t statistics can have fewer degrees of freedom for fixed n.

3.6.3 Exponential Distribution

Another common distribution is the exponential distribution, which is used extensively in the fields of life testing and reliability. It is not needed for any of the following chapters, but it is included here because the life of a manufactured product certainly depends upon its quality.

The equation for the distribution is

$$f(x) = \frac{1}{\theta} e^{-x/\theta} \qquad x > 0 \tag{3.10}$$

where θ is the mean of the distribution. Readers interested in reliability and life testing are referred first to a survey paper by Lawless (1983) and then to books such as Lawless (1982), Nelson (1982), and Mann et al. (1974).

3.6.4 Lognormal Distribution

A distribution that is related to the normal distribution is the lognormal distribution. Specifically, if ln X (read "the natural logarithm of X") were normally distributed, then X would have a lognormal distribution where the equation for the latter is given by

$$f(x) = \frac{1}{\sqrt{2\pi}\sigma} \cdot \frac{1}{x} \exp\left[-\frac{1}{2\sigma^2}(\ln x - \mu)^2\right] \qquad x > 0$$

where σ is the standard deviation of ln X, μ is the mean of ln x, and exp[$\cdot$] represents e raised to the bracketed power. (If we wished to transform the data and work with ln x, the distribution would then be normal rather than lognormal.) Like a normal distribution, the shape of a lognormal distribution depends upon σ. Unlike a normal distribution, however, a lognormal distribution is not symmetric, but a lognormal distribution will be close to a normal distribution when σ is small (say, $\sigma \leq 0.1$).

Ott (1975, p. 4) indicates that a lognormal distribution is often incorrectly assumed when data are unknowingly obtained from a mixture of two normal distributions with different means, with the proportion of data coming from each being considerably different. There is a very important message here. Assume that Figure 3.8 portrays part of a plant layout. There are thus two machines and two machine operators, but the data that are analyzed consist of mixed units from both machines. What if operator #1 is much faster than operator #2, so that most of the units come from Machine #1? If the units produced with each machine have exactly the same distribution [say, normal ($\mu = 50$, $\sigma = 2$)], the fact that there are two different percentages of output coming from each machine will have no effect on the distribution for the units from the two machines combined; it will still be normal ($\mu = 50$, $\sigma = 2$). But what if machine #2 is out of adjustment so that $\mu_2 = 52$? (Assume further that $\sigma_1 \doteq \sigma_2$.) The distribution for each machine might then be as in Figure 3.9.

When the units from the two machines are combined, the single distribution might appear as in Figure 3.10. If the data represented by the distribution in Figure 3.10 were standardized by subtracting the mean and dividing by the standard deviation, the resultant distribution would resemble a standardized lognormal distribution with $\sigma = 0.5$.

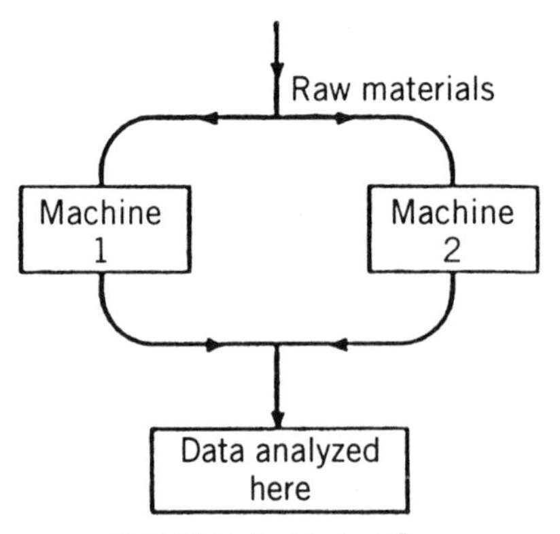

FIGURE 3.8 Product flow.

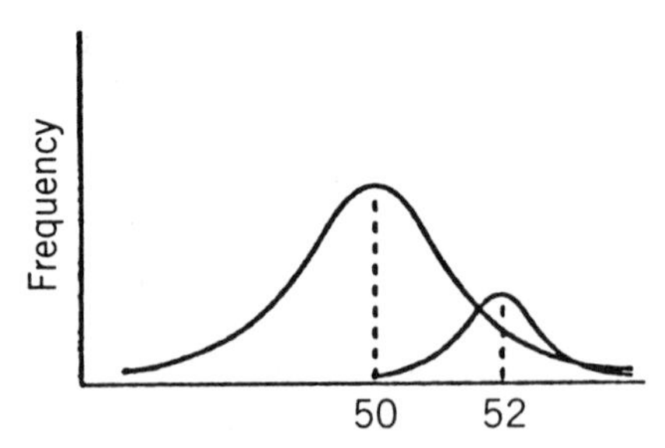

FIGURE 3.9 Distribution of units from machine #1 ($\mu_1 = 50$) and machine #2 ($\mu_2 = 52$).

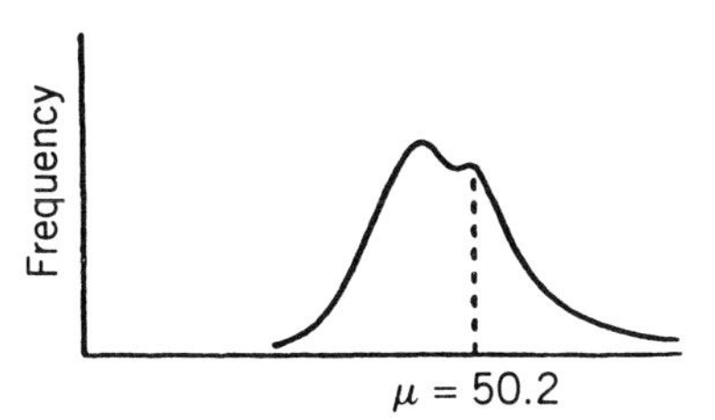

FIGURE 3.10 Distribution of units from the two machines combined.

Thus, the data might appear to have come from a single lognormal distribution, whereas in actuality they came from two normal distributions. For this example, the mixture of the two normal distributions will have a single "hump" (i.e., be unimodal) provided that $\sigma^2 > 32/27$, assuming that $\sigma_1^2 = \sigma_2^2$ and $\mu_1 = 50$ and $\mu_2 = 52$ [see Johnson and Kotz (1970, Vol. 2, p. 89)]. Since the lognormal distribution has only one hump, a mixture of two normals that produces two humps (i.e., in this case if $\sigma^2 \leq 32/27$) could not be mistaken for a lognormal.

This discussion of the lognormal distribution illustrates how easily erroneous conclusions can be drawn if data are not carefully analyzed. In general, data should be collected and analyzed in such a way that different causes of variation in the data (such as different machines) can be easily identified.

The lognormal distribution has found some application in quality improvement work, as is described, for example, in Morrison (1958).

3.6.5 Weibull Distribution

Like an exponential distribution, a Weibull distribution has been used extensively in life testing and reliability. The equation for the distribution is

$$f(x) = \alpha\beta(\alpha x)^{\beta-1}e^{-(\alpha x)^\beta} \qquad x > 0$$

where $\alpha > 0$ and $\beta > 0$ are parameters of the distribution. As with other distributions, the shape of a Weibull distribution depends upon the values of the parameters. When $\beta = 1$, a Weibull reduces to an exponential distribution. References on how a Weibull distribution can be used in quality improvement work include Berrettoni (1964).

3.6.6 Gamma Distribution

The gamma distribution is actually a family of distributions represented by the equation

$$f(x) = \frac{1}{\beta^\alpha \Gamma(\alpha)} x^{\alpha-1} e^{-x/\beta} \qquad x > 0 \tag{3.11}$$

where $\alpha > 0$ and $\beta > 0$ are parameters of the distribution and Γ refers to the gamma function. In the special case in which $\alpha = 1$ we obtain the exponential distribution where β in Eq. (3.11) corresponds to θ in Eq. (3.10).

3.6.7 Chi-Square Distribution

This distribution is a special case of the gamma distribution .Specifically, the chi-square distribution is obtained by letting $\alpha = r/2$ and $\beta = 2$ in Eq. (3.11), where r is the degrees of freedom of the chi-square distribution.

3.6.8 Truncated Normal Distribution

This distribution has not been previously used to any extent in quality improvement work, but it is presented here because it is needed for the chapter on Taguchi methods (Chapter 14).

It results when a random variable has a normal distribution but there is a lower bound and/or an upper bound for values of the random variable. When there is only a lower bound, the distribution is said to be *left truncated*, and when there is only an upper bound, the distribution is *right truncated*. When the truncation point is at the mean, this produces a half-normal distribution. When both bounds exist the distribution is said to be *doubly truncated*. The latter is discussed in detail by Johnson and Kotz (1970), whereas singly truncated normal distributions are discussed, at a more elementary level, by Meyer (1970) as well as in other statistical texts, including Nelson (1982).

Only singly truncated normal distributions will be discussed here. A left-truncated normal distribution is represented by the equation

$$f(x) = \frac{k}{\sigma\sqrt{2\pi}} e^{-1/2[(x-\mu)/\sigma]^2} \qquad x \geq a$$

where

$$k = \left[1 - \Phi\left(\frac{a-\mu}{\sigma}\right)\right]^{-1}$$

a is the truncation point, and $\Phi[(a-\mu)/\sigma]$ is the cumulative area under the normal curve at the value of $(a-\mu)/\sigma$.

It can be shown that the mean of the distribution, $E(x)$, is given by

$$E(x) = \mu + k\sigma f\left(\frac{a-\mu}{\sigma}\right)$$

where k is as previously defined, and

$$f\left(\frac{a-\mu}{\sigma}\right) = \frac{1}{\sqrt{2\pi}} e^{-1/2[(a-\mu)/\sigma]^2}$$

The variance is given by the expression

$$\text{Var}(x) = \sigma^2\{1 + cf(c)[1 - \Phi(c)]^{-1} - [f(c)]^2[1 - \Phi(c)]^{-2}\} \tag{3.12}$$

where $c = (a - \mu)/\sigma$. [See, e.g., Nelson (1982), p. 65.]

For a right-truncated normal distribution with truncation point a', the variance can be found by using Eq. (3.12), where the value of a would be chosen so that $a' - \mu = \mu - a$.

The mean can be found from

$$E(x) = \mu - \sigma\left[\Phi\left(\frac{a' - \mu}{\sigma}\right)\right]^{-1} f\left(\frac{a' - \mu}{\sigma}\right)$$

where

$$f\left(\frac{a' - \mu}{\sigma}\right) = \frac{1}{\sqrt{2\pi}} e^{-1/2[(a' - \mu)/\sigma]^2}$$

The equation for the distribution is given by

$$f(x) = \frac{1}{(\sigma\sqrt{2\pi})\{\Phi[(a' - \mu)/\sigma]\}} e^{-1/2[(x - \mu)/\sigma]^2} \qquad x \leq a'$$

3.6.9 Bivariate and Multivariate Normal Distributions

These distributions also have not been used to any extent in quality improvement work. They are presented here because they are needed for the material on multivariate control charts in Chapter 9.

Recall that the equation for the normal distribution is

$$f(x) = \frac{1}{\sigma\sqrt{2\pi}} e^{-(x - \mu)^2/2\sigma^2} \qquad -\infty < x < \infty$$

If we have two *independent* process characteristics, x_1 and x_2, with means and standard deviations given by μ_1, μ_2 and σ_1, σ_2, respectively, their joint distribution would be given by

$$f(x_1, x_2) = \frac{1}{2\pi\sigma_1\sigma_2} \exp[-(x_1 - \mu_1)^2/2\sigma_1^2 - (x_2 - \mu_2)^2/2\sigma_2^2] \tag{3.13}$$

where, as indicated previously, $\exp[\cdot]$ represents e raised to the bracketed power.

Using matrix notation, Eq. (3.13) becomes

$$f(\mathbf{x}) = \frac{1}{2\pi|\mathbf{\Sigma}|^{1/2}} \exp\left[-\frac{1}{2}(\mathbf{x} - \mu)'\mathbf{\Sigma}^{-1}(\mathbf{x} - \mu)\right]$$

where

$$\mathbf{x} = \begin{bmatrix} x_1 \\ x_2 \end{bmatrix} \qquad \boldsymbol{\mu} = \begin{bmatrix} \mu_1 \\ \mu_2 \end{bmatrix} \qquad \boldsymbol{\Sigma} = \begin{bmatrix} \sigma_1^2 & 0 \\ 0 & \sigma_2^2 \end{bmatrix}$$

$\boldsymbol{\Sigma}^{-1}$ is the inverse of the matrix $\boldsymbol{\Sigma}$ and $|\boldsymbol{\Sigma}|$ denotes the determinant of that matrix. For the general case of p variables the expression is

$$f(\mathbf{x}) = \frac{1}{(2\pi)^{p/2}|\boldsymbol{\Sigma}|^{1/2}} \exp\left[-\tfrac{1}{2}(\mathbf{x}-\boldsymbol{\mu})'\boldsymbol{\Sigma}^{-1}(\mathbf{x}-\boldsymbol{\mu})\right] \qquad (3.14)$$

where $\mathbf{x}$, $\boldsymbol{\mu}$, and $\boldsymbol{\Sigma}$ contain p, p, and p^2 elements, respectively.

The p variables $x_1, x_2, \ldots, x_p$ will generally not be independent, but Eq. (3.14) is the general expression for the *multivariate normal distribution*, regardless of whether the variables are independent.

When $p = 2$ and the variables are not independent, the *bivariate normal distribution* would be written as

$$f(x_1, x_2) = \frac{1}{2\pi\sigma_1\sigma_2\sqrt{1-\rho^2}} \exp\left\{-\frac{1}{2(1-\rho^2)}\left[\left(\frac{x_2-\mu_1}{\sigma_1}\right)^2 - 2\rho\frac{(x_1-\mu_1)(x_2-\mu_2)}{\sigma_1\sigma_2} + \left(\frac{x_2-\mu_2}{\sigma_2}\right)^2\right]\right\}$$

which could also be written in the general matrix form given by Eq. (3.14). (Here ρ designates the correlation between x_1 and x_2.)

Readers requiring additional information on matrix algebra are referred to Searle (1982) and to a multivariate text such as Morrison (1990) for further reading on the multivariate normal distribution.

3.6.10 *F* Distribution

This is another distribution that is not generally used in control charting, but it is needed for a few of the statistical procedures covered in Part III, as well as in Chapter 9.

If two independent chi-square random variables are each divided by their respective degrees of freedom and a fraction formed from these two fractions, the result is a random variable that has an F distribution. That is.

$$F_{\nu_1\nu_2} = \frac{x_{\nu_1}^2/\nu_1}{x_{\nu_2}^2/\nu_2} \qquad (0 < F < \infty)$$

where ν_1 and ν_2 are the degrees of freedom for each of the chi-square random variables and are also the numerator and denominator degrees of freedom, respectively, for the random variable denoted here by F (which has an F distribution).

The shape of the distribution depends upon ν_1 and ν_2, so, as with the other distributions discussed in this chapter, there is not a single F distribution, but rather a family of such distributions.

Table D in the Appendix to the book gives the values of F for different combinations of ν_1 and ν_2 as well as for $\alpha = .01$, .05, and .10, where α is the upper tail area (the upper tail is the only one that is used in most statistical procedures).

3.6.11 Beta Distribution

Although generally of more theoretical than practical interest, the beta distribution is needed in Chapter 9, so it is presented here. The probability density function is given by

$$f(x) = \frac{(r+s-1)!}{(r-1)!(s-1)!}x^{r-1}(1-x)^{s-1} \qquad 0 \le x \le 1$$

and the parameters are r and s. The percentiles of a beta distribution are not obtainable with some commonly used statistical software, so it is sometimes useful to use the relationship between the beta and F distributions. That relationship is as follows. Let $B(\alpha; r, s)$ denote the $1-\alpha$ percentile of a beta distribution with parameters r and s. Then

$$B(\alpha; r, s) = \frac{(2r/2s)F(\alpha; 2r, 2s)}{1 + (2r/2s)F(\alpha; 2r, 2s)}$$

The mean of a random variable that has a beta distribution is $r/(r+s)$, and the variance is $rs(r+s)^{-2}(r+s+1)^{-1}$.

3.6.12 Uniform Distribution

There is a discrete uniform distribution and a continuous uniform distribution. For the former, the probability is the same for each possible value of the random variable, whereas for the latter the probability is the same for intervals of possible values that have equal length. The continuous uniform distribution is covered here since it is used in Section 14.3; for that section it is useful to view the distribution as in Figure 3.11.

It should be intuitively apparent that the mean for the distribution depicted in Figure 3.11 is a, and the variance can be shown to be $\frac{1}{3}h^2$. Thus, if $h = 5$, as in Section 14.3, the variance is $\frac{25}{3}$.

3.7 CHOICE OF STATISTICAL DISTRIBUTION

It is important to realize that the distributions that are given in Section 3.5 and 3.6 are simply models of reality. As stated previously. Geary (1947) pointed out

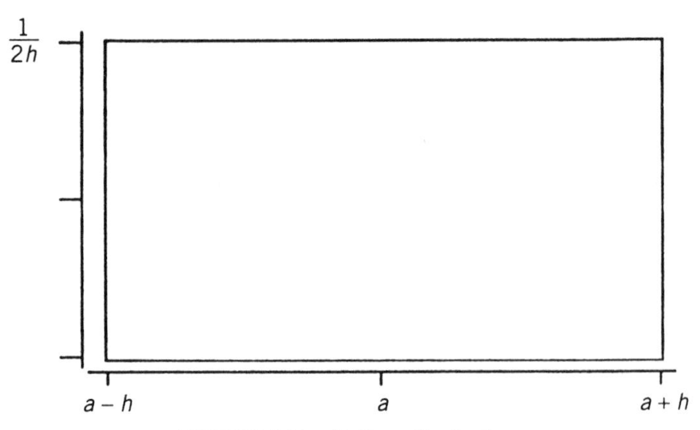

FIGURE 3.11 Uniform distribution.

that there is no such thing as a normal distribution (in practice), and this point is also made in various other books and articles, including Box and Luceño (1997). Even a binomial distribution might not be the appropriate distribution in a practical setting. For example, in studying nonconforming units, even though there are two possible outcomes (conforming and nonconforming), units of production may not be independent in terms of their classification as conforming or nonconforming. If so, then it may be necessary to use a different distribution, such as a beta–binomial distribution (Section 3.5.2).

3.8 STATISTICAL INFERENCE

Various statistical distributions were presented in the preceding two sections. In this section we present methods of statistical inference that can be used for estimating the parameters of those distributions, and which are also needed to thoroughly understand control charts as well as the statistical methods presented in Part III.

3.8.1 Central Limit Theorem

It was stated in the section on the t distribution that $\overline{X} \sim N(\mu, \sigma^2/n)$ when $X \sim N(\mu, \sigma^2)$. When the distribution of X is unknown (the usual case), the distribution of $\overline{X}$ is, of course, also unknown. When the sample size is large, however, the distribution of $\overline{X}$ will be approximately normal. How large must the sample size be? That depends upon the shape of the distribution of X. If the distribution differs very little from a normal distribution (e.g., a chi-square distribution with a moderate number of degrees of freedom), a sample size of 15 or 20 may be sufficient. At the other extreme, for distributions that differ greatly from a normal distribution (e.g., an exponential distribution), sample sizes in excess of 100 will generally be required.

Stated formally, if $X_1, X_2, \ldots, X_n$ constitute a sequence of independent random variables (not necessarily identically distributed) with means $\mu_1, \mu_2, \ldots, \mu_n$ and variances $\sigma_1^2, \sigma_2^2, \ldots, \sigma_n^2$, then

$$Z = \frac{\sum_{i=1}^{n} X_i - \sum_{i=1}^{n} \mu_i}{\sqrt{\sum_{i=1}^{n} \sigma_i^2}}$$

approaches the standard normal distribution [i.e., $N(0, 1)$] as n approaches infinity. We may alternatively express the result in terms of the sample average.

Assume that the X_i are not identically distributed. The theorem applies if there are no dominant effects among the X_i. This often happens in quality improvement work, where a particular effect may result from the summation, loosely speaking, of many small effects. When we take a random sample, the X_i will all have the same distribution.

The central limit theorem has been cited as forming the underlying foundation for many of the control charts presented in subsequent chapters. We will see, however, that the "normal approximations" that are used for determining the control limits for several charts will often be inadequate. This will be observed for the attribute control charts presented in Chapter 6.

3.8.2 Point Estimation

The distributions that were presented in Sections 3.5 and 3.6 all have one or more *parameters* that were, for the most part, represented by Greek letters. These (population) parameters are generally estimated by *sample statistics*. For example, μ is estimated by $\overline{X}$, σ^2 is estimated by S^2, and p is estimated by the sample proportion $\hat{p}$. In control chart methodology, however, the sample statistics are represented by slightly different symbols. This will be explained when the various control charts are presented in subsequent chapters. The important point is that the values of these parameters are generally unknown and must be estimated before control chart analyses and other types of statistical analyses can be performed. A *point estimate* is one type of estimate that can be obtained; in Section 3.8.3 another type of estimate is presented.

3.8.2.1 Maximum Likelihood Estimation

One method of obtaining point estimates, which is discussed later in Section 5.3, is *maximum likelihood* estimation. This is a method for obtaining point estimates such that the probability of observing the data that are in a sample that has been obtained is maximized. When viewed in this manner, it is a very intuitive approach.

We need to distinguish between *point estimators* and *point estimates*. The former refers to the form of the estimator (e.g., $\overline{X}$), whereas the later refers to the numerical value of an estimator (e.g, $\overline{X} = 24.2$).

The *likelihood function* is the probability associated with the sample. Specifically, for a sample of size n this is obtained as $f(x_1, x_2, \ldots, x_n) = f(x_1) \cdot$

$f(x_2)\cdots f(x_n)$. The function for the normal distribution was given in Section 3.6.1, and it was stated in Section 3.8.2 that μ is estimated by $\overline{X}$. We will show that the maximum likelihood estimator of μ for a normal distribution is, in fact, $\overline{X}$. For a normal distribution, the likelihood function $L(\mu, \sigma^2; x_1, x_2, \ldots, x_n)$ is

$$L(\mu, \sigma^2; x_1, x_2, \ldots, x_n) = \left(\frac{1}{\sigma\sqrt{2\pi}}\right)^n \exp\left[-\frac{1}{2\sigma^2}\sum_{i=1}^{n}(x_i - \mu)^2\right]$$

Because of the exponential term, it is convenient to maximize the natural logarithm of L rather than maximizing L. Thus,

$$\frac{\partial \ln(L)}{\partial \mu} = \frac{1}{\sigma^2}\sum_{i=1}^{n}(x_i - \mu)$$

and setting the derivative equal to zero leads to the solution $\hat{\mu} = \overline{X}$ since $\sum_{i=1}^{n}(x_i - \overline{x}) = 0$.

3.8.3 Confidence Intervals

A confidence interval is an *interval estimator* in which an experimenter knows that an interval that he or she is about to construct will contain the unknown value of a parameter with (approximately) a specified probability. For example, a 95% confidence interval for μ is one in which an experimenter is 95% confident that the interval that is about to be constructed will contain μ. (The degree of confidence can only be approximate, however, because the true distribution is unknown.)

The desired degree of confidence determines the width of the interval—an increase in the width will increase the degree of confidence (for a fixed sample size). Increasing the sample size will decrease the width of a confidence interval. Narrow confidence intervals are more meaningful than wide confidence intervals, as narrow intervals reflect less uncertainty in regard to the range of possible values of the parameter. As pointed out by Hahn and Meeker (1993), however, a low level of uncertainty is contingent upon the requisite assumptions being met. Since the assumptions are not likely to be met exactly (especially distributional assumptions), caution should be exercised in interpreting confidence intervals.

Confidence intervals that utilize either t or z are always of the form

$$\hat{\theta} \pm t(\text{or } z)\ s_{\hat{\theta}}$$

where θ is the parameter to be estimated, $\hat{\theta}$ is the *point estimator* (the value of which is called the point estimate) of that parameter, and $s_{\hat{\theta}}$ is the estimated standard deviation of the point estimator.

For example, a large-sample $100(1 - \alpha)\%$ confidence interval for μ would be of the form

$$\overline{x} \pm z_{\alpha/2} s_{\overline{x}}$$

so that a 95% confidence interval would be of the form

$$\bar{x} \pm 1.96 s_{\bar{x}}$$

Confidence intervals are not always symmetric about $\hat{\theta}$. One example is a confidence interval for σ^2 (using the chi-square distribution) where s^2 is not in the middle of the interval.

Confidence intervals can also be one-sided. For example, a lower (large-sample) 95% confidence bound for μ would be of the form

$$\bar{x} - 1.645 s_{\bar{x}}$$

where an experimenter would be 95% confident that μ is greater than or equal to the value that will be obtained for this lower bound.

In general, a 95% confidence interval means that if 100 samples were obtained and 100 intervals constructed, our best guess would be that 95 would contain θ, as this is the expected number. If successive samples were taken, the results might appear as in Figure 3.12. A number of statistics texts contain a table that provides the general form of the confidence interval for a variety of parameters [see, e.g., Montgomery and Runger (1999, inside cover)].

3.8.4 Tolerance Intervals

Confidence intervals should not be confused with *statistical tolerance intervals;* the latter are statements on the *proportion* of the values in a population that will lie within an interval with a certain probability. For example, there is a probability of .90 that at least 95% of the population values for a normal distribution are contained in the interval $\bar{x} \pm 3.018s$, where $\bar{x}$ and s are computed from $n = 10$ observations. The reader is referred to Hahn (1970a,b) and Hahn and Meeker, (1991, 1993) for a discussion of tolerance intervals and how they differ from confidence intervals and prediction intervals.

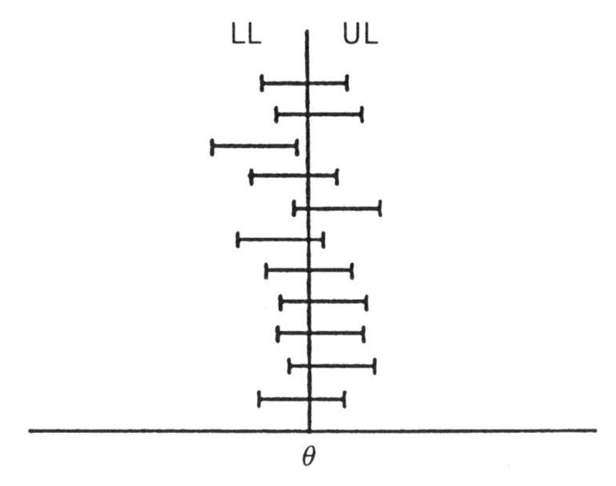

FIGURE 3.12 Confidence intervals (LL, lower limit; UL, upper limit).

3.8.5 Hypothesis Tests

Hypothesis tests are covered in detail in introductory statistics texts and will not be treated extensively here. The general idea is to formulate a hypothesis that is to be tested (e.g., hypothesizing that data have come from a normal distribution) and then use data to test the hypothesis and determine whether or not the hypothesis should be rejected. The hypothesis that is being tested is called the *null hypothesis*, which is tested against an *alternative hypothesis*. Hypothesis tests are used implicitly when control charts are employed.

Hypothesis tests using t or z are of the form

$$t(\text{or } z) = \frac{\hat{\theta} - \theta}{s_{\hat{\theta}}}$$

which is essentially a rearrangement of the components of the corresponding confidence interval.

It is worth noting that, in general, hypotheses that are tested are never true, as has been pointed out by, for example, Tukey (1991). For example, true values of parameters are never equal to hypothesized values. To illustrate, if we hypothesize that the mean of some random variable is 100, the mean is almost certainly not going to be exactly 100. It might be 99.7 or 100.2, but we would not expect it to be 100. Similarly, if we hypothesize that a particular set of data has come from a specified distribution, the hypothesis will almost certainly be false. The assumption of a normal distribution is frequently made (and tested), but as was mentioned in Section 3.6.1, normal distributions do not occur in practice.

We should also recognize that a hypothesis that is being tested is more likely to be rejected for a very large sample size than for a very small sample size. For example, a sample of size 100 might lead to rejection of $\mu = 100$ when the population mean is actually 100.2, whereas a larger difference between the actual mean and the hypothesized mean would be required for rejection when the sample size is much smaller than 100.

The reader is referred to Nester (1996) for practical advice regarding hypothesis tests and related topics.

3.8.5.1 Probability Plots

Although hypothesis tests are of limited usefulness, the use of probability plots to test distributional assumptions is common and does have some value. We *might* view this as a type of hypothesis test. If we construct a *normal probability plot*, we are (apparently) checking to see if the data could have come from a population that could be represented by a normal distribution. Of course, we know that the data could not have come from a normally distributed population since there is no such thing, so what we are actually checking? If the actual, unknown distribution is not much different from a normal distribution, a normal probability plot will generally suggest that a normal distribution provides a reasonable fit to the data.

Thus, it is better to think about a probability plot relative to distribution fitting than viewing it as a form of hypothesis test.

It was mentioned in Section 2.3 that there are various ways to construct a probability plot. One approach is to transform the observations to what they would be if the data could be fit by the distribution in question. Then, when the transformed values are plotted against the raw values, the points will generally form essentially a straight line if the hypothesized distribution is a good representation of the actual distribution. Another approach is to order the values from smallest to largest and then plot the cumulative probability against the ordered values, with the vertical axis being scaled in accordance with the hypothesized distribution.

The following examples illustrate the second approach. Figure 3.13 is a normal probability plot of 50 observations that were generated from a normal distribution with $\mu = 25$ and $\sigma^2 = 5$. The plot was generated using MINITAB, Release 11.

Notice that the points essentially form a straight line. Notice also that statistics are provided in addition to the graph. The value of the Anderson–Darling statistic (Anderson and Darling, 1952) is given, in addition to the corresponding p-value. The latter should be large, generally much larger than .05, when the hypothesized distribution is adequate. We can see that is the case here, so the graph gives the correct message that the assumption of approximate normality is appropriate.

Now consider data from a nonnormal distribution. The chi-square distribution, mentioned briefly in Section 3.6.7, is highly skewed when the degrees of freedom (d.f.) is small. Figure 3.14 shows the distribution when d.f. $= 7$, and Figure 3.15 shows the corresponding normal probability plot of 50 data values from this distribution. Regarding the latter, notice in particular the points that lie below the line in the left portion of the plot. Since the distribution is skewed to the

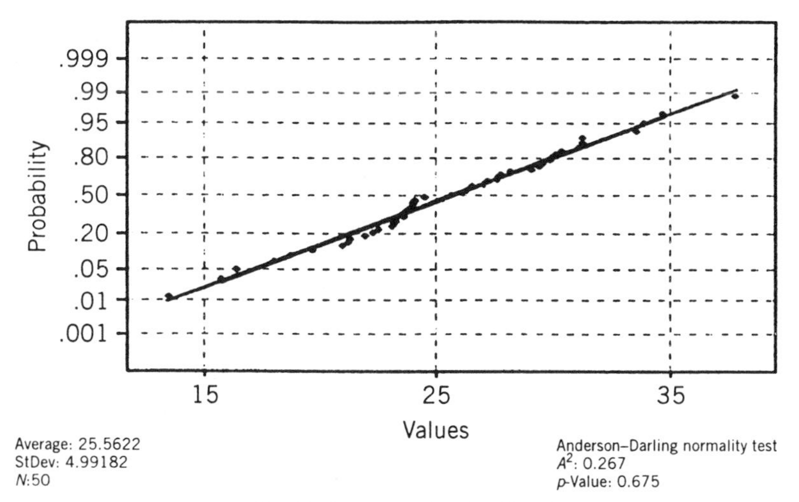

FIGURE 3.13 Normal probability plot.

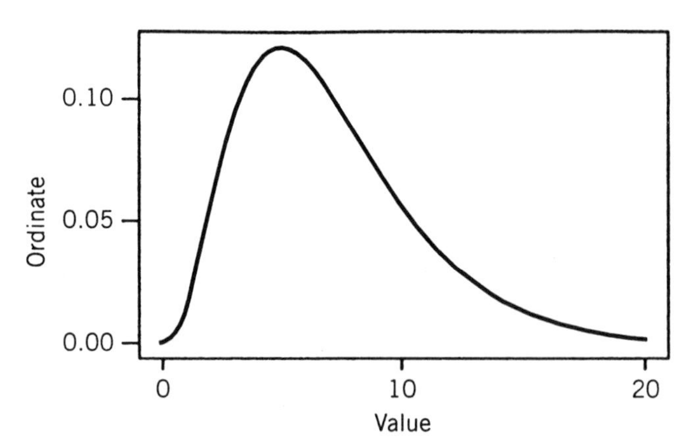

FIGURE 3.14 Graph of χ^2_7 distribution.

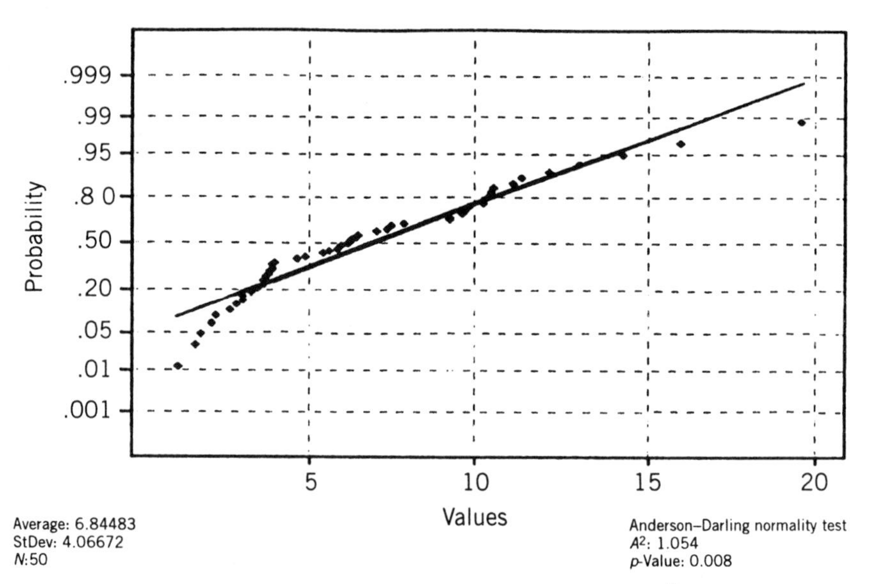

FIGURE 3.15 Normal probability plot of 50 observations from χ^2_7 distribution.

right, there will be large increases in the cumulative probability for small values of the random variable. Therefore, the plotted points will have a steep slope for small values. Conversely, since most of the probability is concentrated at small values, it follows that there will be less probability in the region of large values than is the case for a normal distribution. Therefore, the points at large values plot below the line.

If a chi-square probability plot were constructed for these points, the points would lie approximately on a straight line.

3.8.5.2 Likelihood Ratio Tests

There is a discussion of a specific likelihood ratio test in Section 5.3, so the method is presented here. The general idea is to form the ratio of the likelihood function using the hypothesized value and the likelihood function using an alternative value.

To illustrate, assume that a certain type of defect (nonconformity) can be modeled by a Poisson distribution. A mean value of $\lambda = 7$ has been the norm, but a reduction to $\lambda = 4$ would constitute significant improvement. Therefore, we would like to devise a test for detecting a change from $\lambda = 7$ to $\lambda = 4$.

A *likelihood ratio test* for this scenario would be constructed in the following way. Let H_0: $\lambda = 7$ denote the null hypothesis, and H_1: $\lambda = 4$ represent the alternative hypothesis. We need to first compute the likelihood function under H_0 and under H_1. We will denote these as L_0 and L_1, respectively. Thus,

$$L_0(x) = \frac{e^{-7}7^x}{x!} \quad \text{and} \quad L_1(x) = \frac{e^{-4}4^x}{x!}$$

Then, $\lambda(x) = L_0(x)/L_1(x) = e^{-3}(1.75)^x$, with x denoting the number of nonconformities. We need a decision rule for $\lambda(x)$ such that we have a small probability of rejecting $\lambda = 7$ when it is true and a high probability of accepting $\lambda = 4$ when it is true. But $\lambda(x)$ is not a probability distribution, so we cannot work with it directly. For various likelihood ratio tests, $-2[\log(\lambda)]$ will have approximately a chi square distribution for a large sample size. However, we do not have a sample size, as such, in applications of the Poisson distribution. Rather, we have "areas of opportunity" that might be a week, a square yard, etc. Therefore, we might proceed as follows, recognizing that $\lambda(x)$ is a strictly increasing function of x. We can determine the probability of rejecting the null hypothesis when it is true by working directly with the distribution of X. Assume that we specify a probability of .01 of rejecting H_0 when it is true. Then we would set

$$\sum_{x=0}^{a} \frac{e^{-7}7^x}{x!} = .01$$

and solve for a. This can be easily done using available software. For example, using the INVCDF function in MINITAB, we obtain $x = 1$, which corresponds to a cumulative probability of .0073 (the closest we can come to .01). Thus, we would conclude that λ has become smaller than 7 if we observe either 0 or 1 nonconformities. Notice that this is not a test against $\lambda = 4$, however. Rather, this is actually a test of H_0 against H_1: $\lambda < 7$.

See, for example, Montgomery and Runger (1999), for a detailed explanation of likelihood ratio tests.

3.8.6 Bonferroni Intervals

Bonferroni intervals are illustrated in Chapter 9 relative to multivariate control charts. They are based on a common form of the Bonferroni inequality. When the latter is applied to confidence intervals, the result is as follows. If k $100(1-\alpha)\%$ confidence intervals are constructed, one for each of k parameters, the probability that every interval contains the unknown parameter value that it estimates is at least $1-k\alpha$. Thus, if is desired to have the probability of coverage for all k intervals equal to at least $1-\alpha$, each interval should then be a $100(1-\alpha/k)\%$ confidence interval. This has application in multivariate control charting in which there are p quality characteristics, and p intervals could be constructed to determine which characteristics are causing an apparent out-of-control condition. (The reader will observe in Chapter 9, however, that the Bonferroni approach can produce rather conservative results, especially when there are high correlations among a set of variables.)

See Alt (1982) for additional information on Bonferroni inequalities and intervals.

3.9 ENUMERATIVE STUDIES VERSUS ANALYTIC STUDIES

The basic concepts that have been presented in this chapter are primarily applicable to enumerative studies. [The term is due to Deming (1975).] An enumerative study is conducted for the purpose of determining the "current state of affairs" relative to a fixed frame (population). For example, if a large company wanted to estimate the number of clerical errors per hour made by clerical workers of a particular classification, a random sample from this class might be selected, and the average number of clerical errors made by *those* workers used as the estimate for all of the clerical workers of this type in the company.

By contrast, in an analytic study attention would be focused upon determining the *cause(s)* of the errors that were made with an eye toward reducing the number. Having an estimate of the number is obviously important, but trying to reduce the number is more important. Control charts could be used to monitor performance over time, and other graphical aids could also be employed. Special causes of variation in clerical errors might come to light, and some of the causes could be removable.

Statistical theory is not as easily applied to analytic studies as it is to enumerative studies. Even when an experiment is conducted to identify factors that may significantly affect process yield, so that process yield might be increased in the future, the results of the experiment can be applied, strictly speaking, only to future production in which the conditions are the same as those under which the experiment was conducted. (See Section 13.4 for additional discussion of this point.)

We may summarize by stating that the distinction between enumerative studies and analytic studies is essentially the difference between making inferential and descriptive statements regarding a fixed frame (a fixed list of population elements), versus determining how to improve future performance. Deming (1975) contains a detailed discussion of the difference between these two types of studies. See also Deming (1953), in which the author states that with an enumerative study one attempts to determine "how many," whereas with an analytical study one tries to determine "why." A good discussion of the difference between the two types of studies is given by Hahn and Meeker (1993), who also give various illustrative examples.

REFERENCES

Alt, F. B. (1982). Bonferroni inequalities and intervals. In S. Kotz and N. Johnson, eds. *Encyclopedia of Statistical Sciences*, Vol. 1, pp. 294–300. New York: Wiley.

Anderson, T. W. and D. A. Darling (1952). Asymptotic theory of certain goodness of fit criteria based on stochastic processes. *Annals of Mathematical Statistics 23*: 193–212.

Berrettoni, J. N. (1964). Practical applications of the Weibull distribution. *Industrial Quality Control 21*(1): 71–79.

Box, G. E. P. and A. Luceño (1997). *Statistical Control by Monitoring and Feedback Adjustment*. New York: Wiley.

Deming, W. E. (1953). On the distinction between enumerative and analytic surveys. *Journal of the American Statistical Association 48*(262): 244–255.

Deming, W. E. (1975). On probability as a basis for action. *American Statistician 29*(4): 146–152.

Geary, R. C. (1947). Testing for normality. *Biometrika 34*: 209–242.

Hahn, G. J. (1970a). Statistical intervals for a normal population. Part I. Tables, examples and applications. *Journal of Quality Technology 2*(3): 115–125.

Hahn, G. J. (1970b). Statistical intervals for a normal population. Part II. Formulas, assumptions, some derivations. *Journal of Quality Technology 2*(4): 195–206.

Hahn, G. J. and W. Q. Meeker (1991). *Statistical Intervals: A Guide for Practitioners*. New York: Wiley.

Hahn, G. J. and W. Q. Meeker (1993). Assumptions for statistical inference. *The American Statistician 47*(1): 1–11.

Harvard Computation Laboratory (1955). *Tables of the Cumulative Binomial Probability Distribution*. Cambridge, MA: Harvard University Press.

Johnson, N. L. and S. Kotz (1970). *Distributions in Statistics: Continuous Univariate Distributions — 1*. New York: Wiley.

Lawless, J. F. (1982). *Statistical Models and Methods for Lifetime Data*. New York: Wiley.

Lawless, J. F. (1983). Statistical methods in reliability. *Technometrics 25*(4): 305–316 (Discussion: pp. 316–335).

Mann, N. R., R. E. Schafer, and N. D. Singpurwalla (1974). *Methods for Statistical Analysis of Reliability and Lifetime Data*. New York: Wiley.

Meyer, P. L. (1970). *Introductory Probability and Statistical Applications*, 2nd ed. Reading, MA: Addison-Wesley.

McCullagh, P. and J. A. Nelder (1989). *Generalized Linear Models*, 2nd ed. New York: Chapman and Hall.

Montgomery, D. C. and G. C. Runger (1999). *Applied Statistics and Probability for Engineers*, 2nd ed. New York: Wiley.

Morrison, D. F. (1990). *Multivariate Statistical Methods*. 3rd ed. New York: McGraw-Hill.

Morrison, J. (1958). The lognormal distribution in quality control. *Applied Statistics* 7(3): 160–172.

Nelson, W. (1982). *Applied Life Data Analysis*. New York: Wiley.

Nester, M. R. (1996). An applied statistician's creed. *Applied Statistics 45*, 401–410.

Ott, E. R. (1975). *Process Quality Control*. New York: McGraw-Hill.

Rocke, D. M., G. W. Downs, and A. J. Rocke (1982). Are robust estimators really necessary? *Technometrics 24*(2): 95–101.

Searle, S. R. (1982). *Matrix Algebra Useful for Statistics*. New York: Wiley.

"Student" (1908). The probable error of a mean. *Biometrika 6*(1): 1–25.

Tukey, J. W. (1991). The philosophy of multiple comparisons. *Statistical Science 6*: 100–116.

U.S. Army Materiel Command (1972). Tables of the cumulative binomial probabilities. AMC Pamphlet No. 706–109. Washington, DC: U.S. Army Materiel Command (Second printing: June 1972).

Weintraub, S. (1963). *Tables of Cumulative Binomial Probability Distribution for Small Values of p*. New York: Free Press of Glencoe (Macmillan).

EXERCISES

1. Compute S^2 for the sample 2, 6, 7, 8, 9, and show that the same value of S^2 is obtained for the sample 72, 76, 77, 78, 79. Comment.
2. Compute $\overline{X}$ for the numbers 23, 26, 27, 29, 30.
3. Use Table A in the Appendix of the book produce a random sample of seven two-digit numbers.
4. Use a hand calculator or computer software to determine the probability of observing at most one nonconforming unit in a sample of 100 if 1% nonconforming units are produced during that time period.
5. Determine the following quantities for $X \sim N\ (\mu = 40,\ \sigma = 5)$ and $Z \sim N(0, 1)$:
   (a) $P(Z < 1.65)$
   (b) $P(Z > 1.30)$
   (c) $P(30 < X < 35)$
   (d) $P(30 < X < 50)$
   (e) z_0 where $P(-z_0 < Z < z_0) = .95$
   (f) $P(X = 43)$
6. Construct a 90% confidence interval for μ where $n = 100$, $s = 2$, and $\overline{x} = 20$. (Assume that X is approximately normally distributed.)
7. What is the probability of obtaining a head on a coin and a 6 on a die when a coin is tossed and a die is rolled?
8. Determine $P(t > 2.086 \text{ or } t < -2.086)$ where a t statistic has 20 degrees of freedom.

9. Determine the approximate probability that a sample average ($\bar{x}$) taken from some population with $\mu = 50$ and $\sigma = 4$ exceeds 50.6, where the sample size is 100. Why is the probability approximate?
10. Explain the difference between a confidence interval and a tolerance interval.
11. Determine $\binom{8}{3}$ and explain what it means in words.
12. A sample is obtained from a normal population with $\bar{x} = 22.4$ and $s = 3.1$. We can say with probability .90 that at least 95% of the values in the population will be between what two numbers? (Assume $n = 10$.)
13. For a normal distribution, what percentage of values should be between $\mu - 2\sigma$ and $\mu + 2\sigma$?
14. What is the single parameter in the binomial distribution, and what sample statistic would be used to estimate it?
15. If we knew that a process was producing 0.5% nonconforming units, how many such units would be expect to observe if we inspected 400 units? What should we conclude about the process if we observed 12 nonconforming units?
16. Explain why we would generally not be able to compute σ^2.
17. Assume that the number of scratches per square yard of plate glass can be represented by the Poisson distribution. What is the probability of observing at most two scratches if it is known that the manufacturing process is producing a mean of 0.64 scratches per square yard?
18. Use appropriate software to construct a normal probability plot of 50 observations from a χ^2_6 distribution and interpret the plot.
19. If the lower limit of a 95% two-sided confidence interval for μ is 23.4, will the limit of a lower one-sided 95% confidence interval for μ (using the same data) be less than 23.4 or greater than 23.4? Explain.
20. Assume that all units of a certain item are inspected as they are manufactured. What is the probability that the first nonconforming unit occurs on the 12th inspected item if the probability that each unit is nonconforming is .01? What is the probability that the third nonconforming unit occurs on the 16th inspected item?

PART II

Control Charts and Process Capability

CHAPTER 4

Control Charts for Measurements with Subgrouping (for One Variable)

In this chapter we discuss control charts that can be used when subgroups (samples) of data are formed. We consider control charts that can be used when measurements are made and the values are obtained with sufficient speed to allow subgroups to be formed. Typical measurements are length, width, diameter, tensile strength, and Rockwell hardness.

Subgrouping is an important topic in itself, so subgrouping considerations and basic control chart principles are discussed first.

4.1 BASIC CONTROL CHART PRINCIPLES

Control charts can be used to determine if a process (e.g., a manufacturing process) has been in a state of statistical control by examining past data. This is frequently referred to as retrospective data analysis. We shall also refer to this as Stage 1. More importantly, recent data can be used to determine control limits that would apply to future data obtained from a process, the objective being to determine if the process is being maintained in a state of statistical control. This is termed Stage 2. (Note: Some writers have referred to these two stages as Phase 1 and Phase 2, respectively.)

Control charts alone cannot produce statistical control; that is the job of the people who are responsible for the process. Control charts can indicate whether or not statistical control is being maintained and provide users with other signals from the data. The charts can also be used in studying process capability, as is illustrated in Chapter 7.

Best results will generally be obtained when control charts are applied primarily to *process* variables (e.g., temperature and pressure) than to *product* variables (e.g., diameter and thickness). Bajaria (1994) reports, however, that many companies emphasize the use of control charts applied to product variables rather than to process variables. If the diameter of a product can be affected by temperature at an earlier stage, then certainly it is desirable to monitor temperature.

In general, it is desirable to monitor all process variables that affect important product variables.

As discussed in Chapter 2, control charts are essentially plots of data over time — a desirable way to plot any set of data. Figure 4.1 is an example of a control chart.

The asterisks in Figure 4.1 denote points that are outside the control limits. If control charts are being used for the first time, it will be necessary to determine *trial control limits.* To do so, it is desirable to obtain at least 20 subgroups or at least 100 individual observations (depending upon whether subgroups or individual observations are to be used) from either past data, if available, or current data. If collected from past data, it should be relatively recent data so that the data adequately represent the current process. The points in Figure 4.1 might represent such data. Because there are two points outside the control limits, these points should be investigated since they were included in the calculation of the trial control limits. Whether or not these trial limits should be revised depends upon the outcome of the investigation. If each of the points can be traced to a "special cause" (e.g., a machine out of adjustment), the limits should be recomputed *only* if the cause can be removed. If the cause cannot be removed, it should be regarded (unfortunately) as a permanent part of the process, so the trial limits should not be recomputed. Thus, the trial limits should be recomputed only if the cause of points lying outside the limits can be both detected *and* removed. Occasionally, points will fall outside the limits because of the natural variation of the process. This is much more likely to happen when a *group* of points is plotted rather than the sequential plotting of individual points. This is discussed in detail in the next section.

The question arises as to what should be done when one or more points are outside the new limits that are obtained from deleting points that were outside the old limits. That could happen with the points in Figure 4.1, since deleting points outside the limits will cause the new limits to be closer together. Thus, the points that are just barely inside the old limits could be just barely outside

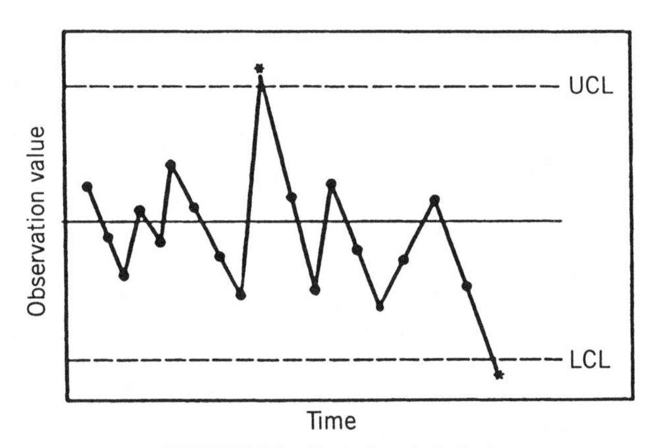

FIGURE 4.1 Typical control chart.

the new limits. Technically, points that are outside the new limits should also be deleted and the limits recomputed if and only if an assignable cause can be both detected and removed, and this cycle should be continued until no further action can be taken.

After a process has been brought into a state of statistical control, a process capability study can be initiated to determine the capability of the process in regard to meeting the specifications. It would be illogical to undertake such a study if the process is not in control, since the objective should be to study the capability of the process after all problematic causes have been eliminated, if possible. (Many companies do use *process performance indices*, however, which show process performance when a process is not in a state of statistical control.) Various methods can be employed for ascertaining process capability; these are discussed in Chapter 7.

4.2 REAL-TIME CONTROL CHARTING VERSUS ANALYSIS OF PAST DATA

Factors to consider in determining the control limits for Stage 1 have not been covered to any extent in the literature. When a set of points is plotted all at once (in Stage 1 or perhaps even in Stage 2), the probability of observing at least one point that is outside the control limits will obviously be much greater than 0.0027, which applies to points plotted individually when 3-sigma limits are used *and* parameters for the appropriate distribution are assumed to be known. For n points, the probability of having at least one point outside the 3-sigma limits when the process is, in fact, in control is given in Table 4.1 for different values of n. The actual probability (assuming a normal distribution, known parameter values, and independent points) can be approximated by $.0027n$.

TABLE 4.1 Probabilities of Points Plotting Outside Control Limits

n	$.0027n$	Actual Probability ($\geq$1 Point Outside Limits)
1	.0027	.0027
2	.0054	.0054
5	.0135	.0134
10	.0270	.0267
15	.0405	.0397
20	.0540	.0526
25	.0675	.0654
50	.1350	.1264
100	.2700	.2369
350	.9450	.6118

It can be observed from Table 4.1 that the approximation works quite well for moderate values of n. The reason for this is given in the Appendix to this chapter.

The important point is that there is a much higher probability of observing *at least* one point that is outside the limits when, say, 15 or 20 points are plotted together than when an individual point is plotted. When points are plotted individually in real time, the .0027 probability applies to each point, so there is indeed a very small probability of one particular point being outside the limits when the process is, in fact, in control. But when trial control limits are established and periodically revised (using a set of observations each time), and when control charting is not performed in real time, the probability of observing one or more points outside the limits when the process is in control is clearly much greater.

This does not mean that we should ignore such points and not look for assignable causes; it simply means that we should not be too surprised if we cannot find them.

Although the use of 3-sigma limits has become quite customary, at least in manufacturing applications, there is no reason why such limits should always be used. If, for example, a particular work situation mandates that 20 points will always be plotted together on a chart whenever the charting is performed, the limits can be adjusted so that, if desired, the probability of observing at least 1 point out of 20 outside the limits when the process is in control is close to .0027.

Using the rule-of-thumb given in Table 4.1, we would use

$$np = .0027$$

so that with $n = 20$

$$20p = .0027$$

$$p = .000135$$

Thus, we would then look up .50000 − .00007 in Table B in the Appendix to the book and observe that .49993 corresponds to approximately $z = 3.81$. We would then use 3.81-sigma limits. This is not to suggest that this should be done, but rather that it *could* be done.

As implied previously, this analysis applies only to the case for which the parameters are assumed to be *known*. When the parameters are unknown, the true probability of at least one of n points plotting outside the limits when the process is in control cannot be determined analytically because the deviations of each of the n points from the control limits are correlated since each deviation contains realizations of common random variables. This is true for both Stage 1 and Stage 2. [See, e.g., Sullivan and Woodall (1996).]

Therefore, an exact probability will not result from calculations such as in the preceding example. The exact probability can be determined only by simulation.

In general, the possible use of k-sigma limits in Stage 1 needs to be addressed. In Stage 1 the user is not concerned with some loss occurring because a process is stopped, but the cost of a false alarm in Stage 1 may not be trivial. Medical practitioners often tend to favor 2-sigma limits. This is reasonable if false alarms can be easily tolerated but there is a need to detect assignable causes as quickly as possible.

4.3 CONTROL CHARTS: WHEN TO USE, WHERE TO USE, HOW MANY TO USE

General information has been given concerning how to use a control chart, and similar information is given in later chapters for each chart. When the decision is made to begin using control charts, various questions must be addressed, such as those indicated above.

It would be impractical to think of using a control chart at every work station in a plant. The nature of the product will often preclude measurements being made at various stages of production. There is also no need to use control charts at a point in a manufacturing process at which it is highly unlikely that the process could ever go out of control. Control charts should be used where trouble is likely to occur. When control charts are first implemented, it is also important that they be used where the potential for cost reduction is substantial. This is desirable so that management can see the importance of the charts and support their continued use in a company.

The number of charts that can be handled in a plant may very well depend upon whether the control charting is to be performed manually or by computer. If it is done manually, more workers will have to understand the fundamentals of control charting than will be the case if the charting is to be handled by computer. Manual control charting could also be a problem if there is a need to use some of the mathematically sophisticated types of charts. There is also the problem of storing and displaying the charts when manual charting is used. The advantages of computerized control charting are discussed briefly in Chapter 10.

4.4 BENEFITS FROM THE USE OF CONTROL CHARTS

The benefits that result from the use of control charts are many and varied. One of the most important benefits is that they result in good record keeping. Whether charts are being used for administrative applications (as in Chapter 10) or for process control (or both), good records are essential if company-wide quality control is to become a reality. The mere maintenance of control charts often leads to a reduction in product variability simply because operators realize that management is placing considerable emphasis on product quality, and this causes them to exercise greater care. Control charts are invaluable as an aid in identifying special causes of variation, which often can be removed only through management action (e.g., replacing a faulty machine). Even if all special causes have been identified and removed, process improvement should still be sought. This is discussed further in Chapter 7.

4.5 RATIONAL SUBGROUPS

An important requirement is that the data that are chosen for the subgroups come from the same population. Ideally, this means that data from different operators, different shifts, and different machines, for example, should not be mixed. Granted, it may be difficult or impractical to use a separate chart for each operator, for example. If data from different populations are mixed, however, the control limits that are constructed will then correspond to a mixture (i.e., aggregated) distribution and could be quite unsuitable when applied to individual data points.

Recall the egg timer discussion in Section 2.3. Assume that a subgroup of four observations is formed by using two times from one half of the timer and two times from the other half. Since the halves differ, the four observations are therefore not all from the same population. Assume that the timer is defective in that one of the halves does not actually measure 3 minutes. This would be analogous to having half of the observations in each subgroup from an out-of-control process caused by a defective timer.

Similarly, if half of the observations in a subgroup were measurements obtained from the work of a well-trained operator and the other half consisted of measurements obtained from the work of a new, poorly trained operator, this would be a case of mixing data from two populations, and the result would be an overestimate of the variability due to operators.

Nelson (1988) discusses the need for rational subgroups and points out that data collected over a short time period will not necessarily constitute rational subgroups. Palm (1992) illustrates the importance of a good sampling plan for control chart construction.

4.6 BASIC STATISTICAL ASPECTS OF CONTROL CHARTS

If we are interested in controlling the process mean, μ, and the limits are given as $\mu \pm 3\sigma_{\overline{x}}$, the total probability outside the limits is .0027 (.00135 on each side) if X has a normal distribution. As explained in Chapter 3, $\overline{X}$ denotes a sample average and $\sigma_{\overline{x}}$ is its standard deviation. Thus, *if* we had exactly a normal distribution and if $\sigma_{\overline{x}}$ were known, the chances would be 27 in 10,000 of observing a plotted value of $\overline{X}$ outside the limits when the mean is at μ. We should not expect to have exactly a normal distribution, however, nor will we know the true process mean, μ, or $\sigma_{\overline{x}}$. Therefore, we should think of the limits as "3-sigma limits" (as they are usually called) rather than probability limits, since the exact probabilities are unknown. Some authors have provided probability limits for certain charts (as is discussed, e.g., in Section 4.7.2), but the limits will not be true probability limits when the actual distribution is unknown (as is the usual case) and when parameters must be estimated. We can also argue that even if we knew μ (which will generally have to be estimated), we would not expect it to remain constant over a long period of time. Thus, when probabilities are applied to the future, they are only approximations—possibly poor approximations.

Nevertheless, if we take samples of at least size 4 or 5, the distribution of $\overline{X}$ will not differ greatly from a normal distribution as long as the distribution of X is reasonably symmetric and bell shaped. This results from the fact that the distribution of $\overline{X}$ will be more normal, in general, than the distribution of X. This is the result of the central limit theorem, mentioned in Chapter 3.

Even if the distribution is highly asymmetric so that the distribution of $\overline{X}$ will also be clearly asymmetric for small samples, data can often be transformed (e.g., log, square root, reciprocal) so that the transformed data will be approximately normal.

The $\overline{X}$ chart has been the most frequently used control chart. We shall first use a hypothetical data set to illustrate the construction and handling of an $\overline{X}$ chart and an R chart. An R (range) chart can be used for controlling the process variability and should generally indicate control before an $\overline{X}$ chart is constructed. The reason for this recommendation is that unless the variability of the process is in a state of statistical control, we do not have a stable distribution of measurements with a single fixed mean.

4.7 ILLUSTRATIVE EXAMPLE

The data in Table 4.2 will be used to illustrate the construction of each chart.

TABLE 4.2 Data in Subgroups Obtained at Regular Intervals

Subgroup	x_1	x_2	x_3	x_4	$\overline{x}$	R	s
1	72	84	79	49	71.00	35	15.47
2	56	87	33	42	54.50	54	23.64
3	55	73	22	60	52.50	51	21.70
4	44	80	54	74	63.00	36	16.85
5	97	26	48	58	57.25	71	29.68
6	83	89	91	62	81.25	29	13.28
7	47	66	53	58	56.00	19	8.04
8	88	50	84	69	72.75	38	17.23
9	57	47	41	46	47.75	16	6.70
10	13	10	30	32	21.25	22	11.35
11	26	39	52	48	41.25	26	11.53
12	46	27	63	34	42.50	36	15.76
13	49	62	78	87	69.00	38	16.87
14	71	63	82	55	67.75	27	11.53
15	71	58	69	70	67.00	13	6.06
16	67	69	70	94	75.00	27	12.73
17	55	63	72	49	59.75	23	9.98
18	49	51	55	76	57.75	27.	12.42
19	72	80	61	59	68.00	21	9.83
20	61	74	62	57	63.50	17	7.33

If the data in Table 4.2 had been real data, they might have been obtained by measuring four consecutive units on an assembly line every 30 minutes until the 20 subgroups are obtained. (In general, 20 or more subgroups with at least four or five observations per subgroup should be obtained initially, so as to have enough observations to obtain a good estimate of the process mean and process variability.) The s is the sample standard deviation introduced in Chapter 3, namely,

$$s = \sqrt{\frac{\sum_{i=1}^{n}(x_i - \overline{x})^2}{n-1}}$$

We could use either R or s in controlling the process variability. The latter is preferable, particularly if a statistical quality control (SQC) operation is computerized, since it uses all the observations in each subgroup, whereas R is calculated from only two observations in each subgroup (the largest minus the smallest). Thus, although the range is much easier to calculate by hand than the standard deviation, it is wasteful of information. This loss of information is relatively inconsequential when the subgroup size is 4 (as in this example), but is much more serious when the subgroup size is somewhat larger. Therefore, when large subgroup sizes are used, the range should not be used to control the process variability. Instead, some statistic should be used that is calculated from all of the observations (such as s).

Another reason for preferring s over R is that other statistical methods that are useful in quality improvement work are generally based upon s (or s^2) rather than R.

A cursory glance at Table 4.2 reveals two comparatively small numbers in subgroup 10 (13 and 10), which cause the average for that subgroup to be much smaller than the other subgroup averages. At this point we might ask the following question: What is the probability of obtaining a subgroup average as small or smaller than 21.25 when, in fact, the process mean is in control at $\overline{\overline{x}}$, the average of the subgroup averages? Recall from Chapter 3 that we can "standardize" $\overline{X}$ as

$$Z = \frac{\overline{X} - \mu}{\sigma/\sqrt{n}}$$

so that $Z \sim N(0, 1)$ if $X \sim N(\mu, \sigma^2)$. The (usually) unknown process mean μ is estimated by $\overline{\overline{x}}$, where

$$\overline{\overline{x}} = \frac{\sum_{i=1}^{k} \overline{x}_i}{k}$$

for k subgroup averages. In this example $k = 20$ and the average of the 20 subgroup averages is 59.44, which is the same as the average of the 80 numbers. The process standard deviation σ could be estimated using either s or R. The latter will be illustrated first since that has been the conventional approach. How might we use the subgroup ranges to estimate σ? Could we use the average of the ranges to estimate σ as we used the average of the subgroup averages to estimate μ? The answer to the latter question is "no." We would want our estimator of σ to be a statistic such that if we took a number k of samples from a large population of values whose standard deviation is σ, we would "expect" the average of the k values of the statistic to be equal to σ. That is, we would want the value of the statistic from a single sample to be an "unbiased" (in a statistical sense) estimate of the value of σ such that the value of the statistic would be our best guess of the value of σ.

This will not happen if we use either the average of the subgroup ranges or the average of the subgroup standard deviations. Fortunately, however, tables have been constructed that allow the average of the ranges or the average of the standard deviations to be divided by a constant such that the resultant statistics are unbiased estimators of σ. Those constants, for different subgroup sizes, are given in Table E in the Appendix to the book as d_2 and c_4. Thus, if we use ranges, we would estimate σ by $\overline{R}/d_2$, where $\overline{R}$ is the average of the ranges. Similarly, if we use standard deviations, we would estimate σ by $\overline{s}/c_4$, where $\overline{s}$ is the average of the standard deviations. (These and other constants are derived in the Appendix to this chapter.) For the data in Table 4.2,

$$\begin{aligned}\hat{\sigma} &= \frac{\overline{R}}{d_2} \\ &= \frac{31.30}{2.059} \\ &= 15.20\end{aligned}$$

where, as in Chapter 3, the "hat" ($\hat{}$) is used to indicate that σ is being estimated. If we were to use s, our estimate would be

$$\begin{aligned}\hat{\sigma} &= \frac{\overline{s}}{c_4} \\ &= \frac{13.90}{.921} \\ &= 15.09\end{aligned}$$

We can see that there is some difference in the two estimates, even though the subgroup size is quite small in this case. If we use the first estimator, we would estimate $\sigma_{\overline{x}}$ as

$$\begin{aligned}\hat{\sigma}_{\overline{x}} &= \frac{\overline{R}/d_2}{\sqrt{n}} \\ &= \frac{15.20}{\sqrt{4}} \\ &= 7.60\end{aligned}$$

Then, *if* we were willing to assume approximate normality for $\overline{X}$, $P(\overline{X} \leq 21.25)$ could be estimated using

$$\begin{aligned} z &= \frac{\overline{x} - \hat{\mu}}{\hat{\sigma}_{\overline{x}}} \\ &= \frac{21.25 - 59.44}{7.60} \\ &= -5.02 \end{aligned}$$

where $\hat{\mu} = \overline{\overline{x}}$. We would estimate $P(Z < -5.02)$ as approximately zero since we cannot look up $z = 5.02$ in Table B in the Appendix to the book. It clearly must be smaller than .00003 since that would be the probability for $z = -3.99$. In fact, the estimated probability can be shown to be 2.588×10^{-7}. (The word "estimated" is used since μ and σ are unknown, and the true distribution is almost certainly not exactly normal. How could we tell if the distribution of $\overline{X}$ is approximately normal? We could perform a test of normality on the individual values, or we might simply construct a normal probability plot of the subgroup averages. Constructing the latter, as the reader is asked to do in exercise 13, shows the average for subgroup 10 to be unusual, but otherwise the plot looks okay.)

Thus, before we even construct the $\overline{X}$ chart, we have strong reason to suspect that the process was out of control (with respect to its mean) at the time the data in subgroup 10 were collected.

Are there other numbers that stand out? We observe that the range in subgroup 5 of Table 4.2 is considerably larger than any of the other ranges and that this is due in large part to the existence of the number 97, which is the largest of the 80 numbers in Table 4.2. What is the probability of obtaining a value for the range that is at least as high as 71 (for this example) when the process variability is in a state of statistical control? This question cannot be answered as easily as the previous question.

4.7.1 *R* Chart

We can, however, construct an *R* chart for these data and see if 71 falls within the control limits. As with the other "standard" control charts that are given in this book, the control limits for an *R* chart are 3-sigma limits. Specifically, the control limits are obtained as

$$\overline{R} \pm 3\hat{\sigma}_R \tag{4.1}$$

The limits given by Eq. (4.1) can be shown to be equal to $D_3\overline{R}$ for the lower control limit (LCL) and $D_4\overline{R}$ for the upper control limit (UCL). (See the Appendix to this chapter for the derivations.) Values of D_3 and D_4 for various sample sizes are contained in Table E in the Appendix to the book. These values are based on the assumption of normality of the individual observations. A normal probability plot suggests that a normal distribution is a plausible model.

Since we found that $\overline{R} = 31.30$ for the data in Table 4.2, the control limits are

$$\begin{aligned} \text{LCL} &= D_3\overline{R} & \text{UCL} &= D_4\overline{R} \\ &= 0(31.3) & &= 2.282(31.3) \\ &= 0 & &= 71.43 \end{aligned}$$

It should be noted that D_3 will always be zero whenever the subgroup size is less than 7. Therefore, some authors would say the LCL for an R chart is zero for such subgroup sizes. But Nelson (1996) defines a control limit as "a limit beyond which a point can lie". If we use this definition, which certainly seems desirable, then we would say that there is no LCL when the subgroup size is less than 7.

The chart is given in Figure 4.2. We can see from Figure 4.2 that the range of 71 for subgroup 5 is (barely) inside the UCL. Nevertheless, we still might wish to investigate the values in subgroup 5 and try to determine why that range is considerably higher than the other ranges. There is also obvious evidence of a downward trend. Since this represents a reduction in variability (which is obviously desirable), we would certainly want to determine the cause (if possible), so that the cause could become a permanent part of the process.

The point to be made is that even though the chart indicates control, there is also evidence that some "detective work" might lead to further improvement. As Ott (1975) mentions in the preface: "Troubleshooting cannot be entirely formalized, and there is no substitute for being inquisitive and exercising ingenuity." Statements such as these apply to every type of control chart and every type

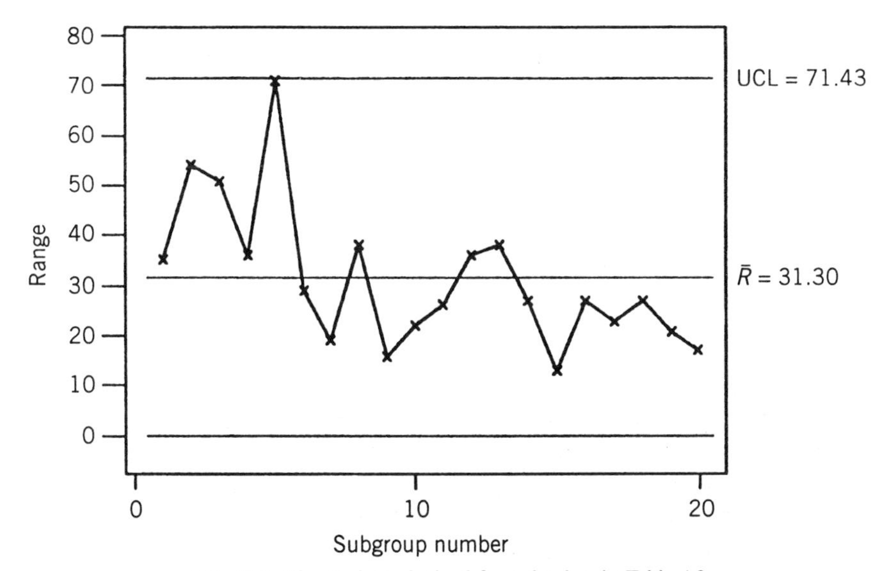

FIGURE 4.2 R chart obtained from the data in Table 4.2.

of statistical procedure that we might use. No statistical quality control system should ever be so formalized that it does not leave room for the exercise of good judgment.

Implicit in the construction of an R chart is the assumption that the population of individual values from which the subgroups are obtained can be adequately represented by a normal distribution. Indeed, the D_3 and D_4 constants are tabulated under the assumption of a normal distribution. In practice, the "good judgment" referred to above should be used to see if the data are approximately normally distributed. Methods for doing so were discussed in Section 3.8.5.1.

Even if the population were normally distributed, however, the distribution of the range is highly asymmetric and, thus, nowhere near a normal distribution. Nevertheless, whenever 3-sigma limits are used on any control chart, there is the implicit assumption that whatever is being charted has a symmetric distribution and that the "tail areas" are thus equal (and small).

The statistically proper approach would thus be to not use 3-sigma limits for an R chart if tables for "probability limits" are readily available. Grant and Leavenworth (1980, p. 292) do give an abbreviated table for this purpose; more extensive tables are given in Harter (1960). Table F in the Appendix to the book contains some of the entries found in the latter.

4.7.2 *R* Chart with Probability Limits

One possibility in constructing an R chart would be to use $D_{.001}$ and $D_{.999}$ from Table F in the Appendix to the book in place of D_3 and D_4. This would give equal tail areas (assuming normality) of .001. The probability limits would then be obtained as

$$\text{UCL} = D_{.999}\left(\frac{\overline{R}}{d_2}\right) \qquad \text{LCL} = D_{.001}\left(\frac{\overline{R}}{d_2}\right)$$

If this approach had been used for the present example, the limits would have been

$$\begin{aligned} \text{UCL} &= 5.31\left(\frac{31.3}{2.059}\right) & \text{LCL} &= 0.20\left(\frac{31.3}{2.059}\right) \\ &= 2.579(31.3) & &= 0.097(31.30) \\ &= 80.72 & &= 3.04 \end{aligned}$$

These limits obviously differ greatly from the 3-sigma limits. In particular, the (.001 and .999) probability limits will always be higher than the 3-sigma limits. Unlike the limits for the conventional R chart, the LCL using the probability limits will always exist since all of the values in the $D_{.001}$ column in Table F are greater than zero. This is a desirable feature since "significantly" small values of R can then show up as being below the LCL. Similarly, the UCL using $D_{.999}$ will always exceed the 3-sigma UCL. (This can be easily verified by calculating

$D_{.999}/d_2$ for each value of n and observing that each of these values is greater than the corresponding value of D_4.)

Admittedly, $D_{.001}$ is used here rather than $D_{.00135}$ (there are no tables that give $D_{.00135}$), so the limits are not totally comparable. Nevertheless, the difference between the probability limits and the 3-sigma limits would not be this great if the distribution of the range was symmetric.

Which set of limits should be used in practice? Both sets are based upon the assumption of a normal distribution (of X) and the assumption that $\sigma = \overline{R}/d_2$. Neither assumption is likely to be met, and the statistical theory does not exist to allow the two approaches to be compared under other conditions. (In particular, the distribution of the range is not widely known, and especially not tabulated, for distributions other than the normal.) Nevertheless, the probability limits do have considerable appeal because the user of such limits is at least attempting to correct for the asymmetry in the distribution of the range.

4.7.3 s Chart

Although an s chart is generally preferable to an R chart when the computations are computerized, the former suffers from some of the same shortcomings as the latter. Specifically, the distribution of s is also asymmetric when the distribution of individual values is normal, although 3-sigma limits are typically used. Specifically, the control limits are

$$\bar{s} \pm 3\hat{\sigma}_s \tag{4.2}$$

where $\bar{s}$ is the average of the subgroup standard deviations and $\hat{\sigma}_s$ is the estimate of the standard deviation of s. It can be shown that Eq. (4.2) leads to

$$\text{UCL} = B_4\bar{s} \qquad \text{LCL} = B_3\bar{s}$$

where B_3 and B_4 are given in Table E in the Appendix to the book. (See the Appendix to this chapter for the derivation.) For the data in Table 4.2 we obtain

$$\begin{aligned} \text{UCL} &= 2.266(13.899) & \text{LCL} &= 0(13.899) \\ &= 31.495 & &= 0 \end{aligned}$$

Like the two different types of limits for the R chart, these limits are also based upon the assumption of a normal distribution. We can also see from Table E that the LCL will not exist when the subgroup size is less than 6. (Recall that the LCL for the conventional R chart will not exist when the subgroup size is less than 7.)

4.7.4 s Chart with Probability Limits

Can we improve upon these limits by using probability limits, as was done for the R chart? The answer is "yes"; the probability limits can be obtained by using the chi-square distribution discussed in Chapter 3, in conjunction with the following well-known theorem:

$$\text{If } X \sim N(\mu, \sigma^2), \quad \text{then } \frac{(n-1)S^2}{\sigma^2} \sim \chi^2_{n-1}$$

where $n - 1$ is the degrees of freedom. It follows from this result that

$$P\left(\chi^2_{.001} < \frac{(n-1)S^2}{\sigma^2} < \chi^2_{.999}\right) = .998$$

and so

$$P\left(\frac{\sigma^2}{n-1}\chi^2_{.001} < S^2 < \frac{\sigma^2}{n-1}\chi^2_{.999}\right) = .998$$

By taking square roots, we obtain

$$P\left(\sigma\sqrt{\frac{\chi^2_{.001}}{n-1}} < S < \sigma\sqrt{\frac{\chi^2_{.999}}{n-1}}\right) = .998$$

Thus, if the process variability is in control at σ, 99.8% of the time the subgroup standard deviation, s, will fall between the endpoints of the interval. [The .998 is roughly equal to the (assumed) area between the 3-sigma limits on a standard chart, which is .9973.]

If an estimate, $\hat{\sigma}$, of σ is available from past experience or past (but relatively recent) data, that estimate could be used to obtain the control limits as

$$\text{LCL} = \hat{\sigma}\sqrt{\frac{\chi^2_{.001}}{n-1}} \qquad \text{UCL} = \hat{\sigma}\sqrt{\frac{\chi^2_{.999}}{n-1}}$$

If σ is to be estimated from data collected in subgroups (as in the present example), an unbiased estimator of σ is $\bar{s}/c_4$ with c_4 given in Table E in the Appendix to the book. The control limits would then be

$$\text{LCL} = \bar{s}/c_4\sqrt{\frac{\chi^2_{.001}}{n-1}} \qquad \text{UCL} = \bar{s}/c_4\sqrt{\frac{\chi^2_{.999}}{n-1}}$$

and the centerline would be $\bar{s}$. Table 4.3 gives the .001 and .999 percentage points of the χ^2 distribution for different values of n.

TABLE 4.3 The .001 and .999 Percentage Points of the χ^2 Distribution[a]

n	$\chi^2_{.001,n-1}$	$\chi^2_{.999,n-1}$
2	1.570×10^{-6}	10.827
3	0.002	13.815
4	0.024	16.266
5	0.091	18.467
6	0.210	20.515
7	0.381	22.457
8	0.598	24.322
9	0.857	26.124
10	1.152	27.877
11	1.479	29.587
12	1.834	31.263

[a]The entries in this table (except for the first entry) were obtained using Minitab, a statistical software package. Minitab is a registered trademark of Minitab, Inc., 1829 Pine Hall Rd., State College, PA 16801-3008. Tel: (814) 238–3280, (800) 448–3555.

For the present example $n = 4$ so the limits would be

$$\text{LCL} = \frac{13.899}{0.921}\sqrt{\frac{0.024}{3}} = 1.350 \qquad \text{UCL} = \frac{13.899}{0.921}\sqrt{\frac{16.266}{3}} = 35.140$$

Notice that these limits differ somewhat from the 3-sigma limits obtained previously (0 and 31.495).

Which set of limits should we use? The same kind of remarks that were made for the two types of R-chart limits also apply to the two types of s-chart limits. Specifically, both sets of limits are based upon the assumption of a normal distribution (of X) and the assumption that $\sigma = \bar{s}/c_4$. Neither assumption is likely to be met exactly, but the probability limits are more appealing than the 3-sigma limits since the area above the UCL for the latter will not be particularly close to the nominal value (0.00135).

This can be demonstrated as follows. The 3-sigma limits given by Eq. (4.2) can be converted from limits that give *nominal* tail areas of .00135 to limits that give nominal tail areas of .00100 by using 3.09 instead of 3.00. (The British actually use 3.09 instead of 3.00 for the various control charts.) The limits would then be obtained by using

$$\bar{s} \pm 3.09\hat{\sigma}_s$$

which is equivalent to

$$\bar{s} \pm 3.09\left(\frac{\bar{s}}{c_4}\right)\sqrt{1 - c_4^2}$$

For the present example the limits would be

$$\begin{aligned}\text{LCL} &= \bar{s} - 3.09\left(\frac{\bar{s}}{c_4}\right)\sqrt{1-c_4^2} \\ &= 13.899 - 3.09\left(\frac{13.899}{.921}\right)\sqrt{1-(.921)^2} \\ &= -4.267\end{aligned}$$

and

$$\begin{aligned}\text{UCL} &= \bar{s} + 3.09\left(\frac{\bar{s}}{c_4}\right)\sqrt{1-c_4^2} \\ &= 13.899 + 3.09\left(\frac{13.899}{.921}\right)\sqrt{1-(.921)^2} \\ &= 32.065\end{aligned}$$

Of course, the LCL would be set equal to zero, but notice that 32.065 differs considerably from the probability limit UCL of 35.140. Thus, *if* $\sigma = \bar{s}/c_4$ and we had exactly a normal distribution, attempting to obtain a nominal probability value of .001 by using $\bar{s} \pm 3.09\hat{\sigma}_s$ does not lead to a UCL that is particularly close to the .001 probability limit UCL.

Another way to view the difference would be to determine the χ^2 value that would cause the probability limit UCL to be equal to $\bar{s} \pm 3\hat{\sigma}_s$ and then compare the resultant upper tail area with .00135. Specifically,

$$\text{UCL} = \bar{s}/c_4\sqrt{\frac{\chi_a^2}{n-1}} = 31.495$$

where χ_a^2 is the value that will result from solving the equation. Substituting in the values of $\bar{s}$, c_4, and n leads to

$$\chi_a^2 = 13.066$$

The value of a can be shown to be .0045. This is 3.33 times the nominal value of .00135.

What does all of this mean from a practical standpoint? It means that the number of "false alarms" will be much greater with 3-sigma limits than what the user would expect. Numerically, a false signal can be expected to occur once every 222 subgroups (1/.0045), whereas the user of the 3-sigma limits will naturally assume that a false signal will be received once every 741 subgroups (1/.00135). This is obviously a considerable difference and could be a major consequence if the cost of looking for assignable causes is high. On the other hand, if the .001 probability limit is used, a false signal would be received once every 500 subgroups (1/.002). (Note that .001 is doubled since there are two limits.)

Here, again, this assumes that $\sigma = \bar{s}/c_4$ and that we have a normal distribution. Nevertheless, the type of data encountered in industry from which control charts are constructed is often reasonably close to a normal distribution, and we would certainly hope that the value of $\bar{s}/c_4$ is a good estimate of σ. Therefore, the probability limits for an s chart certainly have more appeal than the 3-sigma limits.

The reader should understand that a control chart or other statistical procedure is not invalidated just because the assumptions upon which it is based are not likely to be met. Rather, it is a matter of determining how "robust" the procedure is when the assumptions are not met (i.e., how insensitive is the procedure to a violation of the assumptions?). Some procedures will not be seriously affected by a slight-to-moderate departure from the assumptions, whereas other procedures will be seriously affected. These "robustness" considerations are discussed in each of the chapters on control charts.

The s chart will not be displayed here. Instead, the reader will be asked in exercise 1 to construct the chart and to compare the configuration of points with the configuration of points on the R chart.

4.7.5 s^2 Chart

An s^2 chart is another chart that could be used for controlling the process variability. (With this chart the process variance would be controlled instead of the process standard deviation.) As would be expected, the control limits are similar to the control limits for the s chart. The limits are

$$\text{LCL} = \bar{s}^2 \left(\frac{\chi^2_{.001}}{n-1} \right) \qquad \text{UCL} = \bar{s}^2 \left(\frac{\chi^2_{.999}}{n-1} \right)$$

The limits are not quite the same as the square of the limits for the s chart because $\bar{s}^2$ is an unbiased estimator of σ^2, not $(\bar{s}/c_4)^2$, where $\bar{s}^2$ is the average of the s^2 values.

4.7.6 $\overline{X}$ Chart

Regardless of which chart we select to control the process variability (R, s, or s^2), all of the points for the data in Table 4.2 lie within the control limits. Therefore, since the process variability is evidently in a state of statistical control, we can logically proceed to investigate whether or not the process mean is in control. An $\overline{X}$ chart will be used for that purpose.

The control limits for an $\overline{X}$ chart are obtained from

$$\bar{\bar{x}} \pm 3\hat{\sigma}_{\bar{x}}$$

where $\bar{\bar{x}}$ denotes the overall average of the subgroup averages and $\hat{\sigma}_{\bar{x}}$ denotes an estimator of the standard deviation of the subgroup averages. It was established in Chapter 3 that $\sigma_{\bar{x}} = \sigma_x/\sqrt{n}$. Therefore,

$$\hat{\sigma}_{\bar{x}} = \frac{\hat{\sigma}_x}{\sqrt{n}}$$

The usual procedure for the $\overline{X}$ chart is to obtain $\hat{\sigma}_x$ from the subgroup ranges, namely,

$$\hat{\sigma}_x = \frac{\overline{R}}{d_2}$$

The control limits would then be written as

$$\begin{aligned} \bar{\bar{x}} \pm 3\hat{\sigma}_{\bar{x}} &= \bar{\bar{x}} \pm 3\frac{\hat{\sigma}_x}{\sqrt{n}} \\ &= \bar{\bar{x}} \pm 3\frac{(\overline{R}/d_2)}{\sqrt{n}} \\ &= \bar{\bar{x}} \pm A_2\overline{R} \end{aligned}$$

where $A_2 = 3/(d_2\sqrt{n})$.

The control limits for the data in Table 4.2 are

$$\begin{aligned} &\bar{\bar{x}} \pm A_2\overline{R} \\ &\quad = 59.4375 \pm 0.729(31.3) \\ &\quad = 59.4375 \pm 22.8177 \end{aligned}$$

so that

$$\text{LCL} = 36.6198 \qquad \text{UCL} = 82.2552$$

(The values of A_2 are given in Table E in the Appendix to the book.)

There is no reason why the control limits must be obtained from $\overline{R}$, however. In particular, if the control chart user prefers an s chart over an R chart, it would make more sense to estimate σ from $\bar{s}$ instead of from $\overline{R}$. The control limits would then be obtained as

$$\bar{\bar{x}} \pm 3\frac{(\bar{s}/c_4)}{\sqrt{n}} = \bar{\bar{x}} \pm A_3\bar{s}$$

where $A_3 = 3/(c_4\sqrt{n})$ and the values of A_3 are given in Table E.

The control limits for this example would then be

$$\begin{aligned} \bar{\bar{x}} \pm A_3\bar{s} &= 59.4375 \pm 1.628(13.899) \\ &= 59.4375 \pm 22.6276 \end{aligned} \tag{4.3}$$

so that

$$\text{LCL} = 36.8099 \qquad \text{UCL} = 82.0651$$

Thus, we see that there is some difference in the two sets of control limits, although the difference is fairly small.

The $\overline{X}$ chart with the limits obtained from using Eq. (4.3) is given in Figure 4.3. We observe that 1 of the 20 subgroup averages is below the LCL: the one corresponding to subgroup 10. Using the approximation given in Chapter 4, we would estimate the probability of observing at least one point outside the control limits when the process mean is in control as $20(.0027) = .0540$. (The probability of observing exactly one point outside the limits when the process is in control is .0513, assuming normality.) Since this probability is not extremely small (say, less than .05), the process might have been in control at the time the data were collected, and the data may have been recorded correctly. Recall, however, that we previously estimated the probability of observing a subgroup average as small or smaller than 21.25 as being approximately zero. Therefore, by supplementing the control chart information with this extra detective work, we now have reason to believe that it is virtually impossible for the subgroup values to be valid numbers if the process was in control at that point. Perhaps the 13 and 10 should have been recorded as 43 and 40, respectively. If that were the case, the limits could then be recomputed using the correct values.

We should note that the best estimator of σ (in terms of having the smallest variance) when subgroup data are used is not $\overline{s}/c_4$. Rather, the best estimator is $\sqrt{\text{Ave}(s^2)}$, with $\text{Ave}(s^2)$ denoting the average of the subgroup variances. This estimator can be shown to generally have a slightly smaller variance than $\overline{s}/c_4$.

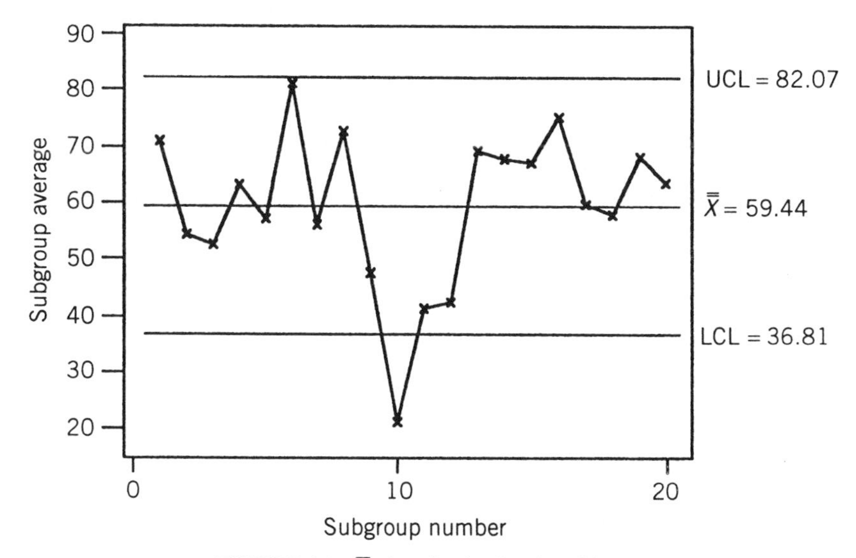

FIGURE 4.3 $\overline{X}$ chart for the data in Table 4.2.

Table 1 of Bissell (1990) contains the ratio of the variance of the optimal estimator to the variance of $\bar{s}/c_4$ for various subgroup sizes. Using the reciprocals of these numbers, we find that $\mathrm{Var}(\bar{s}/c_4) = 1.07$ times the variance of the optimal estimator for $n = 4$, with the multiplier being 1.05 for $n = 5$.

Since the multiplier is thus very close to 1.00 for typical subgroup sizes, we have adopted the more conventional approach (in the quality improvement literature) of using $\bar{s}/c_4$.

4.7.7 Recomputing Control Limits

Assume that the cause was detected as operator error but there is no way to determine the correct value(s). The control limits should then be recomputed using the remaining 19 subgroups. Although this is the proper way to proceed (a point should be removed from the control limit computations only if an assignable cause is both detected and removable), the question arises as to whether or not it will be worth the trouble. (The word "trouble" obviously has different meanings in this context, depending upon whether or not the computations are computerized.)

Some generalizations can be made concerning the extent to which the recomputed limits will differ from the original limits. When the original data set contains n subgroups and one of the subgroup averages is below the LCL, it can be shown that the new UCL must exceed the old UCL by at least $1/(n-1)A_3\bar{s}_{(n)}$, provided that the value of s for the deleted subgroup is less than $\bar{s}_{(n)}$. (Here $\bar{s}_{(n)}$ denotes the average value of s for all n subgroups.) For this example, $s_{10} = 11.35 < \bar{s}_{(20)}$ so the difference must be at least $1/19(1.628)(13.899) = 1.191$.

Similar statements can be made when there is more than one point outside the control limits. In regard to the other control charts, if 30 samples of size 4 are used to compute the control limits for an R chart, and one of the ranges is above the UCL, the smallest possible difference between the original and recomputed UCL values is $.101\overline{R}$. Thus, if $\overline{R} = 50$, the difference would exceed 5. Again, that is not a small difference. (In exercise 2 the reader will be asked to determine the difference for an s chart.)

The message is that not only is it proper to recompute the control limits when an assignable cause can be both detected and removed, but also it can make a considerable difference in the limits.

When the $\overline{X}$ limits are recomputed, we obtain

$$\begin{aligned} \bar{\bar{x}}_{19} \pm A_3\bar{s}_{19} &= 61.477 \pm 1.628(14.033) \\ &= 61.447 \pm 22.846 \end{aligned}$$

where $\bar{\bar{x}}_{19}$ and $\bar{s}_{19}$ indicate that the statistics are computed from the remaining 19 subgroups. The new limits are thus

$$\mathrm{LCL} = 38.601 \qquad \mathrm{UCL} = 84.293$$

We can see that the new limits are considerably different from the old ones. In particular, the new UCL exceeds the old UCL by 2.2279. This is considerably greater than the lower bound on the difference (1.191) since the average of the deleted subgroup was well below the old UCL and, to a lesser extent, because $\bar{s}$ increased slightly.

This completes the analysis of the historical data, which we will hereinafter refer to as the *Stage 1* analysis.

4.7.8 Applying Control Limits to Future Production

We can tell by looking at the remaining 19 subgroup averages that all of them are contained by the new control limits. Thus, we could extend these limits and have them apply to future subgroups as they are plotted individually in real time, or to sets of future subgroups if for some reason real-time plotting is not feasible. The real-time plotting of points, which is generally termed *process monitoring*, will be referred to, as indicated previously, as *Stage 2*, to distinguish it from Stage 1.

For how long should these limits be used? There is no simple answer to this question, but there are some guidelines that can be given. When control charts are implemented for the first time, the process variability usually decreases. This is due not only to the removal of assignable causes, but also to the fact that workers become aware of the importance that management is attaching to quality improvement, and subsequently exercise greater care. Thus, variability that is due to workers will usually decrease even though it may have been "in control" initially.

Accordingly, the control limits should be revised from time to time by repeating the process of computing control limits from at least 20 subgroups. It is particularly important to revise the limits when there is evidence of a reduction in the process variability. The control limits for both $\overline{X}$ and s charts (and R charts) obviously depend greatly upon the process variability. If, for example, an s chart is being used and the limits are not revised (made tighter) when the variability decreases, a subsequent increase in variability (which would indicate that the process is probably out of control) might not be detected very quickly (if at all) if the points are within the (original) control limits. Similarly, a shift in the process average might not be readily detected on an $\overline{X}$ chart if the limits on the chart are based upon a poor, outdated estimate of the current process variability.

4.7.9 Standards for Control Charts

Other writers have discussed the use of "standards" in conjunction with control charts. For example, if a machine can be adjusted so that the length of a bolt should be exactly 2 inches, the centerline of an $\overline{X}$ chart maintained for length could be set at 2 rather than some value of $\bar{\bar{x}}$. Similarly, if the variability has stabilized at σ', we could use that value in obtaining the control limits for an $\overline{X}$ and s (or R) chart rather than estimating σ by either $\bar{s}/c_4$ or $\overline{R}/d_2$.

The control limits for an $\overline{X}$ chart would then be set at $\mu' \pm A\sigma'$, where, for the example just given, $\mu' = 2$ and the values of A are given in Table E in the Appendix to the book. For an s chart the centerline would be $c_4\sigma'$ and the control limits would be obtained from $B_5\sigma'$ and $B_6\sigma'$. For an R chart the centerline would be $d_2\sigma'$ and the control limits would be obtained from $D_1\sigma'$ and $D_2\sigma'$. The values for B_5, B_6, D_1 and D_2 are also given in Table E.

Some caution should be exercised when standards are used, however. For example, it might be desirable to control the process mean at a particular standard value but the current process may be incapable of meeting that standard. If the control limits are calculated from the standard, and, say, the current process mean exceeds the standard, subgroup averages could frequently exceed the UCL even when the process is in control. Another example would be where a company wanted to control the percentage of nonconforming items produced at a particular level (using one of the charts in Chapter 6). If the current percentage of nonconforming items exceeds this standard value by a sizable margin, many points will fall above the UCL, and the "assignable cause" will be simply that there is a considerable difference between the current percentage of nonconforming items and the "hoped-for" value.

Therefore, there can be serious consequences if standards are used unwisely. Even if the process mean and process variability do seem to have stabilized, the standard values do not have to be used. For a stabilized process we would expect $\overline{\overline{x}}$ and $\overline{s}/c_4$ to be very close to μ' and σ', respectively. Consequently, limits obtained using $\overline{\overline{x}}$ and $\overline{s}/c_4$ should differ very little from limits that would be obtained using μ' and σ'. Thus, nothing is lost by using the conventional approach. When in doubt as to whether or not to use standard values, it is best not to use them since control chart limits should be based upon current conditions as reflected by recent data.

4.7.10 Deleting Points

A final word concerning the deletion of points and the recomputation of control limits. For the hypothetical example that was used, all of the points on the R and s charts fell within the limits, so it was not necessary to recompute the limits. What if a point had fallen outside the control limits on the s chart, with the point subsequently discarded and the limits recomputed; should that point also be excluded from the computations for the $\overline{X}$ control chart limits? Should it be included in computing $\overline{\overline{x}}$ but excluded in estimating σ? If a point is excluded on an s chart, it should certainly not be used in estimating σ for the $\overline{X}$ chart, whether σ is estimated from $\overline{R}$ or $\overline{s}$. Technically, an assignable cause that affects σ need not have any effect on μ, and vice versa. Nevertheless, including a point for the estimation of $\mu(\sigma)$ and excluding it for the estimation of $\sigma(\mu)$ is apt to be more confusing than beneficial when charts are maintained by hand, and would make the programming somewhat tricky when a computerized system is used. The exclusion of such points will not make very much difference anyway unless the points are near one of the control limits.

4.7.11 Target Values

It is important to realize that a process can be in a state of statistical control and yet still be performing poorly if the mean of a process characteristic is not equal to the target value. For example, assume that a particular process characteristic should have a value of 12.0 for a product to function optimally, and the process appears to be in statistical control with $\bar{\bar{x}} = 13.3$. Having plotted points fall within 3-sigma limits is simply not good enough when $\bar{\bar{x}}$ is far removed from the target value. Target values are discussed in some detail in Chapter 14. Although the establishment of target values for process characteristics can be quite beneficial, the use of target values to establish control limits is another matter.

4.8 ILLUSTRATIVE EXAMPLE WITH REAL DATA

We will turn our attention to the real data set given in Table 4.4.

The data were originally analyzed by Ott (1949) and can also be found on p. 32 of Ott (1975). The objective was to determine whether the variability (which was considered excessive) in a particular electrical characteristic that was involved in the assembly of electronic units was significant over the 11 ceramic sheets relative to the variability between the 7 strips within each sheet. If the variability between sheets turned out to be significant, inferior sheets could then be discarded.

Ott addressed this question by constructing an $\bar{X}$ and an R chart. The control limits for the R chart were

$$\begin{aligned} \text{LCL} &= D_3\bar{R} & \qquad \text{UCL} &= D_4\bar{R} \\ &= 0.076(2.8) & &= 1.924(2.8) \\ &= 0.2128 & &= 5.3872 \end{aligned}$$

We observe that the range for subgroup 9 exceeds the UCL.

TABLE 4.4 Electrical Characteristics (dB) of Final Assemblies from 11 Strips of Ceramic

16.5	15.7	17.3	16.9	15.5	13.5	16.5	16.5	14.5	16.9	16.5
17.2	17.6	15.8	15.8	16.6	13.5	14.3	16.9	14.9	16.5	16.7
16.6	16.3	16.8	16.9	15.9	16.0	16.9	16.8	15.6	17.1	16.3
15.0	14.6	17.2	16.8	16.5	15.9	14.6	16.1	16.8	15.8	14.0
14.4	14.9	16.2	16.6	16.1	13.7	17.5	16.9	12.9	15.7	14.9
16.5	15.2	16.9	16.0	16.2	15.2	15.5	15.0	16.6	13.0	15.6
15.5	16.1	14.9	16.6	15.7	15.9	16.1	16.1	10.9	15.0	16.8
$\bar{x} = 16.0$	15.8	16.4	16.5	16.1	14.8	15.9	16.3	14.6	15.7	15.8
$R = 2.8$	3.0	2.4	1.1	1.1	2.5	3.2	1.9	5.9	4.1	2.8

Source: E. R. Ott (1949). Variables and control charts in production research. *Industrial Quality Control 6*(3): 30. Reprinted with permission of the American Society for Quality Control.

The limits for the $\overline{X}$ chart can be shown to be

$$\text{LCL} = \overline{\overline{x}} - A_2\overline{R} = 15.81 - 0.419(2.8) = 14.637 \qquad \text{UCL} = \overline{\overline{x}} + A_2\overline{R} = 15.81 + 0.419(2.8) = 16.983$$

It can be observed that the average for ceramic sheet #9 is just outside the LCL.

Thus, there is an apparent problem with subgroup 9 in regard to both average and variability. Closer examination of the data reveals that this is the result of the fact that two values for this ceramic sheet (10.9 and 12.9) are smaller than any of the other values in the table.

In particular, the 10.9 is well below the average value for that sheet number. If that number was excluded, the average and range for that ceramic sheet would fall in line with the averages and ranges of the other ceramic sheets. Consequently, this value is perhaps either an outlier or a value that was recorded incorrectly, rather than indicating that ceramic sheet 9 is inferior to the other sheets.

It is worth noting that if $\hat{\sigma}_{\overline{x}}$ had been obtained using s rather than R, $\overline{x}_9$ would have been slightly above the LCL, as the reader is asked to demonstrate in exercise 3.

It should also be noted that this is an application to past data, as is also done when trial control limits are determined. The difference is that in this example there is no intent to apply the limits to future production.

4.9 DETERMINING THE TIME OF A PARAMETER CHANGE

It would be helpful to know when there has been a change in the mean or variance. For example, if raw materials from some supplier other than the regular supplier(s) were used beginning at a certain time, and working backward from an out-of-control signal we estimate the time of a parameter change as the time at which the raw materials from the other supplier were used, then the change in suppliers would appear to be the problem.

Samuel, Pignatiello, and Calvin (1998a,b) give a method for determining the point at which there is a change in the mean (a) or a change in the variance (b). They show that the maximum likelihood estimator (see Section 3.8.2.1) of the point at which a change in the mean occurred is the maximum over t of $\{(T - t)(\overline{\overline{X}}_{T,t} - \hat{\mu})^2\}$, with T denoting the time at which a signal occurs and $0 \leq t < T$. As the authors state, this result is due to Hinkley (1970). The simulation results of Samuel et al. (1998a) indicate that their procedure performs reasonably well. The probability of correctly detecting the time of a one standard deviation mean shift is indicated to be about .26, and the estimator has a small standard deviation.

Samuel et al. (1998b) use the same general approach for detecting a permanent step change in the variance. Their approach is intended to be used in conjunction with any control chart that is used for monitoring process variability, although they illustrate the methodology using only an S chart. Although the authors do not discuss this, it would be desirable to use their approach for an S chart with probability limits, or in general a chart that has a lower control limit, so that the time of process improvement, as suggested by a point falling below the LCL. can be ascertained.

4.10 ACCEPTANCE SAMPLING AND ACCEPTANCE CONTROL CHART

Acceptance sampling is not a process control technique. Rather, an acceptance sampling plan specifies the sample size that is to be used and the decision criteria that are to be employed in determining whether a lot or shipment should be rejected. Acceptance sampling plans can be for either variables (i.e., measurement data) or attributes (e.g., good or bad). When the former is used, normality is generally assumed, although there are variables sampling plans for nonnormal distributions. Since normality does not exist in practice, serious problems can ensue when a normality-based acceptance sampling plan is used (Montgomery, 1996, p. 655).

Several decades ago, quality control was generally equated with inspection, and acceptance sampling was viewed as a compromise between no inspection and 100% inspection. Today there are more options for quality control and improvement.

Acceptance sampling has been rather controversial. As the prominent industrial statistician Harold F. Dodge (1893–1976) once stated, "You cannot inspect quality into a product." The quality of an item has been determined by the time it reaches the final inspection stage. Thus, 100% final inspection, as has been routinely practiced in many companies both large and small, will not ensure good quality. Studies have shown that only about 80% of nonconforming units are detected during 100% final inspection, so not only is such inspection rather expensive, but also it is not effective in preventing nonconforming units from reaching customers.

What about using inspection at each stage of a manufacturing process? This would have the same weaknesses of 100% final inspection: nonconforming units would be undetected at each stage and the inspection would be too expensive. W. Edwards Deming stated that a company should use either 100% final inspection or no inspection, and that the former will generally be too expensive, although probably desirable for critical parts. Papadakis (1985) examines what he terms the "Deming Inspection Criterion." [See also Chapter 15 of Deming (1986).]

One view of acceptance sampling is the following. If a company is using acceptance sampling plans it should do so only temporarily, as acceptance sampling is simply a defensive measure for preventing product quality from deteriorating any further. It might thus be used for *quality control*, but not for *quality improvement*. The same message can be found in the acknowledged best reference on acceptance sampling, Schilling (1982, p. 5):

> Simultaneous use of acceptance quality control and process quality control should eventually lead to improvement in quality levels to the point that regular application of acceptance sampling is no longer needed.

A somewhat stronger message is contained in Schilling (1983):

> Since we cannot "inspect quality into a product," acceptance sampling plans should always be set up to self-destruct in favor of such procedures as soon as conditions warrant.

Vardeman (1986) similarly supports the use of acceptance sampling until processes can be brought into a state of statistical control. See also Schilling (1991).

Problems with acceptance sampling plans include the fact that the producer's and the consumer's risk can *both* be unacceptably high, and the commonly used plans, which were developed decades ago, are not suitable for the high quality that is expected today. Consequently, it is best to try to avoid the use of acceptance sampling as much as possible. One way for a manufacturing company to do so is to require its suppliers to use process control techniques. This can greatly reduce, if not eliminate, the need to inspect incoming materials.

Most quality control books contain considerable information on acceptance sampling, including Montgomery (1996) and Wadsworth, Stephens, and Godfrey (1986), and the reader interested in acceptance sampling is referred to such sources.

4.10.1 Acceptance Control Chart

Control chart users have employed modified forms of an $\overline{X}$ chart when the specification limits are well beyond the endpoints of the interval $\mu \pm 3\sigma$. The rationale for doing so can be explained as follows. Assume that the specification limits are $\mu \pm 6\sigma$ and that a value outside this interval will result in a nonconforming unit of production. What will be the effect of a shift in the process mean from μ to $\mu + 2\sigma_{\overline{x}}$ for $n = 4$? Before the shift occurs, only about one of every one billion units will be judged nonconforming ($z = 6$) by having exceeded the upper specification limit (assuming approximate normality). Since $\sigma_{\overline{x}} = \sigma/2$ when $n = 4$, it follows that $2\sigma_{\overline{x}} = \sigma$ so that the new mean is at $\mu' = \mu + \sigma$. It then follows that the proportion of nonconforming units will be given by $P(Z > 5)$ where $Z \sim N(0, 1)$. The answer is that slightly less than 3 out of every 10 million units will be nonconforming because of having exceeded the upper specification limit. Such a small increase is not likely to cause any real concern.

With an $\overline{X}$ chart the 2-sigma shift would produce an average run length (ARL) of approximately 3 for the UCL. (See Section 4.14 for an explanation of ARL.) Nevertheless, in this instance the shift would be relatively inconsequential in terms of the increase in the percentage of nonconforming units.

Accordingly, when the specifications are well outside $\mu \pm 3\sigma$, which is the natural variability of the process, an argument could be made for not using the conventional $\overline{X}$ chart. The control limits on an $\overline{X}$ chart could, of course, be increased from 3-sigma to, say, 5-sigma, but it would be far more logical to design a chart for plotting subgroup averages in which the control limits are determined from the specification limits. An *acceptance control chart* is such a chart. Recall, however, that the specification limits should not be displayed on an $\overline{X}$ chart, or, in general, on any control chart on which averages are plotted. Thus, they should not be displayed on an acceptance chart either; they should be used only in obtaining the control limits.

The acceptance control chart was developed by Freund (1957) and reviewed later by Woods (1976). The limits are obtained from the same components that are used in acceptance sampling procedures. They are:

APL (Acceptable Process Level). This is the process level farthest from standard that yields product quality that is to be accepted $100(1 - \alpha)\%$ of the time that the level is attained. There will be two APL values when there is both an upper and lower specification limit.

α (Alpha). This is the probability of receiving a signal when the process level is at the APL.

RPL (Rejectable Process Level). This is the process level closest to the standard that yields product quality that is to be rejected almost all $100(1 - \beta)\%$ of the time it occurs. There will be two RPL values when there is both an upper and lower specification limit.

β (Beta). This is the probability of not receiving a signal when the process level is at the RPL.

p_1. The acceptable percentage of units falling outside of the specifications.

p_2. The rejectable percentage of units falling outside of the specifications.

In general, all six must not be specified. Doing so would most likely result in inconsistent solutions for the acceptance control limits because of the way in which the limits are computed. Specifically, the upper acceptance control limit (UACL) can be computed as

$$\text{UACL} = \text{USL} - z_{p_1}\sigma + z_\alpha \frac{\sigma}{\sqrt{n}} \tag{4.4}$$

where USL is the upper specification limit and z_{p_1} and z_α denote the z-values (for the standard normal distribution) corresponding to tail areas of p_1 and α, respectively. The UACL can also be computed as

$$\text{UACL} = \text{USL} - z_{p_2}\sigma - z_\beta \frac{\sigma}{\sqrt{n}} \tag{4.5}$$

Equality between Eqs. (4.4) and (4.5) would require that $z_{p_2}\sigma + z_\beta\sigma/\sqrt{n} = z_{p_1}\sigma - z_\alpha\sigma/\sqrt{n}$, which will not hold for most choices of p_1, p_2, α, and β.

Similarly, the lower acceptance control limit (LACL) can be computed as

$$\text{LACL} = \text{LSL} + z_{p_1}\sigma - z_\alpha \frac{\sigma}{\sqrt{n}} \tag{4.6}$$

or as

$$\mathrm{LACL} = \mathrm{LSL} \ + z_{p_2}\sigma + z_{\beta}\frac{\sigma}{\sqrt{n}} \tag{4.7}$$

In essence, an acceptance control chart is a modified $\overline{X}$ chart, but it is not necessarily the same as charts obtained from "modified control limits" as discussed in Hill (1956). Those modified charts do not allow for the specification of β and p_2. They are discussed later in the chapter.

4.10.1.1 Acceptance Chart with $\overline{X}$ Control Limits

The acceptance chart can be constructed with the $\overline{X}$ (chart) control limits either shown or not shown on the chart. It would be advantageous to have them displayed on the chart. The reason for this is that the $\overline{X}$ control limits would enable the user to determine whether or not the process is in control at the assumed mean. Evidence of a mean shift would not necessarily cause any action to be taken, however. This use of $\overline{X}$ limits would be for detecting a "tolerable shift" before the shift becomes larger and causes major consequences.

An example will be given to illustrate the determination of the acceptance control limits. In Table 4.2 the smallest value was 10 and the largest was 97. Let us assume that the specification limits are 4 and 108 and that we will specify p_1 and α. Readers familiar with acceptance sampling will recognize that p_1 is the same as the acceptable quality level (AQL). Since arguments against use of the latter were given in Chapter 1, this begs the question of whether the same arguments apply to the use of p_1. They do, but not to the same extent. We can make p_1 as small as we want, but we do not have that luxury with AQL. The smallest possible AQL value that can be used in the various sampling plans is .001. Stating a value for p_1 implies that we are willing to tolerate some nonconforming units, so we would probably want to counteract that by making p_1 very small. If we make it too small, however, the acceptance control limits will be *inside* the $\overline{X}$ limits, which is not in accordance with the way that the chart is designed to function. To illustrate this, we will use $p_1 = .00001$, which means that 1 nonconforming unit of every 100,000 units will be considered an acceptable rate. We will also use $\alpha = .05$, which means that we will run the risk of rejecting a process that has this value of p_1 5% of the time. A value of $p_1 = .00001$ corresponds to $z = 4.265$ and $\alpha = .05$ corresponds to $z = 1.645$. The limits are

$$\begin{aligned}
\mathrm{LACL} &= \mathrm{LSL} + z_{p_1}\hat{\sigma} - z_{\alpha}\frac{\hat{\sigma}}{\sqrt{n}} \\
&= 4 + 4.265(15.202) - 1.645\left(\frac{15.202}{\sqrt{4}}\right) \\
&= 56.333
\end{aligned}$$

and

$$\begin{aligned}\text{UACL} &= \text{USL} \quad - z_{p_1}\hat{\sigma} + z_{\alpha}\frac{\hat{\sigma}}{\sqrt{n}} \\ &= 108 - 4.265(15.202) + 1.645\left(\frac{15.202}{\sqrt{4}}\right) \\ &= 55.667\end{aligned}$$

Thus, not only are both limits well *inside* the $\overline{X}$ limits, but also the LACL is actually larger than the UACL! This occurs, in part, because a small value of p_1 was chosen, but primarily because the specification limits were not very far outside the natural variability of the process. The USL was at $\hat{\mu} + 3.19\hat{\sigma}$ and the LSL was at $\hat{\mu} - 3.65\hat{\sigma}$, where $\hat{\sigma} = \overline{R}/d_2$.

We will now assume that the data in Table 4.2 are "coded" (e.g., by subtracting 1000 from the original observations) so that a negative value for LSL is possible for the coded data. Can we use a very small value for p_1 when the specification limits are at, say, $\hat{\mu} \pm 6\hat{\sigma}$? The answer is "yes" and can be demonstrated as follows. The specification limits would be

$$\begin{aligned}\hat{\mu} \pm 6\hat{\sigma} &= 59.4375 \pm 6(15.202) \\ &= 59.4375 \pm 91.2120\end{aligned}$$

so that USL = 150.6495 and LSL = −31.7745. The acceptance control limits for these specification limits can be shown to be

$$\text{LACL} = 20.5585 \qquad \text{UACL} = 98.3165$$

Notice that these values are well outside the $\overline{X}$-chart limits of 36.6198 and 82.2552 that were given much earlier in the chapter. Recall that the average for subgroup 10 in Table 4.2 is 21.25, which is well below the LCL of 36.6198. It is not, however, below the LACL, so a search for an assignable cause would not necessarily be initiated. The acceptance chart with the $\overline{X}$ limits would appear as in Figure 4.4. Notice in this case that the acceptance control limits are equidistant from the estimated process mean. This is the result of the fact that the specification limits were assumed to be at $\hat{\mu} \pm 6\hat{\sigma}$. Obviously, the specification limits will not always be equidistant from the estimated process mean, and when they are not, the acceptance control limits will not be equidistant, nor will they be equidistant from their respective $\overline{X}$ limits.

These $\overline{X}$ limits essentially serve as "warning limits" on the acceptance chart. Readers interested in acceptance control charts are referred to Woods (1976) for further reading.

4.10.1.2 Acceptance Charts Versus Target Values

The focus for acceptance charts has been on the location of the specification limits relative to the variability of the variable being measured. There has been

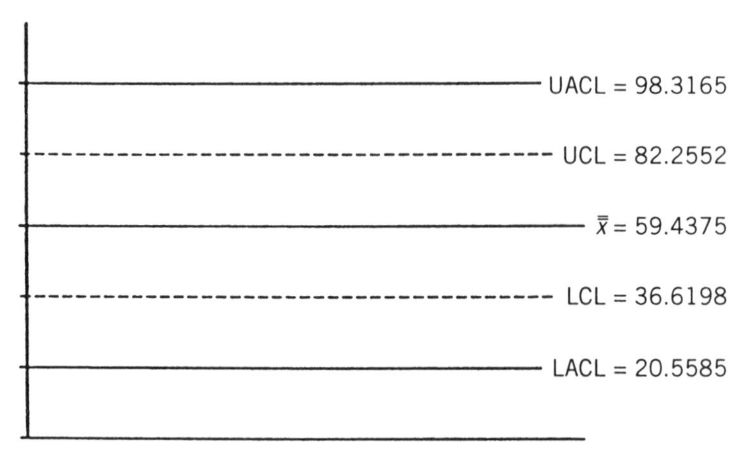

FIGURE 4.4 Acceptance control chart with the $\overline{X}$ limits.

no mention of a "target value" for the mean of that variable. If, on the other hand, a target value does exist and a "loss" of some sort will be incurred if there is even a slight deviation from that target value, then an acceptance chart should not be used. (The reader is referred to Chapter 14 for additional information concerning target values and loss functions.) It is apparent, however, that many companies do not employ target values and are simply interested in staying within specifications. For them an acceptance chart can be useful.

Users of acceptance charts should recognize, however, that the use of acceptance charts is not in accordance with contemporary views on quality improvement. The current objective is to strive to "hit the target" (value) while minimizing variation about the target value. Thus, an acceptance chart really does not have a role in a quality improvement program, and use of the chart should be discouraged.

4.11 MODIFIED LIMITS

The preceding comments also apply if upper and lower modified limits (UML and LML) are used. As stated previously, these do not allow for the use of β and p_2. Accordingly, the control limits would be obtained using Eqs. (4.4) and (4.6). Figure 4.5 illustrates a possible configuration. Here USL $= \mu + 8\sigma$ and LSL $= \mu - 8\sigma$ (assuming for simplicity that μ and σ are known). Since the USL and LSL are well beyond $\mu \pm 3\sigma$, there is considerable room for the distribution of $\overline{X}$ to shift without causing an appreciable increase in the percentage of nonconforming units. If it is desired to use 3-sigma limits relative to the midlines for the two distributions of $\overline{X}$ shown in Figure 4.5 (as is the usual case), the effect is to widen the usual $\overline{X}$ control limits by, in this case, 5σ since $8 - 3 = 5$. In general, if the specification limits were at $\mu \pm k\sigma$, the limits would be widened by $(k - 3)\sigma$. Thus, we would not want to use modified limits unless

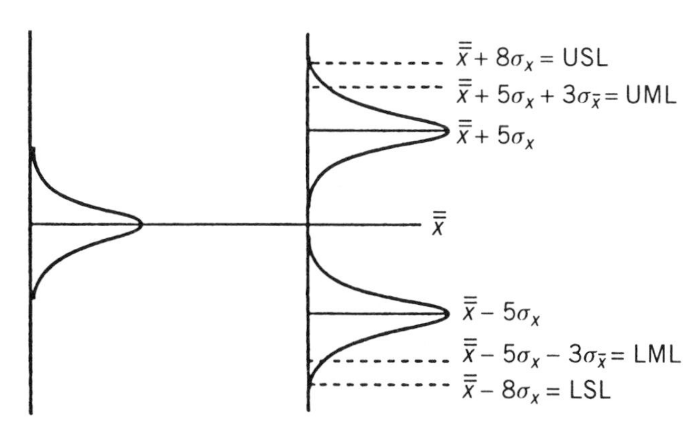

FIGURE 4.5 Illustration of modified control limits (UML and LML).

$k > 3$. (Of course, the specification limits will not necessarily be equidistant from μ or $\bar{\bar{x}}$.)

Modified limits, acceptance control charts, and a modified version of acceptance control charts are discussed in detail by Wadsworth et al. (1986).

4.12 DIFFERENCE CONTROL CHARTS

Two types of difference charts have been proposed. Grubbs (1946) proposed a difference chart for ballistics testing during World War II. The general idea is to separate process instability caused by uncontrollable factors (i.e., humidity) from process instability due to assignable causes. This is accomplished by taking samples (subgroups) from the current production and also samples from what is referred to as a reference lot. The reference lot consists of units that are produced under controlled process conditions except for possibly being influenced by uncontrollable factors. Since samples from the current production and the reference lot are equally susceptible to influence by uncontrollable factors, any sizable differences between subgroup averages should reflect process instability in the current production due to controllable factors. The 3-sigma control limits are obtained from

$$0 \pm A_2\sqrt{\overline{R}_{\mathrm{r}}^2 + \overline{R}_{\mathrm{c}}^2}$$

where A_2 is the same constant that is used for $\overline{X}$ charts; $\overline{R}_{\mathrm{r}}$ and $\overline{R}_{\mathrm{c}}$ designate the average range for the reference lot and the current lot, respectively, for the (20 or more) subgroups that are used in determining the control limits.

Each point that is plotted on the chart represents $\bar{x}_{\mathrm{r}} - \bar{x}_{\mathrm{c}}$ for each subgroup, where $\bar{x}_{\mathrm{r}}$ and $\bar{x}_{\mathrm{c}}$ denote a subgroup average for the reference and current lot, respectively.

This chart would be used in place of an $\overline{X}$ chart. An R chart could be kept on the current production or an R chart of differences could be maintained if it is believed that the uncontrollable factors affect the within-sample variability for the current production. The limits on such a chart would be obtained from

$$
\begin{aligned}
&0 \pm 3\sqrt{\mathrm{Var}(R_{\mathrm{r}} - R_{\mathrm{c}})} \\
&\quad = 0 \pm 3\sqrt{\mathrm{Var}(R_{\mathrm{r}}) + \mathrm{Var}(R_{\mathrm{c}})} \\
&\quad = 0 \pm \frac{3d_3}{d_2}\sqrt{\overline{R}_{\mathrm{r}}^2 + \overline{R}_{\mathrm{c}}^2}
\end{aligned}
$$

and each point plotted on the chart would represent $R_{\mathrm{r}} - R_{\mathrm{c}}$ for each subgroup (d_3 is explained in the Appendix to this chapter).

Readers familiar with statistics will recognize that this difference chart for averages is analogous to a pooled-t test. In fact, when each point is plotted separately and compared with the limits, it is comparable to a pooled-t test for testing the equality of population means with a significance level of .0027 provided that $\overline{x}_{\mathrm{r}} = \overline{x}_{\mathrm{c}}$ is approximately normally distributed. (A pooled-t test does not utilize ranges, however.)

Another type of difference chart was proposed by Ott (1947). With this chart differences are obtained between paired observations within each subgroup, and what is plotted on the chart is the average difference for each subgroup. The control limits are obtained from $0 \pm A_2\overline{R}$, where $\overline{R}$ is the average of the ranges of the differences. (That is, there is one range per subgroup, and that range is the difference between the largest difference between the pairs in a subgroup and the smallest difference.) Notice that the limits are obtained in essentially the same way as they are obtained for an $\overline{X}$ chart, except that the centerline is 0 instead of $\overline{\overline{x}}$.

This type of difference chart is of particular value in calibration work in which test equipment is to be kept in good working order. See Ott (1975, pp. 64–68) for further reading.

Another way to identify possible extraneous effects, such as the (possibly) poor quality of raw materials, is to use a cause-selecting control chart. These charts are discussed in Section 12.8.

4.13 OTHER CHARTS

A median chart is a substitute for an $\overline{X}$ chart and could be used, in particular, when the subgroup size is an odd number (e.g., $n = 5$). The median was defined in Chapter 3 as the middle value in a set of ordered observations (from smallest to largest), or the average of the two middle values when there is an even number of observations. Charts that utilize medians are favored by, for example, Joiner (1994).

The median is easy to determine and easily understood, but it is not as efficient as the average since not all of the observations are being used. References on the median chart include Nelson (1982), which gives the factors for calculating the 3-sigma limits and also gives efficiency figures for the median relative to the average. With a computerized SQC system a chart that uses averages is to be preferred over a median chart, but the latter will probably continue to be used by a number of companies that use manual charting.

Ferrell (1953) proposed a midrange chart for controlling the process mean and provided tables to be used in constructing such charts. (The midrange is defined as the average of the largest and smallest observations in a subgroup.) Midrange charts have been used very little.

4.14 AVERAGE RUN LENGTH

If the parameters were known, the expected number of plotted points before a point plots outside the control limits could be obtained as the reciprocal of the probability of a single point falling outside the limits when each point is plotted individually. This expected value is called the average run length (ARL). It is desirable for the in-control ARL to be reasonably large, so that false alarms will rarely occur. With 3-sigma limits, the in-control ARL would be $1/.0027 = 370.37$ if normality existed. The in-control ARL being equal to the reciprocal of the sum of the tail areas results from the fact that the geometric distribution is the appropriate distribution, and the mean of a geometric random variable was given in Section 3.5.4 as $1/p$.

When a parameter change of an amount that is considered to be consequential occurs, we want to detect the change as quickly as possible. Accordingly, the *parameter-change ARL* should be small. Unfortunately, Shewhart-type charts do not have good ARL properties. For example, assume that a $1\sigma_{\overline{x}}$ increase in the mean occurs when an $\overline{X}$ chart is used. Again assuming that the parameters are known, we may determine the probability of a point falling outside the control limits by first computing two z-values.

Specifically, we obtain $z = [\mu + 3\sigma_{\overline{x}} - (\mu + \sigma_{\overline{x}})]/\sigma_{\overline{x}} = 2$ for the UCL and $z = (\mu - 3\sigma_{\overline{x}} - (\mu + \sigma_{\overline{x}})]/\sigma_{\overline{x}} = -4$. The ARL is then obtained as $1/[P(z > 2) + P(z < 4)] = 43.89$. The ARLs for other mean changes as a multiple of $\sigma_{\overline{x}}$ can be computed by using $k\sigma_{\overline{x}}$ in place of $\sigma_{\overline{x}}$.

The reader should realize that parameters will generally have to be estimated, however, and doing so will inflate both the in-control ARL and the parameter-change ARLs. A partial explanation of this result is as follows. For simplicity, assume that σ is known so that only μ must be estimated. With the assumption of normality, the UCL is as likely to be overestimated by a given amount as it is to be underestimated by the same amount. If the UCL is overestimated by $0.1\sigma_{\overline{x}}$, the reciprocal of the tail area of the normal distribution above the UCL is 1033.41, whereas for underestimation by the same amount the reciprocal is 535.94. The average of these two numbers is 784.67, which is noticeably greater than the nominal value of 740.8. Estimation errors of larger amounts would produce a larger discrepancy.

The determination of ARLs when parameters are estimated has been addressed by several authors. Quesenberry (1993) used simulation to obtain the ARLs and concluded that it is desirable to have the trial control limits based upon at least 100 observations, with "permanent" limits based on at least 300 observations. Thus, if a subgroup of size 5 is being used, the historical data that are used to compute the trial limits should consist of at least 20 subgroups. See also Ghosh, Reynolds, and Hui (1981), who determined the ARLs using numerical integration when σ^2 was estimated and a target value was assumed for μ.

4.14.1 Weakness of the ARL Measure

The use of ARLs has come under considerable fire in recent years. This is because (1) the run length distribution is quite skewed so that the average run length will not be a typical run length, and (2) the standard deviation of the run length is quite large. Consider again the case where the parameters are known so that the geometric distribution applies. Since the standard deviation of a geometric random variable is $(1 - p)^{1/2}/p$, the standard deviation will be close to the mean of $1/p$ whenever p is small. Since $p = .0027$ when normality is assumed, the standard deviation of the run length is $(1 - .0027)^{1/2}/.0027 = 369.87$, which is very close to the ARL of 370.37. Similarly, the ARL and the standard deviation will be close for parameter changes since the standard deviation approaches zero as p approaches 1 and the mean approaches 1.

This problem also occurs when parameters are estimated, with both the mean and the standard deviation being inflated (see Quesenberry, 1993).

Whether the parameters are known or not, the magnitude of the standard deviation of the run length is disturbing. Table 4.5 contains 100 simulated in-control run lengths under the assumptions of (1) normality, (2) known parameter values, and (3) 3-sigma limits. The average of these run lengths is 330.4, the standard deviation is 348.6, and the median is 194.0. Since the standard deviation is quite large, the standard deviation of the average based on 100 run lengths is not small (34.9). Consequently, it is not surprising that the average of the 100 run lengths differs considerably from the theoretical average. Notice also that the minimum run length is 1 and the maximum is 1524, with many of the run lengths being quite small.

The numbers in Table 4.5 suggest that a control chart user should design a control chart scheme so as to perhaps (1) produce a particular median or (2) have a certain maximum probability for the occurrence of a particular run length, such as 100. Alternatively, one might focus attention on the probability of receiving a signal on a single plotted point. For the in-control case this would be called the *false-alarm rate*.

Keats, Miskulin, and Runger (1995) introduced the *average production length*. This is the expected amount of production between the time of a parameter change and the time when a signal is received. The average production length will often be of more practical value than the ARL since the former incorporates the sampling frequency whereas the latter does not. We should note, however, that the average production length should also have a large standard deviation since it is a function of the ARL.

TABLE 4.5 Simulated In-Control Run Lengths

1105	134	75	50	63
229	1041	150	626	1488
543	603	322	124	40
430	161	38	426	76
27	1017	105	1245	22
156	284	1524	117	216
1	365	446	110	304
96	45	71	67	1350
498	694	328	111	53
172	478	378	433	182
23	56	130	655	938
212	1284	61	377	398
304	286	218	652	67
201	722	260	103	390
86	158	360	193	84
182	66	412	921	67
188	178	643	158	351
182	195	330	43	219
140	18	97	22	226
172	92	421	847	31

4.15 DETERMINING THE SUBGROUP SIZE

The subgroup size is generally not determined analytically but rather is determined by convenience. Subgroup size *is* determined when the economic design of control charts is considered (see Chapter 10) and would also be determined when an average production length approach, as described in Section 4.14, is used.

Typically subgroups of four or five observations are used for convenience, and this practice has been followed for decades. The larger the subgroup, the more power a control chart will have for detecting parameter changes. Osborn (1990) surveyed 117 companies in the food manufacturing/processing and metalworking industries. Although the respondents were asked to give the sample (subgroup) size that they typically used, Osborn did not provide any summary statistics involving subgroup size. Instead, for the 61 respondents who provided usable information, Osborn computed the power of detecting a (minimum) shift that the respondent believed was crucial. The average power (i.e., the probability of detecting the shift) was .3426.

From this information we cannot determine the average sample size. Nevertheless, we can gain some insight by proceeding as follows. Assume that a 1-sigma shift is the smallest shift that is important to detect. If the subgroup size is 5 and we assume that μ and σ are known (as Osborn had to assume), then the probability of detecting the shift on the first subgroup (i.e., the "power") is $P(Z > 3 - \sqrt{5}) + P(Z < -3 - \sqrt{5}) = .22$. The power is .45 if a subgroup size of 6 is used, so it seems likely that most respondents used subgroup sizes of 5, or a size very close to 5. Indeed, one operator stated that he always used a subgroup size of 5, regardless of the process, because he had memorized the control chart constants for that subgroup size and could thus compute control limits without having to refer to a table.

It is interesting that 34 of the respondents believed that the power of the chart that they were using was 99.73%. Of course .9973 would be the probability of not receiving a signal when a process is in control and normality is assumed.

These bits of information from Osborn's paper suggest that many users do not have a sound understanding of control charts, and subgroup size is often not determined with control chart properties in mind. This seems to be true in general, not just for the food manufacturing/processing and metalworking industries.

4.15.1 Unequal Subgroup Sizes

Although unequal subgroup sizes frequently occur with charts for nonconformities and nonconforming units (discussed in Chapter 6), they should occur less frequently for the procedures presented in this chapter since the subgroup sizes are typically small and the sampling is usually performed in real time. Nevertheless, missing data can occur. The presence of unequal subgroup sizes can be handled for an $\overline{X}$, R, or s chart by using an approach such as that given by Burr (1969). The problem has also been addressed by Nelson (1989), who discusses the use of standardized plotting statistics for various charts, including $\overline{X}$ and R charts. See also Nelson (1990).

There is also the matter of purposeful variation of the subgroup size. Costa (1994) gives the properties of an $\overline{X}$ chart where the size of a subgroup depends on the observations in the preceding subgroup. Specifically, two subgroup sizes are used, where the smaller of the two sizes is used for the next subgroup if the mean of the current subgroup falls within "warning limits," with the larger size used if the mean is between one of the warning limits and the corresponding action (control) limit. Such procedures are referred to in the literature as variable sample size (VSS) procedures. We would expect a VSS $\overline{X}$ chart to be superior to a regular $\overline{X}$ chart in detecting small parameter changes since for a small parameter change it is more likely that a subsequent subgroup average will plot between a warning limit and an action limit than plot outside an action limit, with the advantage going to zero as the parameter change becomes larger. Of course, a comparison of the two types of charts in terms of detection capability is meaningful only when the warning limits are chosen in such a way that the in-control ARL is the same for each procedure. See Costa (1994) for details.

4.16 OUT-OF-CONTROL ACTION PLANS

In order for the control charts presented in this chapter and in subsequent chapters to be useful tools that will facilitate process improvement, out-of-control action plans (OCAPs) must be devised and implemented. Such plans are discussed in an excellent article by Sandorf and Bassett (1993). They describe how Philips Semiconductors, headquartered in Eindhoven, The Netherlands, has used OCAPs since 1986. Phillips Semiconductors is Europe's leading manufacturer of integrated circuits (Roes and Does, 1995).

An OCAP is a flow chart that has three main components: activators, checkpoints, and terminators. An *activator* is an out-of-control signal. The discussion of Sandorf and Bassett (1993) suggests that Philips uses runs rules in addition to the signal resulting from a point outside the control limits. The inadvisability of using runs rules in conjunction with Shewhart charts is discussed in Section 8.2.1. A *checkpoint* is a potential assignable cause. The checkpoints are numbered in accordance with their likelihood of causing an out-of-control signal, with the lowest-numbered checkpoint being the most probable cause and/or the most easily investigated. The *terminator* is the action that is taken to resolve the out-of-control condition.

An important feature of an OCAP is that the searching for an assignable cause and the action taken to remove an assignable cause are taken by the person who maintains each chart. (At Philips this is usually the operator.) Thus, the operator has complete control over a process, rather than having engineering or maintenance personnel attempting to repair an out-of-control process.

As discussed by Sandorf and Bassett (1993), the use of an OCAP is an iterative process. At Philips, a process action team (PAT) is responsible for designing control charts, calculating control limits, and constructing the OCAP that accompanies each control chart. Interestingly, a control chart is never implemented unless there is an accompanying OCAP.

A Pareto chart is constructed by the PAT after a certain period of time, and this is used in assigning numbers to the terminators. These numbers should of course change over time as certain assignable causes are eliminated.

A control chart is useful in signaling the apparent existence of an assignable cause. Process improvement will result only from the removal of assignable causes. Palm and DeAmico (1995) state, "any chart in operations without such plan is not likely to be a useful process improvement tool." See also the discussion in Roes and Does (1995).

4.17 ASSUMPTIONS FOR THE CHARTS IN THIS CHAPTER

This section should be studied by those readers who are interested in the statistical foundation of the control charts presented in this chapter and in control chart modifications that should be used under certain conditions.

There are various assumptions that must be (at least) approximately satisfied before the control chart methodologies given in this chapter can be considered valid. These are not often stated and are checked even less frequently, but they are very important. The model that is assumed is a "white-noise model": $Y_i = \mu + \varepsilon_i$. That is, the mean is assumed fixed (under the null hypothesis), and the variation in the observations (the Y_i) is assumed to be a random error term (ε_i). This is the model that is assumed when the process is in control.

We shall now consider the other assumptions for each control chart presented in this chapter.

4.17.1 Normality

For the $\overline{X}$, R, s, and s^2 charts the basic assumptions are that the individual observations are independent and (approximately) normally distributed, since the tabled constants that are used are constructed under the assumption of normality. We have seen that the usual 3-sigma limits can be off considerably from the actual probability limits due to the fact that the distributions of R, s, and s^2 differ considerably from a normal distribution. The charts for R and s that were illustrated with probability limits will also have incorrect probability limits, however, when the individual observations are not normally distributed.

There are many process characteristics whose distribution will not be well approximated by a normal distribution. In particular, James (1989) points out that quality characteristics such as diameter, roundness, mold dimensions, and customer waiting time will be nonnormal; Lee and Matzo (1998) explain that leakage from a fuel injector has a very right-skewed distribution; and Gunter (1991) points out that characteristics such as flatness, runout, and percent contamination will have skewed distributions.

Burr (1967) investigated the effect of nonnormality on the control limits of an R chart (and an $\overline{X}$ chart) and concluded that nonnormality is not a serious problem unless there is a considerable deviation from normality. He provides constants that should be used (instead of D_3 and D_4) for 28 (nonnormal) distributions in the family of Burr distributions. These constants, however, simply facilitate the construction of 3-sigma limits in the presence of nonnormality. The resultant limits are not probability limits and the probability of a point falling outside the limits will, in general, be unknown. This is due to the fact that the distribution of the range is known (or at least tabulated) only for a few nonnormal distributions. Consequently, since tables of percentage points of the distribution of the range are unavailable for general nonnormal distributions, probability limits cannot be determined analytically.

The same general problem exists with s and s^2 charts since the distributions of s and s^2 are not known for nonnormal distributions, in general, and are certainly not tabulated. As Hahn (1970) points out, confidence intervals and hypothesis tests on σ or σ^2 are highly sensitive to slight-to-moderate departures from normality, and an s chart and an s^2 chart correspond to hypothesis tests on σ and σ^2, respectively. [See also Box (1953) for a discussion of the sensitivity of tests on σ and σ^2 to the normality assumption.]

Since there is no general way to obtain probability limits for an R, s, or s^2 chart for every type of nonnormal data, what can a user do who would like to have control limits such that the probability of a point falling outside one of the control limits when the process is in control is approximately the same for each limit, and a good estimate can be obtained of each of the two probabilities? A logical approach would be to transform the data such that the transformed data are approximately normally distributed and then use the probability-limit approach to obtain the limits on the desired chart. Such limits would then be approximate probability limits. The variability of the transformed variable would then be controlled, not the variability of the original variable. Transformed variables will often have physical meaning when simple transformations are used (e.g., log, reciprocal, square root). Simple transformations will not necessarily lead to normality, whereas more sophisticated transformations that produce approximate normality will often produce new variables that do not have any physical meaning. For example, what is the physical interpretation of $X^{2.6}$ when X is measured in feet? The user might thus be faced with the choice between approximate probability limits for a less-than-desirable transformed variable and false probability limits for the original variable where the limits may result in highly different tail probabilities.

This choice need be made, however, only if there is evidence that the distribution may be more than slightly nonnormal. This can be checked in various ways, as was discussed briefly in Chapter 2. These methods are described in detail in Shapiro (1980). Suggested methods include plotting all of the observations on normal probability paper and observing if the points form approximately a straight line. (Alternatively, a systematic sample of every ith observation might be plotted if the data set is large.) Another commonly used test is the Shapiro–Wilk W test. The mechanics of normal probability plots are discussed and illustrated in Section 11.4.

The assumption of independent observations can be checked by calculating autocorrelation coefficients of various lags [see, e.g., Box, Jenkins, and Reinsel (1994)]. The control limits should be adjusted when there is strong evidence of correlated observations. This is discussed later in the chapter.

For an $\overline{X}$ chart it is assumed that the observations are independent and the *averages* are approximately normally distributed. Thus, the individual observations need not be approximately normally distributed, as is the case for an R, s, or s^2 chart. It is often assumed, however, that the subgroup averages are approximately normally distributed for subgroup sizes of 4 or 5, regardless of the way the individual observations are distributed. This implicit assumption can be traced to the work of Walter A. Shewhart, who took samples of size 4 from various nonnormal distributions and constructed empirical distributions of the sample averages. The results, reported in Shewhart (1931), indicate that the distribution of the sample averages does not differ greatly from a normal distribution, even when the observations come from a highly nonnormal distribution such as a right triangular distribution. The right triangular distribution used by Shewhart was obtained from a population that consisted of 820 chips, with 40 chips marked -1.3, 39 chips marked -1.2, 38 chips marked -1.1, and so on, with the pattern continued so that one chip is marked 2.6. The mean of this distribution can be shown to be zero, the coefficient of skewness, α_3, is 0.5661, and the coefficient of kurtosis, α_4, is 2.3978. These are measures of the asymmetry and "peakedness" of a distribution. For a normal distribution $\alpha_3 = 0$ (since the distribution is symmetric) and $\alpha_4 = 3.0$.

Our interest, however, should be in α_3 and α_4 for $\overline{X}$, not X, and to see the extent to which these differ from 0 and 3.0, respectively. Measures of skewness and kurtosis are perhaps more frequently given as β_1 and β_2, respectively, where $\beta_1 = \alpha_3^2$ and $\beta_2 = \alpha_4$. For the distribution of $\overline{X}$ we have

$$\beta_{1:\,\bar{x}} = \frac{\beta_{1:\,x}}{n} \qquad \beta_{2:\,\bar{x}} = \frac{\beta_{2:\,x} - 3}{n} + 3$$

For the right triangular distribution, $\beta_{1:\,\bar{x}} = 0.0801$ and $\beta_{2:\,\bar{x}} = 2.84945$ when $n = 4$. Since these are very close to 0 and 3, respectively, Shewhart's sampling distribution of $\overline{X}$ obtained from 1000 samples of size 4 from the population of 820 chips was expectedly very close to a normal distribution in shape.

Samples drawn from a less asymmetric, "mound-shaped" distribution could be expected to have values of $\beta_{1:\,\bar{x}}$ and $\beta_{2:\,\bar{x}}$ that are closer to 0 and 3, respectively, than the values for the right triangular distribution.

Our attention, however, should be focused on the effect that the nonnormal distributions have on the probabilities outside the 3σ limits on an $\overline{X}$ chart. These "tail probabilities" have been examined by Schilling and Nelson (1976), who showed that the sum of the two tail probabilities differs very little from the nominal value (.0027) even when the distribution of X differs considerably from a normal distribution.

The individual tail probabilities have been examined by Moore (1957) in addition to Schilling and Nelson (1976). A much different picture emerges when the tail probabilities are examined separately rather than combined. Examination of the individual tail areas can reveal how unequal the tail probabilities are.

Moore examined, in particular, what happens to the tail probabilities when the data are drawn from either a chi-square distribution with four degrees of freedom or a chi-square distribution with eight degrees of freedom (the former is more asymmetric than the latter). Such an analysis, however, begs the question, "Are these distributions likely to arise in practice when control charts are constructed?" As Moore (1957, p. 171) points out, data received from Shewhart resulting from work at Bell Labs had considerable skewness and kurtosis, roughly equal to that of a chi-square distribution with eight degrees of freedom.

Although Moore examined the true tail probabilities for samples of size 5 from the latter distribution when normality is assumed, the smallest nominal tail probability given in his appropriate table is .005, which corresponds to 2.575σ limits. Therefore, it is necessary to derive the results for 3σ limits.

The derivation will now be given. It can be easily shown that if

$$X \sim \chi_r^2$$

then $\overline{X}_n \sim (1/n)\chi_{nr}^2$. Thus, for samples of size 5 drawn from a chi-square distribution with r degrees of freedom, the distribution of the sample average will be

$\frac{1}{5}$ times a chi-square variable with $5r$ degrees of freedom. In this case $r = 8$, so $\overline{X}_5 \sim (\frac{1}{5})\chi^2_{40}$. It follows that $E(\overline{X}_5) = 8$ and $\text{Var}(\overline{X}_5) = 3.2$ since the mean and the variance of a chi-square random variable are the degrees of freedom and twice the degrees of freedom, respectively. Also, the variance of a constant times a random variable is equal to the constant squared times the variance of the random variable. The 3σ control limits would then be obtained from $8 \pm 3\sqrt{3.2}$, so that UCL = 13.3666 and LCL = 2.6334. It is then a matter of determining $P(\overline{X}_5 > 13.3666)$ and $P(\overline{X}_5 < 2.6334)$. Since $5\overline{X}_5 \sim \chi^2_{40}$, this is equivalent to determining the tail area of χ^2_{40} above 66.8328 and below 13.1672. These probabilities can be shown to be .0049 and .000019, respectively. These probabilities follow the general pattern of probabilities reported in Moore (1957). Specifically, for small nominal tail probabilities the actual tail probabilities differ greatly, with the lower tail probability being virtually zero. Furthermore, the actual upper tail probability in this case is almost 4 times the nominal upper tail probability (.00135) and is thus almost double the nominal probability for both tails. [This differs somewhat from the results reported by Schilling and Nelson (1976) for the distributions they considered.]

The practical significance of these results is that if $X \sim \chi^2_8$, the expected frequency of "false alarms" above the UCL is actually 1 per 204 subgroups instead of the assumed expected frequency of 1 per 741 subgroups. Thus, if samples of size 5 are selected every 30 minutes, a false signal will be received about once every 13 work days (assuming an 8-hour day), instead of the assumed rate of once every 46 work days. If it is costly to search for assignable causes, this would obviously increase those costs considerably.

How can this problem be avoided? *If* we knew the actual distribution from which we are sampling, we could obtain the actual limits from that distribution such that the tail areas are 0.00135. For the present example the actual limits would be (approximately)

$$\text{UCL} = \tfrac{1}{5}(72.20) = 14.44 \qquad \text{LCL} = \tfrac{1}{5}(18.38) = 3.676$$

Notice that these differ considerably from the 3σ limits.

We will not generally know the type of distribution from which we are sampling, however. One possible solution would be to fit a distribution to the data (such as a member of the Pearson family of distributions) and then use the percentage points of that distribution to obtain control limits for equal tail probabilities (.00135 or some other value). Another possible solution, as mentioned earlier, would be to transform the data so that the transformed data seem to be approximately normally distributed and then follow the basic procedure for controlling the transformed variable. For a detailed discussion of these two suggested approaches, the reader is referred to Chapter 21 of Cowden (1957).

See also Ryan and Howley (1999) and Yourstone and Zimmer (1992) for a discussion and illustration of the problem of nonnormality.

4.17.2 Independence

Another possible problem is lack of independence of observations made in sequential order. Is the weight of a ball bearing coming off an assembly line correlated with the weight of the preceding ball bearing, for example? In particular, is that ball bearing likely to be nonconforming because the preceding one was nonconforming? The distorting effect that correlated data can have on the control limits of $\overline{X}$, R, and other control charts has received increased attention during the past 10 years.

We should note that the estimation of $\sigma_{\overline{x}}$ by $\hat{\sigma}_x/\sqrt{n}$ is appropriate only when the data are independent. If consecutive observations are correlated, $\sigma_{\overline{x}}$ will expectedly be underestimated, perhaps by a considerable amount, if the correlation is positive (and overestimated if ϕ_1 is negative) when $\hat{\sigma}_x/\sqrt{n}$ is used. The process standard deviation, σ_x, might also be poorly estimated by $\overline{R}/d_2$ in the presence of autocorrelation. The appropriate expression for $\sigma_{\overline{x}}$ can be determined from the time series model that is fit to the data, and $\hat{\sigma}_{\overline{x}}$ would then be obtained by substituting sample estimates for the parameters in the expression for $\sigma_{\overline{x}}$.

Vasilopoulos and Stamboulis (1978) showed that the limits for an $\overline{X}$ chart, in particular, can be off by a wide margin when the correlation structure of the data is ignored. To remedy this problem, they provided factors that should be used with an s chart or an $\overline{X}$ chart when the data can be fit with either a first- or second-order autoregressive model [AR(1) or AR(2)]. Figure 4.6 illustrates what can happen when an obvious trend in the data is ignored when determining control limits for an $\overline{X}$ chart. If subgroups are obtained in such a way that the n observations in each subgroup all happen to be at either the top or the bottom of the curve (as indicated by the vertical lines), σ^2 would be poorly estimated since there would be hardly any variability within each subgroup. Although we would usually not expect any trend in the data to be quite as obvious as in Figure 4.6, the intent here is to show that ignoring trended data can cause serious problems.

A simple example will now be given to illustrate this point. Assume that data obtained for some quality characteristic can be represented by a *first-order*

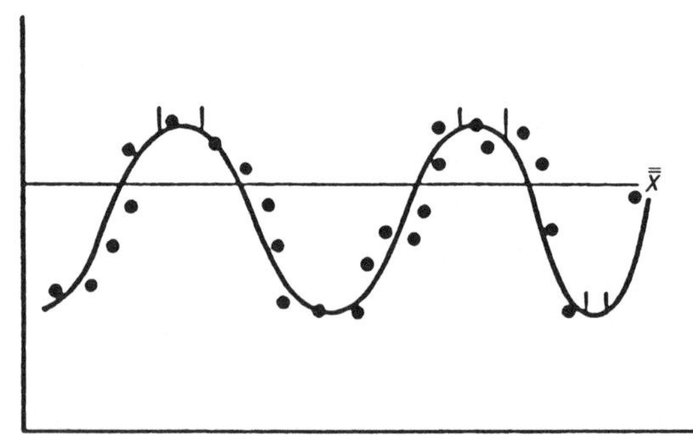

FIGURE 4.6 Nonrandom data.

autoregressive process of the form

$$X_t = \phi_1 X_{t-1} + \varepsilon_t$$

where $-1 < \phi_1 < 1$ and $\varepsilon_t \sim \text{NID}(0, \sigma_\varepsilon^2)$. (The requirement on ϕ_1 is necessary for the process to be stationary, and NID indicates that the errors are normally and independently distributed.) Successive observations will obviously be uncorrelated if $\phi_1 = 0$ but will be highly correlated if $|\phi_1|$ is close to 1.

One hundred consecutive values generated from an AR(1) process with $\phi_1 = 0.5$ are displayed in Table 4.6, and a time-sequence plot of that data is given in Figure 4.7. If subgroups of size 4 or 5 are formed from these data (preserving the time order of the data), σ will be underestimated if subgroup ranges are used. For example, if subgroups of size 5 are formed and σ is estimated by $\overline{R}/d_2$, the estimate is 0.9190 compared to the known value of 1.1547. (See Chapter 5 for explanation of the latter.) A better estimate would be obtained if the data were not put into subgroups, and this applies, in general, to trended data. This is discussed at greater length in Chapter 5.

Although the data in Figure 4.7 exhibit a very obvious trend, even stronger trends have been encountered in practice. A somewhat classic article on this subject is McCoun (1949), which was reprinted in the October 1974 issue of *Quality Progress*. The article describes an actual study of a tooling problem in which the individual values graphed similar to Figure 4.7. Subgroups were

TABLE 4.6 100 Consecutive Values from AR(1) [$\phi_1 = 0.5$, $\varepsilon \sim$ NID(0, 1)]

1.30	0.06	−1.63	1.28	0.52
1.59	−1.46	0.03	0.48	−0.29
0.17	−1.75	0.52	−0.50	2.22
0.01	−1.46	1.21	0.99	1.21
0.07	0.19	0.87	1.00	−0.20
−1.18	−0.60	1.23	−0.05	0.59
−3.36	−0.67	0.78	−1.55	0.42
−3.35	1.10	0.86	−0.88	0.63
0.50	1.15	2.71	0.79	0.52
−0.26	1.30	1.68	1.04	−0.89
−2.07	1.61	−0.52	2.52	−0.50
−2.02	0.12	−1.40	1.79	−0.99
−1.77	1.11	−0.59	3.72	−2.02
−0.12	0.76	−1.15	2.65	−1.40
0.32	−0.21	−0.48	1.92	0.39
−1.16	−0.73	0.13	0.90	1.36
−0.89	0.17	0.33	−0.60	1.24
−0.80	0.45	0.07	−1.70	−1.10
0.08	0.27	0.34	−0.75	−1.98
1.74	−0.78	0.78	0.76	−0.41

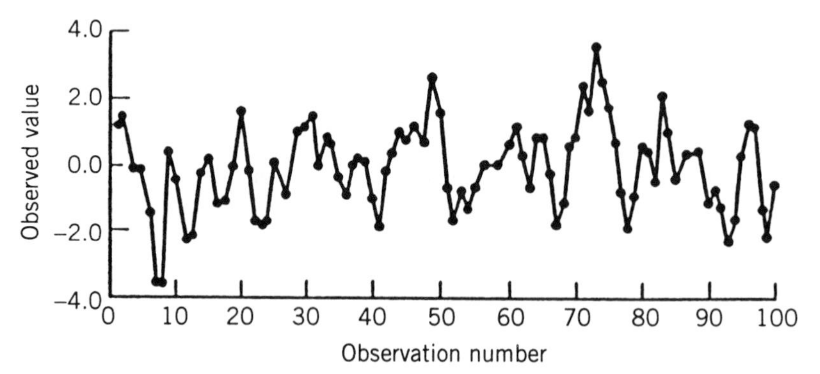

FIGURE 4.7 Time-sequence plot of the data in Table 4.6.

unknowingly taken at the bottom of each cycle and σ was badly underestimated because $\overline{R}$ was close to zero. Nevertheless, in spite of the fact that the resultant control limits were very narrow, the subgroup averages when plotted provided evidence that the process was in control. This was because the averages also differed very little because of the way that the data were obtained. In spite of this control chart message, many of the individual values did not even meet engineering specifications. The fact that the trend was fouling up the $\overline{X}$-chart limits was not detected until the individual values were plotted and the trend became apparent.

It should be pointed out that Vasilopoulos and Stamboulis (1978) did not consider the possibility that σ could be poorly estimated as a result of the serial correlation. Instead, they just considered the appropriate expression for $\sigma_{\overline{x}}$ for an AR(2) process and contrasted what the control limits would be for that expression with the usual $\overline{X}$-chart limits obtained from using $3(\sigma/\sqrt{n})$.

Berthouex, Hunter, and Pallesen (1978) found that the use of a standard Shewhart or CUSUM chart (see Chapter 8) for certain environmental data was unsatisfactory in that the charts gave too many false alarms. This was due to the fact that the data were correlated. The problem was remedied by fitting a time-series model to the data and using the *residuals* from the model in monitoring the process. (A residual is the difference between an observed value and the corresponding predicted value. Residuals are discussed and illustrated in Chapters 12 and 13.) Using residuals to monitor a process is thus an alternative to the use of adjustment factors such as those given by Vasilopoulos and Stamboulis (1978), and is a more general approach. Obviously it would be much easier to fit an adequate model to a set of data than to construct tables of adjustment factors for every type of model that might be fit to data from some process. If an $\overline{X}$ chart of residuals begins to exhibit a lack of control, this would be evidence that the fitted model is no longer appropriate, and, in particular, the mean of the measured variable is no longer at the assumed value.

There are drawbacks, however. A major problem is a residuals chart can have poor ARL properties, depending on the magnitude of the correlation. This is discussed further in Section 10.3. Another problem is that it is harder to relate to a residual than to an $\overline{X}$ value. Consequently, if residuals are used, the $\overline{X}$ values should probably also be plotted (without control limits). Readers interested in fitting time series models to data should consult time series books such as Box et al. (1994).

As far as the other charts are concerned, the assumptions of normality and independence are also important for the acceptance chart, the two types of difference charts that were presented, and the median and midrange charts that were mentioned.

4.18 MEASUREMENT ERROR

A possible impediment to the effective use of control charts is measurement error. Consequently, it is important that measurement systems be in a state of statistical control. How is this accomplished?

We can view the important role of a measurement system in the following way. The variance in the recorded observations is comprised of three components: the variance of the product (process characteristic), the short-term variability in the measurement system, and the long-term variability in the measurement system that is caused by changes in the measurement conditions. The latter two components are generally referred to as *repeatability* and *reproducibility*, respectively. Thus, $\sigma^2_{\text{observations}} = \sigma^2_{\text{product}} + \sigma^2_{\text{repeatability}} + \sigma^2_{\text{reproducibility}}$, if we assume that measurement variability is independent of product variability.

Studies of repeatability and reproducibility are commonly referred to as gauge capability studies or *gauge R& R* studies.

If the last two variance components in the expression for $\sigma^2_{\text{observations}}$ can be estimated, then the product variance can be found by subtraction and compared to the variance of the observations. Clearly control charts will not be particularly effective if the variance of the observations is much greater than the variance of the product. Using the variance of the observations as an estimator of the product variance might lead to a false conclusion that the product variance is out of control when it is actually the measurement system that is out of control.

Readers familiar with analysis of variance (ANOVA), which is covered in Chapter 13, will perhaps see an analogy with one-factor ANOVA, as we might view repeatability as corresponding to the variability that results when the level of the single factor is fixed and reproducibility as corresponding to the variability that results when the factor varies over its levels. Accordingly, it is not surprising that designed experiments are used in gauge R& R studies. The conditions that are expected to change over time are systematically varied in a designed experiment, and reproducibility is determined by the performance of the measurement process under these changing conditions. If none of the factors has a statistically significant effect on the variation of the measurements, then the measurement precision is considered to be adequate, and the reproducibility variance component is considered to not differ significantly from zero. Buckner, Chin, and Henri

(1997) present a designed experiment that was performed to assess reproducibility and none of the factors were found to be significant.

See Montgomery and Runger (1993a,b) for a detailed discussion of gauge R&R studies using designed experiments.

It might also be of interest to study reproducibility by using a control chart, as discussed by Leitnaker et al. (1996, p. 390). Here no attempt would be made to try to isolate effects due to different factors. Rather, this would be an attempt to determine if reproducibility is in a state of statistical control. The control chart might spotlight specific problems that could be hidden when an experimental design is used. A weakness of this approach, however, is that any judgment regarding whether the reproducibility variability was large would be subjective. Therefore, a control chart for reproducibility might be a valuable adjunct to a designed experiment but would not be a satisfactory substitute for a designed experiment.

The variance component for repeatability should be estimated using at least a moderately large number of measurements made back-to-back under identical measurement conditions. Czitrom (1997) suggests that at least 30 measurements be taken. This might be adequate if the reproducibility variance component is not large; otherwise, a much larger number of measurements may have to be taken, since the variance of a sample variance is a function of the square of the variance that is being estimated. Specifically, $\text{Var}(s^2) = 2\sigma^4/(n-1)$ when normality is assumed.

How large must the variance components for reproducibility and repeatability be relative to the product variance component before the measurement system is considered to be a problem? Various rules-of-thumb have been given. Montgomery (1996, p. 457) states that $6\hat{\sigma}_{\text{gauge}}$ should be at most one-tenth of USL − LSL, with $\sigma^2_{\text{gauge}} = \sigma^2_{\text{repeatability}} + \sigma^2_{\text{reproducibility}}$. Note that for normally distributed data this is equivalent to saying that σ^2_{gauge} should be at most one-tenth of $\sigma^2_{\text{observations}}$. Czitrom (1997), however, states that the ratio should be less than 0.3, rather than at most 0.1. In the automotive industry, gauge variability that is 15% of the total variability is considered tolerable, although when 15% does exist, the objective should be to reduce it to 10% (Linna, 1998). Mittag (1995) shows that measurement errors could have a considerable effect on the performance of Shewhart-type control charts.

Experimental designs for measurement system variation are discussed briefly in section 13.16.5. See Leitnaker et al. (1996, Chapter 9) and Wheeler and Lyday (1989) for detailed discussions of measurement systems, and see Chapters 1–5 of Czitrom and Spagon (1997) for examples of gauge studies.

4.19 SUMMARY

The control chart user has many different types of control charts to select from when subgroups can be formed. Standard Shewhart-type charts were presented in this chapter, in addition to some lesser-known charts and modifications of the standard charts. The main problem with the conventional R chart is that the

3σ limits will result in highly unequal tail probabilities since the distribution of the sample range is highly asymmetric even when the individual observations are normally distributed. This problem can be overcome by using an R chart with probability limits, provided that the original or transformed observations are approximately normally distributed. An s chart with probability limits would be a better choice under such conditions, however, since an s chart is superior to an R chart even for small sample sizes. An acceptance chart can be used in place of an $\overline{X}$ chart when the specification limits are much wider than the inherent process variability and a target value is not being specified. A difference chart for paired data can be used in calibration work, and the "grouped" difference chart can be used to adjust for the influence of extraneous factors.

Although Shewhart charts are the most frequently used charts, other process control schemes have generally superior properties. See Chapter 8 for some of these procedures.

There is also the question of what to do when the control chart assumptions are violated. Suggested alternative procedures for use when the independence assumption is violated can be found in Section 10.3.

APPENDIX

Derivation of Control Chart Constants

d_2

Assuming $Y \sim N(\mu, \sigma^2)$, it is well known that

$$E(R) = \int_{-\infty}^{\infty} R f(R)\, dR = d_2 \sigma_y$$

where $R = (y_{(n)} - y_{(1)})$ with $y_{(n)}$ the largest value in a sample of size n and $y_{(1)}$ the smallest value, $f(R)$ represents the probability density function for the range, and d_2 depends upon n. Thus, since

$$\mu_R = d_2 \sigma_y$$

it follows that we would estimate σ_y as

$$\hat{\sigma}_y = \frac{\hat{\mu}_R}{d_2} = \frac{\overline{R}}{d_2}$$

where $\overline{R}$ is the average of the ranges for a set of subgroups.

c_4

Again, assuming $Y \sim N(\mu, \sigma^2)$, we can utilize the result given in this chapter; that is,

$$\frac{(n-1)s_y^2}{\sigma_y^2} \sim \chi^2(n-1)$$

It then follows that $s\sqrt{n-1}/\sigma$ has a chi distribution with $(n-1)$ degrees of freedom, so that

$$E(s) = \frac{\sigma}{\sqrt{n-1}} E(\chi_{n-1})$$

Since the expected value of a chi random variable with $(n-1)$ degrees of freedom can be shown to be equal to

$$\sqrt{2}\frac{\Gamma(n/2)}{\Gamma[(n-1)/2]}$$

where $\Gamma(n) = \int_0^\infty y^{n-1}e^{-y}\,dy$, it follows that

$$\begin{aligned} E(s) &= \sqrt{\frac{2}{n-1}}\frac{\Gamma(n/2)}{\Gamma[(n-1)/2]}\sigma \\ &= c_4\sigma \end{aligned}$$

where c_4 is given in Table E in the Appendix to the book for different values of n. Thus, $\hat{\sigma} = \bar{s}/c_4$.

D_3 and D_4

Assuming a normal distribution, it can be shown that the control limits for the R chart are $D_3\overline{R}$ and $D_4\overline{R}$, by starting with the expression

$$\overline{R} \pm 3\hat{\sigma}_R$$

that produces the control limits. As with $E(R)$, it is also well known that the standard deviation of R can be written as $\sigma_R = d_3\sigma_y$, where d_3 depends on n. It then follows that

$$\begin{aligned} \hat{\sigma}_R &= d_3\hat{\sigma}_y \\ &= d_3(\overline{R}/d_2) \end{aligned}$$

so that the 3-sigma limits could be written as

$$\overline{R} \pm 3(d_3/d_2)\overline{R}$$

but are generally written as $D_3\overline{R}$ and $D_4\overline{R}$, where $D_3 = 1 - 3(d_3/d_2)$ and $D_4 = 1 + 3(d_3/d_2)$.

Values of D_3, D_4, and d_3 are given in Table E for different values of n.

B_3 and B_4

It was shown in the discussion of c_4 above that $E(s) = c_4\sigma$. Since the variance of s equals $E(s^2) - [E(s)]^2$, it follows that

$$\begin{aligned} \text{Var}(s) &= E(s^2) - [E(s)]^2 \\ &= \sigma^2 - (c_4\sigma)^2 \\ &= \sigma^2(1 - c_4^2) \end{aligned}$$

With σ^2 estimated as $(\bar{s}/c_4)^2$, it follows that an estimate of the standard deviation of s is given by

$$\hat{\sigma}_s = \frac{\bar{s}}{c_4}\sqrt{1 - c_4^2}$$

so that the 3-sigma limits for the s chart could be written as

$$\bar{s} \pm 3\frac{\bar{s}}{c_4}\sqrt{1 - c_4^2}$$

but more conveniently as $B_3\bar{s}$ and $B_4\bar{s}$, where $B_3 = 1 - (3/c_4)\sqrt{1 - c_4^2}$ and $B_4 = 1 + (3/c_4)\sqrt{1 - c_4^2}$.

ARL Calculations

The probability of obtaining the first out-of-control message on the nth subgroup is $p(1-p)^{n-1}$ $n = 1, 2, \ldots$, which is the general form for the geometric distribution. Thus, with $f(n) = p(1-p)^{n-1}$ $n = 1, 2, \ldots$, we need the expected value of n. Specifically, $E(n) = \sum_{n=1}^{\infty} np(1-p)^{n-1} = p\sum_{n=1}^{\infty} n(1-p)^{n-1}$. Using the fact that $\sum_{n=1}^{\infty}(1-p)^{n-1}$ is an infinite geometric series so that

$$\sum_{n=1}^{\infty}(1-p)^{n-1} = \frac{1}{p}$$

we can take the derivative of each side of this last equation and then multiply by p to obtain

$$p\sum_{n=1}^{\infty}(n-1)(1-p)^{n-2} = \frac{1}{p} = p\sum_{n=1}^{\infty} n(1-p)^{n-1}$$

so that the expected subgroup number is $1/p$, where p is the probability associated with a single subgroup.

REFERENCES

Bajaria, H. J. (1994). Quality execution strategies for world-class status. In *ASQC Quality Congress Proceedings*, pp. 813–820. Milwaukee, WI: American Society for Quality Control.

Berthouex, P. M., W. G. Hunter, and L. Pallesen (1978). Monitoring sewage treatment plants: Some quality control aspects. *Journal of Quality Technology 10*(4): 139–149.

Bissell, A. F. (1990). How reliable is your capability index? *Applied Statistics 39*(3): 331–340.

Box, G. E. P. (1953). Non-normality and tests on variances. *Biometrika 40*: 318–335.

Box. G. E. P., G. M. Jenkins, and G. C. Reinsel (1994). *Time Series Analysis: Forecasting and Control.* Englewood Cliffs, NJ: Prentice-Hall.

Buckner, J., B. L. Chin, and J. Henri (1997). Prometrix RS35e gauge study in five two-level factors and one three-level factor. In V. Czitrom and P. D. Spagon, eds. *Statistical Case Studies for Industrial Process Improvement,* Chapter 2. Philadelphia: American Statistical Association and Society for Industrial and Applied Mathematics.

Burr, I. W. (1967). The effect of non-normality on constants for $\overline{X}$ and R charts. *Industrial Quality Control 23*(11): 563–569.

Burr, I. W. (1969). Control charts for measurements with varying sample sizes. *Journal of Quality Technology 1*(3): 163–167.

Costa, A. F. B. (1994). $\overline{X}$ charts with variable sample size. *Journal of Quality Technology 26*(3): 155–163.

Cowden, D. J. (1957). *Statistical Methods in Quality Control.* Englewood Cliffs, NJ: Prentice-Hall.

Czitrom, V. (1997). Introduction to gauge studies. In V. Czitrom and P. D. Spagon, eds. *Statistical Case Studies for Industrial Process Improvement,* Chapter 1. Philadelphia: American Statistical Association and Society for Industrial and Applied Mathematics.

Czitrom, V. and P. D. Spagon, eds. (1997). *Statistical Case Studies for Industrial Process Improvement.* Philadelphia: American Statistical Association and Society for Industrial and Applied Mathematics.

Deming, W. E. (1986). *Out of the Crisis.* Cambridge, MA: Massachusetts Institute of Technology, Center for Advanced Engineering Study.

Duncan, A. J. (1986). *Quality Control and Industrial Statistics,* 5th ed. Homewood, IL: Irwin.

Ferrell, E. B. (1953). Control charts using midranges and medians. *Industrial Quality Control 9*(5): 30–34.

Freund, R. A. (1957). Acceptance control charts. *Industrial Quality Control 14*(4): 13–23.

Ghosh, B. K., M. R. Reynolds, Jr., and Y. V. Hui (1981). Shewhart $\overline{X}$-charts with estimated process variance. *Communications in Statistics—Theory and Methods 10*(18): 1797–1822.

Grant, E. L. and R. S. Leavenworth (1980). *Statistical Quality Control,* 5th ed. New York: McGraw-Hill.

Grubbs, F. E. (1946). The difference chart with an example of its use. *Industrial Quality Control 3*(1): 22–25.

Gunter, B. (1991). The use and abuse of C_{pk} revisited (response to Steenburgh). *Quality Progress 24*(1): 93–94.

Hahn, G. J. (1970). How abnormal is normality. *Journal of Quality Technology 3*(1): 18–22.

Harter, H. L. (1960). Tables of range and studentized range. *Annals of Mathematical Statistics 31*(4): 1122–1147.

Hill, D. (1956). Modified control limits. *Applied Statistics 5*(1): 12–19.

Hinkley, D. V. (1970). Inference about the change-point in a sequence of random variables. *Biometrika 57*: 1–17.

James, P. C. (1989). C_{pk} equivalencies. *Quality 28*(9): 75.

Joiner, B. L. (1994). *Fourth Generation Management.* New York: McGraw-Hill. (in collaboration with Sue Reynard).

Keats, J. B., J. D. Miskulin, and G. C. Runger (1995). Statistical process control scheme design. *Journal of Quality Technology 27*(3): 214–225.

Lee, C. and G. A. D. Matzo (1998). An evaluation of process capability for a fuel injector process using Monte Carlo simulation. In R. Peck, L. D. Haugh, and A. Goodman, eds. *Statistical Case Studies: A Collaboration Between Industry and Academe*. Philadelphia: Society of Industrial and Applied Mathematics and American Statistical Association.

Leitnaker, M. G., R. D. Sanders, and C. Hild (1996). *The Power of Statistical Thinking: Improving Industrial Processes*. Reading, MA: Addison-Wesley.

Linna, K. (1998). Personal communication.

McCoun, V. E. (1974). The case of the perjured control chart. *Quality Progress 7*(10): 17–19. (Originally published in *Industrial Quality Control*, May 1949.)

Mittag, H.-J. (1995). Measurement error effect on control chart performance. In: *ASQC Quality Congress Proceedings*, pp. 66–73. Milwaukee, WI: American Society for Quality Control.

Moore, P. G. (1957). Normality in quality control charts. *Applied Statistics 6*(3): 171–179.

Montgomery, D. C. (1996). *Introduction to Statistical Quality Control*, 3rd ed. New York: Wiley.

Montgomery, D. C. and G. C. Runger (1993a). Gauge capability and designed experiments: Part I: Basic methods. *Quality Engineering 6*(1): 115–135.

Montgomery, D. C. and G. C. Runger (1993b). Gauge capability analysis and designed experiments: Part II: Experimental design models and variance component estimation. *Quality Engineering 6*(2): 289–305.

Nelson, L. S. (1982). Control chart for medians. *Journal of Quality Technology 14*(4): 226–227.

Nelson, L. S. (1988). Control charts: Rational subgroups and effective applications. *Journal of Quality Technology 20*: 73–75.

Nelson, L. S. (1989). Standardization of Shewhart control charts. *Journal of Quality Technology 21*(4): 287–289.

Nelson, L. S. (1990). Setting up a control chart using subgroups of varying sizes. *Journal of Quality Technology 22*: 245–246.

Nelson, L. S. (1996). Answers to a true-false test in quality control and statistics. *Journal of Quality Technology 28*(4): 480–482.

Osborn, D. P. (1990). Statistical power and sample size for control charts — survey results and implications. *Production and Inventory Management Journal 31*(4): 49–54.

Ott, E. R. (1947). An indirect calibration of an electronic test set. *Industrial Quality Control 3*(4): 11–14.

Ott, E. R. (1949). Variables control charts in production research. *Industrial Quality Control 6*(3): 30–31.

Ott, E. R. (1975). *Process Quality Control*. New York: McGraw-Hill.

Palm, A. C. (1992). Some aspects of sampling for control charts. *ASQC Statistics Division Newsletter 12*: 20–23.

Palm, A. C. and R. L. DeAmico (1995). Discussion (of paper by Roes and Does). *Technometrics 37*(1): 26–29.

Papadakis, E. P. (1985). The Deming inspection criterion for choosing zero or 100 percent inspection. *Journal of Quality Technology 17*(3): 121–127.

Quesenberry, C. P. (1993). The effect of sample size on estimated limits for $\overline{X}$ and X control charts. *Journal of Quality Technology* *25*(4): 237–247.

Roes, K. C. B. and R. J. M. M. Does (1995). Shewhart-type charts in nonstandard situations. *Technometrics 37*(1): 15–24 (discussion: 24–40).

Ryan, T. P. and P. P. Howley (1999). Comment on Janacek and Meikle (with reply). *The Statistician 48*: 441–444.

Samuel, T. R., J. J. Pignatiello, Jr., and J. A. Calvin (1998a). Identifying the time of a step change with $\overline{X}$ control charts. *Quality Engineering, 10*(3): 521–527.

Samuel, T. R., J. J. Pignatiello, Jr., and J. A. Calvin (1998b). Identifying the time of a step change in a normal process variance. *Quality Engineering, 10*(3): 529–538.

Sandorf, J. P. and A. T. Bassett III (1993). The OCAP: Predetermined responses to out-of-control conditions. *Quality Progress 26*(5): 91–95.

Schilling, E. G. (1982). *Acceptance Sampling in Quality Control.* New York: Marcel Dekker.

Schilling, E. G. (1983). Two new ASQC acceptance sampling standards. *Quality Progress 16*(3): 14–17.

Schilling, E. G. (1991). Acceptance control in a modern quality program. *Quality Engineering 3*: 181–191.

Schilling, E. G., and P. R. Nelson (1976). The effect of non-normality on the control limits of $\overline{X}$ charts. *Journal of Quality Technology 8*(4) 183–188.

Shapiro, S. S. (1980). How to test normality and other distributional assumptions. In *Basic References in Quality Control: Statistical Techniques*, Vol. 3. Milwaukee, WI: American Society for Quality Control.

Shewhart, W. A. (1931). *Economic Control of Quality of Manufactured Product.* New York: Van Nostrand.

Sullivan, J. H. and W. H. Woodall (1996). A comparison of multivariate control charts for individual observations. *Journal of Quality Technology 28*(4): 398–408.

Vardeman, S. B. (1986). The legitimate role of inspection in modern SQC. *The American Statistician 40*: 325–328.

Vasilopoulos, A. V. and A. P. Stamboulis (1978). Modification of control chart limits in the presence of data correlation. *Journal of Quality Technology 10*(1): 20–30.

Wadsworth, H. M., K. S. Stephens, and A. B. Godfrey (1986). *Modern Methods for Quality Control and Improvement.* New York: Wiley.

Wheeler, D. J. and R. W. Lyday (1989). *Evaluating the Measurement Process*, 2nd. ed. Knoxville, TN: SPC Press.

Woods, R. F. (1976). Effective, economic quality through the use of acceptance control charts. *Journal of Quality Technology 8*(2): 81–85.

Yourstone, S. A. and W. J. Zimmer, (1992). Non-normality and the design of control charts for averages. *Decision Sciences 23*: 1099–1113.

EXERCISES

1. Construct the s chart, without probability limits, for the data in Table 4.2. Compare the configuration of points with the R chart given in Figure 4.2.

2. Deleting points and recomputing the control limits can cause the limits to change considerably. Show this for the s chart with 3-sigma limits by determining the smallest possible difference between the original and recomputed UCL values for $n = 4$ when 1 point out of 30 plots above the UCL and is subsequently discarded.

3. For the data in Table 4.4 show that $\bar{x}_9$ would have been inside the control limits if σ had been estimated using subgroup standard deviations rather than subgroup ranges. Comment.

4. Explain the purpose of an acceptance control chart, and indicate one or more conditions under which it should not be used.

5. Discuss the alternatives that an experimenter has when constructing an $\overline{X}$ chart for correlated data.

6. Twenty subgroups of size 5 are obtained for the purpose of determining trial control limits for an $\overline{X}$ and an R chart. The data are as follows.

Subgroup Number	$\overline{X}$	R	Subgroup Number	$\overline{X}$	R
1	23	5	11	26	5
2	22	3	12	21	4
3	24	2	13	22	4
4	20	4	14	20	4
5	18	3	15	23	3
6	17	4	16	21	6
7	24	4	17	20	5
8	10	3	18	18	4
9	16	5	19	15	3
10	20	4	20	17	2

(a) Determine the trial control limits for each chart.

(b) Explain why there are so many subgroup averages outside the control limits for the $\overline{X}$ chart in spite of the fact that the averages do not vary greatly.

(c) What should be done with those subgroups whose average is beyond the limits?

(d) Since the number of points outside the control limits on the $\overline{X}$ chart is quite high relative to the number of points that are plotted, what might this suggest about the type of distribution for the population from which the data came?

(e) Significant nonnormality cannot be investigated for the individual observations, since all of the data are not given. Nonnormality of the subgroup averages *can* be investigated, however, by constructing a normal probability plot for the averages. Use appropriate computer software to do so and interpret the plot. What do you conclude about the relatively high number of points that are outside the control limits?

7. Construct an R chart with probability limits for the data in exercise 6 and compare with the conventional R chart. Would you be reluctant to use an R chart with probability limits based only on what is given in that exercise?
8. Construct an R chart and an $\overline{X}$ chart for the following data and interpret:

Subgroup Number	Observations			
1	3.1	3.2	3.6	3.8
2	3.3	3.4	3.1	3.7
3	3.0	2.9	3.6	3.5
4	3.2	3.2	3.1	3.5
5	3.7	3.8	3.4	4.0
6	3.3	3.5	3.2	3.0
7	3.4	3.1	3.6	3.0
8	3.3	3.9	3.7	3.6
9	3.5	3.8	3.1	3.3
10	4.1	3.7	3.5	3.2
11	3.2	4.6	2.9	3.4
12	3.0	3.5	3.1	3.3
13	3.2	3.7	3.4	3.1
14	3.4	3.1	2.8	3.5
15	3.6	3.3	3.4	3.9
16	3.8	3.2	3.4	3.5
17	3.7	3.5	3.2	3.3
18	3.5	3.4	3.1	3.0
19	3.6	3.7	3.9	3.2
20	3.4	3.8	3.5	3.1

9. Construct a normal probability plot using the subgroup averages in Table 4.2 and comment on the plot.
10. Explain why an out-of-control action plan is highly desirable.
11. Assume that μ and σ are known to be approximately 25 and 5, respectively, so that they need not be estimated from data. If 50 historical data points are plotted to determine if a process was previously in a state of statistical control, what should be the value of k in k-sigma limits such that the probability of at least one point falling outside the limits is (a) .05 (b) .001, and (c) .0027? What value of k would you select and why?
12. Critique the following statements: We have often found that $\hat{\sigma}^2_{\text{gage}}$ is more than 35% of $\hat{\sigma}^2_{\text{observations}}$, with each estimate based on 25 observations since measurements are expensive. We are reluctant to rely on control chart results because of our high ratio of gauge variability to observation variability.
13. Table 4.5 provides evidence of run length variability when parameters are assumed to be known. Explain how you would perform a similar study when the parameters are *not* assumed to be known.

14. A company that makes rollers is having some quality problems relating to the flatness of the end of one type of roller. For a selected measure of flatness, three rollers are measured every 15 minutes. The average of the three measurements is plotted on a control chart, but the limits are the specification limits (as discussed in Section 4.10.1), which happen to be at $\mu \pm 3\sigma$. Let X represent the (coded) measure of flatness with the assumption of $X \sim N(0, 1)$ being a reasonable approximation. Using this approximation, what percentage of the averages would you expect to be outside the control limits that the company is using if the mean increases from $\mu = 0$ to $\mu + \sigma = 1$? What would be your recommendation to this company?

15. In a recent article in *Quality Engineering* ("Variations in conventional control charts" by C. E. Noble, Vol. 10, No. 4, 1998, pp. 705–711), the author discusses and illustrates some variations of standard control chart usage for applications in the paper industry. The author gave the following data for the tensile strength of each of two types of paper, A and B, with the data collected over a 10-hour period, with samples (subgroups) of size 3 obtained each hour.

Type A				Type B			
Hour	Observations			Hour	Observations		
1	17	19	18	1	26	20	24
2	23	19	19	2	23	25	21
3	22	22	22	3	23	24	22
4	19	21	18	4	24	23	24
5	22	18	19	5	21	22	24
6	19	20	18	6	22	19	21
7	21	17	25	7	24	23	23
8	19	17	22	8	25	24	23
9	20	20	21	9	20	23	24
10	20	21	19	10	25	19	25

(a) The author stated that the use of only 10 hourly samples was just for the sake of illustration, with 100 hourly samples generally used in setting up the control charts. What could be expected if only 10 hourly samples were used in practice, and do you consider 100 samples to be sufficient?

(b) The tensile strength for type A paper is supposed to be 20 pounds, and the tensile strength of type B paper is supposed to be 23 pounds. The author used these two values in determining $\overline{X}$ chart limits for each paper type and used the average range for the two machines combined in estimating the standard deviations, since they were assumed to be about the same. Do you agree with this approach? Construct $\overline{X}$ charts for each paper type (1) using these standard values and (2) not using the standard values. Also, estimate σ using (1) the range method and (2) the standard deviation method. Compare the results for the three sets of charts.

(c) Since each paper type was run on only one of two machines, there is the possibility that the machine could be out of control rather than the "pulp furnish" that relates to the tensile strength. To overcome this problem, the author combined the data on each machine so that the subgroup size became 6. But $\sigma_{\overline{x}}$ was estimated using the same value of $\overline{R}$ (i.e., 3.15) as was used for the control chart of each paper type. Thus, $\sigma_{\overline{x}}$ is estimated as $3.15/(d_2\sqrt{6})$ using the value of d_2 for a subgroup size of 6. Do you agree with this approach? Comment.

CHAPTER 5

Control Charts for Measurements without Subgrouping (for One Variable)

The control charts given in the preceding chapter can be used when subgroups can be formed. This is not always possible or practical, however. Items coming off an assembly line may be produced at such a slow rate so as to preclude the forming of subgroups. If items are produced every 30 minutes, it would take 2.5 hours to form a subgroup of size 5; by then the process might already have gone out of control. Variables such as temperature and pressure could also not be charted with subgrouping since, for example, a "subgroup" of five temperature readings made in quick order would likely be virtually the same. Thus, nothing would be gained. Clerical and accounting data would also have to be charted using individual numbers rather than subgroups. (This will be discussed further in Chapter 10.)

5.1 INDIVIDUAL OBSERVATIONS CHART

Charts based upon individual observations are not as sensitive as an $\overline{X}$ chart for detecting a mean shift. This can be demonstrated as follows. Assume that there is a shift in the process mean of $a\sigma$, where $a > 0$. The control limits on an individual observations chart are at (estimates of) $\mu \pm 3\sigma$. Assuming for the moment that the parameters are known, the probability of an individual observation plotting above the upper control limit (UCL) is equal to $P(X > \mu + 3\sigma)$ given that the actual mean is equal to $\mu + a\sigma$. For the $\overline{X}$ chart we need to determine $P(\overline{X} > \mu + 3\sigma_{\overline{X}})$ when the mean is $\mu + a\sigma$. The two z-values are

$$z = \frac{\mu + 3\sigma - (\mu + a\sigma)}{\sigma} = 3 - a$$

and

$$z = \frac{\mu + 3\sigma_{\overline{X}} - (\mu + a\sigma)}{\sigma_{\overline{X}}} = 3 - a\sqrt{n}$$

(The latter results from the fact that $\sigma_{\overline{X}} = \sigma/\sqrt{n}$.) Clearly, $P(Z > 3 - a\sqrt{n})$ is greater than $P(Z > 3 - a)$, so for an $a\sigma$ increase in μ, we would expect to observe a subgroup average above its UCL before we would observe an individual observation above its UCL. (The same type of result would hold when $a < 0$.)

This is not only intuitive but also in line with general statistical theory, which holds that the "power" of a hypothesis test increases as the sample size increases. (Here we are comparing a sample size of 1 with a sample size of $n > 1$. Of course this "power analysis" is based on the supposition that all of the subgroup measurements are made at essentially the same point in time.)

The important point is that when conditions are such that either an X (individual observations) chart or an $\overline{X}$ chart *could* be used, it would be unwise to use an X chart.

This does not mean, however, that the two charts could not be used together. A subgroup average can be above the UCL on an $\overline{X}$ chart because of the presence of one exceedingly large observation. The individual observations that comprise that subgroup could be plotted on an X chart, and the magnitude of that one large observation can be assessed relative to the control limits for the individual observations. A subsequent investigation might reveal that the value was recorded in error. An X chart can be used in conjunction with an R, s, or s^2 chart for the same purpose.

5.1.1 Control Limits for the *X* Chart

The control limits for an X chart are obtained from

$$\hat{\mu} \pm 3\hat{\sigma}$$

For Stage 1, the estimate of the mean should be obtained from at least 100 historical observations. [Quesenberry (1993) suggests that the "permanent" limits should be based on at least 300 observations. These would be the limits that would be used in Stage 2—process monitoring.] The estimation of σ is another matter. Obviously subgroup ranges cannot be used since we have no subgroups. The most frequently used procedure is to "create" ranges by taking differences of successive observations (second minus first, third minus second, etc.) and dropping the sign of the difference when it is negative. The average of these "moving ranges" of size 2 is then used in the same manner that $\overline{R}$ is used in estimating σ for an $\overline{X}$ chart. Specifically,

$$\hat{\sigma} = \frac{\overline{\mathrm{MR}}}{d_2}$$

where $\overline{\text{MR}}$ denotes the average of the moving ranges. As when σ is estimated using ranges with subgroup data, the constant d_2 is used so as to make the estimator unbiased.

This approach is reasonable for Stage 1, but not for Stage 2. In Stage 1 the process that is to be monitored is probably out of control (since control charts will not have been used, unless the limits are simply being revised), and the estimate of sigma based on the moving range is more robust (i.e., insensitive) than the estimate based on the standard deviation relative to the type of out-of-control conditions that might have occurred in Stage 1 (Harding, Lee, and Mullins, 1992).

When an X chart is used, there is the tacit assumption that the data have come from a population whose distribution is very close to being normal. (Regardless of what the actual distribution is, the data that are used to compute the control limits for Stage 2 must have all come from the same in-control distribution; otherwise, the control limits really have no meaning, as was also emphasized in Section 4.5).

If that distribution is a normal distribution, Cryer and Ryan (1990) showed that it is much better to estimate σ as $\hat{\sigma} = s/c_4$, where c_4 is the tabled constant (see Table E in the Appendix to the book) that makes the estimator unbiased. Specifically, $\text{Var}(\overline{\text{MR}}/d_2)/\text{Var}(s/c_4) \doteq 1.65$, regardless of the number of historical observations that are used for Stage 2. This has important ramifications because it means that if μ were known and we used the moving range approach rather than estimating σ using the standard deviation, we would need, roughly speaking, about 65% more observations in order for the X chart to have the ARL properties that it would have if the standard deviation were used. Since it is necessary to use a large number of observations for an X chart in order for the chart to have approximately the ARL properties that the chart would have in the parameters-known case, there is no need to make matters worse by using an inferior estimator.

5.1.2 X-Chart Assumptions

The assumptions for an X chart are the same as the assumptions for an $\overline{X}$ chart — normality and independence. The normality assumption is far more important when individual observations are plotted than it is when averages are plotted, since there is no central limit theorem–type effect with individual observations. Hence, even a slight departure from normality can cause the in-control ARL to be far less than the normal-theory value. For example, a chi-square distribution with 30 degrees of freedom does not differ greatly from a normal distribution, but the in-control ARL (179.04) is less than half of the normal-theory value (370.37).

The consequences of the independence assumption depend considerably on whether σ is assumed to be known. Assume that data have come from an AR(1) process with $\phi = 0.75$. Table 3 of Zhang (1997) shows that the in-control ARL is 516.58 and the ARL for detecting a 1-sigma change in the mean is 76.89. The latter is almost 80% greater than the normal-theory value, and the former is about 40% greater than the normal-theory value. If 2.90-sigma limits are used rather than 3-sigma limits, the in-control ARL is 373.52, and the ARL for detecting a 1-sigma shift is 64.73.

The picture is quite different if σ is not assumed to be known, with the effect of autocorrelation depending on the estimator of σ that is used. Whereas σ should not be estimated using the moving range approach (in Stage 2) for independent data, the situation is much worse when the data are autocorrelated. As shown by Cryer and Ryan (1990), $E(\overline{\text{MR}}/d_2) = \sigma\sqrt{1-\rho_1}$, with ρ_1 denoting the correlation between consecutive observations [$= \phi$ for an AR(1) model]. (This is also derived in the chapter Appendix.) Thus, the control limits will be much too narrow when ρ_1 is close to 1, and the in-control ARL could be much less than 100. [See also Maragah and Woodall (1992), who considered the effect of the moving-range estimator when data are from either an AR(1) process or an MA(1) process, a first-order moving-average process.]

Since $E(S/c_4) \to \sigma$ as $n \to \infty$ for an AR(1) process, using the sample standard deviation to estimate sigma will not be a serious problem for large sample sizes.

We may conclude by saying that normality is a more serious problem than autocorrelation when individual observations are plotted. Autocorrelation can be compensated for by adjusting the control limits and by using $\hat{\sigma} = S/c_4$ and also by having the estimate based on a large number of observations. (Recall from Section 4.14 that at least 100 observations should be used in computing the Stage 1 control limits, with the "permanent" limits preferably computed from at least 300 observations.) Increasing the number of observations does not help the nonnormality problem, however, as the distribution for the population from which the data are obtained is obviously unaffected by the number of observations that are obtained from that population.

5.1.3. Illustrative Example: Random Data

Two sets of simulated data and two sets of real data will be used to illustrate an X chart and the other charts presented in this chapter.

Assume that the random numbers in Table 5.1 constitute 50 consecutive observations from a manufacturing process. (The consecutive observations are obtained by reading down each column.) In this instance we know that the data are normally distributed, as the numbers were generated from a normal distribution. Nevertheless, the normal probability plot is given in Figure 5.1 since a test for normality should be the first step in a typical application. It should be observed that the points do not form exactly a straight line even though the data came from a normal distribution, but neither does it depart greatly from a straight line. (We should not expect exactly a straight line, however. See the discussion in Section 13.9.)

The plot exhibits a noticeable departure from a straight line, but not enough to cause the p-value for the normality test to lead to rejection of normality. If we

TABLE 5.1 Random Numbers from $N(\mu = 25, \sigma^2 = 9)$

28.30	17.89	26.45	24.69	24.18
23.64	28.38	26.83	24.35	24.54
26.92	22.71	23.75	16.79	21.35
30.53	24.35	24.26	24.00	27.09
26.68	23.80	27.60	25.66	24.30
25.87	30.80	25.74	30.70	29.54
24.26	25.54	29.16	21.27	27.11
20.03	24.85	25.86	21.27	21.50
21.58	23.58	27.03	26.43	23.89
22.24	24.14	24.28	28.01	24.33

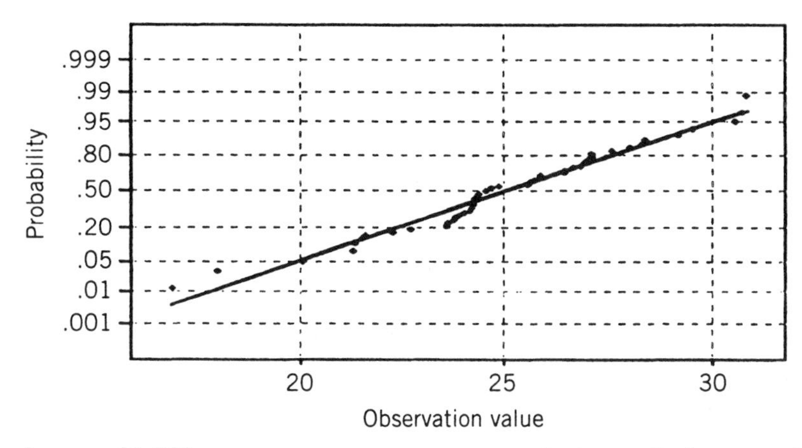

FIGURE 5.1 Normal probability plot of Table 5.1 data.

constructed at least a moderate number of normal plots for normally distributed data, some of the plots would deviate greatly from a straight line, as Daniel (1976) illustrated. If we wanted to construct a normal probability plot with decision lines, we could use the approach described by Nelson (1989). Then if points fell outside the decision lines, we would reject the hypothesis of normality. Alternatively, we could use software that provides the value of one of more test statistics for testing normality in addition to the plot.

For the sake of illustration, σ (which we know is 3) will be estimated using both s and $\overline{\text{MR}}/d_2$ for a moving range of size 2. It can be shown that $s/c_4 = 2.9872/0.9949 = 3.0025$ and $\overline{\text{MR}}/d_2 = 3.1484/1.128 = 2.7911$. Thus, the two estimates differ noticeably, even though the observations are normally distributed random numbers.

The control limits using s would be

$$\bar{x} \pm 3\frac{s}{c_4} = 24.96 \pm 3(3.0025)$$

so that UCL = 33.9675 and LCL = 15.9525. The control limits using the moving range would be

$$\bar{x} \pm 3\left(\frac{\overline{\text{MR}}}{d_2}\right) = 24.96 \pm 3(2.7911)$$

so that UCL = 33.3333 and LCL = 16.5867. Thus, the two sets of limits differ somewhat, and this is due to the comparatively poor estimate of σ obtained by using the moving range. The X control chart is displayed in Figure 5.2, and it can be observed that all of the points lie within the limits (which are the limits using s).

5.1.4 Example with Particle Counts

Chou, Halverson, and Mandraccia (1998) use data on equipment-generated particle counts to illustrate the construction of an X chart. The data are given in Table 5.2, and the authors indicate it has been determined that there are no bad data points. One disadvantage of looking at real data is we do not know the true state of nature. More specifically, we do not know an appropriate distribution for the data. We can see from the normal probability plot in Figure 5.3, however, that even approximate normality seems quite implausible since the plot indicates that the data are strongly right skewed. When symmetric, 3-sigma limits are constructed for right-skewed data, multiple points will fall above the UCL. The reader is asked to show in exercise 9 that this is exactly what happens for these

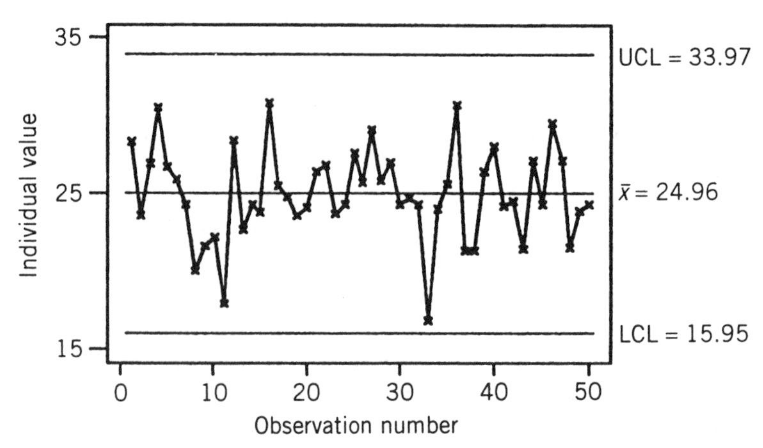

FIGURE 5.2 X chart for Table 5.1 data.

TABLE 5.2 Fifty Consecutive Values from AR(1) [$\phi_1 = 0.5$, $\varepsilon \sim \text{NID}(0, 1)$]

Value	Moving Range ($n = 2$)	Value	Moving Range ($n = 2$)	Value	Moving Range ($n = 2$)
1.30		−0.80		−0.21	
	0.29		0.88		0.52
1.59		0.08		−0.73	
	1.42		1.66		0.90
0.17		1.74		0.17	
	0.16		1.68		0.28
0.01		0.06		0.45	
	0.06		1.52		0.18
0.07		−1.46		0.27	
	1.25		0.29		1.05
−1.18		−1.75		−0.78	
	2.18		0.29		0.85
−3.36		−1.46		−1.63	
	0.01		1.65		1.66
−3.35		0.19		0.03	
	3.85		0.79		0.49
0.50		−0.60		0.52	
	0.76		0.07		0.69
−0.26		−0.67		1.21	
	1.81		1.77		0.34
−2.07		1.10		0.87	
	0.05		0.05		0.36
−2.02		1.15		1.23	
	0.25		0.15		0.45
−1.77		1.30		0.78	
	1.65		0.31		0.08
−0.12		1.61		0.86	
	0.44		1.49		1.85
0.32		0.12		2.71	
	1.48		0.99		1.03
−1.16		1.11		1.68	
	0.27		0.35		
−0.89		0.76			
	0.09		0.97		

data. Chou et al. (1998) were able to transform the data to approximate normality, and the transformed data were all within 3-sigma limits.

5.1.5 Illustrative Example: Trended Data

Limits obtained from the two approaches can be expected to differ greatly when the data contain a nonremovable trend. This will be illustrated by using part of the data for the AR(1) process from Chapter 4.

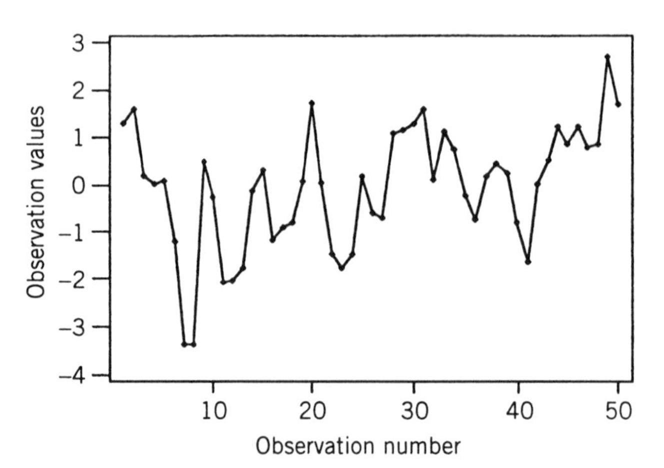

FIGURE 5.3 Time-sequence plot of Table 5.2 data.

Recall that the data (in Table 4.6) were generated using

$$x_t = \phi_1 x_{t-1} + \varepsilon_t$$

where ϕ_1 was set equal to 0.5 and we assumed $\varepsilon \sim$ NID (0, 1). The first 50 of those 100 values are listed in Table 5.2 with the corresponding moving ranges.

Although data from an AR(1) with $\phi_1 = 0.5$ should not show as strong a trend as would be the case if $0.5 < \phi_1 < 1.0$, there is nevertheless some evidence of a trend when the data in Table 5.2 are graphed in Figure 5.3.

It is well known (see, e.g., p. 58 of Box and Jenkins, 1976) that the variance of an AR(1) process is

$$\sigma_x^2 = \frac{\sigma_\varepsilon^2}{1 - \phi_1^2}$$

Thus, with $\sigma_\varepsilon^2 = 1$ and $\phi_1 = 0.5$, $\sigma_x^2 = 1.33$. Therefore, we *know* the actual process variance, and with a sample consisting of a large number of consecutive values we would expect the sample variance, s_x^2, to be reasonably close to 1.33. For the 50 observations in Table 5.2 it can be determined that $s_x^2 = 1.679$.

We would expect that the estimator of σ_x based upon moving ranges of size 2 would underestimate σ_x, however. In particular, successive values will not differ greatly, in general, when ϕ_1 is close to 1, which will cause the moving ranges to be close to zero. Consequently, the average moving range will also be close to zero, and since $d_2 = 1.128$ for $n = 2$, the estimate of σ_x will differ only slightly from the average moving range.

For fixed σ_ε, $\sigma_x = \sigma_\varepsilon / \sqrt{1 - \phi_1^2}$ will increase rapidly as ϕ_1 approaches 1, but the estimate of σ_x obtained using moving ranges will not track σ_x. For example, if $\sigma_\varepsilon = 1$ and $\phi_1 = 0.99$, then $\sigma_x = 7.09$, but the estimate obtained using the moving range approach should be close to 0.71. (See the Appendix to this Chapter for details.)

In this example $\phi_1 = 0.5$ and the average moving range is 0.85, so the estimate of σ_x is thus

$$\begin{aligned}\hat{\sigma}_x &= \frac{\overline{\text{MR}}}{d_2} \\ &= \frac{0.85}{1.128} \\ &= 0.754\end{aligned}$$

Thus, $\sigma_x = \sqrt{1.33} = 1.155$ is estimated by 0.754 using the moving range, and by $s_x = \sqrt{1.679}/0.9949 = 1.296/0.9949 = 1.302$. Since the control limits for the X chart are generally obtained using the moving range, the limits in this instance will be much too narrow. In general, limits that are too narrow can give a false signal that a process is out of control when, in fact, the process is actually in control.

This is exactly what happens in this case, as can be seen from Figure 5.4.

Three of the 50 values are outside the control limits even though the process that generated the values is a stationary (i.e., in-control) process. With the nominal tail areas of .00135 we would expect 50(0.0027) = .135 values to be outside the control limits. (In other words, we would expect all of the values to be within the control limits.)

If the limits had been obtained from $\bar{x} \pm 3s/c_4$, however, they would have been

$$\text{UCL} = \bar{x} + 3\frac{s}{c_4} = -0.0462 + 3\left(\frac{1.296}{0.9949}\right) = 3.8617$$

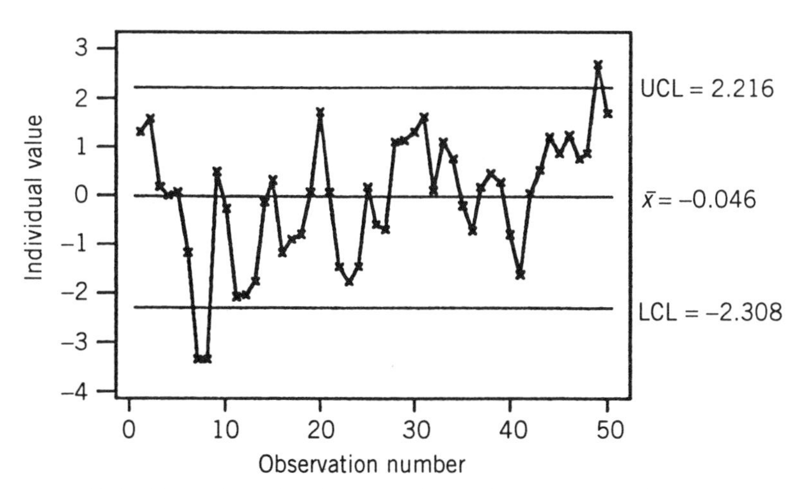

FIGURE 5.4 *X* chart of Table 5.2 data; limits obtained using moving ranges.

and

$$\mathrm{LCL} = \bar{x} - 3\frac{s}{c_4} = -0.0462 - 3\left(\frac{1.296}{0.9949}\right) = -3.9541$$

Inspection of Table 5.2 reveals that all of the values lie between these limits. This can be seen more easily by viewing the X chart with the new limits, as shown in Figure 5.5. (Note: Instead of 3-sigma limits, we might use 2.98-sigma limits, as suggested by Table 3 of Zhang (1997), but obviously this will make only a slight difference in the limits for this example.)

The important point is that the use of the moving range to estimate σ for an X chart is inadvisable when data contain a nonremovable trend. On the other hand, the use of s to estimate σ is also inadvisable if s is calculated from data that contain a trend that is due to an assignable cause that is (immediately) removable. Thus, considerable caution should be exercised in determining the limits for an X chart. The problem in determining trial limits when an X chart is first used is that it may be difficult to determine whether or not a trend is removable.

5.1.6 Trended Real Data

The fact that what was discussed in the preceding section can happen with real data can be illustrated by analyzing the data in Table 5.3. The data are measurements of color for some actual chemical data. The control chart with σ estimated using moving ranges is shown in Figure 5.6. We can observe that 7 of the 102 points are outside the control limits. In particular, 5 of those points are above the upper control limit. We would not expect this from a process that is in control.

A histogram of the data (not shown) reveals that the data are slightly skewed—there are more large values than small values. This coupled with the fact that the data are obviously autocorrelated (and hence σ is underestimated) causes so

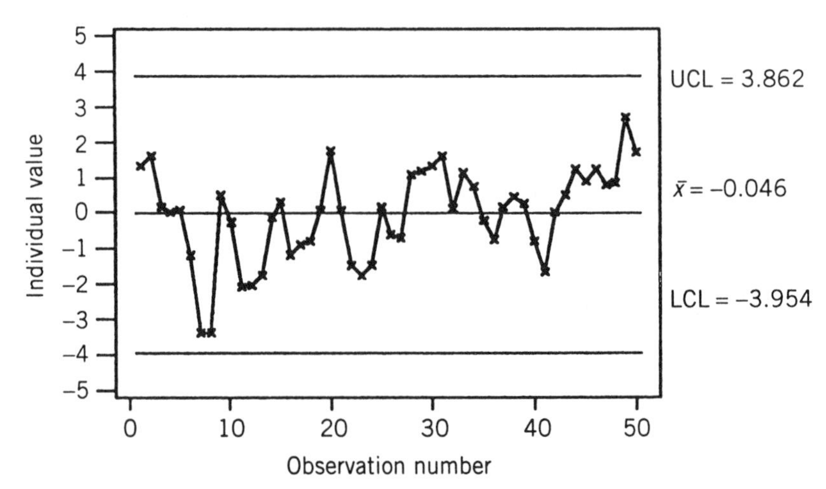

FIGURE 5.5 X chart with limits obtained from $\bar{x} \pm 3s/c_4$.

TABLE 5.3 Trended Real Data

Observation Number	Value	Observation Number	Value	Observation Number	Value
1	0.67	35	0.67	69	0.77
2	0.63	36	0.73	70	0.76
3	0.76	37	0.69	71	0.77
4	0.66	38	0.73	72	0.71
5	0.69	39	0.73	73	0.74
6	0.71	40	0.74	74	0.66
7	0.72	41	0.71	75	0.65
8	0.71	42	0.65	76	0.68
9	0.72	43	0.67	77	0.69
10	0.72	44	0.68	78	0.66
11	0.83	45	0.71	79	0.71
12	0.87	46	0.64	80	0.74
13	0.76	47	0.66	81	0.74
14	0.79	48	0.71	82	0.72
15	0.74	49	0.69	83	0.76
16	0.81	50	0.73	84	0.73
17	0.76	51	0.77	85	0.73
18	0.77	52	0.78	86	0.78
19	0.68	53	0.70	87	0.76
20	0.68	54	0.67	88	0.77
21	0.74	55	0.77	89	0.66
22	0.68	56	0.77	90	0.70
23	0.69	57	0.78	91	0.66
24	0.75	58	0.77	92	0.73
25	0.80	59	0.80	93	0.74
26	0.81	60	0.79	94	0.85
27	0.86	61	0.80	95	0.66
28	0.86	62	0.75	96	0.70
29	0.79	63	0.85	97	0.66
30	0.78	64	0.74	98	0.73
31	0.77	65	0.74	99	0.74
32	0.77	66	0.71	100	0.85
33	0.80	67	0.74	101	0.57
34	0.76	68	0.76	102	0.62

many points to be above the upper control limit. (The fact that the data are autocorrelated can be seen from the runs of points above and below the midline.) The company's process engineers believed that the process was not operating properly, but it was also known that the company's chemical data were naturally autocorrelated because of the nature of the chemical process. Thus, there may be some question as to whether or not the process was really out of control.

In any event, since $s = 0.0574$, the 3-sigma control limits obtained using s would be 0.9052 and 0.5608, and none of the points are outside these limits. Thus, whether or not the process is out of control for these 102 values depends

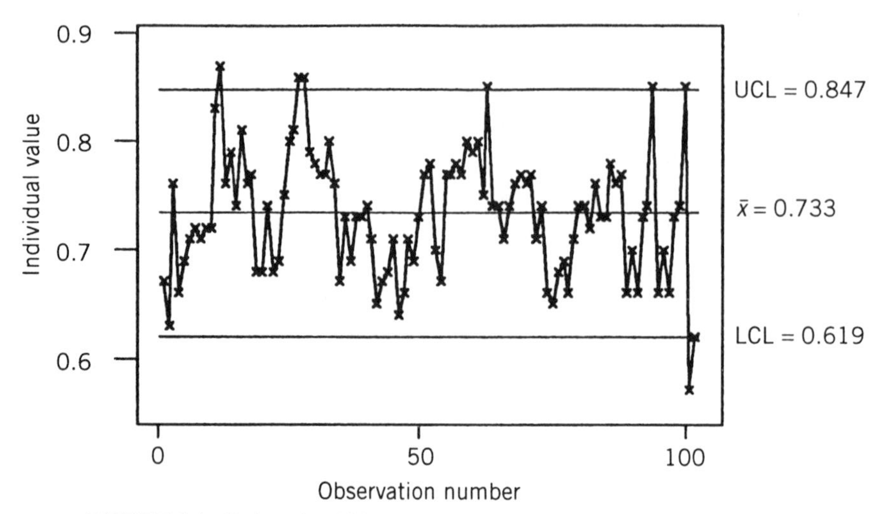

FIGURE 5.6 *X* chart for Table 5.3 data; limits obtained using moving ranges.

on whether or not successive values should be correlated. (Note that we could have adjusted the limits, in the spirit of Zhang (1997), but since the first-order autocorrelation is only 0.421, the appropriate adjustment limits would be to use 2.98-sigma limits instead of 3-sigma limits. Since s is quite small, however, the 2.98-sigma limits would be practically the same as the 3-sigma limits.)

The treatment of autocorrelated data is discussed in detail in Section 10.3.

5.2 MOVING AVERAGE CHART

Whereas moving averages have been used by many practitioners to "smooth out data," a moving average control chart suffers from some of the same deficiencies as a moving range chart. In particular, the plotted points are correlated even when the individual observations are independent. Another problem, if the chart is used in Stage 2, is that σ is generally estimated used moving ranges.

The control limits are generally obtained from

$$\bar{x} \pm \frac{3}{\sqrt{n}} \frac{\overline{\text{MR}}}{d_2}$$

where $\bar{x}$ represents the average of the individual observations, n is the number of observations from which each moving average is computed, and $\overline{\text{MR}}/d_2$ is the estimate of σ (with d_2 obtained using $n = 2$). The moving averages that are plotted on the chart could be obtained as follows. If averages of size 5 are to be used, the first average would be the average of the first five observations, the second average would be the average of observations 2–6, the third would be the average of observations 3–7, and so on. One problem with this approach, however, is that

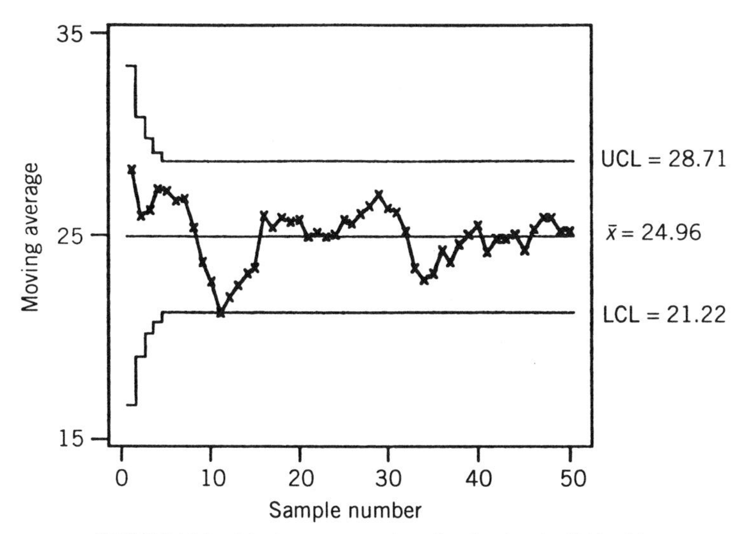

FIGURE 5.7 Moving average chart for the data in Table 5.1.

any of the first four observations could indicate that the process is out of control, but the first four data values are not used until the fifth data value is available. Consequently, it would be much more reasonable to plot some statistic at each time period when the data are collected. An obvious alternative would be to plot the first observation against the control limits for an individual observation and then to compute control limits based on i observations for $i = 2, 3, 4$. Of course, this means that the control limits would be different for the first five plotted points.

Obviously this is a Stage 1 problem because in Stage 2 the historical data would be utilized so that a constant moving subgroup size could be maintained.

For the data in Table 5.1 the moving average chart is given in Figure 5.7. (Recall that the data are ordered sequentially by columns.)

Notice that the graph indicates somewhat of a trend despite the fact that the data were generated from a stable process. Furthermore, the 11th plotted point, which is the average of observations 7–11, is 21.20, which is slightly below the LCL. This occurs despite the fact that all of the observations were generated from the same distribution. Nelson (1983) illustrates how moving averages can be quite deceiving. Figure 5.7 is a perfect example of this. Despite the lure of moving averages in general, a moving average chart is not a very good chart to use for controlling a process mean when only individual observations are available.

5.3 CONTROLLING VARIABILITY WITH INDIVIDUAL OBSERVATIONS

A moving range chart has often been used in conjunction with an X chart for controlling the process variability, with the moving ranges computed as described in Section 5.1.1. The control limits are obtained from $D_3\overline{\text{MR}}$ and $D_4\overline{\text{MR}}$, with

D_3 and D_4 as given in Section 4.7.1, for a subgroup size of 2. Various authors, including Roes, Does, and Schurink (1993) and Rigdon, Cruthis, and Champ (1994), have recommended that an X chart be used for monitoring both the process mean and process variability. Why?

The points that are plotted on a moving range chart are correlated, which can make interpretation somewhat difficult. It is for the latter reason that Nelson (1982) recommends that the moving ranges not be plotted. [It can be shown using results given by Cryer and Ryan (1990) that the correlation between consecutive moving ranges is .229].

In addition to problems with interpretation, a moving range chart simply does not contribute very much when combined with an X chart, as shown in Table 7 of Roes et al. (1993) and by Rigdon et al. (1994). This is intuitive because an increase in σ will cause points on the X chart to be a greater distance from the midline on the chart, and if the increase in σ is sufficiently great, then we would expect points on the X chart to begin falling outside the limits of the chart.

This can be illustrated as follows. The first 30 observations in Table 5.4 were generated from $N(50, 1)$, whereas the next 20 observations were generated from $N(50, 4)$. Assume that the in-control distribution is $N(50, 1)$, so that the control limits are LCL $= 47$ and UCL $= 53$. It can be observed that the first point from the out-of-control distribution is outside the control limits, as are three of the other points. Thus, the change in σ was detected very quickly on a chart that is ostensibly for controlling the mean.

In discussing an alternative to the use of an X chart and a moving range chart in Stage 1, Sullivan and Woodall (1996) also showed that little is gained by using a moving range chart in addition to an X chart. They also showed, however, that it is possible to do considerably better by using a completely different approach. Specifically, a likelihood ratio test approach (Section 3.8.5.2) is used. A short description of the method follows.

Under the assumption of normality, the n historical data values are partitioned into two groups, with the first partition containing n_1 observations, with $n_1 = 2, 3, \ldots, n-2$ and the second partition containing $n_2 = n - n_1$ observations.

TABLE 5.4 Random Numbers from $N(\mu = 50, \sigma^2 = 1)$ and $N(\mu = 50, \sigma^2 = 4)$

51.1692	49.6146	49.9113	54.3343	46.9921
49.5848	50.4418	48.8756	48.9897	49.2840
50.5234	49.7623	50.9346	50.2957	53.2407
51.2021	51.5255	48.8860	48.5961	45.8092
49.2391	51.2715	51.6562	48.2164	50.8160
49.3705	49.0103	49.0290	53.9226	52.5341
49.3705	48.9178	50.9294	51.9704	49.9224
49.0180	52.3443	50.5536	49.3518	50.4417
48.7330	49.1353	49.4807	49.6224	47.8571
49.6408	49.7464	49.6370	48.6043	51.6544

All possible partitions are examined, with the data sequence maintained. The log of the likelihood function for each partition is computed, using the maximum likelihood estimators for μ and σ^2 in each partition. The two log-likelihood functions are then added. Sullivan and Woodall (1996) used l_a to denote this sum. The log-likelihood is then computed without any partitions (say, l_0), and the maximum value of $r = -2(l_0 - l_a)$ is found by computing the difference in the log-likelihoods for all of the partitions. If a sustained change in either the mean or the variance has occurred, the value of n_1 at which the maximum occurs is the maximum likelihood estimate of the change point.

A control chart can be constructed by dividing r by its in-control expected value for each value of n_1. This is necessary because the expected value is not constant over n_1. The UCL is then determined from simulations or approximated as described by Sullivan and Woodall (1996). When an out-of-control signal is received, the statistic r can be broken down into two components that sum to r—one that is zero when the means in each partition are equal and the other that is zero when the variances are equal. Thus, the relative magnitude of the two components can suggest whether the mean or variance has changed.

It is also worth noting that with this type of control chart the point at which a change is detected will generally be much closer to the time that the change actually did occur than is the case when a combined X–MR chart is used. The superiority of the likelihood ratio test (LRT) control chart over the combined use of an X and moving range chart is expected since the test statistics are computed using both historical and very recent data, whereas only the current data point is plotted when a Shewhart chart is used.

As discussed by Sullivan and Woodall (1996), the LRT control chart is less effective when there is only a temporary change in the mean or variance, however. Therefore, they suggest that the LRT chart be combined with either a stalactite analysis or an X chart. (The former is discussed in Atkinson and Mulira (1993) and a modification is given by Sullivan and Woodall (1996).)

5.4 SUMMARY

Although not used as frequently as control charts for subgroup data, charts for individual observations are often used. Because of the sensitivity of an X chart to the normality assumption, it is important to try to determine if there is evidence of more than slight nonnormality. If there is evidence of autocorrelation, k-sigma control limits with k slightly different from 3 can be used to approximately reproduce the properties of the chart for independent data. If the autocorrelation is extreme (and positive), there is not much that can be done other than adjusting the limits and settling for properties that may be less than desirable (see Section 10.3).

A moving range chart is controversial, but the evidence suggests that very little is gained by using it. More sophisticated methods might be used for controlling both the process mean and variability, as indicated in Section 5.3.

APPENDIX

We may easily derive $E(\overline{\text{MR}}/d_2)$ under the assumption that individual observations adhere to an AR(1) process. From the Chapter 4 Appendix we have that the expected value of a range is $d_2\sigma$, with d_2 depending on the number of observations used in computing each range and σ representing the standard deviation of *independent* observations. Since $d_2 = 2/\sqrt{\pi}$ for a range based on two observations (and, thus, also for a "moving range"), we have $E(R) = E(\text{MR}) = d_2\sigma$, which is the same as $E(\overline{\text{MR}})$ since the average of $d_2\sigma$ is obviously $d_2\sigma$. Here σ is for independent observations. It is well known, and also follows from Section 5.1.5, that the corresponding (white noise) standard deviation for an AR(1) process is $\sigma\sqrt{1-\rho_1}$. Using this expression then produces

$$E\left(\frac{\overline{\text{MR}}}{d_2}\right) = \sigma\sqrt{1-\rho_1}$$

REFERENCES

Atkinson, A. C. and H. M. Mulira (1993). The stalactite plot for the detection of multivariate outliers. *Statistics and Computing 3*: 27–35.

Box, G. E. P. and G. M. Jenkins (1976). *Time Series Analysis: Forecasting and Control*, rev. ed. San Francisco: Holden-Day.

Chou, Y.-M., G. D. Halverson, and S. T. Mandraccia (1998). Control charts for quality characteristics under nonnormal distributions. In R. Peck, L. D. Haugh, and A. Goodman, eds. *Statistical Case Studies: A Collaboration Between Academe and Industry*, Chapter 8. Philadelphia: Society of Industrial and Applied Mathematics and American Statistical Association.

Cryer, J. D. and T. P. Ryan (1990). The estimation of sigma for an X chart: $\overline{\text{MR}}/d_2$ or S/c_4? *Journal of Quality Technology* 22: 187–192.

Daniel, C. (1976). *Applications of Statistics to Industrial Experimentation*. New York: Wiley.

Finison, L. J., M. Spencer, and K. S. Finison (1993). Total quality measurement in health care: Using individual charts in infection control. In *ASQC Quality Congress Transactions*, pp. 349–359. Milwaukee, WI: American Society for Quality Control.

Harding, A. J., K. R. Lee, and J. L. Mullins (1992). The effect of instabilities on estimates of sigma. *ASQC Quality Congress Transactions*, 1037–1043.

Maragah, H. D. and W. H. Woodall (1992). The effect of autocorrelation on the retrospective X-chart. *Journal of Statistical Computation and Simulation 40*: 29–42.

Nelson, L. S. (1982). Control charts for individual measurements. *Journal of Quality Technology 14*(3): 172–173.

Nelson, L. S. (1983). The deceptiveness of moving averages. *Journal of Quality Technology 15*(2): 99–100.

Nelson, L. S. (1989). A stabilized normal probability plotting technique. *Journal of Quality Technology 21*(3): 213–215.

Quesenberry, C. P. (1993). The effect of sample size on estimated limits for $\overline{X}$ and X control charts. *Journal of Quality Technology 25*: 237–247.

Rigdon, S. E., E. M. Cruthis, and C. W. Champ (1994). Design strategies for individuals and moving range control charts. *Journal of Quality Technology 26*: 274–287.

Roes, K. C. B., R. J. M. M. Does, and Y. Schurink (1993). Shewhart-type control charts for individual observations. *Journal of Quality Technology 25*(3): 188–198.

Sullivan, J. H. and W. H. Woodall (1996). A control chart for preliminary analysis of individual observations. *Journal of Quality Technology 28*: 265–278.

Zhang, N. F. (1997). Detection capability of residual control chart for stationary process data. *Journal of Applied Statistics 24*(4): 475–492.

EXERCISES

1. Reorder the data in Table 5.2 in the following manner: Put the data in ascending order grouped by the last digit, ignoring the sign of the number. (Thus, the first 12 numbers are 0.50, −0.60, −0.80, 1.10, 1.30, 1.30, 0.01, −0.21, 1.11, 1.21, 1.61, and 2.71. Put 0.12 before −0.12.)
   (a) Would you expect the estimate of σ obtained from $\overline{\text{MR}}/d_2$ to be satisfactory, or should σ be estimated using s/c_4?
   (b) Construct the X chart and display the 3-sigma limits obtained using both $\overline{\text{MR}}/d_2$ and s/c_4 as estimates of σ.
   (c) Comment on the difference in the two sets of control limits.
2. Assume that there has been a 1.6σ upward shift in the process mean. How many individual observations would we expect to obtain before finding one that is above the UCL? Compare this with the number of subgroups of $n = 4$ that we would expect to observe before a subgroup average plots above its UCL.
3. List the objections to the use of moving range and moving average charts.
4. The model that is implied by the use of a conventional X chart is $y_i = \mu + \varepsilon_i$, where $\varepsilon \sim \text{NID}(0, \sigma^2)$ and the y_i are the values that are charted. If, however, the data actually come from an AR(1) process with $\phi_1 = 0.6$, by what amount would we expect σ_y^2 to be underestimated when $\overline{\text{MR}}/d_2$ is used to obtain the estimate?
5. Explain why a control chart user should select an $\overline{X}$ chart over an X chart if a process is such that either could be used.
6. Construct an X chart for the following data set using s/c_4 as the estimate of σ. (The data are ordered in columns.)

14.2	14.1	14.5	14.1	15.7
15.3	13.9	14.8	14.7	15.1
14.4	13.7	14.0	13.5	13.9
13.6	14.4	13.9	13.8	14.2
13.2	15.1	13.5	14.2	14.3

7. The following 50 observations have been generated from a normal distribution with $\mu = 75$ and $\sigma = 10$:

78.2	92.6	74.7	83.9	54.3
67.6	75.2	67.4	71.7	77.3
54.8	79.2	75.1	88.1	83.8
67.2	87.0	56.8	83.6	78.5
85.4	65.4	91.8	74.1	66.8
64.4	75.5	84.8	72.0	86.4
69.2	69.1	71.0	87.2	90.5
79.7	59.2	96.8	86.3	85.5
80.7	58.4	75.7	80.0	75.8
64.3	100.1	75.0	77.5	79.2

Construct the set of 46 moving averages of size 5. Note the trend in the moving averages, even though the individual observations are randomly generated. (The data are ordered in columns.)

8. The following 50 observations have been generated from an AR(1) process with $\phi_1 = 0.4$:

13.0	10.2	10.4	10.2	11.2
8.9	9.9	9.6	10.9	8.8
7.6	11.1	9.2	10.0	9.0
8.0	10.2	9.2	9.6	9.3
10.9	11.5	11.9	10.3	10.3
10.2	11.9	11.4	11.1	9.9
10.2	10.9	10.0	10.2	9.1
10.5	12.3	11.3	10.3	9.6
9.4	10.9	9.7	11.6	8.7
8.8	9.5	8.1	10.6	9.3

Construct the set of 46 moving averages of size 5, after first plotting the individual observations. Compare the trend in the moving averages with the trend in the individual observations. Merging this result with the outcome from the preceding problem, what does this suggest about the effect created by moving averages, whether the individual observations are independent or not? (The data are ordered in columns.)

9. Construct an X chart for the data in Table 5.3 (using s) and comment on the results relative to what Figure 5.6 shows.

10. The following are actual data on days between infections at a small hospital (Finison et al., 1993):

31, 19, 33, 129, 102, 19, 10, 4, 10, 17, 169, 283, 99, 75, 5, 3, 1, 4, 4, 5, 1, 1, 18, 9, 12.

If an X chart is to be constructed for these data, should the data be transformed first? Why or why not?

11. Explain why it is difficult to control variability when individual observations are obtained rather than subgroups.

CHAPTER 6

Control Charts for Attributes

It is not always possible or practical to use measurement data in quality improvement work. Instead, *count data* (often referred to as *attribute data*) are used where, for example, the number of nonconforming parts for a given time period may be charted instead of measurements charted for one or more quality characteristics. Although automatic measuring devices have greatly simplified the measurement process, it is still often easier to classify a unit of production as conforming or nonconforming than to obtain the measurement for each of many quality characteristics. Furthermore, attribute control charts are used in many applications, such as clerical operations, for which count data occur naturally, not measurement data.

Before studying and implementing charts for the number of nonconforming units, it is important to remember that, ideally, nonconforming units should not be produced. Consequently, attempting to control the number of nonconforming units at a particular level would generally be counterproductive. The objective should be, of course, to continually reduce the number of such units. When used for that purpose, such charts can be of value in indicating the extent to which the objective is being achieved over time. Thus, such charts might be used in *conjunction* with measurement charts, rather than simply in place of them. If either type of chart could be used in a particular situation, it would be wasteful of information to use an attribute chart by itself. Accordingly, such charts should be used alone only when there is no other choice.

Several different types of attribute charts are presented in this chapter. Before examining these charts in detail, there is the question of which terminology to adopt. For example, should a chart that displays the fraction or percentage of defective units be called a chart for fraction rejected, fraction defective, or fraction of nonconforming units? The latter will be adopted in accordance with national and international standards (see, e.g., ANSI/ASQC A1–1987). In particular, *nonconforming unit* will be used in place of *defective*. A unit could also have a number of defects (e.g., scratches) that might not be severe enough to cause it to be classified as a nonconforming unit, but a chart of each type of defect might be maintained. *Defect* will not be used; it will be replaced by *nonconformity*.

6.1 CHARTS FOR NONCONFORMING UNITS

When interest centers on nonconforming units, the user can employ a chart for either the fraction or number of such units. These are labeled the p chart and np chart, respectively. When these charts are constructed using 3-sigma limits, there is the tacit assumption that the normal approximation to the binomial distribution, alluded to in Chapter 3, is adequate. It is also assumed that the nonconforming units occur independently; that is, the occurrence of a nonconforming unit at a particular point in time does not affect the probability of a nonconforming unit in the time periods that immediately follow. Extra binomial variation can occur when this assumption is violated; this is considered in Section 6.1.7. Letting X represent the number of nonconforming units in a sample of size n and $\hat{p} = X/n$ denote the proportion of such units,

$$\frac{X - np}{\sqrt{np(1-p)}} \tag{6.1}$$

and

$$\frac{\hat{p} - p}{\sqrt{\dfrac{p(1-p)}{n}}} \tag{6.2}$$

will be approximately distributed as $N(\mu = 0, \sigma = 1)$, when n is at least moderately large and p does not differ greatly from .5. In practice, the first requirement is generally met; in fact, n has often been equal to an entire day's production. We would certainly hope that the second requirement is never met, however, since anything close to 50% nonconforming units would be disastrous.

6.1.1 *np* Chart

As mentioned in Chapter 3, the rule-of-thumb generally advanced is that for Eqs. (6.1) and (6.2) to be approximately $N(0, 1)$, both np and $n(1-p)$ should exceed 5. We would certainly hope that p will be quite small in virtually any type of application, especially manufacturing applications, so it is the first requirement that is of interest. A company embarking upon a statistical quality control program might well have p approximately equal to .10 for some products, however. For $p = .10$, the rule-of-thumb would require n greater than 50. Assume that $n = 400$ and $p = .10$ so that the requirement is easily satisfied. In general, the control limits for an np chart would be obtained (if p were known) from

$$np \pm 3\sqrt{np(1-p)} \tag{6.3}$$

where p will usually have to be estimated. For this example we assume p to be known so the limits would be

$$400(.10) \pm 3\sqrt{400(.1)(.9)} = 40 \pm 18$$

Thus, UCL = 58 and LCL = 22. *If* the normal approximation were quite good, we would expect $P(X > 58)$ and $P(X < 22)$ to be close to .00135 (recall that this is the area in each tail for an $\overline{X}$ chart when normality is assumed). Unfortunately, the probabilities differ greatly even though $np = 40$. Specifically, $P(X > 58) = .0017146$ and $P(X < 22) = .0004383$, using the binomial distribution. Thus, although the first probability is fairly close to the nominal value, the second one is not. This shows that the frequently cited rules-of-thumb regarding the adequacy of the normal approximation to the binomial distribution do not apply to control charts. The problem is that the adequacy of the approximation depends primarily on the value of p, as has been discussed, for example, by Schader and Schmid (1989).

The practical implication of this is that when the actual lower tail area is very close to zero, sizable reductions in p will not be immediately reflected by points falling below the LCL. For example, $P(X < 22) = .02166$ when $p = .08$ and $P(X < 22) = .00352$ when $p = .09$. Thus, for a 10% reduction in the percentage of nonconforming units, we expect to have to obtain $1/.00352 = 284$ samples of size 400 before observing a value of X that is below the LCL of 22. Similarly, for a 20% reduction, to $p = .08$, we would expect to observe $1/.02166 = 46$ samples before observing a value of X less than 22. This means that the LCL will have very little value in many practical situations as a benchmark for indicating quality improvement. The problem can be even more acute when p is smaller than in the preceding example.

The fact that the actual upper tail area is close to the nominal value is virtually irrelevant for manufacturing applications because that would have meaning only if an np chart were being used for *defensive* purposes, that is, to keep the quality from deteriorating any further. (Recall from Section 4.10 that this is essentially the objective of acceptance sampling.) The UCL will have value when an np chart is used in nonmanufacturing applications, however.

Thus, if an np chart is to be used in a quality *improvement* program, it is certainly desirable to have an LCL value such that points that begin appearing below that value may reflect a significant reduction in p; conversely, a significant reduction in p will cause points to begin immediately appearing below that value.

It is worth noting, however, that points falling below the LCL may not be indicative of quality improvement but may actually suggest some type of problem. As discussed by Box and Luceño (1997, p. 39), small plotted values for the number or proportion of nonconformities could be caused by faulty test equipment and/or procedures. Similarly, Montgomery (1996, p. 264) points out that inspectors could be poorly trained, pass nonconforming units, and/or report fictitious data. So the assignable cause of points plotting below the LCL may not be a favorable one.

Another problem is that there will frequently not be an LCL since $np - 3\sqrt{np(1-p)}$ can easily be negative. For example, with $n = 400$, $np - 3\sqrt{np(1-p)}$ will be negative if $p < .022$. In general, it can be stated that $np - 3\sqrt{np(1-p)}$ will be negative if $p < 9/(9+n)$, which is equivalent to

$n < 9(1-p)/p$. Since it will typically be easier to change n than to change p, it would be wise to select n so that the LCL > 0.

6.1.2 *p* Chart

In the preceding section attention was focused upon the number of nonconforming units in each sample. Alternatively, a chart could be used that shows the proportion of nonconforming units in each sample. Such a chart is called a *p chart*. Neither type of chart is superior to the other from a statistical standpoint. Some users may wish to plot the number of nonconforming units per sample (which would be reasonable if the sample size is constant), whereas others may prefer to see the proportion of such units.

The control limits for a p chart would be obtained (with p known) from

$$p \pm 3\sqrt{\frac{p(1-p)}{n}} \tag{6.4}$$

Notice that if these limits are multiplied by n, the result will be equal to the limits given by Eq. (6.3), and if the points that are plotted on a p chart are also multiplied by n, the numerical values would be equal to the values of the points plotted on an np chart. Thus, a p chart is simply a scaled version of an np chart, with the same configuration of points on each chart. Thus, there is no need to use both charts.

Since a p chart is simply a scaled version of an np chart, it is subject to the same general weakness as an np chart. Namely, the tail probabilities will be the same as for the np chart.

6.1.3 Stage 1 and Stage 2 Use of *p* Charts and *np* Charts

If one of these charts is to be used, a set of historical data would be used to establish the trial control limits and to estimate p, analogous to what was discussed in Chapter 4 for measurement data. If we first assume that n is constant over the historical data, p would be estimated as the total number of nonconforming units divided by the total number of items that were inspected.

If historical data exist, the data may or may not be adequate for estimating p. To obtain a good estimate of p, we initially need a rough estimate of p, or at least a lower bound on p, so that we can select a sample size large enough to ensure that each sample will contain at least a few nonconforming units. If, for example, we believed that p was approximately .005, it would be unwise to use samples of size $n = 100$ since we would expect most of the samples to have zero nonconforming units.

The construction of a p chart and an np chart can be illustrated using the fictitious data in Table 6.1. An estimate of p would be obtained as

$$\overline{p} = \frac{\text{total number of nonconforming units}}{\text{total number inspected}}$$

TABLE 6.1 Number of Nonconforming Transistors Out of 1000 Inspected Each Day During the Month of April

Day	Numbers of Nonconforming Units
1	7
2	5
3	11
4	13
5	9
6	12
7	10
8	10
9	6
10	14
11	9
12	13
13	8
14	11
15	12
16	10
17	9
18	12
19	14
20	12
21	13
22	7
23	9
24	12
25	8
26	14
27	12
28	12
29	11
30	13

where $\overline{p}$ will be used instead of $\hat{p}$ to designate the estimate of p, in accordance with standard notation. The data in Table 6.1 produce

$$\overline{p} = \frac{318}{30,000} = .0106$$

so that 1.06% of the units examined were nonconforming. The standard 3-sigma limits for the np chart are

$$n\overline{p} \pm 3\sqrt{n\overline{p}(1-\overline{p})} = 1000(.0106) \pm 3\sqrt{1000(.0106)(1-.0106)}$$
$$= 10.6 \pm 9.715$$

so that the UCL = 20.315 and the LCL = 0.885.

The corresponding limits for the *p* chart would be obtained by dividing the *np*-chart limits by 1000. Thus, the *p*-chart limits would be UCL = 0.0203 and LCL = 0.0009. The *p* and *np* charts are shown in Figures 6.1 and 6.2, respectively. We can see that the configuration of points is the same for the two charts, as will

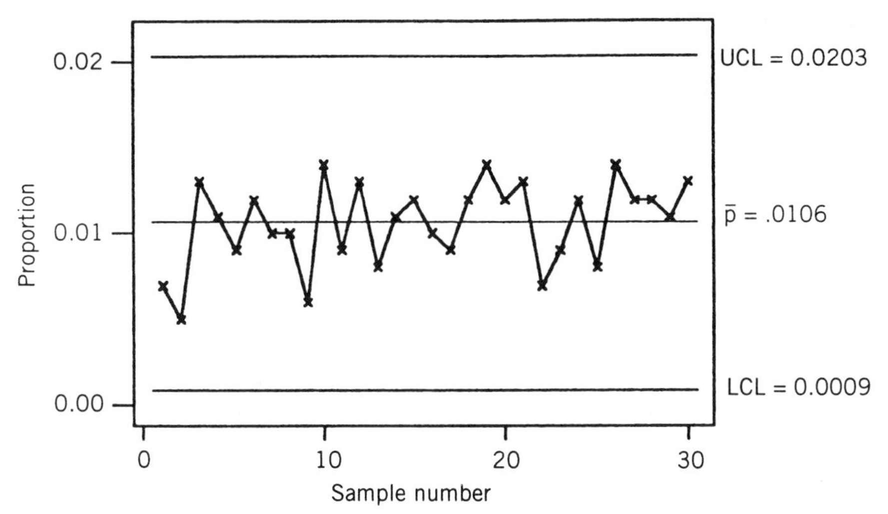

FIGURE 6.1 The *p* chart for data in Table 6.1.

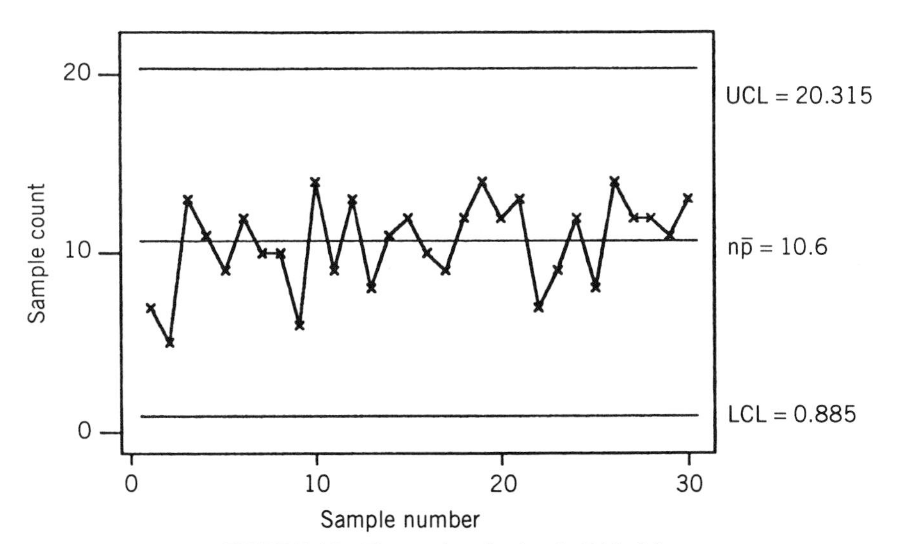

FIGURE 6.2 The *np* chart for data in Table 6.1.

always be the case. All of the points are within the control limits on each chart, although the last five points being above the midline could indicate (with real data) that the number (percentage) of nonconforming units has increased. If any points had plotted outside the limits, we would search for assignable causes and proceed in the same general way that we would if we were using an $\overline{X}$ chart. Since all of the points are within the limits for the current example, the limits would be extended and applied to future samples provided that $n = 1000$.

The point that was made in Chapter 4 should be borne in mind when a set of samples is used to determine trial control limits. Namely, the probability of observing at least one of k points outside the control limits is approximately kp, where p is the sum of the two-tail probabilities. Thus, using a nominal value of $p = .0027$, $kp = .081$ so that at least one point could easily fall outside the limits due strictly to chance.

6.1.4 Alternative Approaches

Although 3-sigma limits have historically been used for p and np charts, such an approach will be unwise for most applications, since the LCL will generally be too small. Thus an alternative to the use of 3-sigma limits is desirable for each chart.

Several alternatives have been proposed. One such method would be to transform the binomial data so that the transformed data are approximately normally distributed and then use 3-sigma limits for the transformed variable. Alternatively, the control limits could be adjusted to compensate for the nonnormality. These various approaches are discussed in subsequent sections.

6.1.4.1 Arcsin Transformations

Some (almost identical) *arcsin transformations* have been proposed. Johnson and Kotz (1969) state that

$$y = \sin^{-1}\sqrt{\frac{x+\frac{3}{8}}{n+\frac{3}{4}}} \tag{6.5}$$

will be approximately normally distributed with mean $\sin^{-1}\sqrt{p}$ and variance $1/4n$, where $\sin^{-1}$ denotes the inverse of the sine function in trigonometry (generally referred to as the arcsin). Thus, for each sample of size n, one would plot the value of y on a control chart with the midline at $\sin^{-1}\sqrt{p}$ (if p is known) and the control limits given by

$$\sin^{-1}\sqrt{p} \pm 3\sqrt{\frac{1}{4n}} = \sin^{-1}\sqrt{p} \pm \frac{3}{2\sqrt{n}} \tag{6.6}$$

For the example with $n = 400$ and $p = .10$, the control limits would be

$$\sin^{-1}\sqrt{0.10} \pm \frac{3}{2\sqrt{400}} = 0.32175 \pm 0.075$$

so that the LCL = 0.24675 and the UCL = 0.39675.

A simple way to compare the worth of this approach for this particular example would be to determine the value of x such that $P(X \leq x) \doteq .00135$ (from the binomial distribution) and then see if that value of x produces a value of y that is quite close to the LCL of 0.24675. Here, $P(X \leq 23) = .00168$ and for $x = 23$, $y = 0.2439$. Thus, since y is indeed almost equal to the LCL, we can see the value of obtaining the control limits by using the arcsin transformation — at least for this example.

Comparing this arcsin result with the result obtained using the conventional approach, the LCL using the arcsin transformation would correspond to $x = 23$ (to the nearest integer), whereas the conventional approach yielded $x = 22$ for the LCL. Thus, the conventional LCL is one unit too low. These values will differ by a somewhat larger amount, however, when p is much smaller than .10, as is shown later in Table 6.3.

Notice in Table 6.2 that the actual tail areas for the 3-sigma limits are off considerably from .00135, although we should not expect that value to be met exactly since the binomial random variable is discrete instead of continuous. Of particular concern, however, is the fact that the lower tail areas are quite close to zero when p is around .02 for $n \leq 1000$. In particular, when $n = 600$ and $p = .02$ the lower tail area is .00007. As indicated earlier, the problem this creates is that a considerable reduction in p could go undetected for quite some time when an np chart (or p chart) is used in the conventional manner. For example, if there is a subsequent reduction in p from .02 to .015 (a 25% improvement), the probability of a point falling below the LCL obtained using $p = .02$ is only .0012, so it would require, on the average, 856 samples before a point would fall below the LCL. (Notice that this is even a higher ARL value than would exist with a nominal tail area of .00135 and no change in p.) Similarly, a drop to $p = .01$ would produce a tail area of .01698 and an ARL value of approximately 59.

In general, by comparing Table 6.2 with Table 6.3, we can see that the LCL values for the 3-sigma limits differ from the probability limit LCL values by either one or two units in all cases, but for the UCL values the two differ by only one unit in almost all cases. Specifically, the 3-sigma limit LCL is too low by two units roughly half the time (and by one unit the rest of the time), and the UCL is generally one unit too low.

The limits obtained from use of the arcsin transformation fare somewhat better. The UCL values on the original scale agree exactly (in terms of the possible integer values of X) for 21 of the 35 combinations and differ by 1 unit for the other 14. For the LCL values there is exact agreement for 19 of the combinations and a difference of 1 unit for the other 16. We should not always expect exact agreement, however, since the use of 3-sigma limits with the arcsin transformation is designed to hit the .00135 probability limits almost exactly, *if* they existed. We should not expect them to exist for a discrete random variable, however. For these combinations of n and p some of the tail probabilities are not particularly close to .00135 (see, e.g., the LCL for $n = 600$ and $p = .02$), even though they are the *closest* to .00135 for each combination. Notice, however, that when the tail probability is very close to .00135, the arcsin-based limit for the LCL is almost

TABLE 6.2 *np* Chart Control Limits

		3-Sigma Limits		Actual Tail Areas	
n	p	LCL	UCL	LCL	UCL
100	.12	2.250	21.750	.00030	.00341
	.10	1.000	19.000	.00003	.00198
	.09	0.415	17.586	.00008	.00342
200	.10	7.272	32.728	.00048	.00292
	.08	4.490	27.510	.00027	.00276
	.06	1.924	22.076	.00006	.00221
	.05	0.753	19.247	.00004	.00266
300	.10	14.412	45.588	.00057	.00242
	.08	9.903	38.097	.00028	.00196
	.06	5.660	30.340	.00024	.00246
	.05	3.675	26.325	.00016	.00257
	.04	1.818	22.182	.00006	.00247
	.03	0.136	17.864	.00011	.00464
400	.10	22.000	58.000	.00044	.00171
	.05	6.923	33.077	.00020	.00207
	.03	1.765	22.235	.00007	.00261
600	.10	37.955	82.045	.00058	.00169
	.05	13.984	46.016	.00031	.00189
	.03	5.464	30.536	.00028	.00288
	.02	1.712	22.288	.00007	.00275
800	.10	54.544	105.456	.00081	.00190
	.05	21.507	58.493	.00057	.00226
	.03	9.525	38.475	.00037	.00257
	.02	4.121	27.879	.00036	.00374
1000	.10	71.540	128.460	.00086	.00184
	.05	29.324	70.676	.00072	.00233
	.03	13.817	46.183	.00035	.00210
	.02	6.718	33.282	.00023	.00243
	.01	0.561	19.439	.00004	.00329
2000	.10	159.751	240.249	.00094	.00162
	.05	70.760	129.240	.00076	.00178
	.03	37.113	82.887	.00084	.00245
	.02	21.217	58.783	.00067	.00263
	.01	6.650	33.349	.00024	.00256
	.007	2.814	25.186	.00009	.00252

identical to the probability limit, even when the decimal places are considered. For example, for the LCL with $n = 1000$ and $p = .05$ there is hardly any difference between 31 and 31.027.

All factors considered, the superiority of the alternative approaches over the traditional approach should be apparent.

Other forms of the arcsin transformation have also been used. One is

$$y_1 = \sin^{-1}\sqrt{\frac{x}{n}}$$

TABLE 6.3 ***np*** **Chart Probability Limits and Arcsin Transformation (Values of** ***x*** **That Give Areas Closest to .00135)**

n	p	LCL	Tail Area[a]	UCL	Tail Area[a]	LCL[b]	UCL[b]
100	.12	4	.00145	22	.00151	3.750	23.100
	.10	3	.00194	20	.00081	2.568	20.432
	.09	2	.00087	18	.00140	2.017	19.058
200	.10	9	.00139	33	.00154	8.820	34.180
	.08	6	.00099	28	.00137	6.108	29.042
	.06	4	.00184	23	.00101	3.612	23.688
	.05	2	.00040	20	.00116	2.476	20.899
300	.10	16	.00127	46	.00141	15.951	47.049
	.08	12	.00184	39	.00108	11.513	39.636
	.06	7	.00080	31	.00128	7.341	31.959
	.05	5	.00069	27	.00127	5.391	27.984
	.04	4	.00199	23	.00115	3.568	23.882
	.03	2	.00111	19	.00086	1.921	19.604
400	.10	24	.00168	59	.00105	23.534	59.466
	.05	9	.00172	34	.00110	8.635	34.740
	.03	4	.00206	23	.00123	3.546	23.979
600	.10	40	.00166	83	.00112	39.482	83.518
	.05	16	.00158	47	.00111	15.692	47.683
	.03	7	.00092	31	.00153	7.243	32.282
	.02	4	.00214	23	.00131	3.525	24.075
800	.10	56	.00125	106	.00134	56.068	106.930
	.05	23	.00112	59	.00143	23.212	60.163
	.03	11	.00095	39	.00147	11.301	40.223
	.02	6	.00128	29	.00100	5.932	29.668
1000	.10	73	.00127	129	.00134	73.061	129.939
	.05	31	.00128	71	.00154	31.027	72.348
	.03	16	.00172	47	.00125	15.591	47.934
	.02	9	.00193	34	.00133	8.528	35.072
	.01	2	.00048	20	.00150	2.405	21.270
2000	.10	161	.00123	241	.00128	161.266	241.734
	.05	72	.00111	130	.00131	72.458	130.917
	.03	39	.00139	84	.00115	38.884	84.641
	.02	23	.00128	60	.00107	23.024	60.576
	.01	8	.00075	34	.00141	8.493	35.182
	.007	5	.00176	26	.00125	4.667	27.030

[a] These values for tail areas are binomial probabilities such that $P(X < \text{LCL}) =$ tail area and $P(X > \text{UCL}) =$ tail area.
[b] Values of X in the arcsin transformation $y = \sin^{-1}\sqrt{(x + 3/8)/(n + 3/4)}$ that cause y to equal the upper and the lower limits obtained from $\sin^{-1}\sqrt{p} \pm 3\sqrt{1/4n}$.

[see, e.g., Ehrenfeld and Littauer (1964, p. 310), Brownlee (1960, p. 115), or Hald (1952, p. 685)]. Another variation is

$$y_2 = \frac{1}{2}\left(\sin^{-1}\sqrt{\frac{x}{n+1}} + \sin^{-1}\sqrt{\frac{x+1}{n+1}}\right)$$

The latter is generally referred to as the average angular transformation and is discussed, for example, in Nelson (1983). It was originally suggested in Freeman and Tukey (1950), in which the $\frac{1}{2}$ was also omitted. The reason for not using the $\frac{1}{2}$ is that without it the variance of the transformed variable is $1/n$, which is considered by some to be more convenient than a variance of $1/4n$.

If either of these two transformations were to be used instead of Eq. (6.5), the control limits would still be obtained from Eq. (6.6).

Since the three transformations differ only slightly, the results obtained from using any of the transformations should be virtually the same.

It is also possible that the LCL could be zero when using the arcsin transformation, but this will happen for a smaller range of values for p and n than with the traditional approach. This is illustrated in Table 6.4.

The minimum sample size needed using the arcsin approach is thus roughly one-fourth the size needed for the traditional approach. Although an entire day's production has often constituted the sample size, it would generally be impractical to inspect 22,499 items for each sample. Therefore, regardless of the approach that is used, a very small percentage of nonconforming units will cause the LCL to be zero for reasonable values of n.

But for many values of p, Table 6.4 could be a useful guide. For example, if p was approximately .005, we might wish to use samples of size 500 or larger so as to have some assurance that the LCL will be positive (if an arcsin transformation is used). This should also provide at least one or two nonconforming units in each sample since the expected number is 2.5.

TABLE 6.4 Comparison of Minimum Sample Size Necessary to Make LCL > 0 for Traditional Approach and Arcsin Transformation Approach

	Value n Must Exceed	
p	Arcsin Method	Traditional Method
.10	21	81
.08	27	103
.06	36	141
.05	44	171
.04	55	216
.03	74	291
.02	111	441
.01	224	891
.005	449	1,791
.002	1,124	4,491
.001	2,249	8,991
.0001	22,499	89,991
.00001	224,999	899,991

6.1.4.2 *Q Chart for Binomial Data*

Another approach to the problem is to use a Q chart, as developed by Quesenberry (1991 a,b,c). See also Hawkins (1987). Quesenberry (1997) contains an extensive discussion of the various types of Q charts.

The general idea of a Q chart for binomial data is as follows. Let X be as previously defined, and let $u_i = B(x_i; n, p)$ denote $P(X \leq x_i)$. If 3-sigma limits are used, the statistics $Q_i = \Phi^{-1}(u_i)$ are plotted against control limits of LCL $= -3$ and UCL $= 3$, where Φ denotes the normal cumulative distribution function and p is here assumed to be known. In comparing the performance of the Q chart against the arcsin transformation approach, Quesenberry (1991b) states that the Q chart will generally have tail probabilities that are less than .00135 for the lower tail and greater than .00135 for the upper tail. Indeed, this is true for every combination of n and p considered by Quesenberry and must be true by definition. Quesenberry subsequently concluded that the arcsin approach gives a better approximation to the nominal lower tail area, and the Q chart provides a better approximation to the nominal upper tail area. Since in industrial applications we would want a p chart or an np chart to show process improvement in the form of points plotting below the LCL, this comparison favors the arcsin approach for such applications.

6.1.4.3 *Regression-Based Limits*

A computer such as a PC will be needed for the Q chart and at least a hand calculator will be needed for each plotted point using the arcsin approach. This will put both approaches beyond the reach of many control chart users. Another shortcoming of both approaches is that there is no attempt made to come as close to the nominal tail probabilities as possible. Both approaches rely upon a nonlinear normal approximation that, while being far superior to the linear normal approximation upon which the p and np charts are based, does leave room for improvement.

Another obvious shortcoming of both approaches is that a transformed statistic is plotted, rather than the statistic of interest, as mentioned for the Q chart by Quesenberry (1991a). It should be borne in mind that numbers on a transformed scale will, in general, not be as easy to relate to as numbers on the original scale. In particular, the arcsin transformation will not produce numbers that have a direct physical interpretation. If this is of concern to the user, it would be preferable to use a regular p or np chart with the actual tail probabilities used as a guide in determining the control limits, rather than automatically using the 3-sigma limits. Box and Luceño (1997, p. 41) provide a table, produced by interpolating in the binomial distribution, that can be used for producing the control limits for either chart when the normal approximation is deemed inadequate.

Another approach, which would undoubtedly be preferred by many control chart users, would be to use a slightly modified form of the 3-sigma limits that would essentially emulate the limits that would be produced by direct use of binomial tables.

Ryan and Schwertman (1997) gave regression-bassed limits for an *np* chart as

$$\begin{aligned} \text{UCL} &= 0.6195 + 1.00523np + 2.983\sqrt{np} \\ \text{LCL} &= 2.9529 + 1.0195np - 3.2729\sqrt{np} \end{aligned} \tag{6.7}$$

with the corresponding *p*-chart limits obtained by dividing each of the limits produced by Eqs. (6.7) by n. The reader will observe that these limits differ from the corresponding 3-sigma limit expressions in that the latter do not contain a constant, and the regression-based limits do not contain a factor of $1 - p$ in the third term. The latter is due to the fact that the regression-based limits were first constructed for a c chart (see Section 6.2.1), with the limits for the p and *np* charts obtained as a byproduct.

The general idea is to obtain control limit expressions by regressing the optimal limits that would be determined with a computer search against the terms in the control limit expression. (Regression analysis is covered in Chapter 12.) For an *np* chart, these terms could be *np* and $\sqrt{np(1-p)}$, although in most applications $\sqrt{1-p}$ should be very close to 1, so that factor might be dropped. The term "optimal" implies the use of some criterion for determining the best limits. Since control chart properties and performance are determined by the reciprocal of the sum of the two tail areas and by the reciprocals of the individual tail areas, it seems preferable to seek closeness to the reciprocals of the nominal tail areas rather than closeness to those tail areas. Accordingly, a reasonable objective would be to minimize $(|1/\text{LTA} - 1/.00135| + |1/\text{UTA} - 1/.00135|)$, with LTA and UTA denoting the lower tail area and upper tail area, respectively. Notice that this criterion also reflects the objective of having ARLs for parameter shifts in one direction that are close to the ARLs for shifts in the other direction.

Ryan and Schwertman (1997) stated that the control limits based on Eqs. (6.7) will generally produce the optimal limits when $p \doteq .01$. When p is larger than, say, .03, the limits could be obtained using the program given in Schwertman and Ryan (1997). Alternatively, an approach given by Shore (1999) might be used.

When p is smaller than .01, a p or *np* chart could be a poor choice. This is because there will be a concentration of probability at $X = 0$ when p is very small. For example, assume that $p = .00003$ and $n = 5000$. There will be no practical LCL since $P(X = 0) = .86$. Both the arcsin and Q-chart approaches produce a UCL of $X = 1$, as does an *np* chart with 3-sigma limits. The tail area above the UCL is .01018—almost eight times the nominal value. The best UCL is $X = 2$, which produces a tail area of .00050—about one-third of the nominal value. Thus, when p is very small, it will not be possible to construct a meaningful LCL, and it could also be difficult to produce a good UCL. One solution to this problem is to use a chart based on the geometric distribution (Section 3.54) or the negative binomial distribution (Section 3.5.5), but charts based on these distributions do have one troublesome flaw, as is discussed in Section 6.1.6.

Another problem with a p or *np* chart is that even if p is not so small as to preclude the use of a LCL, a very large sample size may be needed to detect small changes in p. This can be seen by examining Table 1 of Ryan and Schwertman (1997). Very large values of n may be impractical, so Schwertman and Ryan (1999) provide a modified approach.

6.1.4.4 ARL-Unbiased Charts

Acosta–Mejia (1999) introduced the concept of *ARL-unbiased* control charts. These are charts where the control limits are such that the in-control ARL is larger than any of the parameter change ARLs. Clearly this is desirable, but Shewhart charts and many other types of control charts do not have this property when applied to skewed distributions. For skewed distributions the ARL will *increase* initially as one moves from the in-control parameter value in the direction of the skewness. For example, for a p chart with the in-control value of p equal to .01, the ARL will be much higher at values of p slightly less than .01 that at .01. This problem is caused by the skewness of the binomial distribution — the smaller the value of p, the greater the skewness and the greater the problem.

One solution to this problem is to force the in-control value of the parameter to have the largest ARL. This can be accomplished in the following general way. For some random variable Y, obtain the appropriate function that represents the probability that Y plots outside the control limits when the process is in control. (Obviously the function will depend on LCL and UCL.) Then differentiate this function with respect to the appropriate parameter(s) and set the derivative(s) equal to zero. Using the in-control parameter value(s) and solving for LCL and UCL produces a pair of control limits for an ARL-unbiased chart, if the solution exists. Since the solution might not exist, a practitioner may have to settle for a chart that is nearly ARL-unbiased.

Since a computer program must be used to produce ARL-unbiased or nearly ARL-unbiased charts, the question arises as to when these limits will be different from the limits produced using the Ryan and Schwertman (1997) approach, which is much easier to use. Acosta–Mejia (1999) reports that "extensive numerical work" showed that the limits given by Eqs. (6.7) result in ARL-unbiased (or *nearly* ARL-unbiased) performance when $p < 0.03$. Accordingly, when this condition is met, and p is not so small that the LCL becomes meaningless, Eqs. (6.7) can be used instead of a computer program, provided that the user is satisfied with an in-control ARL of approximately 370.

When p is extremely small, we would not want to use a chart based on the binomial distribution anyway (for the reasons cited in Section 6.1.4.3), so the range of p for which Eqs. (6.7) produces ARL-unbiased or near ARL-unbiased control limits may be fairly small.

Another possible approach, discussed by Acosta–Mejia (1999) and Nelson (1997), is to use a runs test in place of the LCL; the motivation given in Nelson (1997) is that this is desirable when the calculated LCL is negative. A superior approach in some instances may be to simply forsake the binomial distribution and instead plot run lengths between nonconforming units. This approach is discussed in Section 6.1.6.

6.1.5 Variable Sample Size

The number of units that are examined may not be constant. If each sample is comprised of a full-day's production, we would certainly not expect production to be constant from day to day. (Of course, this would be 100% inspection, which would be impractical for very long.) On the other hand, management might later decide that it is too expensive to inspect 1000 units in each sample. Or perhaps there is a decision to inspect more than 1000 units.

Whatever the reason, it would be inadvisable to use an *np* chart when n varies, but a p chart could be used. The reason for this is as follows. With an *np* chart both the limits and the centerline will vary with n , whereas with a p chart only the limits will vary. Thus, with a "jagged centerline" on an *np* chart it would be difficult to compare adjacent points to see, for example, if there is evidence of a downward trend.

The (variable) limits for the p chart would be obtained from

$$\overline{p} \pm 3\sqrt{\frac{\overline{p}(1-\overline{p})}{n_i}} \tag{6.8}$$

where n_i is the size of the ith sample. Thus, the limits will vary as the n_i vary. This means that the limits would have to be computed for each different sample size. To circumvent this problem, some authors have suggested using the average sample size ($\overline{n}$) in place of n_i in Eq. (6.8) so as to produce constant limits, provided that the n_i vary only slightly. Such "approximate" limits can be a time saver if the charting is performed by hand but would be unnecessary if a computer is used. If such limits are used, it would be highly desirable to compute the exact (variable) limits and compare a point with those limits whenever the point plots close to one of the approximate limits. (It should be noted that $\overline{p}$ would be obtained by adding the number of nonconforming units for the different samples and dividing by the sum of the sample sizes.)

One suggested approach to the problem of variable sample size is to use a *standardized p chart*, with the plotted quantity being

$$Z_i = \frac{\hat{p}_i - p}{\sqrt{\frac{p(1-p)}{n_i}}}$$

with p typically estimated from historical data. (A standardized *np* chart would produce the same plotted quantity.)

Thus, the chart user would plot z-scores, with the control limits being -3 and $+3$. Although such transformed data are easier to relate to than the data produced by most transformations, there is naturally some loss of interpretability. Furthermore, the standardized chart does not overcome the problem of the LCL frequently not existing (whether n is constant or variable) with the untransformed data. That is, if an LCL does not exist on a p chart or an *np* chart, it will not be possible for Z_i to be less than -3, as the reader is asked to show in exercise 12. The use of a p chart with variable control limits seems preferable, provided that the LCL exists for each n_i.

Standardized p charts have also been advocated for short production runs (see Section 10.2), although their use in such situations also seems questionable.

6.1.6 Charts Based on the Geometric and Negative Binomial Distributions

As stated in Section 6.1.4.3, the use of a p chart when p is extremely small is inadvisable. Accordingly, for very small values of p it is preferable to plot the number of plotted points until k nonconforming units are observed. If $k = 1$, then the appropriate distribution is the *geometric distribution* (see Section 3.5.4).

It is inadvisable to use 3-sigma limits for a control chart based on the geometric distribution because the distribution is highly skewed. It is easy to obtain probability-limit forms for the LCL and UCL, and these were given by Quesenberry (1995), who suggested the use of a geometric control chart when the probability of observing a nonconforming unit is so small as to render a p chart useless. These limits are

$$\text{LCL} = 1 + \frac{\log(.99865)}{\log(1-p)} \qquad \text{UCL} = \frac{\log(.00135)}{\log(1-p)}$$

Although these limits correct for the extreme skewness of the geometric distribution by making the tail areas equal, a control chart based on these limits will not be remotely close to being ARL unbiased.

In an effort to overcome the skewness problem, Bourke (1991) suggested plotting the cumulative probability $\alpha_N = 1 - (1-p)^{N+1}$, $N = 0, 1, 2,$ associated with each observed value of the random variable and termed this an α chart. Here N denotes the number of units observed *between* nonconforming units. The α_N will have approximately a uniform (0, 1) distribution since the percentiles of any cumulative mass function are equally likely.

There are some nuances that should be understood, however. For example, since N is discrete, α_N cannot be a continuous random variable, which means that probability limits for a desired probability are not possible. Furthermore, there will not be a LCL unless $p \leq LTA^*$, with LTA* denoting the desired LTA (lower-tailed area) value. If, for example, $p = .0015$, the smallest possible value for the LTA is $p = .0015$. Failure to recognize this relationship could result in an in-control ARL that is much shorter than desired, so the relationship might be used in establishing a range of p values for which an α chart would be used.

An α chart has some intuitive appeal, but it might be best to use the chart as a supplementary chart to which runs rules and other tests could be applied. Management would naturally be more interested in knowing the number of units observed between nonconforming units than to know the cumulative probability associated with the observed value.

Another solution to the problem—which seems preferable—is to create an ARL-unbiased chart based on the geometric distribution using the methods of Acosta–Mejia (1999). As was stated in Section 6.1.4.4, however, such an approach requires the use of a computer program, so a practitioner might wish to consider using N much greater than 1, since the skewness decreases as N increases.

Another reason for using $N > 1$ is as follows. For $p > .00101$, the optimal LCL will be 2 and the LTA $= p$. Thus, for values of p much greater than .001 the in-control ARL could be deemed too short, as the probability of a false signal that p has decreased increases linearly with p. (Of course, if p is an order of magnitude larger than .001, then a p chart could be used.) Thus, for such values of p in the vicinity of, say, .005, it would be preferable to use a control chart for the number of units inspected when the kth nonconforming unit is observed, with $k > 1$. Another motivation for using $k > 1$ is that there will be better sensitivity in detecting changes in p relative to a chart using $k = 1$.

If we use $k > 1$, the appropriate distribution is the *negative binomial distribution* (see Section 3.5.5). How large should k be? The coefficient of skewness for the negative binomial distribution is $(2 - p)/\sqrt{k(1 - p)}$. Unlike the binomial distribution, the skewness *increases* as p increases from a small value for a fixed value of k. The skewness will not be sensitive to changes in p when p is small, but it will be sensitive to the value of k. Clearly, when p is very small, the skewness coefficient will be approximately $2/\sqrt{k}$. Therefore, the skewness can be decreased considerably by increasing the value of k. We would want k to be large enough so that optimal limits could be determined in the spirit of Ryan and Schwertman (1997) and still have a chart that is ARL unbiased, or at least approximately ARL unbiased.

Acosta–Mejia (1999) gave an example for which the ARL-unbiased approach and the Ryan–Schwertman approach produce the same limits. In that example $n = 1000$ and $p = .01$. The coefficient of skewness for the binomial distribution is $(1 - 2p)/\sqrt{np(1 - p)}$, so the skewness equals 0.31 for this combination of n and p. In order to have approximately the same skewness for the negative binomial distribution, we would have to set $k = 42$, which would be impractical unless a computer program were being used. For a given value of k, a regression approach similar to that described in Ryan and Schwertman (1997) could be used to determine the control limits.

For this example, the expected number of nonconforming units per sample is 10, so when the processes are in control, no point would be plotted for most of the samples. Many practitioners could find this discomforting. If so, such practitioners could use the same general approach as given in Acosta–Mejia (1999).

6.1.7 Overdispersion

One important point to keep in mind when constructing charts for nonconforming units, regardless of the chart that is used, is that *overdispersion* may exist. This means that the actual variance of X (or of $\hat{p}$) is greater than the variance obtained using the binomial distribution.

Overdispersion can result from various causes, including (1) a nonconstant p and/or (2) autocorrelation. The latter has been discussed extensively in the literature for measurement data but has received comparatively little attention for attribute data. Would we expect the probability of a unit of production being nonconforming to be independent of whether the preceding unit was nonconforming? Probably not.

If p is not constant, but rather has its own variance, σ_p^2, the variance of X is not the binomial variance $np(1-p)$, but rather is $np(1-p)+n(n-1)\sigma_p^2$ [see the discussion in Box and Luceño (1997, p. 43) or Box et al. (1978, p. 136)]. Obviously this will be more of a problem for large sample sizes than for small sample sizes. A simple way to see if there is evidence of a nonconstant p is to compare $n\overline{p}(1-\overline{p})$ with $\sum(X-\overline{X})^2/n$, with the two computations performed over at least a few dozen samples. The two values will almost certainly not be equal but should not differ by an order of magnitude.

A similar approach, given by Snedecor and Cochran (1974, p. 202), is to compute

$$X = \sum_{i=1}^{c} \frac{(p_i - \overline{p})^2}{\overline{p}\,\overline{q}/n} = \frac{(c-1)s_p^2}{\overline{p}\,\overline{q}/n}$$

with $\overline{q} = 1 - \overline{p}$. This statistic, which is obviously $c-1$ times the sampling variance of p divided by the binomial variance, has approximately a chi-square distribution with $c-1$ degrees of freedom

Of course, we can also use the two standard control charts that have been discussed previously in this chapter to determine if p is nonconstant—that being the purpose of the charts! The control limits will be inflated if p varies, however, which could make it difficult to determine all instances of a nonconstant p.

In general, the binomial distribution could be inappropriate for control charting for various reasons. Ramirez and Cantell (1997) and Randall, Ramirez, and Taam (1996) consider the overdispersion caused by spatial considerations. See also Quarshie and Shindo (1995), Luceño and de Ceballos (1995), and Lai, Govindaraju, and Xie (1998).

6.2 CHARTS FOR NONCONFORMITIES

As stated at the beginning of the chapter, a unit of production can have one or more *nonconformities* without being labeled a nonconforming unit. Similarly, nonconformities can occur in nonmanufacturing applications of control charts. An example of a nonconformity that would not necessarily produce a nonconforming unit would be a small scratch on a plate of glass.

As with charts for nonconforming units, there are multiple charts available for use in trying to control (and preferably reduce) nonconformities and different methods that can be used for obtaining the control limits for each chart. These methods are discussed in subsequent sections.

6.2.1 *c* Chart

A c chart can be used to control the number of nonconformities per inspection unit, where the latter may be comprised of one or more than one physical unit.

For example, the inspection unit may consist of a single bolt or a container of bolts. The chart can be used for controlling a single type of nonconformity or for controlling all types of nonconformities without distinguishing between types. (Another chart that can be used for controlling multiple types of nonconformities is the demerit chart, D chart, which is discussed at the end of this chapter.)

As mentioned earlier in the chapter, a physical unit could have one or more (generally minor) nonconformities without being labeled (and discarded) as a nonconforming unit. Examples would include a minor blemish on a tire and wrapping on a food item that is not fastened properly. Ideally, such nonconformities should not occur, but elimination of all types of imperfections might be too expensive to be practical. [This relates to Deming's contention that product quality should be just good enough to meet the demands of the market place and also relates to Six Sigma programs (see Section 17.3), which are driven in part by customer requirements.]

The standard approach is to use 3-sigma limits. The 3-sigma limits for a c chart, where c represents the number of nonconformities, are obtained from

$$\overline{c} \pm 3\sqrt{\overline{c}} \tag{6.9}$$

where the adequacy of the normal approximation to the Poisson distribution is assumed, in addition to the appropriateness of the Poisson distribution itself. (Here, c is assumed to have a Poisson distribution.) Specifically, the probability of observing a nonconformity in the inspection unit should be small, but a large number of nonconformities should be theoretically possible. The size of the inspection unit should also be constant over time.

The mean and variance of the Poisson distribution are the same, and $\overline{c}$ is the estimate of each. The "± 3" implies the use of a normal approximation, as was the case with the p and np charts. With those two charts, the normal approximation to the binomial distribution was assumed to be adequate, although it was shown that this will often not be true. The normal approximation to the Poisson distribution has been assumed to be adequate when the mean of the Poisson distribution is at least 5, but that really applies to estimating Poisson probabilities rather than to the related, but distinct, problem of determining control limits.

When applied to the c chart, this requires that $\overline{c}$ should be at least 5. Although this requirement will often be met in practice, control limits obtained from Eq. (6.9) should not automatically be used when it is met. In particular, when $5 \leq \overline{c} < 9$, zero would be used for the lower limit since $\overline{c} - 3\sqrt{\overline{c}}$ would be negative. Thus, there would not be a lower limit that could be used as a possible indicator of a reduction in the number of nonconformities. (Recall that this is the same type of problem that can exist with p and np charts.) The tail areas beyond the control limits can also be off considerably from the assumed (nominal) areas of .00135. For example, when $\overline{c} = 10$, the lower limit is approximately 0.5, but $P(X = 0) = .000045$ when the Poisson mean is 10, which is 1/30th of the assumed area. If the objective was to come as close as possible to .00135, the lower limit would be set at 2 and the tail area would be .00050.

This type of problem can be avoided by using probability limits obtained from tables such as those given Kitagawa (1952) and Molina (1973) or by using one of the methods given in the following sections.

6.2.2 Transforming Poisson Data

A transformation analogous to the transformations discussed for the p chart can be used. One such transformation is to let $Y = 2\sqrt{c}$, which will be, for a large sample, approximately normally distributed with mean $2\sqrt{\lambda}$ and variance 1, where λ is the mean of the Poisson distribution. Here, Y would be plotted on the chart and the control limits would be obtained as $\overline{y} \pm 3$. Similar transformations have been proposed by Anscombe (1948) and Freeman and Tukey (1950). When applied to a c chart, the transformations would be $y_1 = 2\sqrt{c + 3/8}$ and $y_2 = \sqrt{c} + \sqrt{c+1}$, respectively, and the control limits would be obtained as $\overline{y}_1 \pm 3$ and $\overline{y}_2 \pm 3$, respectively.

There is obviously very little difference between these three transformations. The following considerations should be made in choosing between the use of probability limits and some type of transformation. Probability limits allow the user to work with the data on the original scale, but they do require special tables. Limits obtained with transformed variables are easier to construct, but they are for data on a different scale.

6.2.3 Illustrative Example

The regular (untransformed) c chart and one of the three transformations will be illustrated using the data in Table 6.5. It can be determined that $\overline{c} = 189/25 = 7.56$, so the control limits for the regular c chart are obtained from

$$7.56 \pm 3\sqrt{7.56} = 7.56 \pm 8.25$$

The control limits are thus 0 and 15.81. The chart is shown in Figure 6.3. It can be observed that one point is above the UCL, and there is evidence of a downward trend. Notice, however, that there is no LCL (since $\overline{c} < 9$), so other methods would have to be used to determine if there has indeed been a true downward shift in the number of nonconformities.

It is interesting to note that if any of the three transformations had been used, the fifth point would have been inside the UCL, but the twenty-third point (bolt 1 on November 19) would have been below the LCL. The results for each of the transformations are given in Table 6.6. In particular, there is very little difference between the values for the last two transformations.

Can the use of one of these transformations instead of the regular c chart be justified in this instance? The answer is yes. If interest had centered upon determining the control limits in such a way that they are as close as possible to being .00135 probability limits, the limits would have been set in such a way that $c \geq 18$ would fall above the UCL and $c < 1$ (i.e., $c = 0$) would be below

TABLE 6.5 Nonconformity Data

Number of Nonconformities (c)	Bolt Number	Date
9	1	November 9
15	2	
11	3	
8	4	
17	5	
11	1	November 10
5	2	
11	3	
13	1	November 16
7	2	
10	3	
12	4	
4	5	
3	1	November 17
7	2	
2	3	
3	4	
3	1	November 18
6	2	
2	3	
7	4	
9	5	
1	1	November 19
5	2	
8	3	

Source: Ford Motor Co. (1985). *Continuing Process Control and Process Capability Improvement*, p. 46a. Plymouth, MI: Ford Motor Co., Statistical Methods Publications. Copyright © 1985.

TABLE 6.6 Nonconformity Data from Table 6.5 Using Transformations

Transformation	Value of 5th Point	Value of 23rd Point	LCL	UCL
$y = 2\sqrt{c}$	8.246	2.0	2.257	8.257
$y = 2\sqrt{c + 3/8}$	8.337	2.345	2.415	8.415
$y = \sqrt{c} + \sqrt{c+1}$	8.366	2.414	2.459	8.459

the LCL. It can be determined using the results in Table 6.6 that the UCL will be exceeded for each of the transformations when $c \geq 18$, although $c \leq 1$ will be below the LCL. It can be shown that $P(C \geq 16) \doteq .005$ when the mean of the Poisson distribution is 7.56. Therefore, the actual probability is roughly four times the nominal probability. Thus, although the requirement of $\overline{c} \geq 5$ is met

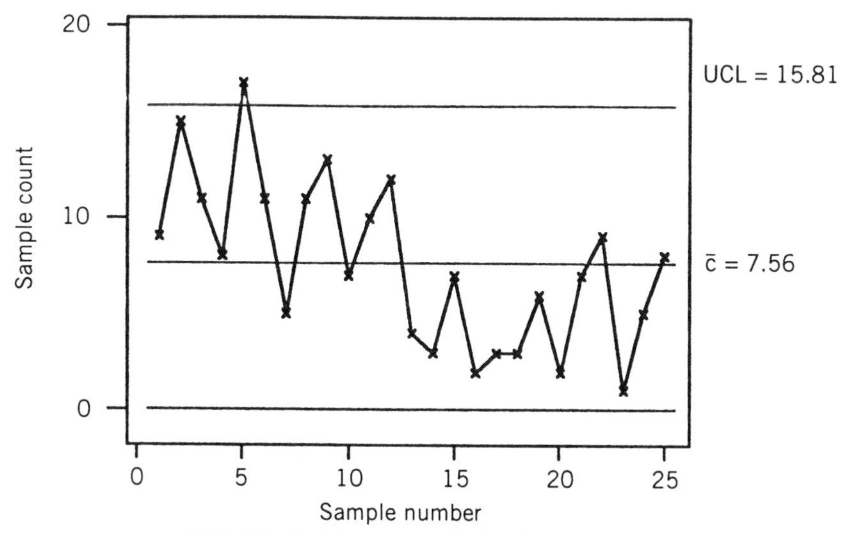

FIGURE 6.3 The c chart with data from Table 6.5.

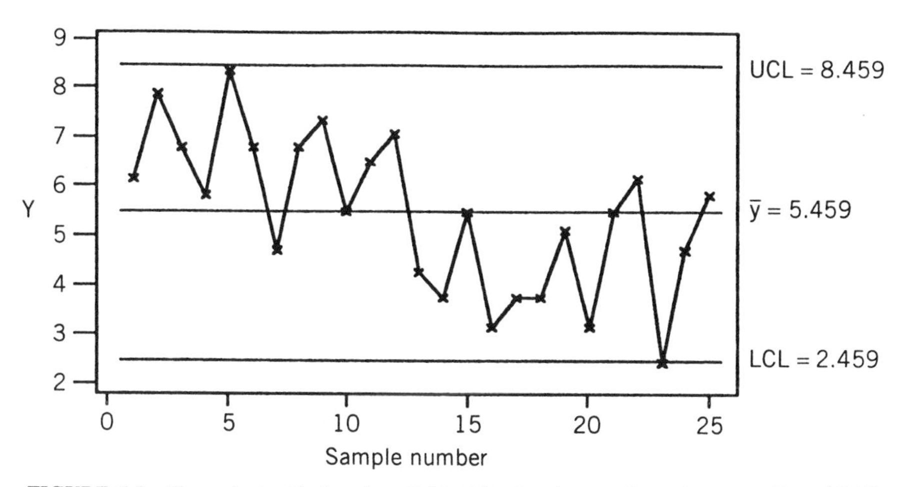

FIGURE 6.4 The c chart with data from Table 6.5 using the transformation $y = \sqrt{c} + \sqrt{c+1}$.

for this example, the UCL is nevertheless off by about two units relative to the probability limit. (Recall that this is the same type of problem that occurred with the p chart.) In general, any of the transformations can be expected to produce control limits that are closer to the desired probability limits than those obtained from use of the regular c chart. The c chart obtained from using the transformation $y = \sqrt{c} + \sqrt{c+1}$ is shown in Figure 6.4. The configuration of points is essentially the same as for the regular c chart in Figure 6.3, as should be expected.

One way to avoid the need for transformations or adjustment factors would be to define the sample size so that it would consist of more than one inspection unit. This will naturally cause $\bar{c}$ to be larger than it would be if the sample size equaled one inspection unit. The chart user would then have to be satisfied with charting the number of nonconformities per sample, however, and $\bar{c}$ would have to be quite large before the tail areas would be close to .00135. Consequently, the use of a transformation or adjustment factor seems preferable.

Some authors (see, e.g., Grant and Leavenworth, 1980, p. 256) have pointed out that there are many instances in which the requirements for using the Poisson distribution will not be strictly met. For example, the area of opportunity for the occurrence of nonconformities may not be constant over time, or a distribution similar to the Poisson might be more appropriate. When the first requirement is violated, a *u* chart can be used (to be discussed later). When a *c* chart is being maintained, where *c* is the count of different types of nonconformities, *c* will not necessarily have a Poisson distribution. [See Grant and Leavenworth (1980, p. 259) for an extended discussion.] In particular, if different types of nonconformities with different opportunities for occurrence comprise the value of *c*, it can be shown that *c* will not have a Poisson distribution, even if each type of nonconformity has a Poisson distribution.

6.2.4 Regression-Based Limits

Ryan and Schwertman (1997) also gave *c* chart control limits that were based on the same loss function as was mentioned in Section 6.1.4.3. The control limits that minimize this loss function for selected values of λ are given in Table 6.7. [Although these are the "optimal limits," it could be shown that most of the

TABLE 6.7 Optimal *c*-Chart Limits for Selected Values of λ

λ	LCL	UCL
5	1	12
6	1	14
7	1	16
8	2	17
9	2	19
10	3	20
11	3	22
12	4	23
13	4	24
14	5	26
15	6	27
20	9	34
25	12	41
30	16	47

in-control ARLs differ considerably from the nominal value of 740.7(= 1/.00135) since the Poisson random variable is discrete.] Since λ could have any value along a continuum, it is desirable to seek an approximation that will reproduce the values in Table 6.7 and that can also be used for non-integer values.

When the optimal control limits given in a more extensive version of Table 6.7 [with $\lambda = 5(.25)50$] are regressed against λ and $\sqrt{\lambda}$, the following control limit expressions are produced:

$$\begin{aligned} \text{UCL} &= 0.6195 + 1.0052\lambda + 2.983\sqrt{\lambda} \\ \text{LCL} &= 2.9529 + 1.01956\lambda - 3.2729\sqrt{\lambda} \end{aligned} \tag{6.10}$$

These expressions reproduce all of the control limits given in Table 6.7, and is the best known method for producing c-chart limits when λ is between 5 and 50. Of course, λ will generally have to be estimated, and the value of $\bar{c}$ would be its estimate.

For control limits produced by Eqs. (6.10), there will be no LCL when $\lambda <$ 4.07 since the computed value for the LCL will round to at most zero for such values of λ. When λ is only slightly larger than 4.07, the LTA will be more than 10 times the nominal value. Specifically, this occurs for $\lambda = 4.1, 4.2, 4.3$. Thus, even though the LCL is the optimal LCL in terms of the loss function that we are using, the discreteness of the Poisson random variable creates major problems when λ is small. Therefore, when λ is only slightly larger than 4.07, it would be wise for the control chart user to define the sample size in such a way as to avoid such small values. Examples of this are given in Ryan and Schwertman (1997).

Increasing the sample size will also increase the likelihood that control limits obtained from Eqs. (6.10) will produce an ARL-unbiased chart. If not, the approach of Mejia (1999) could be applied to a c chart, if desired.

If we apply the control limits given by Eqs. (6.10) to the data in Table 6.5, acting as if $\lambda = 7.56$, we obtain LCL = 1.66 and UCL = 16.42. We may observe that the application of these limits to the Table 6.5 data differs slightly from the results shown in Table 6.6 when any of the transformations are used. Specifically, although the decision regarding point 23 is the same, point 5 is slightly above the regression-based UCL but is slightly below the UCL when any of the square root transformations are used.

The superiority of the regression-based approach was demonstrated in Ryan and Schwertman (1997). This example shows that the results can differ for actual data. Which limits lead to the correct interpretation? The actual control chart is shown in Ford Motor Company (1985), and on the chart it is noted that the use of heavy-gauge blue material was initiated at bolt 5. There was a change at that point and we want the limits to indicate that. Thus, the regression-based limits are superior to the limits for any of the square root transformations in this example, and the former are also clearly superior to the 3-sigma limits since the downward trend eventually produces a point below the regression-based LCL, whereas there is no 3-sigma LCL.

6.2.5 *u* Chart

When the area of opportunity for the occurrence of nonconformities does not remain constant, a u chart should be used rather than a c chart, where $u = c/n$ and n is the number of inspection units from which c is obtained. When n is constant, either a u chart or a c chart can be used, but when n varies, a u chart must be used. (It can be noted that u does not have a Poisson distribution, however, so it does differ from c in that respect.)

The control limits for a u chart are related to the limits for a c chart. Specifically, $\text{Var}(c/n) = (1/n^2)\text{Var}(c)$, which is estimated by $(1/n^2)\bar{c}$. The latter is obviously equal to $(\bar{c}/n)/n$, which equals $\bar{u}/n$. Thus, the control limits are obtained from

$$\bar{u} \pm 3\sqrt{\frac{\bar{u}}{n}} \tag{6.11}$$

where $\bar{u} = \bar{c}/n$ and n is assumed to be constant. When the latter is true, the u-chart limits are simply the c-chart limits divided by n.

When the sample size varies, n_i would replace n in Eq. (6.11) and $\bar{u}$ would be calculated as $\bar{u} = \sum c_i / \sum n_i$. The control limits will thus vary as n_i varies, but the center line will still be constant. When the n_i do not vary greatly, the average sample size, $\bar{n}$, might be used in place of n. One suggested rule-of-thumb is that this approach can be used as long as no individual sample size differs from the average by more than 25% (Ford Motor Company, 1985). As was discussed for a p chart, when a point falls near a control limit based on $\bar{n}$, the exact limit using n_i should be calculated to see if the point falls inside or outside the exact limit.

A u chart can thus be used under each of the following conditions:

1. As a substitute for a c chart when the (constant) sample size contains more than one inspection unit and there is a desire to chart the number of nonconformities per inspection unit.
2. When the sample size varies so that a c chart could not be used. The control limits would then be
   a. variable limits using individual sample sizes or
   b. constant limits using the average sample size when the sample sizes do not differ greatly.

It should be noted that a u chart could be produced when a transformation is used for c. For example, if $y = \sqrt{c} + \sqrt{c+1}$ and n is constant, the control limits would then be obtained from

$$\frac{\bar{y}}{n} \pm \frac{3}{n}$$

where $\bar{u} = \bar{y}/n$ is the average number of "transformed nonconformities" per inspection unit. Similarly, when the sample size varies, the exact variable limits would be obtained from

$$\frac{\sum y}{\sum n_i} \pm \frac{3}{n_i}$$

The motivation for using transformations in conjunction with a u chart is the same as the motivation for using them with a c chart. Specifically, if the normal approximation is not adequate for nonconformity data, it is not going to be adequate for nonconformities per unit.

6.2.5.1 Regression-Based Limits

Regression-based limits can also be produced for a u chart. Therefore, the control limits for a u chart that utilize the improved c-chart limits would have to be obtained as follows. Let the UCL for a c chart using this approximation be represented by $\bar{c} + k_1\sqrt{\bar{c}}$ and the LCL be represented by $\bar{c} - k_2\sqrt{\bar{c}}$. The values of k_1 and k_2 can be easily solved for after the limits are computed using the approximations. Then, for variable n_i, the control limits for the u chart would be given by $\text{UCL} = \bar{u} + k_1\sqrt{\bar{u}/n_i}$ and $\text{LCL} = \bar{u} - k_2\sqrt{\bar{u}/n_i}$.

Consider the following example. A company has been having difficulties with a minor tuning problem for a particular type of radio, with 24 (small) radios in each box that is shipped by the manufacturer. Assume that quality assurance personnel typically record the number of nonconformities per 100 boxes sampled, but we wish to chart the number per box. Assume that 20,000 boxes were sampled over a 6-month period at times when the process was considered to be performing typically (i.e., "in control"), and 1264 tuning problems were recorded.

If a c chart were constructed, we would have $\bar{c} = 1264/200 = 6.32$. Using this as our estimate of λ, the regression-based limits would be $\text{UCL} = 14.47$, and $\text{LCL} = 1.17$. Performing the appropriate algebra produces $k_1 = 3.24$ and $k_2 = 2.04$. It then follows that the u-chart limits are $\text{UCL} = 6.32/100 + 3.24\sqrt{(6.32/100)/100} = 0.89$ and $\text{LCL} = 6.32/100 - 2.04\sqrt{(6.32/100)/100} = 0.47$.

When the n_i are equal, this has the effect of multiplying each of the c chart limits by $1/n$, as stated previously. Accordingly, the accuracy of the u chart limits will depend upon $1/n$ and the accuracy of the corresponding c chart limits. This suggests that n be chosen to be as large as is practical, which in turn will make λ large. As shown in Ryan and Schwertman (1997), the approximation improves as λ increases.

6.2.6 Overdispersion

Just because nonconformities are to be charted does not necessarily mean that the Poisson distribution should be used in producing the control limits. Friedman (1993) discusses the use of u charts in integrated circuit fabrication and shows that the in-control ARL for a u chart can be less than 10 due to clustering. Consequently, it is important that the assumption of a Poisson process be checked (see Friedman, 1993, for additional details).

A simple, but somewhat crude, method for comparing the observed sample variance with the Poisson variance was given by Snedecor and Cochran (1974, p. 198). See also Hawkins and Olwell (1998, p. 119). Using a goodness-of-fit approach, $X = \sum_{i=1}^{k}(c_i - \bar{c})^2/\bar{c}$ is distributed approximately as chi-square with

$k - 1$ degrees of freedom, where k is the number of independent samples, with c_i denoting the count in each sample. Since the sample variance of c is $s_c^2 = \sum_{i=1}^{k}(c_i - \overline{c})^2/(k-1)$, and $\overline{c}$ is the estimator of both the Poisson mean and Poisson variance, it follows that $X = (k-1)$ (sample variance)/(estimator of Poisson variance). To illustrate, if a sample of $k = 10$ counts is 5, 8, 7, 2, 4, 5, 3, 2, 5, 6, then $\overline{c} = 4.70$, $s_c^2 = 4.01$, and $X = 9(4.01)/(4.70) = 7.68$. Since $P(X < 7.68) = .4332$, using the chi-square approximation, there is no reason to doubt that the sample variance can be represented by the Poisson variance. How good is this approximation? Well, the approximation is based on the assumed adequacy of the normal approximation to the Poisson distribution. That approximation is generally assumed to be adequate when the Poisson mean is at least 5, as stated previously in Section 6.2.1, although we have seen that for control chart purposes this rule-of-thumb is inadequate. Furthermore, here the mean is only 4.70, but since the mean and variance differ very little, a formal test seems unnecessary.

In other instances, however, the appropriate conclusion may not be quite so obvious. Other tests for Poisson overdispersion are described by Bohning (1994) and Lee, Park, and Kim (1995).

6.2.7 *D* Chart

It was previously stated that a c chart can be used to chart a single type of nonconformity or to chart the sum of different types of nonconformities. If the latter is being used, each type of nonconformity is implicity assigned the same weight. This will be unsuitable in some applications because some nonconformities could be very minor. Consequently, many companies have used demerit charts (D charts) in which different weights (demerits) can be assigned to nonconformities of differing severity. The idea is due to Dodge (1928).

Assume, for example, that there are three different types of nonconformities, expressed as 1, 2, and 3, with the number of nonconformities of each type expressed as c_1, c_2, and c_3. If the corresponding weights are expressed as w_1, w_2, and w_3, and D represents the number of demerits per inspection unit, we then have

$$D = w_1 c_1 + w_2 c_2 + w_3 c_3$$

Assuming that $w_i \geq 1$, $i = 1, 2, 3$, D will not have a Poisson distribution unless all $w_i = 1$, which could be rather meaningless. Nevertheless, control chart limits can be easily constructed (recall that u does not have a Poisson distribution, either).

In general, if we assume that there are k different types of nonconformities, and the nonconformities are independent, we then have

$$\text{Var}(D) = \sum_{i=1}^{k} w_i^2 \lambda_i$$

where λ_i is the mean of c_i. This variance would be estimated by $\sum_{i=1}^{k} w_i^2 \overline{c}_i$, where $\overline{c}_i$ is obtained by averaging over the n inspection units (which may or may not equal the number of samples). Specifically, if we let c_{ij} represent the number of nonconformities of type i in inspection unit j, then

$$\overline{c}_i = \frac{\sum_{j=1}^{n} c_{ij}}{n}$$

The 3-sigma control limits are obtained from

$$\overline{D} \pm 3\sqrt{\sum_{i=1}^{k} w_i^2 \overline{c}_i}$$

where

$$\overline{D} = \frac{\sum_{j=1}^{n} \sum_{i=1}^{k} w_i c_{ij}}{n} = \frac{\sum_{j=1}^{n} D_j}{n}$$

and $D_j = \sum_{i=1}^{k} w_i c_{ij}$ is the number of demerits for the jth inspection unit.

This would be the counterpart to a c chart. If each sample contains more than one inspection unit and it is desired to chart the number of demerits per inspection unit, then the counterpart to the u chart would be produced. This idea is due to Dodge and Torrey (1956). Specifically,

$$D_u = \sum_{i=1}^{k} w_i u_i$$

where $u_i = c_i / n_l$ is the number of nonconformities of type i per inspection unit in a sample that contains n_l such units.

If m samples are available for computing the control limits, it follows that we would compute

$$\overline{D}_u = \sum_{i=1}^{k} w_i \overline{u}_i$$

where

$$\overline{u}_i = \frac{\sum_{l=1}^{m} c_{il}}{\sum_{l=1}^{m} n_l}$$

Here c_{il} denotes the number of nonconformities of type i in sample l, which contains n_l inspection units.

Assuming again that the k nonconformities are independent, we have

$$\mathrm{Var}(D_u) = \frac{1}{n_l^2} \sum_{i=1}^{k} w_i^2 \lambda_i$$

Since c_i is assumed to have a Poisson distribution with mean (and variance) of λ_i, a weighted average of the c_{il} (assuming that the n_l differ) would be used in estimating λ_i.

Thus,

$$\bar{c}_i = \frac{\sum_{l=1}^{m} n_l c_{il}}{\sum_{l=1}^{m} n_l}$$

where the average would be an unweighted average if the n_l are equal.

The 3-sigma control limits are obtained from

$$\overline{D}_u \pm \frac{3}{n_l}\sqrt{\sum_{i=1}^{k} w_i^2 \bar{c}_i}$$

6.2.7.1 *Probability-Type D-Chart Limits*

As discussed by Jones, Woodall, and Conerly (1999), the normal distribution may not be an appropriate model, depending on the values of the λ_i and the w_i. Accordingly, it would be safer to base the control limits on the exact distribution of D. Unfortunately, however the distribution of D is not easy to determine. Jones et al. (1999) used a characteristic function approach and numerically evaluated the distribution of D. [See, e.g., Lukacs (1970) for information on characteristic functions.] They assumed that the sample size was constant.

In an example given by Dodge and Torrey (1956), the in-control ARL was found to be only 104 when 3-sigma limits are used. Thus, the use of 3-sigma limits will probably often have a very short in-control ARL due to the skewness of the distribution of D.

Unfortunately, tables could not be easily constructed since the distribution of D depends on the λ_i, the w_i, and the sample size. Consequently, the user of a D-chart with probability limits would need to numerically evaluate the distribution for a given combination of the weights and Poisson means, and it would be highly desirable for the sample size to be constant. The possibility of having a very short in-control ARL from using 3-sigma limits should motivate such effort, however. The assumption of the appropriateness of the Poisson distribution should, of course, be assessed before proceeding.

6.3 SUMMARY

Several different types of control charts for attribute data were presented in this chapter. It was observed that the usual Shewhart charts will often not have a lower control limit and will often have tail areas that differ greatly.

Some new charts were proposed (for the first time in a textbook) to remedy these problems. In particular, regression-based limits and ARL-unbiased charts are among the most useful approaches.

We need to keep in mind, however, that the use of these charts will not help improve the quality of products and processes. They will merely show the extent to which progress is being achieved. Obviously this is an important function, however.

See Woodall (1997) for a recent bibliography of papers on control charts for attribute data.

REFERENCES

Acosta-Mejia, C. A. (1999). Improved *P* charts to monitor process quality. *IIE Transactions 31*(6): 509–516.

Anscombe, F. J. (1948). The transformation of Poisson, binomial, and negative binomial data. *Biometrika 35:* 246–254.

Bohning, D. (1994). A note on a test for Poisson overdispersion. *Biometrika 81*(2): 418–419.

Box, G. and A. Luceño (1997). *Statistical Control by Monitoring and Feedback Adjustment.* New York: Wiley.

Box, G. E. P., W. G. Hunter, and J. S. Hunter (1978). *Statistics for Experimenters.* New York: Wiley.

Bourke, P. D. (1991). Detecting a shift in fraction nonconforming using run-length control charts with 100% inspection. *Journal of Quality Technology 23:* 225–238.

Brownlee, K. A. (1960). *Statistical Theory and Methodology in Science and Engineering.* New York: Wiley.

Burr, I. W. (1979). *Elementary Statistical Quality Control.* New York: Dekker.

Dodge, H. F. (1928). A method of rating a manufactured product. *Bell System Technical Journal 7:* 350–368.

Dodge, H. F. and M. N. Torrey (1956). A check inspection and demerit rating plan. *Industrial Quality Control 13*(1) (reprinted in *Journal of Quality Technology* 9(3): 146–153, 1977).

Duncan, A. J. (1986). *Quality Control and Industrial Statistics*, 5th ed. Homewood, IL: Irwin.

Ehrenfeld, S. and S. B. Littauer (1964). *Introduction to Statistical Method.* New York: McGraw-Hill.

Ford Motor Company (1985). *Continuing Process Control and Process Capability Improvement.* (Available from Ford Motor Company, Statistical Methods Publications, P.O. Box 1000, Plymouth, MI 48170.)

Freeman, M. F. and J. W. Tukey (1950). Transformations related to the angular and the square root. *Annals of Mathematical Statistics 21:* 607–611.

Friedman, D. J. (1993). Some considerations in the use of quality control techniques in integrated circuit fabrication. *International Statistical Review 61:* 97–107.

Grant, E. L. and R. S. Leavenworth (1980). *Statistical Quality Control*, 5th ed. New York: McGraw-Hill.

Hahn, G. J. (1981). Tolerance intervals for Poisson and binomial variables. *Journal of Quality Technology 13*(2): 100–110.

Hald, A. (1952). *Statistical Theory with Engineering Applications.* New York: Wiley.

Hawkins, D. M. (1987). Self-starting CUSUM charts for location and scale. *The Statistician 36:* 299–315.

Hawkins, D. M. and D. H. Olwell (1998). *Cumulative Sum Charts and Charting for Quality Improvement.* New York: Springer-Verlag.

Johnson, N. L. and S. Kotz (1969). *Distributions in Statistics, Vol 1: Discrete Distributions*. New York: Wiley.

Johnson, N. L. and F. C. Leone (1977). *Statistics and Experimental Design in Engineering and Physical Sciences*, Vol. 1, 2nd ed. New York: Wiley.

Jones, L. A., W. H. Woodall, and M. D. Conerly (1999). Exact properties of demerit control charts. *Journal of Quality Technology 31*(2): 207–216.

Kitagawa, T. (1952). *Tables of Poisson Distribution.* Tokyo, Japan: Baifukan.

Kittliz, R. G. (1979). Poisson distribution and textile mill problems. *ASQC Technical Conference Transactions*, pp. 126–133.

Lai, C. D., K. Govindaraju, and M. Xie (1998). Effects of correlation on fraction nonconforming statistical process control procedures. *Journal of Applied Statistics 25*(4): 535–543.

Lee, S., C. Park, and B. S. Kim (1995). Tests for detecting overdispersion in Poisson models. *Communications in Statistics—Theory and Methods 24*(9): 2405–2420.

Lukacs, E. (1970). *Characteristic Functions*. London: Griffen.

Lucas, J. M. (1985). Counted data CUSUM's. *Technometrics 27*(2): 129–144.

Lucas, J. M. and R. B. Crosier (1982). Fast initial response for CUSUM quality control schemes: Give your CUSUM a head start. *Technometrics 24*(5): 199–205.

Luceño, A. and F. de Ceballos (1995). Describing extra-binomial variation with partially correlated models. *Communications in Statistics—Theory and Methods 24*: 1637–1653.

Molina, E. C. (1973). *Poisson's Exponential Binomial Limit*. Huntington, NY: Krieger.

Montgomery, D. C. (1996) *Introduction to Statistical Quality Control*, 3rd ed. New York: Wiley.

Nelson, L. S. (1983). Transformations for attribute data. *Journal of Quality Technology 15*(1): 55–56.

Nelson, L. S. (1997). Supplementary runs tests for *np* control charts. *Journal of Quality Technology 29*(2): 225–226.

Quarshie, L. B. and H. Shindo (1995). A control chart for attributes and its applications. *ASQC Quality Congress Transactions*, 378–388.

Quesenberry, C. P. (1991a). SPC Q charts for start-up processes and short and long runs. *Journal of Quality Technology 23:* 213–224.

Quesenberry, C. P. (1991b). SPC Q charts for a binomial parameter: Short or long runs. *Journal of Quality Technology 23:* 239–246.

Quesenberry, C. P. (1991c). SPC Q charts for a Poisson parameter λ: Short or long runs. *Journal of Quality Technology 23:* 296–303.

Quesenberry, C. P. (1995). Geometric Q-charts for high quality processes. *Journal of Quality Technology 27:* 304–315.

Quesenberry, C. P. (1997). *SPC Methods for Quality Improvement.* New York: Wiley.

Ramírez, J. G. and B. Cantell (1997). An analysis of a semiconductor experiment using yield and spatial information. *Quality and Reliability Engineering International 13:* 35–46.

Randall, S. C., J. G. Ramírez, and W. Taam (1996). Process monitoring in integrated fabrication using both yield and spatial statistics. *Quality and Reliability Engineering International 12:* 195–202.

Ryan, T. P. and N. C. Schwertman (1997). Optimal limits for attributes control charts. *Journal of Quality Technology 29*(1): 86–98.

Schader, M. and F. Schmid (1989). Two rules of thumb for the approximation of the binomial distribution by the normal distribution. *The American Statistician 43:* 23–24.

Schwertman, N. C. and T. P. Ryan (1997). Implementing optimal attributes control charts. *Journal of Quality Technology 29*(1): 99–104.

Schwertman, N. C. and T. P. Ryan (1999). Using dual *np*-charts to detect changes. *Quality and Reliability Engineering International 15*: 317–320.

Shore, H. (1999). General control charts for attributes. *IIE Transactions* (to appear).

Snedecor, G. W. and W. G. Cochran (1974). *Statistical Methods*, 7th ed. Ames, IA: Iowa State University Press.

Wadsworth, H. M., K. S. Stephens, and A. B. Godfrey (1986). *Modern Methods for Quality Control and Improvement.* New York: Wiley.

Woodall, W. H. (1997). Control charts based on attribute data: Bibliography and review. *Journal of Quality Technology 29*(2): 172–183.

EXERCISES

1. An experimenter has collected 20 samples of size 200 for the purpose of constructing trial control limits for a p chart and finds that $\overline{p} = .023$. Determine the control limits using the transformation

$$y = \sin^{-1}\sqrt{(x + \tfrac{3}{8})/(n + \tfrac{3}{4})}$$

2. Consider the data in Table 6.1. For each of the last 5 days the number of nonconforming units plots above the midline on the np chart in Figure 6.2. Does this provide evidence that p has increased? Can runs criteria be easily used here? If so, what would the use of runs criteria suggest?

3. Explain why an experimenter should not attempt to construct a p chart with 3-sigma limits when $n = 100$ and p is assumed to be approximately .001. Should the regression-based approach be used for determining the limits? Why or why not? If not, what approach should be used?

4. A soft drink company wishes to control the number of nonconformities per bottle of its product. The data on nonconformities for 30 days are as follows, where 100 inspection units (bottles) constitute the sample size and the bottles are all the same size:

6	7	3	5	2	2
3	8	4	6	4	9
4	5	5	8	4	2
3	3	8	3	10	5
8	4	7	2	6	6

(a) Could either a c chart or u chart be used here? Explain.

(b) If a c chart is used, will the LCL exist?

(c) Compute the regression-based limits for the c chart using $\bar{c}$ in place of λ and construct the chart.

5. Could we determine the largest value of $\bar{c}$ that would make the LCL zero using the transformation $y = \sqrt{c} + \sqrt{c+1}$? Why or why not?

6. Construct a u chart for the following data. Use the conventional approach, but also compute the regression-based limits and compare the results.

c	n	c	n	c	n
12	125	15	145	14	135
13	140	12	130	19	195
15	150	14	150	16	180
10	115	18	175	15	160
11	135	17	180	13	155
14	145	15	155	12	115

7. The following 50 numbers are random values generated from a binomial distribution with $n = 500$ and $p = 0.008$:

1	4	5	3	3	3	3	5	8	4
4	3	5	5	2	3	3	4	7	8
3	1	1	3	6	2	4	1	3	4
5	5	6	2	2	3	2	5	6	5
7	8	6	3	2	3	3	5	8	2

Construct the p chart using the standard approach and also compute the limits using the regression-based approach. Compare the two sets of limits. (Successive values are in columns.)

8. The following 40 numbers have been produced as follows. The first 30 numbers were generated from a Poisson distribution with $\lambda = 9$ and the next 10 were generated from a Poisson distribution with $\lambda = 12$.

$\lambda = 9$						$\lambda = 12$	
7	9	10	12	9	9	11	8
8	7	11	8	12	13	14	11
9	5	8	14	6	9	11	13
11	9	12	9	7	11	18	14
7	11	8	9	9	9	12	13

Use the first 30 values (successive values are in columns) to determine the control limits using the following:

(a) The standard c-chart approach, first ignoring the known value of λ and then using the known value.

(b) The regression-based limits, using the known value of λ and then using $\bar{c}$ in place of λ.

Compare the results for the four sets of limits in terms of detecting the shift to $\lambda = 12$.

(c) Use the Poisson distribution to determine the expected number of plotted points before the shift should be detected for each of the four pairs of limits. Comment.

9. Assume that a demerit chart is to be produced and the numbers of demerits assigned to each of three types of nonconformities are 5, 4, and 8, respectively. The data for 20 inspection units are as follows, where c_1, c_2, and c_3 denote the number of nonconformities of each of the three types:

Unit Number	c_1	c_2	c_3	Unit Number	c_1	c_2	c_3
1	3	5	2	11	2	3	1
2	2	4	5	12	4	4	2
3	1	2	3	13	3	1	0
4	2	4	1	14	2	0	1
5	3	2	3	15	3	2	3
6	4	3	2	16	4	1	2
7	2	1	2	17	5	3	4
8	2	0	3	18	2	3	6
9	3	2	2	19	1	2	4
10	0	3	1	20	3	1	1

(a) Construct the d chart (that corresponds to a c chart) and test for control.

(b) Would three separate c charts have indicated statistical control? [Note that this can be easily determined since the three $\bar{c}$ values were computed in part (a).]

10. What advantage, if any, does a standardized p chart have over a p chart constructed using 3-sigma limits? Would you expect a standardized p chart to be superior to a p chart with regression-based limits? Explain.

11. Compute the regression-based limits for an np chart using the data in Table 6.1, using $\bar{p}$ in place of p. Compare these limits with the limits shown in Figure 6.2.

12. Show that a point cannot plot below the LCL for the standardized p chart in Section 6.1.5 if the LCL does not exist for the regular p chart.

CHAPTER 7

Process Capability

The concept of *process capability* was mentioned briefly in Section 4.1. What is process capability and how is it measured? Process capability generally relates to whether an in-control process is functioning in such a manner that the variables being measured have distributions that lie almost completely within the specification limits, with the latter frequently determined from engineering tolerances and/or customers' needs. A "capable" process is one for which the distributions of the process characteristics lie almost entirely within the engineering tolerances. Numerical measures of capability are obviously needed and are presented in this chapter.

Much attention has been devoted to process capability indices in recent years, with some writers suggesting that they not be used (Nelson, 1992; Pignatiello and Ramberg, 1993; Singpurwalla, 1998), while many researchers have concurrently worked at developing new indices. In general, process capability indices have been quite controversial. Nevertheless, Kotz and Lovelace (1998) believe that process capability indices are here to stay.

Despite various admonitions regarding the use and interpretation of these indices, it is highly probable that they will continue to be used because of their simplicity. It is this simplicity that causes problems, however, as we will see later in the chapter.

There are essentially two phases involved in a process capability study: (1) determining how the data are to be collected and then collecting the data, and (2) selecting one or more indices and performing the computations.

7.1 DATA ACQUISITION FOR CAPABILITY INDICES

The careful selection of the data that are to be used in calculating one or more capability indices is perhaps more important than the selection of which index to use. The data must have come from an in-control process, the sample must be representative of the population, and the sample size must be large enough to provide good estimates of the parameters and to assess the extent of the nonnormality. Process capability indices should not be constructed using data from processes that are not in a state of statistical control. Certainly some measure of performance will generally be needed before a process can be brought under statistical control, however, so *process performance indices* are also used. These are discussed in Section 7.11.

Historical data are used in calculating process capability indices, and it is important that only historical data be used for which there was clear evidence that the process was in control. If a process was not in control for the entire time period covered by the sample, then the process capability would be underestimated (unless "out of control" meant that the variability had decreased).

7.1.1 Selection of Historical Data

It would be unwise to attempt to determine process capability unless there is a considerable amount of very recent historical data that have been charted and analyzed with control charts.

Why is it that we cannot select a moderate-sized sample from the current process and perform the necessary computations? First, a very large sample will generally be necessary, especially if the data are highly nonnormal, and we would not necessarily know if the data have come from an in-control process if the data have not been charted.

Assume that either an X chart or an $\overline{X}$ chart has been used in monitoring a particular process. If the charts indicate control, then the data can be used to estimate the process mean and process standard deviation. If there is evidence that the process may have been out of control at certain points in time, it will be necessary to have information regarding points outside the control limits. If an assignable cause were detected, then the point(s) should not be used in computing process capability unless the cause is nonremovable. If an assignable cause is not detected or suspected, then the point(s) should be used in the computations. (The specifics of the computations will be given in later sections.)

Runger (1993) emphasizes that a large sample will not necessarily be sufficient if the sample comes from a short production run, since it is imperative that the sample capture the long-term variability. The author stresses that it is also a good idea to design a process capability study in such a way that noise factors are varied over the range of levels that are likely to occur in production. [Here we assume that a noise factor is a factor that cannot be controlled (without difficulty, at least) during normal manufacturing operations but can be controlled during an experiment.]

7.2 PROCESS CAPABILITY INDICES

A process capability index should be easy to compute and understand, and if it is based on the assumption of a normal distribution, it should not be undermined by slight-to-moderate departures from normality. If the sampled data are strongly nonnormal, an approach that recognizes the nonnormality, such as fitting a distribution to the data, would have to be used.

The evolution of capability indices has seen them progress from very simple indices to indices that were designed to correct some of the shortcomings of their predecessors.

7.2.1 C_p

For example, one of the first capability indices was C_p, which is defined as

$$C_p = \frac{\text{USL} - \text{LSL}}{6\sigma}$$

with USL and LSL representing the upper and lower specification limits, respectively, and σ is the process standard deviation. The term "six sigma" is often used (by Motorola, Inc., in particular) to represent acceptable quality. A process is said to have six-sigma capability if the difference between the specification limits is at least equal to 12σ, with the mean at all times no closer than 4.5σ to the nearest specification limit. This suggests that an objective might be to work to obtain $C_p = 2.0$.

But consider the following. Let $\mu = 180$ and $\sigma = 25/6$, with USL $= 100$ and LSL $= 75$. If values outside the specification limits define nonconforming units of production and a normal distribution is assumed, then almost all of the units will be nonconforming even though $C_p = 1.0$. If the process mean is later shifted, by appropriate process adjustments, to 87.5, C_p is still equal to 1.0, but now only 0.27% will be nonconforming (still assuming normality). The second process might be considered to have fair capability, but the first process clearly is not capable.

Thus, C_p is a poor measure of process capability since it is meaningless to consider the difference between the specification limits as a multiple of 6σ while ignoring the process mean, or a target value for the mean. Clearly the value of any process capability index should depend on the location of the distribution of the process characteristic relative to the specification limits.

As discussed by Boyles (1991) and others, C_p measures just the *potential* process capability that would be realized only if the process mean were centered between the specification limits. [We could similarly speak of the "potential percent nonconforming," assuming normality, as Littig and Lam (1993) do. This is discussed in Section 7.7.]

7.2.2 C_{pm}

Hsiang and Taguchi (1985) gave a capability index that was also labeled C_p but which will herewith be denoted as C_p^* so as to avoid confusion. With this notation the index is given by

$$C_p^* = \frac{\text{USL} - \text{LSL}}{6\tau}$$

with $\tau^2 = E(X - T)^2$ and T denoting the target value for the process mean. This index was proposed independently by Chan, Cheng, and Spiring (1988), who labeled it C_{pm}. Since $E(X - T)^2 = \sigma^2 + (\mu - T)^2$, C_{pm} is written as

$$C_{pm} = \frac{\text{USL} - \text{LSL}}{6\sqrt{\sigma^2 + (\mu - T)^2}}$$

$$= \frac{(\text{USL} - \text{LSL})/6\sigma}{(1/\sigma)\sqrt{\sigma^2 + (\mu - T)^2}}$$

$$= \frac{C_p}{\sqrt{1 + [(\mu - T)/\sigma]^2}} \tag{7.1}$$

This index is clearly both similar and superior to C_p since it reflects the extent to which the process mean differs from the target value for the mean. Johnson (1992) shows how C_{pm} can be viewed in terms of a squared error loss function, so that values of C_{pm} may be interpreted in terms of the percentage loss that results when $\mu \neq T$ (relative to the maximum worth of the product when $\mu = T$). In general, the selection of a particular process capability index should probably correspond to the form of the loss function, if a particular loss function is assumed.

As with various other indices, C_{pm} can produce unsatisfactory results when the tolerances are not symmetric about the target value. This is because the value of C_{pm} does not depend on the relationship between μ and the specification limits. Assume that we have two processes, and for the first process USL = 110, LSL = 70, $\sigma = 5$, $T = 80$, and $\mu = 75$, and the second process has USL = 60, LSL = 20, $\sigma = 5$, $T = 40$, and $\mu = 45$. The value of C_{pm} is 0.94 for each process, but assuming normality, the first process will produce 15.87% nonconforming units, whereas the second process will produce about 0.1% nonconforming units.

Thus, although C_{pm} is an improvement over C_p since it is a function of both μ and T, it does not necessarily reflect process performance relative to the process characteristic staying within the tolerances.

7.2.3 C_{pk}

The most frequently used process capability index is probably C_{pk}, which is defined as

$$C_{pk} = \tfrac{1}{3}Z_{\min}$$

where $Z_{\min}$ is defined as the minimum of $Z_1 = (\text{USL} - \mu)/\sigma$ and $Z_2 = (\mu - \text{LSL})/\sigma$. (Also, C_{pk} is frequently defined as the minimum of CPU and CPL, with CPU = $(\text{USL} - \mu)/3\sigma$, and CPL = $(\mu - \text{LSL})/3\sigma$.)

In words, if the USL and LSL are equidistant from μ, $Z_{\min}$ can be viewed as the number of standard deviations that the specification limits are from μ. If the distances are not the same, $Z_{\min}$ will then equal the number of standard deviations that the closest specification limit is to μ.

In particular, if the limits are at $\mu \pm 3\sigma$, then $Z_{\min} = 3$ and $C_{pk} = 1.0$, which can be seen from simple substitution. In general, if the limits are at $\mu \pm j\sigma$, then $Z_{\min} = j$ and $C_{pk} = j/3$. Thus, process capability for equidistant limits could be expressed in terms of either $\mu \pm j\sigma$ or $C_{pk} = j/3$. The use of C_{pk} is clearly preferable when the limits are not equidistant, however, as will often be the case.

Since C_{pk} is a function of the process mean, it is superior to C_p, but it can be shown, as the reader is asked to do in exercise 1, that $C_{pk} = C_p$ if μ is equidistant from the specification limits.

Neither C_{pk} nor C_p is useful as a measure of process capability when there is a target value for the process mean. Specifically, the value of each index will approach infinity as $\sigma \to 0$ for any fixed value of $\mu - T$. Therefore, it would be possible to have the value of each index increase because σ is decreasing and yet have the center of the process distribution moving farther away from the target value. (By comparison, C_{pm} can approach infinity only if *both* $\sigma \to 0$ and $\mu \to T$.)

A related problem with C_{pk} is there is not a one-to-one correspondence between C_{pk} and the percentage of nonconforming units. This point is frequently misunderstood (see, e.g., Rodriguez, 1992; Littig and Lam, 1993; or Kaminsky, Dovich, and Burke, 1998). That there is not a one-to-one correspondence should be apparent since C_{pk} is a function of $\min(Z_1, Z_2)$ rather than being a function of both Z_1 and Z_2. That is, C_{pk} is a measure of the capability on one side of the distribution — the side for which the larger of the two proportions nonconforming will result.

In general, comparing C_{pk} values over time could easily create a false impression in regard to the percentage of nonconforming units. For example, assume that LSL $= 20$, $\mu = T = 30$, USL $= 40$, and $\sigma = \frac{10}{3}$. It follows that $Z_1 = Z_2 = 3$, so $C_{pk} = 1$. Assuming a normal distribution, 0.27% will be nonconforming since $P(|Z| > 3) = .0027$ with $Z \sim N(0, 1)$. If the distribution of the process characteristic drifts so that the new mean is $\mu_1 = 35.17$ but the variability is reduced so that $\sigma_1 = \frac{5}{3}$, $Z_{\min} = Z_1 = 2.9$ so that $C_{pk} = 0.97$. Thus, C_{pk} has declined slightly, thus suggesting lower process capability, but the percentage of nonconforming units has *declined* from 0.27 to 0.187% — a reduction of almost one-third.

Therefore, C_{pk} values displayed in a time sequence will not necessarily give a clear picture of process capability relative to the specification limits, vis-á-vis nonconforming units.

Examples such as this one can be constructed to show that it is questionable whether C_{pk} is measuring anything meaningful relative to process capability. Furthermore, there is the tacit assumption that a change in the process mean that is offset by a compensating change in the process standard deviation (so that $Z_{\min}$ remains unchanged) has not changed the process capability. Even in the absence of a target value this would seem to frequently be an unrealistic assumption.

Gunter (1989) discusses shortcomings of C_{pk} in a series of short articles, and the following statement can be found in the last article (p. 86):

> The greatest abuse of C_{pk} that I have seen is that it becomes a kind of mindless effort that managers confuse with real statistical process control efforts. Instead of helping them better understand and improve their processes, it functions as a kind of numerical hurdle that is either met or forgotten, or finessed via negotiations, well-chosen samples, and biased inspection procedures. In short, rather than fostering never-ending improvement, C_{pk} scorekeeping kills it.

[See also Gunter (1991) for a similar perspective.]

This particular criticism is in regard to the way that C_{pk} is misused; it is not really a criticism of the index itself.

But shortcomings in the index have been noted in this section, and Gunter also notes in that article that considerable measurement error could prevent a goal value for C_{pk} from being reached, with the measurement error perhaps being unknown. [See Herman (1989) for a related discussion.]

7.2.4 C_{pmk}

An obvious modification of C_{pm} would be to replace C_p in Eq. (7.1) with C_{pk}. This produces an index that we will write as

$$C_{pmk} = \min\left(\frac{\mu - \text{LSL}}{3\sqrt{\sigma^2 + (\mu - T)^2}}, \frac{\text{USL} - \mu}{3\sqrt{\sigma^2 + (\mu - T)^2}}\right) \tag{7.2}$$

This index was proposed by Pearn, Kotz, and Johnson (1992). What is gained by using C_{pmk} instead of C_{pm}? Although the latter is a function of $|\mu - T|$, it does not depend on the distance that μ is from the specification limits.

If $\mu = \frac{1}{2}(\text{LSL} + \text{USL})$, then $C_{pm} = C_{pmk}$, regardless of the value of $|\mu - T|$. But if the mean is not equidistant from the specification limits, then $C_{pmk} < C_{pm}$. Thus, C_{pmk} is a better indicator of process capability relative to the specification limits than is C_{pm}.

It is clear that C_{pmk} is also a better measure of process capability than the other indices since it measures process capability in what appears to be an appropriate manner in the presence of asymmetrical specification limits, and it is also a function of $|\mu - T|$. It is equivalent to C_{pk} when $\mu = T$ and, as stated, is equivalent to C_{pm} when the specification limits are symmetric about μ. Lastly, C_{pmk} is equivalent to C_p when there are symmetric specification limits and $\mu = T$. Thus, C_{pmk} will reduce to each of the other three indices under certain conditions and is thus a more general and flexible measure of process capability than the other three indices.

We examine C_{pmk} further in Section 7.6.1.

7.2.5 Other Capability Indices

Several other capability indices have been proposed, including other modifications of C_p and C_{pk} that were suggested by Kane (1986), and the $C_{pg}(= C_{pm}^{-2})$ index proposed by Beazley and Marcucci (1988).

Furthermore, Vännman (1995) defined a family of process capability indices as

$$C_p(u, v) = \frac{d - u|\mu - M|}{3\sqrt{\sigma^2 + v(\mu - T)^2}}$$

with $d = (\text{USL} - \text{LSL})/2$, $M = (\text{USL} + \text{LSL})/2$, and u and v are constants to be selected. The capability indices C_p, C_{pk}, C_{pm}, and C_{pmk} are produced by letting $(u, v) = (0, 0)$, $(1, 0)$, $(0, 1)$, and $(1, 1)$, respectively.

Vännman (1995, 1997a) and Vännman and Kotz (1995a,b) have studied $C_p(u, v)$ and derived statistical properties for $\hat{C}_p(u, v)$, in addition to considering suitable choices for u and v. They suggest consideration of $(u, v) = (0, 3)$ or $(0, 4)$ when the normality assumption is not seriously violated and the target value is midway between the USL and LSL. Vännman (1998) claims that these choices of u and v produce indices that are superior to both C_{pm} and C_{pk}.

Spiring (1997) also gave a general form that includes C_p, C_{pm}, and C_{pk} as special cases. Specifically,

$$C_{pw} = \frac{\text{USL} - \text{LSL}}{6\sqrt{\sigma^2 + \omega(\mu - T)^2}}$$

with ω representing a weight function. Obviously $C_{pw} = C_p$ when $\omega = 0$ and $C_{pw} = C_{pm}$ when $\omega = 1$.

Some of the other (special-purpose) indices are discussed in subsequent sections of this chapter, and multivariate process capability indices are presented in Section 9.14.

7.3 ESTIMATING THE PARAMETERS IN PROCESS CAPABILITY INDICES

The indices that were given in Sections 7.2.1–7.2.4 contain parameters that must be estimated. Since process stability must first be established before the value of an index is calculated, a logical approach would be to compute parameter estimates using control chart data.

This raises the question of how many data points should be used. We may view the numerical value for a given index as providing an estimate of the unknown value of the index at a given point in time (which depends on the unknown parameter values) and also more generally providing an estimate of the process capability.

Accordingly, it is desirable to use a large sample size so that the parameter estimators in the index and therefore the index itself will have a reasonably small variance. Large sample sizes will not overcome problems caused by nonnormality, however; they simply provide us better insight into the shape of the actual distribution than do small sample sizes.

7.3.1 X Chart

Assume that an X chart has been used for process monitoring. An estimate of σ is needed for each of the four capability indices that have been presented, with an estimate of μ not needed for C_p. If there is evidence that the process mean and process standard deviation have been in control for the time period that is to be used in estimating the parameters, one reasonable approach would be to use $\hat{\sigma} = s/c_4$ and $\hat{\mu} = \bar{x}$.

Although these are unbiased estimators, the use of these estimators will not cause the resultant estimators of the process capability indices to be unbiased since ratios are involved.

For example, with $\hat{C}_p = (\text{USL} - \text{LSL})/6\sigma$ and $\hat{\sigma} = s/c_4$, $\hat{C}_p$ will be slightly biased, with the bias approaching zero as the sample size approaches infinity.

Pearn et al. (1992) gave the bias expressions for the various indices, under the assumption that $\hat{\sigma} = s$. Therefore, we may use their results to determine unbiased estimators of each of the indices when $\hat{\sigma} = s/c_4$.

For example, it may be easily shown (as the reader is asked to do in exercise 2) that $(1/c_4^2)[(n-2)/(n-1)]\hat{C}_p$ is an unbiased estimator of C_p.

It is considerably more complicated to obtain the unbiasing constants for C_{pk}, C_{pm}, and C_{pmk}; the interested reader might derive the appropriate constants from the results given in Pearn et al. (1992). [See also Kotz, Pearn, and Johnson (1993) and Rodriguez (1992) for results on C_{pk}.]

7.3.2 $\overline{X}$ Chart

If an $\overline{X}$ chart is being used rather than an X chart, it would be reasonable to use $\hat{\mu} = \overline{\overline{x}}$ and $\hat{\sigma} = \overline{s}/c_4$ (or $\overline{R}/d_2$), with the subgroups that are used in the computations being those for which there is evidence that the process was in control at each of those time points.

7.3.3 Case Study

Pearn and Chen (1997) describe a case study that can be used in illustrating the computation of the various process capability indices. The study involved Bopro, a manufacturer and supplier of audio-speaker components in Taipei, Taiwan. The authors studied the weight of the rubber edge, which is one of the key components that reflect the sound quality of driver units. The data on rubber edge are given here in Table 7.1.

A normal probability plot does not indicate a clear departure from approximate normality, as the reader is asked to show in exercise 12.

The USL, LSL, and target value were 8.94, 8.46, and 8.70, respectively, the target value being set equal to the midpoint between the specification limits. The company had been using C_p and C_{pk} as process capability indicators.

Computing $\hat{C}_p = (\text{USL} - \text{LSL})/6\hat{\sigma}$, with $\hat{\sigma} = s/c_4$ and $c_4 \doteq 4(n-1)/(n-3)$ used to approximate the value of c_4 for $n = 80$, we obtain $\hat{C}_p = 1.53$. If we used the unbiased estimator of C_p given in Section 7.3.1, we would obtain 1.52.

TABLE 7.1 Process Capability Data

8.63	8.65	8.57	8.73
8.65	8.58	8.64	8.70
8.57	8.65	8.63	8.65
8.57	8.67	8.57	8.56
8.54	8.67	8.61	8.66
8.69	8.65	8.59	8.65
8.63	8.69	8.56	8.66
8.64	8.66	8.71	8.68
8.59	8.62	8.53	8.62
8.61	8.63	8.51	8.54
8.60	8.59	8.72	8.67
8.66	8.65	8.58	8.62
8.65	8.64	8.64	8.54
8.50	8.64	8.69	8.62
8.61	8.52	8.64	8.66
8.61	8.69	8.75	8.56
8.63	8.66	8.59	8.60
8.67	8.66	8.61	8.62
8.54	8.61	8.58	8.61
8.62	8.55	8.65	8.66

As stated by Pearn and Chen (1997), there are three estimators that have been proposed for C_{pk}, including the Bayesian-type estimator that they gave. Another estimator is the one given by Kotz et al. (1993), which results from using $\hat{\mu} = \overline{x}$ and $\hat{\sigma} = s$. We will use s/c_4, but since $c_4 \doteq 0.997$ when $n = 80$, this will make very little difference. It can be shown that $\overline{x} = 8.62$ and $s = 0.05$. Using these values, we obtain $\hat{C}_{pk} = 1.04$. [Pearn and Chen (1997) obtained 1.07 as the estimate using their Bayesian-type estimator.]

The company's requirement was $1.00 \leq C_{pk} < 1.33$ (note the upper bound). Even though $\hat{C}_{pk}$ exceeds the lower bound on acceptable values of C_{pk}, the lower bound is just barely exceeded, so we know there is a substantial probability that the true value of C_{pk} is less than 1.0. Pearn and Chen (1997) performed a hypothesis test using their estimator and concluded that the process is not capable.

The company also considered the quality to be unacceptable and subsequently began using some Taguchi parameter designs (see Section 14.8) in an attempt to identify factors that were significantly affecting the quality of the rubber edge. Machine settings were subsequently adjusted, and when a new sample of 80 observations was obtained, $\overline{x} = 8.67$ and $s = 0.05$. Since the new sample mean is much closer to the midpoint between the specification limits than was the previous sample mean, the value of $\hat{C}_{pk}$ will be considerably greater. It can be shown that $\hat{C}_{pk} = 1.40$. Since this exceeds the upper bound on acceptable values of C_{pk}, we would expect to conclude that the true value is probably greater than the lower bound. As an approximation, we could use the approximate 95% lower confidence bound given in Eq. (7.3) in Section 7.5.2. This gives a lower bound of 1.17. Since this lower bound is well below the company's upper bound of 1.33, and since the value of $\hat{C}_{pk}$ is not much greater than 1.33, there is no strong evidence that the true value is greater than the upper bound.

Therefore, the logical conclusion is that the company's requirement now appears to be met.

7.4 DISTRIBUTIONAL ASSUMPTION FOR CAPABILITY INDICES

When distributional properties are given for process capability indices [such as those given by Pearn et al. (1992) and Kotz et al. (1993)], there is the stated assumption that the observations have come from a normal distribution. A normal distribution is also assumed when the capability indices are used by practitioners.

As discussed by Nelson (1992), process capability indices that utilize s will not be robust to nonnormality since the distribution of s is not robust to nonnormality. Accordingly, normality should be tested before a process capability index is used. A histogram or normal probability plot might be constructed and/or a goodness-of-fit test could be performed.

Methods that can be used when there is clear evidence of nonnormality are discussed in Section 7.5.6.

7.5 CONFIDENCE INTERVALS FOR PROCESS CAPABILITY INDICES

It would be unwise to use only a point estimate of a capability index unless the sample size was so large that the capability index estimator had a very small variance. Since such a sample size is not likely to be used very often, if at all, it is generally desirable to also report a confidence interval for the index.

Unfortunately, when companies began using process capability indices extensively in the 1980s, only point estimates were used, and the point estimates were used as if they were the true values, completely overlooking the sampling variability associated with each capability index estimator. Kaminsky et al. (1998) indicate that some companies have switched to confidence intervals, citing the Chrysler Corporation in particular, but it seems almost certain that most companies are not doing so. Indeed, Vännman (1997b) and Deleryd (1998) claim that practitioners still generally interpret estimates of process capability indices as true values, although their conclusion may be based only on their experiences with European companies.

Companies that use process capability indices have established a threshold value for the indices that they use, such as 1.33 for C_{pk}. Therefore, a lower confidence interval would be more appropriate than a two-sided confidence interval. A user would not likely be interested in the upper limit, especially when the estimate exceeds the threshold value.

Methods for constructing lower confidence intervals have been given by Chou, Owen, and Borrego (1990), Bissell (1990), Franklin and Wasserman (1991, 1992a,b), and Kushler and Hurley (1992).

In Sections 7.5.1–7.5.4 we assume that individual observations will be used in computing the parameter estimates (as would be the case if the observations came from X charts, for example). Necessary adjustments that should be made when the data are in subgroups are discussed in Section 7.5.5.

7.5.1 Confidence Interval for C_p

Obtaining a confidence interval for C_p is relatively straightforward since only the denominator of C_p contains a random variable and that random variable is s. If the individual observations are assumed to have a normal distribution, then $(n-1)s^2/\sigma^2$ has a chi-square distribution with $n-1$ degrees of freedom.

It then follows that

$$P\left(\chi^2_{n-1,\alpha/2} \le \frac{(n-1)s^2}{\sigma^2} \le \chi^2_{n-1,1-\alpha/2}\right) = 1-\alpha$$

so that

$$P\left(\frac{1}{s}\sqrt{\frac{\chi^2_{n-1,\alpha/2}}{n-1}} \le \frac{1}{\sigma} \le \frac{1}{s}\sqrt{\frac{\chi^2_{n-1,1-\alpha/2}}{n-1}}\right) = 1-\alpha$$

and

$$P\left(\frac{\text{USL}-\text{LSL}}{6s}\sqrt{\frac{\chi^2_{n-1,\alpha/2}}{n-1}} \le \frac{\text{USL}-\text{LSL}}{6\sigma} \le \frac{\text{USL}-\text{LSL}}{6s}\sqrt{\frac{\chi^2_{n-1,1-\alpha/2}}{n-1}}\right) = 1-\alpha$$

so the endpoints of the interval are the confidence limits. We note that this differs slightly from the expression given by Spiring (1997). The latter used the maximum likelihood estimator for σ^2, which produces n in the radicand rather $n-1$. The latter seems preferable, although there will be hardly any difference for large n.

If only a lower bound were desired, this would be obtained by replacing $\chi^2_{n-1,\alpha/2}$ in the lower bound by $\chi^2_{n-1,\alpha}$.

7.5.2 Confidence Interval for C_{pk}

Obtaining a confidence interval for C_{pk} is much more complicated since C_{pk} is defined as the minimum of two functions (CPU and CPL), each of which is a fraction with one random variable ($\overline{x}$) in the numerator of a fraction and a second random variable (s) in the denominator.

These complications have resulted in various forms of an approximate confidence interval being proposed, not all of which give good results. For example, the approach for obtaining a lower 95% confidence limit given by Chou et al. (1990) is based on the assumption that CPU = CPL, which will hold when the process mean is centered between the two specification limits. When this assumption is not met, the lower limit will be conservative in the sense that the limit is lower than it should be (see Chou et al. 1992). [The conservative nature of the lower bound has been noted by Kushler and Hurley (1992) and Franklin and Wasserman (1992a).]

Bissell (1990) provided two general methods for obtaining approximate confidence intervals for process capability indices (one for two-sided indices and the other for one-sided indices). When the appropriate method is used to produce an approximate $100(1-\alpha)\%$ lower confidence limit for C_{pk}, the limit has the form

$$\text{lower limit} = \hat{C}_{pk} - z_\alpha \sqrt{\frac{1}{9n} + \frac{\hat{C}_{pk}^2}{2a}} \tag{7.3}$$

where $a = n - 1$ when individual observations are used. Since z_α is the standard normal variate, this form is based on the assumption that $\hat{C}_{pk}$ is approximately normally distributed. It is also tacitly assumed that the individual observations are from a normal distribution, and the general form in Eq. (7.3) also results from the use of an approximation of the coefficient of variation of two random variables (see Bissell, 1990, p. 335).

Because of this approximation and the assumption that $\hat{C}_{pk}$ is approximately normally distributed, it is desirable to investigate this approach. Numerical results given by Kushler and Hurley (1992) suggest that this approach will provide good results. [As noted, however, by Kotz et al. (1993), the variance approximation employed by Bissell uses only the smaller of $\hat{Z}_1/3$ and $\hat{Z}_2/3$, and the variance expression also uses a biased estimator of C_{pk}. Kotz et al. (1993) note that $\text{Var}(\hat{C}_{pk})$ will be less than the variance considered by Bissell (1990) when $\overline{X}$ is closer to LSL than to USL and that it is $\text{Var}(\hat{C}_{pk})$ that is relevant. The results of Kotz et al. (1993) also show that the exact variance expression for the variance form utilized by Bissell (1990) produced results noticeably different from those obtained by Bissell for the special case of $\mu = \left(\frac{1}{2}\right)(\text{LSL} + \text{USL})$.]

The most accurate approach, however, is to obtain the lower confidence limit by solving the equation

$$\alpha = Q_f\left(Z_1\sqrt{n},\ 3L\sqrt{n};\ \frac{6L\sqrt{n-1}}{Z_1+Z_2},\ \infty\right) - Q_f\left(-Z_2\sqrt{n},\ -3L\sqrt{n};\ \frac{6L\sqrt{n-1}}{Z_1+Z_2},\ \infty\right)$$

where $f = n - 1$, L denotes the lower limit that is to be solved for, and Q_f is the function that was defined by Owen (1965) as

$$Q_f(t, \delta; a, b) = \frac{\sqrt{2\pi}}{\Gamma(f/2)2^{(f-2)/2}} \int_a^b \Phi\left(\frac{tx}{\sqrt{f}} - \delta\right) x^{f-1}\phi(x)\,dx$$

where Φ denotes the cumulative standard normal distribution function and ϕ denotes the standard normal density function.

Guirguis and Rodriguez (1992) provide a computer program for this approach, which uses the Bissell approximation as a starting point.

Spiring (1997) gives the expression for the confidence interval for C_{pk} based on the assumption that μ and σ are known. This approach might be used as an approximation if $\hat{C}_{pk}$ is computed from an extremely large amount of data.

7.5.3 Confidence Interval for C_{pm}

Obtaining a confidence interval for C_{pm} is complicated by the fact that the denominator of C_{pm} contains both σ^2 and μ. Kushler and Hurley (1992) examined several approximate methods that have been proposed and conclude that the approximation given by Boyles (1991) works best.

That approach provides a lower confidence bound of $\sqrt{\chi^2_{\hat{v}}/\hat{v}}\ \hat{C}_{pm}$, with $\hat{v} = (n + \hat{\lambda})^2/(n + 2\hat{\lambda})$ and $\hat{\lambda} = n(\overline{X} - T)^2/s^2$. Spiring (1997) and Kotz and Johnson (1993) give the form of a confidence interval for C_{pm}, but the interval is a function of a chi-square noncentrality parameter that in turn is a function of μ and σ^2.

7.5.4 Confidence Interval for C_{pmk}

There apparently has not been a published confidence limit approach for this capability index that was proposed by Pearn et al. (1992). One approach would be to assume approximate normality of $\hat{C}_{pmk}$ and use the complicated expression for $\text{Var}(\hat{C}_{pmk})$ that was given by Pearn et al. (1992).

7.5.5 Confidence Intervals Computed Using Data in Subgroups

The methods discussed in Sections 7.5.1–7.5.4 apply directly when individual observations are used in computing the parameter estimates. Some modifications are necessary when data in subgroups are used. For example, if an $\overline{X}$ chart is used for controlling the process mean instead of an X chart, it would be reasonable to estimate μ using the plotted points when there is evidence that the process was in control. Similarly, σ could be estimated from an s or R chart.

For data collected in subgroups, it was stated in Section 4.7.6 that the best way to estimate σ is $\hat{\sigma} = \sqrt{\text{Ave}(s^2)}$, where $\text{Ave}(s^2)$ denotes the average of the subgroup variances. Bissell (1990) gives, in his Table 1, multiplicative factors that should be used to provide the effective number of degrees of freedom.

Assume that 50 subgroups of size 5 are to be used in constructing a confidence interval for C_{pk} using the Bissell approach, with subgroup means and standard deviations used for parameter estimation. If $\hat{\sigma} = \sqrt{\text{Ave}(s^2)}$, the a in Eq. (7.3) is replaced by $50(5-1) = 200$; if $\hat{\sigma} = \bar{s}/c_4$, then $0.949(200) = 189.8$ (rounded to 190) is used, and if $\hat{\sigma} = \overline{R}/d_2$, then $0.906(200) = 181.2$ (or 181) would be used.

Assume that $\hat{C}_{pk} = 1.50$ and a 95% lower confidence limit is desired. The three lower limits under each of these scenarios would be 1.225, 1.224, and 1.223, respectively. Thus, the effect of using inefficient estimators of σ is to cause the lower bound to be slightly less than what it would be if σ were estimated as the average of the subgroup variances. (Here the differences are very slight because the use of a large number of subgroups was assumed. If 20 subgroups had been assumed instead of 50, the differences in the lower limits would have been about three times as large, and such differences would more clearly illustrate the value of using the most efficient estimator of σ.)

The multiplicative factors given by Bissell only apply when a normal approximation is used, so they would not be used in conjunction with the confidence interval approaches for C_p, C_{pm}, and C_{pkm} that have been presented.

7.5.6 Nonparametric Capability Indices and Confidence Limits

Most of the suggested indices and methods for constructing confidence limits are based on the assumption of normality, but certainly many process characteristics will not be approximately normally distributed. If a normal distribution is not a suitable proxy for the actual distribution of a process characteristic, then the value of the capability index that is used could be meaningless.

In preceding sections we have observed a number of shortcomings with C_p and C_{pk}, in particular, and nonnormality presents an additional problem. It is well known that process capability indices are not robust to nonnormality in the individual observations, this resulting in part from the fact that the distribution of s is not robust to nonnormality. Thus, nonnormality for the process characteristic of interest will translate into nonnormality for the capability index that is used, and this could cause the lower confidence bound to be off considerably.

Assume that a normal distribution does not provide an adequate fit to a set of process data. There are essentially four distinct approaches that have been suggested: (1) use a robust capability index, (2) fit a distribution to a set of data and use the percentiles of the fitted distribution in a capability index, (3) attempt to transform the data to approximate normality, and (4) use a resampling procedure to generate samples from the n sampled observations and in doing so generate a sampling distribution for the capability index that is to be used.

We consider each of these possible approaches in Sections 7.5.6.1–7.5.6.4.

7.5.6.1 Robust Capability Indices

Pearn et al. (1992) suggest the possibility of using $C_\theta = (\text{USL} - \text{LSL})/\theta\sigma$, where θ would be chosen so that $P[\mu - \theta\sigma < X < \mu + \theta\sigma] \doteq 0.99$. They indicate that

$\theta = 5.15$ would be a reasonable choice for chi-square distributions with various degrees of freedom, the inference being that 5.15 should perhaps be adequate for other types of distributions as well.

Notice that this differs very little from the C_p statistic (which has $\theta = 6$), so it should perform similarly to the C_p statistic for a normal distribution. Unfortunately, we have seen that C_p is lacking as a process capability index, so it would be better to modify one of the other capability indices, such as C_{pmk}.

This idea could also be extended to the C_{pm} index. For example, if $\theta = 5.15$ is a good choice for a robust C_p index, then 6 would similarly be replaced by 5.15 in the C_{pm} index.

It would be difficult to try to similarly modify C_{pk} and C_{pmk}, however, since these are one-sided indices rather than two-sided indices. To illustrate this, assume that the relevant distribution has considerable positive skewness (i.e., the "tail" is to the right) and that $\overline{x}$ is closer to LSL than to USL. If we had a normal distribution (or, in general, a symmetric distribution), this would imply that the process capability was the worst for the smallest possible values of the process characteristic, but the process capability in the neighborhood of the smallest values could be worse than the capability at the other extreme when there is considerable positive skewness.

Thus, this idea of a robust capability index would not work for one-sided indices, and the shortcomings of C_p and C_{pm} were noted in earlier sections. Consequently, it would be preferable to consider the approaches discussed in the next two sections.

7.5.6.2 Capability Indices Based on Fitted Distributions

Clements (1989) and Rodriguez (1992) discuss process capability indices based on the percentiles of the fitted distribution. Letting P_α denote the 100α percentile of the fitted distribution, a percentile-based C_{pk} statistic would be written as

$$C_{pk}^* = \min\left(\frac{\text{USL} - P_{0.5}}{P_{0.9987} - P_{0.5}}, \frac{P_{0.5} - \text{LSL}}{P_{0.5} - P_{0.0013}}\right)$$

Notice that C_{pk}^* reduces to C_{pk} for a normal distribution since $P_{0.5} = \mu$, $P_{0.9987} = \mu + 3\sigma$, and $P_{0.0013} = \mu - 3\sigma$.

A C_{pmk}^* index could similarly be defined, with C_{pk} in C_{pmk} replaced by C_{pk}^*. Here, C_p and C_{pm} might be similarly modified, but this will not be pursued here because of the previously noted shortcomings of these indices.

The success of this general approach depends upon the skill exhibited in identifying and fitting an appropriate distribution. Many of the common probability distributions are members of families of distributions such as the Pearson, Johnson, and Burr families, and data will often be well fit by a distribution from one of these families. [The approach advocated by Clements (1989) was based on using the Pearson family.] There are some potential pitfalls, however; these are discussed by Rodriguez (1992).

7.5.6.3 Data Transformation

Frequently data can be transformed so that the transformed data will be approximately normally distributed. For example, if a process characteristic were known to have a lognormal distribution, the (base e) logarithm of that characteristic would have a normal distribution. Therefore, if there was evidence that the lognormal distribution would provide a good fit to a set of data (which might be seen from using lognormal probability paper), then the logarithm of the characteristic might have a distribution that is close enough to normality for the normal distribution to be an adequate model.

The primary impediment to transforming data, however, is that the use of transformed data is often not appealing to practitioners.

7.5.6.4 Capability Indices Computed Using Resampling Methods

Resampling methods such as the bootstrap have been used for such purposes as approximating sampling distributions and obtaining confidence intervals when no assumption is to be made of the distribution of the random variable whose realizations are in a sample of size n.

For example, the sampling distribution of $\overline{X}$ might be approximated by generating several thousand samples using the original sample as the population, computing $\overline{X}$ for each sample, and then constructing the empirical sampling distribution of these averages. This distribution might be used for various purposes, such as in determining $\overline{X}$-chart probability limits.

Bootstrapping is one type of resampling, and there are various bootstrapping techniques. For an original sample of size n, one method is to generate bootstrap samples of size n with the sampling done with replacement from the original sample. This method is called the *naive bootstrap* (Rodriguez, 1992).

It should be recognized, however, that resampling methods in general can give poor results under certain conditions, such as when a distribution is highly skewed.

Simulations performed by Franklin and Wasserman (1992b) suggest that lower confidence intervals for C_p, C_{pk}, and C_{pm} that are based on the assumption of normality can be expected to perform poorly for distributions that are either highly skewed or symmetric but heavy tailed. These simulations also show that the lower confidence limits constructed using each of three bootstrap methods (standard bootstrap, percentile bootstrap, and biased-corrected percentile bootstrap) perform poorly relative to the known values of each of the three capability indices.

This suggests that bootstrap methods probably cannot be relied on to produce a lower confidence limit for a capability index in the presence of considerable nonnormality. This is not surprising since it is known that bootstrap methods perform poorly for highly skewed distributions, so we might suspect a poor performance for other types of nonnormality.

This is almost a moot issue, however, because the standard capability indices clearly should not be constructed when there is evidence that the distribution is markedly nonnormal.

7.6 ASYMMETRIC BILATERAL TOLERANCES

It was suggested in Section 7.2.4 that C_{pmk} seems to perform in a desirable manner in the presence of asymmetrical specification limits (tolerances). We will compare that index with one given by Kane (1986), who gives a modified version of C_{pk} and illustrates its use for an example with asymmetrical tolerances.

We shall denote this modified version by C'_{pk}; it is defined as

$$C'_{pk} = \max\left\{\min\left\{\frac{T - \mathrm{LSL}}{3\sigma}\left(1 - \frac{|T-\mu|}{T - \mathrm{LSL}}\right),\right.\right.$$
$$\left.\left.\frac{\mathrm{USL} - T}{3\sigma}\left(1 - \frac{|T-\mu|}{\mathrm{USL} - T}\right)\right\},\ 0\right\} \qquad (7.4)$$

Notice that C'_{pk} cannot be negative, whereas C_{pk} would be negative if the USL were less than μ or the LSL were greater than μ—a rather unlikely possibility.

One obvious deficiency in C'_{pk} is that it is not a function of the difference between μ and the tolerances. Therefore, it could be a poor measure of process capability relative to the proportion of nonconforming units. In selecting a capability index, a practitioner must decide what the index should measure. Thus, a practitioner should either have a loss function in mind or at least have general priorities regarding the importance attached to deviations from a target relative to the importance of minimizing the percentage of nonconforming units.

The following quote from Littig and Lam (1993) is relevant:

> A factor which makes calculation and interpretation of capability indices difficult for bilateral nonsymmetric tolerances is the fact that the primary intent of the design engineer is not simply to minimize the proportion of nonconforming units. Instead, the engineer wants to minimize proportion nonconforming under the additional constraint that the majority of units be produced near the target value.

Other researchers have considered capability indices in the presence of asymmetric tolerances, including Vännman (1997b).

7.6.1 Examples

We will start with an example similar to the one used by Littig and Lam (1993) to illustrate how the various capability indices perform for asymmetric tolerances.

Assume that we have three normal distributions, A, B, and C, each with $\sigma = 1$ and $\mu_A = 50$, $\mu_B = 56$, and $\mu_C = 62$. Assume also that $T = 59$, with LSL $= 50$ and USL $= 62$.

Since C_p is not a function of the process mean, $C_p = 2$ for each distribution, with $C_{pk} = 0$ for the distributions with means of 50 and 62 and $C_{pk} = 2$ for the distribution with $\mu = 56$. The C_{pm} values are 0.22, 0.63, and 0.63, respectively, and the C_{pmk} values are 0, 0.63, and 0, respectively. Finally, $C'_{pk} = 0$ for each distribution.

On the basis of this example we would likely argue against the use of C'_{pk} because the three distributions seem to not be equally undesirable. In particular, the first and last distributions will have 50% nonconforming, whereas distribution B will have essentially zero percent nonconforming, so the distribution simply needs to be shifted so that $\mu = 59$. Doing so, however, will cause 0.14% of the units to be nonconforming, so this would have to be offset by a gain in quality that results from centering the distribution at the target value in order for the shift to be defensible.

To see why C'_{pk} seems to perform somewhat peculiarly, we may write the expression within the inner brackets in Eq. (7.4) equivalently as

$$\min\left\{\frac{T - \text{LSL} - |T-\mu|}{3\sigma}, \frac{\text{USL} - T - |T-\mu|}{3\sigma}\right\}$$

so C'_{pk} will be zero if the difference between at least one of the tolerances and the target value is the same as the (absolute value of) the difference between the target value and the mean.

If these two differences are equal and quite small, $C'_{pk} = 0$ would then indicate that the process mean was too close to at least one of the tolerances, so that a sizable percentage of nonconforming units would result.

On the other hand, if the two differences are equal and large, relative to σ, $C'_{pk} = 0$ would indicate that the center of the distribution differed considerably from the target value, although, as we have seen, the percentage of nonconforming units could be practically zero.

Consequently, a small C'_{pk} value could indicate poor capability relative to at least one of the tolerances, or that the distribution needed to be shifted by more than a small amount, or neither since the (approximately) equal differences could obviously be neither large nor small.

The user of C'_{pk} would thus have to understand that there is not a clear message associated with a particular value of the index.

As indicated previously, a practitioner who selects a particular capability index should have a loss function in mind so that the index essentially serves as a proxy for the loss function. Even if such matching could be done, it seems questionable whether a practitioner should use any single capability index, since there will not be a simple interpretation of the value of an index that is a function of both the tolerances and the extent to which the process mean deviates from the target value.

This is discussed further in Section 7.7.1.

7.7 CAPABILITY INDICES THAT ARE A FUNCTION OF PERCENT NONCONFORMING

Since capability indices are (erroneously) assumed to have a one-to-one correspondence with proportion nonconforming, it seems desirable to use a capability index that *is* a direct function of the proportion nonconforming and then report both the value of this index and the extent to which the center of the process distribution differs from the target value.

Lam and Littig (1992) propose capability indices that are a function of the actual and potential proportion nonconforming. Specifically, they recommend that C_{pp} be reported, where

$$C_{pp} = \tfrac{1}{3}\Phi^{-1}\left(1 - \tfrac{1}{2}p\right)$$

where Φ denotes the cumulative normal distribution function and p is the proportion nonconforming. (Thus, normality is assumed.)

They also suggest that C_{p^*} and k be reported, with

$$C_{p^*} = \tfrac{1}{3}\Phi^{-1}\left(1 - \tfrac{1}{2}p^*\right)$$

and p^* defined as the *potential proportion nonconforming*. (It would be somewhat better to label p^* as the "potential minimum proportion nonconforming," however.) We may regard p^* as the proportion nonconforming that would result if the process distribution were centered between the specification limits.

We may think of C_{p^*} relative to C_{pp} similar to the way in which we view C_p relative to C_{pk}. That is, C_{p^*} provides the process capability based upon the potential minimum percent nonconforming relative to the capability based upon the actual percent nonconforming (C_{pp}), just as C_p provides the potential process capability if the mean were centered, and C_{pk} reflects the process capability based upon the position of the mean.

The expression for k, which is due to Kane (1986), is given by

$$k = \frac{|T - \mu|}{\min\{T - \text{LSL}, \text{USL} - T\}}$$

[As explained by Kane (1986), k is essentially a scaled distance measure.]

Confidence intervals have not been proposed for these indices, but Wang and Lam (1996) gave two approximate methods for determining a lower confidence limit on the proportion of *conforming* items, which they claim to be less conservative than an approach given previously. Their approach is based on the assumption of a normal distribution. Obviously their approach could be used to indirectly estimate the proportion *nonconforming*.

7.7.1 Examples

To illustrate the utility of these indices, assume that the process characteristic has a normal distribution and .00135 is the actual proportion nonconforming, with this proportion resulting from the combination of $\mu = 44.5$, $\sigma = 2$, USL = 50.5, and LSL = 33.5. Here, C_{pp} is computed as $C_p = \frac{1}{3}\Phi^{-1}(1 - 0.00135/2) = \frac{1}{3}\Phi^{-1}(0.999325) = \frac{1}{3}(3.205) = 1.068$.

Notice that in this example C_{pp} is very close to the value of C_{pk} (which equals 1.0), since the percent nonconforming at the lower end is essentially zero. Although this process capability might seem adequate, we can see that μ is not centered between USL and LSL, so the capability can be improved. That is, C_{p^*} must be larger than C_{pp}. We can observe that p^* will be minimized (and hence C_{p^*} will be maximized), when $\mu = 42$.

How do we interpret a value of C_{pp}, and in particular, can we interpret it relative to the value of C_{pk}?

Because of the manner in which C_{pp} is computed, it provides the value of C_{pk} that *would result* if there were, for this example, (.00135)/2 percent nonconforming for each tail of the distribution. The only way this could happen would be if the process distribution were centered between the specification limits.

Thus, for a centered distribution we would have $C_p = C_{pk} = C_{pp} = C_{p^*}$. Therefore, in the case of a centered distribution there would be no need to resort to using C_{pp} and C_{p^*}. Furthermore, the values of C_p and C_{pk} would be interpreted in the same way that one would interpret C_{pp}. If the process mean is not centered in this way, however, then the values of C_{pk} and C_p would not translate into proportion nonconforming, whereas C_{pp} would so translate. In this example the process distribution is not centered, however, so the value of C_{p^*} will exceed the value of C_{pp}. We can see that p^* will be minimized when $\mu = 42$, and by using a hand calculator or other type of computer, it can be determined that $p^* = 0.000022$. Thus, $C_{p^*} = \frac{1}{3}\Phi^{-1}(1 - .000022/2) = \frac{1}{3}\Phi^{-1}(.999911) = \frac{1}{3}(3.748) = 1.249$.

If we assume a target value (T), then we may also compute k. If we assume $T = 42$, then $k = 2.5/8.5 = 0.294$.

Notice that k will equal zero if and only if $T = \mu$, regardless of the relationship of T and μ to the specification limits. Therefore, k primarily measures closeness of the process mean to the target value. The values of C_{p^*} and C_{pp} do not depend upon T, so they are unaffected by the relationship between T and μ.

In the preceding example, $T = 42$ and $\mu = 44.5$. If $\mu = 42$ and $T = 44.5$ and the specification limits are as before, then $C_{pp} = C_{p^*} = 1.249$, and the latter is the smallest possible value. As discussed by Lam and Littig (1993), a process engineer is interested in doing more than just minimizing the proportion of nonconforming units, however, as "hitting the target" is also very important.

For this last example, should the process distribution be centered at 44.5 rather than at 42, even though this would increase the percentage of nonconforming units? This *should* invoke the idea of some type of a loss function approach. For the case of specification limits not being equidistant from the target value (as we have here), Littig and Lam (1993) introduced the idea of a *modified actual proportion nonconforming* and *modified potential proportion nonconforming*. We will designate the latter as p_m^*; it is defined as

$$p_m^* = \Phi\left(-\frac{\text{USL} - T}{c_2\sigma}\right) + \Phi\left(-\frac{T - \text{LSL}}{c_1\sigma}\right)$$

with $c_1 = \max\{(T - \text{LSL})/(\text{USL} - T), 1\}$ and $c_2 = \max\ \{(\text{USL} - T)/(T - \text{LSL}), 1\}$.

The modified actual proportion nonconforming will be denoted as p_m, and it is defined as

$$p_m = \Phi\left(-\frac{\text{USL} - \max\{T, \mu\}}{c_2\sigma} - \frac{\max\{T - \mu, 0\}}{c_1\sigma}\right) + \Phi\left(-\frac{\min\{T, \mu\} - \text{LSL}}{c_1\sigma} - \frac{\max\{\mu - T, 0\}}{c_2\sigma}\right)$$

These modified expressions, which would be used in computing $C_{p_m p_m}$ and $C_{p^*_m}$, are based upon the following assumption. Let two processes be denoted by A and B, and let the means of the processes be denoted by μ_A and μ_B. Assume that $\mu_A < T$ and $\mu_B > T$. The two processes would be equally capable if $(T - \mu_A)/(T - \text{LSL}) = (\mu_B - T)/(\text{USL} - T)$. That is, the processes have equal capability if the respective deviations of the means from the target values account for equal proportions of the distances from the nearest specification limit to the target value. If the two processes were not equally capable, process $A(B)$ would be the superior process if $(T - \mu_A)/(T - \text{LSL})$ is less than (greater than) $(\mu_B - T)/(\text{USL} - T)$.

Relative to the preceding example, let A represent the process with $\mu = 42$. If μ is shifted from 42 to 44.5, the second ratio, $(\mu_B - T)/(\text{USL} - T)$, will be zero, and the first ratio will be non-zero. Thus, although $\mu_B = T$, so that the conditions described above are not strictly met, can we still compute p_m and p^*_m and the values of the associated capability indices and let these values determine if a shift should be made?

We obtain $c_1 = \frac{11}{6}$ and $c_2 = 1$, so

$$\begin{aligned} p_m &= \Phi\left(-\frac{50.5 - 44.5}{(1)(2)} - \frac{2.5}{(11/6)(2)}\right) + \Phi\left(-\frac{42 - 33.5}{(11/6)(2)} - \frac{0}{(1)(2)}\right) \\ &= \Phi(-81/22) + \Phi(-51/22) \\ &= .0001 + .0102 \\ &= .0103 \end{aligned}$$

Therefore, $C_{p_m p_m} = \frac{1}{3}\Phi^{-1}(1 - \frac{1}{2}p_m) = \frac{1}{3}\Phi^{-1}(.9897) = 0.7717$.

Similarly,

$$\begin{aligned} p^*_m &= \Phi\left(-\frac{50.5 - 44.5}{(1)(2)}\right) + \Phi\left(-\frac{44.5 - 33.5}{(11/6)(2)}\right) \\ &= \Phi(-3) + \Phi(-3) \\ &= .0027 \end{aligned}$$

so that $C_{p^*_m} = \frac{1}{3}\Phi^{-1}(1 - \frac{1}{2}p^*_m) = \frac{1}{3}\Phi^{-1}(.9993) = 1.0649$.

We should note that, unlike p and p^*, p_m and p_m^* do not have a simple interpretation. Rather, their utility is simply in comparing processes. We should especially note that p_m^* does *not* give the proportion nonconforming that would result if the process distribution were shifted so that $\mu = T$. (It may be easily seen that in this example the proportion nonconforming would be approximately .00135, not .0027, if $\mu = T$.) Rather, p_m^* is essentially giving twice the proportion nonconforming on whichever end of the distribution the capability is worst. Similarly, p_m does not give the actual proportion nonconforming, either.

Now assume that the process distribution is shifted so that $\mu = T = 44.5$. Of course, $C_{p_m^*}$ will still be 1.0649 since the index is not a function of μ. We may note from the expression for p_m that the latter simplifies to p_m^* when $\mu = T$. Thus, $C_{p_m p_m} = C_{p_m^*}$ and the "potential capability" is realized.

These calculations suggest that we should have, in general, $\mu = T$. Littig and Lam (1993) assume that the process engineer "wants to minimize proportion nonconforming under the additional constraint that the majority of units be produced near the target value." For a normal distribution this suggests that μ be set equal to T.

We must ask, however, if increasing the proportion of units produced that is a given distance from the target value more than offsets the increase in the proportion nonconforming. This question is not addressed in the Littig and Lam (1993) approach. Here setting the process mean equal to the target value increases the proportion nonconforming from .000022 to approximately .0014. If a loss function approach is not explicitly used, management would have to decide if a sizable increase in the proportion nonconforming from 22 per million to 14 per ten thousand is acceptable.

Since C_{pmk} performed well for the examples in Section 7.6.1, it is of interest to see how it performs for the examples in this section. If $\mu = T = 44.5$, then $C_{pmk} = 1$, whereas with $\mu = 42$ and $T = 44.5$, $C_{pmk} = 0.88$. If $\mu = T = 42$, then $C_{pmk} = 1.417$, whereas if $\mu = 44.5$ and $T = 42$, $C_{pmk} = 0.62$.

These numbers suggest that it is much more important to center the process distribution at the target value when the target is in the center of the specification limits than it is to do so when the target is not in the center. This is as it should be because when the target is equidistant from the specification limits, centering the process distribution at the target will give us the best of both worlds: We maximize the percentage of units that are a given distance from the target (assuming normality), and we also minimize the proportion of nonconforming units. When the target is not in the center, setting $\mu = T$ is questionable, especially when the target is not far from one of the tolerances.

If $T \geq 47.7$ instead of 44.5, shifting μ from 42 to T would cause C_{pmk} to *decrease*. We would certainly hope that C_{pmk} would decrease if we think of moving T close to USL. But at $\mu = 47.7$, 8.1% of the units produced will be nonconforming, so we would certainly want to receive a signal that we should not set $\mu = T$ for values of T less than 47.7. Thus, C_{pmk} does not perform as we would hope for this scenario.

There is a somewhat similar problem with $C_{p_m p_m}$ in that if, for example, $T > \mu$, then $C_{p_m^*} > C_{p_m p_m}$, which suggests that regardless of the relationship between T and USL, μ should be set equal to T. Thus, if $C_{p_m^*}$ is regarded as representing the potential capability, then the user would always set $\mu = T$ so as to have $C_{p_m p_m} = C_{p_m^*}$. Thus, whereas $C_{p_m p_m}$ and its relationship to $C_{p_m^*}$ may have value for the use illustrated by Littig and Lam (1993), it obviously has limited usefulness for other purposes.

7.8 MODIFIED *k* INDEX

Littig and Lam (1993) also recommend that the k index given by Kane (1986) be replaced by $k_N = \max\{(T - \mu)/(T - \text{LSL}), (\mu - T)/(\text{USL} - T)\}$. For the example with $\mu = 42$ we would have $k_N = 0.227$; if μ were shifted to 44.5, then, of course, $k_N = 0$. By comparison, $k = 0.417$ with $\mu = 42$, and $k = 0$ with $\mu = 44.5$.

The difference in the two indices is that k_N utilizes the distance between the process mean and the target value for the process mean, *relative* to the distance that the target value is from each specification limit. The k index, however, effectively ignores whether μ is greater than T or less than T.

Consider the following example. First assume that LSL = 15, $T = 25$, USL = 75, and $\mu = 20$. Then assume that μ changes to 30. Has the change in μ improved the process capability? Clearly improvement has occurred, but $k = 0.5$ for each value of μ. The improvement is reflected in the value of k_N, however, which changes from 0.5 for $\mu = 20$ to 0.1 for $\mu = 30$. In general, k_N is the superior measure of process capability since it is more closely aligned with the proportion of nonconforming units.

Thus, whereas it seems clear that k_N is a better capability indicator than k, the capability indices $C_{p_m p_m}$ and $C_{p_m^*}$ cannot be used to compare processes without an assumption of the "loss" that occurs when the value of a process characteristic differs from T by a certain amount, relative to the loss that occurs when a nonconforming unit is produced. (And we have also seen that $C_{p_m^*}$ cannot necessarily be used as a guide for the optimal value of $C_{p_m p_m}$.)

7.9 OTHER APPROACHES

Kaminsky et al. (1998) recommend that process capability indices not be used. Instead, they suggest that a probability distribution (e.g., a normal distribution) be fit to a set of data, and the proportion nonconforming determined "by integrating the density function over the appropriate regions." The authors suggest that the variability in the proportion nonconforming could then be determined by using the variance of the percent nonconforming for a binomial distribution.

Although this approach is clearly superior to blindly assuming normality, such an approach will likely be quite sensitive to the difference between the actual distribution and the assumed distribution. (Remember that "all models are wrong; some are useful," as has been emphasized by G. E. P. Box.) For example, assume that the USL is 65 and there is no LSL, the actual distribution is a chi-square distribution with 31 degrees of freedom, and the fitted distribution is a chi-square distribution with 30 degrees of freedom. Now there is hardly any difference between these two distributions, but the number nonconforming for the fitted distribution is 22 per 100,000 and 33 per 100,000 for the actual distribution. Thus, the percent nonconforming is underestimated by 50%, despite the fact that the two distributions are practically indistinguishable. We should expect the model we use to be much further from the true model than is the difference between these two distributions, so any distribution-fitting approach will likely provide a quality practitioner with only a very rough estimate of the actual proportion nonconforming.

Therefore, a distribution-fitting approach might not work. Furthermore, if a target value exists, we should be interested in some index that reflects, in part, departures from the target value. Therefore, we will frequently want a process capability index to give us more than a representation of the proportion nonconforming.

7.10 PROCESS CAPABILITY PLOTS

One problem with interpreting the value of a process capability index that is a function of a target value, such as $\hat{C}_{pmk}$, is that the value does not tell the user whether the value is due primarily to the difference between $\bar{x}$ and T or to the value of s^2. Assume, for example, that a user is satisfied with a particular value of an index. The large value may be caused by a small variance, with a process still off-target by an amount that can be easily corrected. Of course this could be identified just by comparing $\bar{x}$ with T, but it would obviously be helpful to see the contribution of the two components to the value of the index.

Vännman (1998) and Deleryd and Vännman (1998) provide *process capability plots* that can be used for this purpose, and the reader is referred to these sources for details.

7.11 PROCESS CAPABILITY INDICES VERSUS PROCESS PERFORMANCE INDICES

Whereas a process capability index measures the capability of an in-control process, a *process performance index* measures the (recent) performance of a process, without regard to whether the process was in control or out of control at a given point in time.

Pignatiello and Ramberg (1993) mention that some practitioners have criticized the editing of data (as in deleted points that can be identified as corresponding to out-of-control situations) before computing a capability index value. Certainly if a process is frequently out of control, then the value of a capability index will not adequately reflect process performance. Nevertheless, the words "process capability" do suggest that what is being measured is the *capability* of a process, not the performance of a process.

One possible approach would be to use both a performance index and a corresponding capability index and compare the values. If the values differed greatly, then this would suggest that the process is frequently out of control and is thus not realizing its capability.

Performance indices were discussed by McCoy (1991), and an effort to produce a standard approach for performance indices (and capability indices) has been led by Bigelow (1993).

One obvious use of a performance index would be when process monitoring is initiated on a particular process. At that point in time it is highly probable that the process is out of control, so a performance index would show how the process has performed without it being monitored.

One question that must be addressed is how should the parameters be estimated for process performance indices. Bigelow (1993) makes the point that s should be used in estimating σ (instead of $\overline{R}/d_2$ or $\overline{s}/c_4$) so that both common cause and special cause variation is captured. If a process has been frequently out of control in the recent past and the process mean has appeared to change by more than small amounts, the use of $\overline{R}/d_2$ or $\overline{s}/c_4$ could indeed produce a considerable underestimate of σ.

The guidelines must necessarily be somewhat different when individual observations are being used rather than subgroups. Here it is appropriate to use $\hat{\sigma} = s$ for process monitoring (i.e., for Stage 2). In computing the value of a process performance index, however, there is only Stage 1, so that s would be computed using all of the individual observations collected over a fixed time period.

We would similarly estimate μ by using the average of all of the observations in the stated time period, regardless of whether control is suggested or not. (Notice that in estimating μ, the question does not arise as to which estimator should be used.)

Bigelow (1993) defines performance indices analogous to their capability index counterparts. That is, P_{pk} is the process performance index that corresponds to C_{pk}. Specifically, P_{pk} is actually $\hat{C}_{pk}$ with μ and σ always estimated in the indicated manner. Thus, unlike the capability indices, there is not both a process performance index and the estimator for that index.

7.12 PROCESS CAPABILITY INDICES WITH AUTOCORRELATED DATA

Comparatively little attention has been given to estimating process capability indices when data are autocorrelated. Shore (1997) demonstrates with a simulation study that C_p and C_{pk} are overestimated when the true model is a third-order autoregressive model [an AR(3)] but independence is assumed. The problem is not great when $n \geq 100$, however, and certainly at least this many observations should be used in estimating process capability indices.

In general, some care must be exercised so as to not underestimate σ, and there are other considerations. See also Zhang (1998), who discusses the use of C_p and C_{pk} in the presence of autocorrelated data.

7.13 SUMMARY

Process capability indices have been used extensively since the mid-1980s. Practitioners should keep in mind that the value of a capability index is analogous to a point estimate of a population parameter, and the variability of the estimator of the true value of an index at a particular point should ideally be used in some manner. Such use might consist of a confidence interval, or perhaps even a control chart. The use of the latter might be an effective aid in enabling decision-makers to see the variability in index estimators.

There also should not be an attempt to seek a particular value of a process capability index. Rather, the attempt should certainly be to reduce variability as much as possible. Problems with the proposed capability indices were discussed in this chapter. It is important to understand that a capability index does not necessarily measure what it might seem to measure (as was seen in Section 7.2.3 for C_{pk}). Many people believe that process capability indices should not be used. This is evident from the title of the Pignatiello and Ramberg (1993) paper, and similar views were expressed by Nelson (1992) and Singpurwalla (1998). If process capability indices are used, the limitations of the commonly used process capability indices should be recognized.

Only univariate process capability indices were presented in this chapter; multivariate capability indices are covered in Section 9.14.

Process performance indices have been proposed as an alternative to process capability indices. The former purport to show the current performance of a process, not the capability of a process if the process were never out of control. Therefore, the computation of a process performance index may include data collected when a process was out of control.

REFERENCES

Beazley, C. E. and M. D. Marcucci (1988). Capability indices: Process performance measures. *ASQC Annual Quality Congress Transactions*, 516–523.

Bigelow, J. S. (1993). *Standard Method for Calculating Process Capability and Performance Measures, Revision 1.0*, ANSI-Z1 Committee on Quality Assurance, James S. Bigelow, chair.

Bissell, A. F. (1990). How reliable is your capability index? *Applied Statistics 39*: 331–340.

Boyles, R. A. (1991). The Taguchi capability index. *Journal of Quality Technology 23*(1): 17–26.

Chan, L. K., S. W. Cheng, and F. A. Spiring (1988). A new measure of process capability: C_{pm}. *Journal of Quality Technology 20*: 162–175.

Chou, Y., D. B. Owen, and S. A. Borrego A. (1990). Lower confidence limits on process capability indices. *Journal of Quality Technology 22*: 223–229.

Chou, Y. M., D. B. Owen, and S. A. Borrego A. (1992). Corrigenda. *Journal of Quality Technology 24*(4): 251.

Clements, J. A. (1989). Process capability calculations for non-normal distributions. *Quality Progress 22*(9): 95–100.

Deleryd, M. (1998). On the gap between theory and practice of process capability studies. *International Journal of Quality and Reliability Management 15*: 178–191.

Deleryd, M. and K. Vännman (1998). Process capability plots—a quality improvement tool. Research Report 1998:1, Division of Quality Technology and Statistics, Luleå University of Technology, Sweden.

Franklin, L. A. and G. Wasserman (1991). Bootstrap confidence interval estimates of C_{pk}: An introduction. *Communications in Statistics—Simulation and Computation 20*: 231–242.

Franklin, L. A. and G. Wasserman (1992a). A note on the conservative nature of the tables of lower confidence limits for C_{pk} with a suggested correction. *Communications in Statistics—Simulation and Computation 21*(4): 926–932.

Franklin, L. A. and G. Wasserman (1992b). Bootstrap lower confidence limits for capability indices. *Journal of Quality Technology 24*(4): 196–210.

Guirguis, G. H. and R. N. Rodriguez (1992). Computation of Owen's Q function applied to process capability analysis. *Journal of Quality Technology 24*(4): 236–246.

Gunter, B. (1989). The use and abuse of C_{pk}: Parts 1–4. *Quality Progress 22*(1): 72–73; *22*(3): 108–109; *22*(5): 79–80; *22*(7): 86–87.

Gunter, B. (1991). The use and abuse of C_{pk} revisited (response to Steenburgh). *Quality Progress 24*(1): 93–94.

Herman, J. T. (1989). Capability index—enough for process industries? *ASQC Annual Quality Congress Transactions*, 670–675.

Hsiang, T. C. and G. Taguchi (1985). A tutorial on quality control and assurance—the Taguchi methods. Unpublished presentation given at the Annual Meeting of the American Statistical Association, Las Vegas, NV.

Johnson, T. (1992). The relationship of C_{p_m} to squared error loss. *Journal of Quality Technology 24*(4): 211–215.

Kaminsky, F. Ċ., R. A. Dovich, and R. J. Burke (1998). Process capability indices: Now and in the future. *Quality Engineering 10*(3): 445–453.

Kane, V. E. (1986). Process capability indices. *Journal of Quality Technology 18*: 41–52 (corrigenda, p. 260).

Kotz, S. and N. L. Johnson (1993). *Process Capability Indices*. New York: Chapman and Hall.

Kotz, S. and C. R. Lovelace (1998). *Introduction to Process Capability Indices: Theory and Practice*. London: Arnold.

Kotz, S., W. L. Pearn, and N. L. Johnson (1993). Some process capability indices are more reliable than one might think. (A comment on Bissell 1990).) *Applied Statistics 42*(1): 55–62.

Kushler, R. H. and P. Hurley (1992). Confidence bounds for capability indices. *Journal of Quality Technology 24*(4): 188–195.

Lam, C. T. and S. J. Littig (1992). A new standard in process capability analysis. Technical Report No. 92-23, Department of Industrial and Operations Engineering, University of Michigan.

Littig, S. J. and C. T. Lam (1993). Case studies in process capability measurement. In *ASQC Annual Quality Congress Transactions*, pp. 569–575. Milwaukee, WI: ASQC.

McCoy, P. F. (1991). Using performance indexes to monitor production processes. *Quality Progress 24*(2): 49–55.

Nelson, P. R. (1992). Editorial. *Journal of Quality Technology 24*(4): 175.

Owen, D. B. (1965). A special case of a bivariate non-central *t*-distribution. *Biometrika 52*: 437–446.

Pearn, W. L. and K. S. Chen (1997). A practical implementation of the process capability index C_{pk}. *Quality Engineering 9*(4): 721–727.

Pearn, W. L., S. Kotz, and N. L. Johnson (1992). Distributional and inferential properties of process capability indices. *Journal of Quality Technology 24*(4): 216–231.

Pignatiello, J. J., Jr. and J. S. Ramberg (1993). Process capability indices: Just say "no." *ASQC Annual Quality Congress Transactions*, 92–104.

Rodriguez, R. N. (1992). Recent developments in process capability analysis. *Journal of Quality Technology 24*(4): 176–187.

Runger, G. C. (1993). Designing process capability studies. *Quality Progress 26*(7): 31–32.

Shore, H. (1997). Process capability analysis when data are autocorrelated. *Quality Engineering 9*(4): 615–626.

Singpurwalla, N. D. (1998). The stochastic control of process capability indices. *Test 7*(1): 1–33 (discussion: pp. 33–74).

Spiring, F. A. (1997). A unifying approach to process capability indices. *Journal of Quality Technology 29*(1): 49–58.

Vännman, K. (1995). A unified approach to capability indices. *Statistica Sinica 5*: 805–820.

Vännman, K. (1997a). Distribution and moments in simplified form for a general class of capability indices. *Communications in Statistics—Theory and Methods 26*: 159–179.

Vännman, K. (1997b). A general class of capability indices in the case of asymmetric tolerances. *Communications in Statistics—Theory and Methods 26*: 2049–2072.

Vännman, K. (1998). Process capability plots—a complement to capability indices. *Proceedings of the VIth International Workshop on Intelligent Statistical Quality Control*, Würzburz, Germany, pp. 270–283.

Vännman, K. and S. Kotz (1995a). A superstructure of capability indices—distributional properties and implications. *Scandinavian Journal of Statistics 22*: 477–491.

Vännman, K. and S. Kotz (1995b). A superstructure of capability indices—distributional properties and implications. *International Journal of Reliability, Quality, and Safety Engineering 2*: 343–360.

Wang, C. M. and C. T. Lam (1996). Confidence limits for proportion of conformance. *Journal of Quality Technology 28*(4): 439–445.

Zhang, N. F. (1998). Estimating process capability indexes for autocorrelated data. *Journal of Applied Statistics 25*(4): 559–574.

EXERCISES

1. Show that $C_{pk} = C_p$ when μ is equidistant from USL and LSL.

2. Use

$$E(\chi_{n-1}^{-1}) = \Gamma\left(\frac{n-2}{2}\right) \Big/ \left(\sqrt{2}\,\Gamma\left(\frac{n-1}{2}\right)\right)$$

and

$$c_4 = \sqrt{\frac{2}{n-1}}\,\Gamma\left(\frac{n}{2}\right) \Big/ \Gamma\left(\frac{n-2}{2}\right)$$

to show that $(1/c_4^2)[(n-2)/(n-1)]\hat{C}_p$ is an unbiased estimator of C_p when $\hat{\sigma} = s/c_4$.

3. A manufacturing process has specification limits of 0.99 and 1.02 and a standard deviation of 0.005.

(a) What is the value of C_{pk} if the distribution of measurements is centered at the target value of 1.00?

(b) Will C_{pk} increase or decrease if the mean shifts to 1.005 (i.e., halfway between the specification limits)?

(c) Assuming a normal distribution, will the proportion of nonconforming units increase or decrease with a mean shift to 1.005? What is the proportion before and after the shift?

(d) If, rather than being normal, the distribution is actually highly asymmetric with most of the area under the curve lying to the right of the mean, will the answer to part (c) still hold? What does this imply about the steps that should be followed when trying to improve process capability?

4. An $\overline{X}$ chart exhibits control for the control limits of UCL = 30.26 and LCL = 26.34. If subgroup sizes of $n = 4$ were used and the specification limits are 20 and 36, what is the value of $\hat{C}_{pk}$? What would you recommend to this company?

5. An X chart exhibits control for 50 observations with UCL = 75.8 and LCL = 64.4. To what value will $\hat{\sigma}$ have to be reduced to produce a $\hat{C}_{pk}$ of 1.33 if the specification limits are 76 and 62 and $\hat{\mu}$ remains unchanged? Could a $\hat{C}_{pk}$ value of at least 1.33 be obtained by centering the distribution so that $\hat{\mu} = 69$? Comment.

6. Assume that $\mu = 75.6$, $\sigma = 3$, USL = 95, and LSL = 60. Compute the values of C_{pp} and C_{p^*} and explain why they are similar.

7. A process engineer wishes to compare several processes in terms of process capability. What capability index (or indices) would you recommend if (a) the processes do not have target values, and (b) the processes *do* have target values?

8. Explain the difference between a process capability index and a process performance index.
9. If there is strong evidence that a process characteristic has a markedly nonnormal distribution, what options does the user have in selecting a capability index?
10. Explain why the value of a process capability index could not necessarily be computed from a set of recently constructed control charts for a process characteristic.
11. Consider the scenario in exercise 5 and construct a 95% lower confidence limit for C_{pk} using the expression in Eq. (7.3). Do you consider 50 observations to be sufficient for obtaining a lower confidence limit (and a point estimate) of C_{pk}? Comment.
12. Construct a normal probability plot for the data in Table 7.1 and comment.
13. Construct a 95% confidence interval for C_p using the data in Table 7.1 and comment on the usefulness of this interval in light of the numerical values that were given for this data set in Section 7.3.3.
14. Explain why there is not a monotonic relationship between $\hat{C}_{pk}$ and the percentage of nonconforming units.

CHAPTER 8

Alternatives to Shewhart Charts

The Shewhart and Shewhart-type charts that were presented in Chapters 4–6 are the most commonly used control charts. Some have even argued that Shewhart charts are superior to other types of control charts/procedures and should be used exclusively. In particular, Deming (1993, p. 180) states: "The Shewhart charts do a good job under a wide range of conditions. No one has yet wrought improvement."

Times change, however, and charts with superior properties have been developed. This is to be expected; how many people would want to drive a car that was made in 1924, the year that Shewhart sketched out the idea for a control chart? These superior charts are discussed in considerable detail in this chapter.

See Woodall and Montgomery (1999) and Montgomery and Woodall (1997) for the same general view regarding the need to consider and use alternative control chart procedures. In particular, the latter state: "In many cases the processes to which SPC is now applied differ drastically from those which motivated Shewhart's methods."

8.1 INTRODUCTION

We would certainly want whatever procedure we use for controlling the process mean μ to enable us to detect a shift in the mean that is of any consequence as quickly as possible, but not produce a high rate of false signals. For example, we want to quickly detect a shift that would cause nonconforming units to be produced, but not to frequently receive signals suggesting a mean shift when in fact there is no shift.

Assume for the moment that our only objective in process control is to prevent the production of nonconforming units and that this objective will determine our control chart construction and usage.

We will assume that we have a process in which the process characteristic under scrutiny has parameter values $\mu = 75$ and $\sigma = 5$ and the specifications are 55 and 95. The specifications cannot be altered and values outside these specifications will result in the unit being nonconforming. If we assume that the process characteristic has virtually a normal distribution, the fact that the specifications are four standard deviation units from the mean will result in about 6 of every 100,000 units produced being nonconforming. Idealistically, even that number is too high, as at most a small number of nonconforming parts per million is considered acceptable, but this might be the best performance that the current process can attain.

If an $\overline{X}$ chart is being used to control the process mean, we will assume that LCL $= 67.5$ and UCL $= 82.5$. (This is what the limits would be if $\overline{\overline{x}} = \mu$, $\hat{\sigma} = \sigma$, and $n = 4$.) If there is a $2\sigma_{\overline{x}}$ upward shift in the process mean (to 80), it would certainly be of interest to determine the effect that this will have on the number of nonconforming units that will be produced, and also how long it should take to detect this shift with an $\overline{X}$ chart.

The effect of the mean shift will result in a smaller percentage of units falling below the lower specification (3 out of 10 million instead of 3 out of 100,000), but a much higher percentage falling above the upper specification (1 out of 1000 instead of 3 out of 100,000). The net effect is to increase the (expected) number of nonconforming units from 6 out of 100,000 to 135 out of 100,000. If the product is a high-volume item such as a small steel ball, this shift would result in a considerable increase in the number of nonconforming units produced per day. Obviously it would be desirable to detect this shift as soon as possible. How long will it take to detect this shift with an $\overline{X}$ chart?

This question can be answered by determining $P(\overline{X} > 82.5)$ when $\mu = 80$. Thus,

$$z = \frac{82.5 - 80}{2.5}$$
$$= 1$$

and $P(Z > 1) = .1587$. [We can ignore $P(\overline{X} < 67.5)$ when $\mu = 80$ since its value is virtually zero.] The concept of average run length (ARL) was introduced in Section 4.14, and some similar calculations were given there. For subgroup data the ARL is the expected (i.e., average) number of subgroups that would be obtained before an out-of-control signal is received. For example, if with an $\overline{X}$ chart the process mean is in control at $\overline{\overline{x}}$, the probability that a subgroup average will fall outside the control limits is .0027 when the subgroup averages are plotted one at a time. The ARL is then $1/.0027 = 370.37$. Thus, we would "expect" to observe 371 subgroups before receiving an out-of-control signal. (Obviously we want this number to be high when the process is in control. The justification for this ARL calculation is given in the Appendix to Chapter 4.)

Thus, for the mean shift to $\mu = 80$, ARL $= 1/0.1587 = 6.30$. If samples (subgroups) are taken every 30 minutes, we would expect to detect the shift 3.5 hours after the shift had occurred. Accordingly, if 100,000 steel balls were produced per 8-hour day, we would expect about 60 nonconforming units to be produced before the shift is detected. We can improve considerably on the ability to detect small mean shifts by using cumulative sum (CUSUM) procedures.

8.2 CUMULATIVE SUM PROCEDURES: PRINCIPLES AND HISTORICAL DEVELOPMENT

Cumulative sum charts were first proposed by Page (1954), and a number of modifications have resulted since then. As the name implies, sums are accumulated, but an observation (which may be a single reading or a statistic obtained from a sample) is "accumulated" only if it differs from the goal value (e.g., the estimate of the process mean) by more than k units. Not all CUSUM procedures require the use of charts, but charts can nevertheless be used with any CUSUM procedure.

Some CUSUM procedures are easy to learn (and to program) and are also quite intuitive, while others are somewhat more involved. The emphasis in this section will be on the former.

Accordingly, V-mask CUSUM schemes will not be covered. These schemes, if used manually, require the use of a special cutoff or transparency (in the form of a V) rotated 90° counterclockwise so that the mask could be repositioned after each sample. Otherwise, the mask would have to be redrawn each time, which would be unduly laborious. The V-mask approach could also be computerized so that the mask would automatically appear in the proper position after each sample. A signal results when a point plots outside the "arms" of the mask. But the time of the signal is not the time for a point that is outside the mask; rather it is the time that corresponds to the vertex of the mask. This is potentially confusing. It is questionable how often the V-mask approach is still used. V-mask control schemes are covered in various texts, including Van Dobben de Bruyn (1968).

If quality control personnel are to make the transition from $\overline{X}$ charts to CUSUM procedures, it seems reasonable to assume that they will be drawn toward procedures that are easy to understand and that bear some relationship to an $\overline{X}$ chart. The methods proposed by, for example, Lucas (1982) and Lucas and Crosier (1982a) meet both requirements and will now be examined extensively.

8.2.1 CUSUM Procedures Versus $\overline{X}$ Chart

Could the shift be detected any faster by using a substitute for an $\overline{X}$ chart? The answer is "yes." With one of the CUSUM procedures to be presented in this section the ARL = 3.34, and with a slightly different CUSUM procedure, the ARL = 3.226. With either procedure the shift would be detected within 2 hours as compared to 3.5 hours with an $\overline{X}$ chart. For this example about 26 fewer nonconforming units would be produced before the shift is detected.

Historically, the $\overline{X}$ chart has been considered effective in quickly detecting large mean shifts. For example, a 3-sigma shift (in this case to $\mu = 82.5$) would have an ARL $= 1/.5 = 2$, and a 4-sigma shift would have an ARL $= 1/.8413 =$ 1.19. With the first CUSUM procedure alluded to the corresponding numbers are 2.19 and 1.71, and for the second CUSUM procedure the numbers are 1.922 and 1.322.

Thus, for detecting large mean shifts it is difficult to improve upon the performance of an $\overline{X}$ chart, but it is possible to design a CUSUM procedure that will be quite competitive (if not superior) to an $\overline{X}$ chart in detecting such shifts. For detecting small, sustained shifts (around 1-sigma), CUSUM procedures are far superior to an $\overline{X}$ chart.

The following example with simulated data should serve to illustrate how a small mean shift can be "camouflaged" by an $\overline{X}$ chart. Figure 8.1 is an $\overline{X}$ chart that consists of averages of four observations generated from a normal distribution. Initially, the distribution is $N(0, 1)$, but beginning at some point the data are generated from a distribution that is $N(0.5, 1)$. Can you determine visually the point at which the shift occurs?

Since $\sigma = 1$, $\sigma_{\overline{x}} = 1/\sqrt{4} = 0.5$, so the shift is a 1-sigma shift. Yet all of the points are well within the control limits, and it may seem at first glance as though the chart contains a random configuration of points. Closer examination, however, reveals that 10 consecutive averages are above the midline (11–20). The probability of that happening by chance if the mean is at $\mu = 0$ is (using the binomial distribution again) $\left(\frac{1}{2}\right)^{10} = .001$. Since this is a very small probability, we would certainly suspect that a shift had occurred by subgroup number 20. This is an application of what is referred to as *runs criteria* in conjunction with an $\overline{X}$ chart. See, for example, Champ and Woodall (1987). Criteria include seven consecutive points above or below the midline and a number of consecutive points outside of 1-sigma limits or 2-sigma limits. Such criteria could easily be established so that the probability of meeting any of the criteria is quite small when the process is in control. Thus, if any of the criteria are met, we would have reason to suspect that there has been a shift in the process mean.

If the control charts are being maintained by hand, this would require that an operator look not only at the 3-sigma limits but also at 1- and 2-sigma limits,

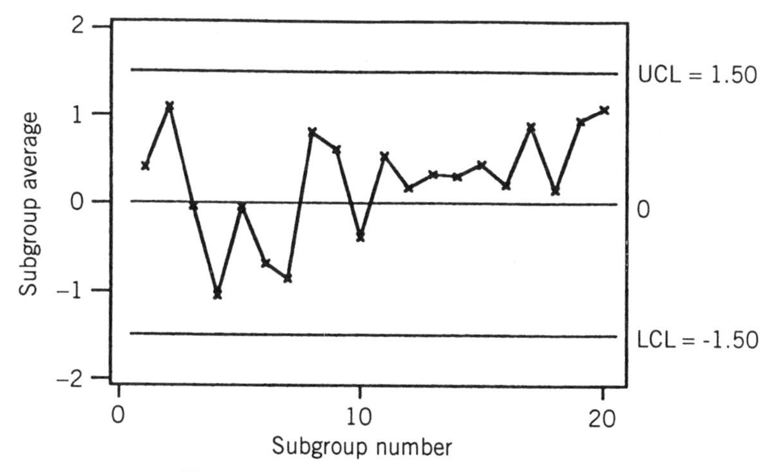

FIGURE 8.1 $\overline{X}$ chart for data generated from $N(0, 1)$ and $N(0.5, 1)$; $n = 4$.

and look for runs of points. Such an approach could, of course, be computerized, which would require that one set of criteria be selected from a number that have been proposed. The choice of criteria would unquestionably be somewhat subjective, however, and it would be much easier to use some simple, straightforward procedure in place of runs criteria.

In addition to being somewhat impractical, a Shewhart chart with runs rules will more importantly have a very short in-control ARL. Specifically, if all of the Western Electric runs rules (as they are called) are used in addition to the Shewhart limits, the in-control ARL is only 91.75, as shown by Champ and Woodall (1987), who also gave the in-control ARLs for various combinations of the runs rules. Stated differently, the false alarm rate is more than four times the false alarm rate of a Shewhart chart, assuming normality. Thus, a Shewhart chart with runs rules is not a viable alternative to CUSUM procedures. [See, e.g., Montgomery (1996, p. 148) for information on the Western Electric runs rules.]

Returning to the example, where did the shift occur? The random numbers from $N(0.5, 1)$ were generated starting with subgroup 13. We will see how quickly the shift would be detected with the suggested CUSUM procedures.

For an $\overline{X}$ chart a subgroup average will be outside the control limits if z is either greater than 3 or less than -3, where

$$z = \frac{\overline{x} - \overline{\overline{x}}}{\hat{\sigma}_{\overline{x}}} \tag{8.1}$$

This follows from the fact that if we solve Eq. (8.1) for $\overline{x}$, we obtain

$$\overline{x} = \overline{\overline{x}} + z\hat{\sigma}_{\overline{x}} \tag{8.2}$$

Notice that the right-hand side of Eq. (8.2) gives the UCL of an $\overline{X}$ chart if $z = 3$ and the LCL if $z = -3$. Thus, if the value of $\overline{x}$ in Eq. (8.1) produces $z = 2$, then $\overline{x} = \overline{\overline{x}} + 2\hat{\sigma}_{\overline{x}}$ is inside the UCL by the amount $\hat{\sigma}_{\overline{x}}$. If the value of $\overline{x}$ produces $z = 4$, then $\overline{x} = \overline{\overline{x}} + 4\hat{\sigma}_{\overline{x}}$ is outside the UCL by the amount $\hat{\sigma}_{\overline{x}}$.

Therefore, a worker who wanted to know whether or not a subgroup average would fall outside the control limits on an $\overline{X}$ chart could simply calculate the z-value and would not have to know the control limits. This might seem an awkward alternative to an $\overline{X}$ chart, but it is important to recognize that a list of z-values would indicate whether the corresponding subgroup averages are inside or outside the control limits. A list of such z-values is an integral part of the method proposed by Lucas (1982).

The pair of cumulative sums used is

$$S_{Hi} = \max[0, (z_i - k) + S_{Hi-1}] \quad S_{Li} = \min[0, (z_i + k) + S_{Li-1}] \tag{8.3}$$

where the first is for detecting mean increases and the second is for detecting mean decreases. We note that some authors, such as Lucas (1982) and Lucas and Crosier (1982a), define the lower sum as $S_{Li} = \max[0, (-z_i - k) + S_{Li-1}]$. We adopt the convention given here for two reasons: (1) as Hawkins and Olwell (1998) state, it seems more reasonable to define the lower sum in such a way that it decreases when there is a mean decrease, and (2) using Eq. (8.3) allows both sums to be plotted on one chart. The value of k is usually selected to be one-half of the mean shift (in z units) that one wishes to detect. We have seen that an $\overline{X}$ chart is effective in detecting a large mean shift such as a 3- or 4-sigma shift. Therefore, there would be no point in setting $k = 1.5$ or $k = 2.0$. The usual choice is $k = 0.5$, which is the appropriate choice for detecting a 1-sigma shift. Two sums, S_{Hi} and S_{Li}, are computed for each z_i, where z is as defined in Eq. (8.3) and i designates the ith subgroup. The sums start with $S_{Ho} = 0$ and $S_{Lo} = 0$. The sums are generally reset if an assignable cause is subsequently discovered and removed after an out-of-control signal is received, but the sums do not have to be reset to zero. This will be explained later.

There obviously must be some "threshold" value such that when either sum exceeds that value an out-of-control signal will be given. This is generally designated by h in the literature. Typically, h is chosen to be either 4 or 5. The logic behind those choices will be discussed later.

We shall now illustrate this procedure with the simulated data that were graphed in Figure 8.1. The data are given in Table 8.1 and the cumulative sums are given in Table 8.2.

TABLE 8.1 Simulated Data Charted in Figure 8.1

From N(0,1)

1.54	0.86	−0.89	−1.88	−1.85	−2.53
−0.09	0.57	0.21	−0.43	2.03	−0.59
1.75	1.17	−1.23	−0.42	−0.64	0.60
−1.58	1.82	1.77	−1.45	0.31	−0.22
$\bar{x} = 0.41$	1.11	−0.04	−1.04	−0.04	−0.68
−0.74	2.10	0.56	−1.53	0.53	−0.81
−1.25	1.48	1.78	0.99	−0.52	0.67
−0.40	0.86	−0.81	−2.38	1.71	0.42
−1.01	−1.19	0.97	1.41	0.43	0.46
$\bar{x} = -0.85$	0.81	0.62	−0.38	0.54	0.18

From N(0.5,1)

0.84	0.22	2.30	2.14	1.03	−0.90	1.56	1.28
−0.71	1.27	−0.33	0.51	0.30	1.71	−0.70	0.98
0.27	0.64	0.19	−1.65	0.55	−1.08	2.06	1.29
0.93	−0.83	−0.38	−0.14	1.65	0.93	0.88	0.81
$\bar{x} = 0.33$	0.32	0.44	0.22	0.88	0.16	0.95	1.09

TABLE 8.2 CUSUM Values

Subgroup Number	Subgroup Average	z	S_H	S_L
1	0.41	0.82	0.32	0
2	1.11	2.22	2.04	0
3	−0.04	−0.08	1.46	0
4	−1.04	−2.08	0	−1.58
5	−0.04	−0.08	0	−1.16
6	−0.68	−1.36	0	−2.02
7	−0.85	−1.70	0	−3.22
8	0.81	1.62	1.12	−1.10
9	0.62	1.24	1.86	0
10	−0.38	−0.76	0.60	−0.26
11	0.54	1.08	1.18	0
12	0.18	0.36	1.04	0
13	0.33	0.66	1.20	0
14	0.32	0.64	1.34	0
15	0.44	0.88	1.72	0
16	0.22	0.44	1.66	0
17	0.88	1.76	2.92	0
18	0.16	0.32	2.74	0
19	0.95	1.90	4.14	0
20	1.09	2.18	5.82[a]	0

[a] Exceeds $h = 5$.

By studying Table 8.2, we can practically "see" how the $\overline{X}$ chart would look without having constructed it. For example, if we look at the first seven values for S_H and S_L, we see a string of nonzero numbers for S_H (and zeros for S_L), followed by a string of nonzero numbers for S_L (and zeros for S_H). This tells us that the first few subgroup averages are considerably above μ (or $\overline{\overline{x}}$, in general) and the next few averages are considerably below μ. The fact that the sums reach 2.04 for S_H and −3.22 for S_L indicates that at least one average had to be far above μ and at least one average had to be far below μ. [If every average was within $\frac{1}{2}$ sigma ($\sigma_{\overline{x}}$) of μ, all of the values for S_H and S_L would be zero.] Of course, we can see this more easily by looking at the averages in Table 8.2, but the important point is that whenever there is a trend it will be reflected by the values of S_H and/or S_L.

The trend begins to appear shortly after subgroup 7, and an out-of-control signal is received at subgroup 20, using $h = 5$. Recall that the shift actually began with subgroup 13 so the shift was detected after eight subgroups.

The cumulative sums need not be displayed graphically, but they can be. A chart could be constructed that has a horizontal line at the value of h, and the values of S_H and S_L could be displayed on the same chart or on separate charts. They are shown on the same chart in Van Dobben de Bruyn (1968) and are similarly shown on the same chart for the Table 8.1 data in Figure 8.2. This seems preferable.

(Note that although Figure 8.2 was produced using Release 12 of Minitab, the latter produces CUSUM charts with the decision lines at $\pm h\hat{\sigma}_{\overline{x}}$ for subgroup data, rather than using h. Decision lines at $\pm h$ in Figure 8.2 were produced by dividing the observations by $\hat{\sigma}_{\overline{x}}$.)

How long would we expect the shift to go undetected using this approach? The answer is contained in Table 8.3, which gives the ARL values for various mean shifts with $h = 4$ and $h = 5$. We can see from Table 8.3 that we would expect the shift to be detected after the tenth subgroup (beyond 10 since the average is 10.4). These values are based on the assumption that both S_H and S_L are zero at the time of the shift. In our example, however, $S_H = 1.04$ at the time of the shift. If S_H had been equal to zero at subgroup 12, then S_H would have been 4.78 at subgroup 20 and the shift would have probably been detected after another one to three subgroups.

How long would it have taken on average to detect this shift using an $\overline{X}$ chart without runs criteria? The answer is 43.89 subgroups. (It is 43.96 if the lower tail area is ignored.) Thus, if subgroups were obtained every 15 minutes and a workday consists of 8 hours, it would take more than one full day longer when an $\overline{X}$ chart is used in this manner than with this particular CUSUM procedure. This applies for *every* $\overline{X}$ chart—not just for this particular example.

For this example $h = 5$ was used, but this should not be taken to imply that $h = 5$ will always be the best choice. Indeed, a fractional value might be used.

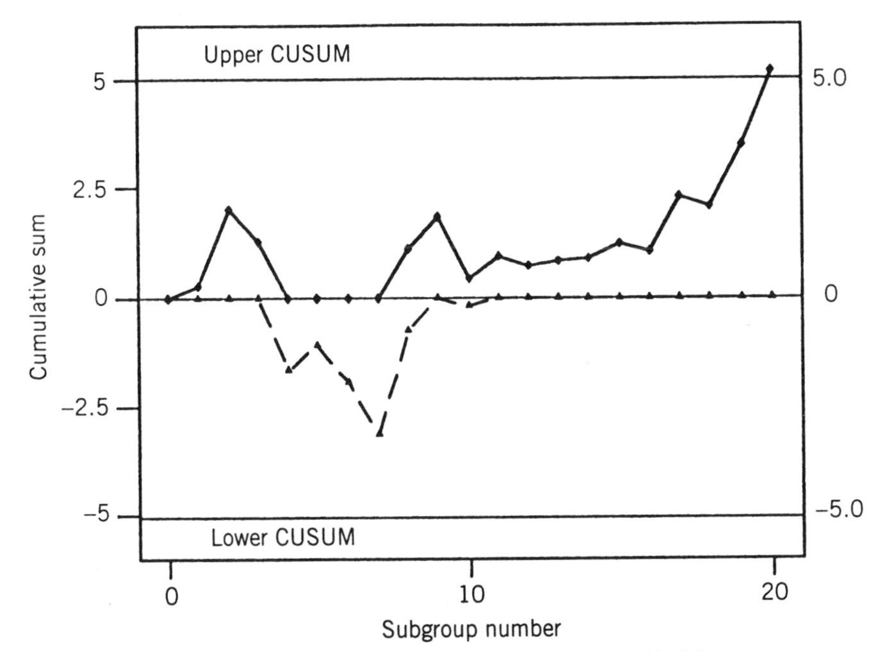

FIGURE 8.2 Chart for S_H and S_L for the data in Table 8.1.

TABLE 8.3 Average Run Lengths for a CUSUM Procedure with $h = 4$ or $h = 5$ ($k = 0.5$)

Mean Shift (in standard units)	$h = 4$	$h = 5$
0	168	465
0.25	74.2	139
0.50	26.6	38.0
0.75	13.3	17.0
1.00	8.38	10.4
1.50	4.75	5.75
2.00	3.34	4.01
2.50	2.62	3.11
3.00	2.19	2.57
4.00	1.71	2.01
5.00	1.31	1.69

Source: Adapted from Table 1 of Lucas and Crosier (1982a).

As Table 8.3 indicates, there is a considerable difference between the in-control ARL values for $h = 4$ and $h = 5$ (168 vs. 465). This is offset to some extent, however, by the fact that the ARL values are larger using $h = 5$ for every mean shift. For example, for a 1-sigma shift about two extra subgroups are required. Thus, if subgroups are obtained every 30 minutes, it would require about 1 hour longer (with $h = 5$) to detect the shift. If the shift could cause a considerable increase in the number of nonconforming units (as was illustrated previously), that extra delay might be deemed critical. Conversely, if the cost of searching for assignable causes when none exists is substantial, the difference between the in-control ARL values might be of critical importance. Thus, the choice should be made after careful consideration of all the relevant factors—both statistical and nonstatistical. In practice, CUSUM schemes with $k = 0.5$ and $h = 5$ have been frequently used. Woodall (1986) reviewed the construction of CUSUM charts, and Gan (1991) provided graphs that can be used for implementing an optimal CUSUM scheme, where the latter is defined as the CUSUM scheme that gives the minimum ARL for the amount of mean shift that a user is most interested in detecting, for a selected in-control ARL.

8.2.2 Fast Initial Response CUSUM

The data in Table 8.2 illustrate how an out-of-control signal is obtained after a mean shift. When such a signal is received, the CUSUM user would search for an assignable cause just as he or she would when a subgroup average plots outside the limits on an $\overline{X}$ chart. After a search has been conducted, the sums could either be reset or left as is. If a cause was not found, the sums might be left as is to see if the appropriate sum stays above the threshold value. If it does, that might mean that a cause does exist but it was simply not detected. Another search might then be initiated. The sums should definitely be reset, however, when an assignable cause is both detected and removed. The sums *could* be reset to zero. It might be, however, that the process was out of control because of multiple causes, and all of them might not have been removed. If the process is still out of control, we would certainly want to detect that as soon as possible. The detection can be accomplished faster by using a "headstart" value as presented in Lucas and Crosier (1982a). They recommend a headstart value of $h/2$. (With our form for the lower sum, the headstart value for the lower sum would be $-h/2$.) Their method is generally referred to as fast initial response (FIR) CUSUM. As with any new CUSUM scheme, the objective should be to achieve a small ARL for mean shifts, especially for the order of magnitude that one wishes to detect, without significantly reducing the in-control ARL. The FIR CUSUM does have this property. For example, with $h = 5$ the ARL for a 1-sigma shift is reduced from 10.4 to 6.35, but the in-control ARL is reduced only from 465 to 430. In fact, this FIR CUSUM with $h = 5$ is far superior to the CUSUM with $h = 4$ without the headstart value since the in-control ARL is considerably longer and all but the $\frac{1}{4}$- and $\frac{1}{2}$-sigma shifts are shorter [See Lucas and Crosier (1982a) for the complete tables.]

The effect of the headstart value is illustrated in Table 8.4, which is produced from additional random numbers generated from $N(0.5, 1)$. We will continue to assume that $h = 5$ is being used and that Table 8.4 is a continuation of Table 8.2. Since in Table 8.2 the out-of-control signal is received at subgroup 20, we will assume that the cause was detected and removed at that point, so the sums are reset before subgroup 21.

TABLE 8.4 Additional Data Generated from $N(0.5, 1)$ to Illustrate the Effect of FIR CUSUM versus CUSUM without FIR ($k = 0.5$, $h = 5$)

			With FIR		Without FIR	
Subgroup	$\bar{x}$	z	S_H	S_L	S_H	S_L
(Reset)	—	—	2.5	−2.5	0	0
21	−0.08	−0.16	1.84	−2.16	0	0
22	0.57	1.14	2.48	−0.52	0.64	0
23	0.80	1.60	3.58	0	1.74	0
24	0.23	0.46	3.54	0	1.70	0
25	0.08	0.16	3.20	0	1.36	0
26	1.33	2.66	5.36[a]	0	3.52	0
27	1.23	2.46			5.48[a]	0

[a] Indicates out-of-control signal.

There are several important points to notice about Table 8.4. We see that with the FIR feature the fact that the process was still out of control was detected after six subgroups, whereas it took seven subgroups to detect this without the FIR feature. (The ARL values are 6.35 and 10.4, respectively, as can be seen later from Table 8.6.) Notice also that the headstart value was used for both S_H and S_L even though we "knew" in this case that there was a positive shift and not a negative shift. In general, however, just because an assignable cause that resulted in a positive shift was removed does not preclude the possibility that the process might now be out of control in the other direction. For example, a machine could be overadjusted. If the shift is still positive the headstart value for S_L will not likely have any effect, as Table 8.4 illustrates.

The question arises as to the effect of the headstart value when the cause has been removed so that the process is subsequently in control. As indicated previously, it causes only a slight decrease in the in-control ARL values; the decrease is from 465 to 430 for $h = 5$ (see Table 8.6 below).

Additional data were generated from $N(0, 1)$ and we can see in Table 8.5 that for this particular set of data, the headstart values have no effect. The effect of the headstart values quickly dissipates to a difference of 1.00 by subgroup 23, and it can be seen that there is a constant difference of 1.00 between S_H with FIR and S_H without FIR from that point until subgroup 32. The headstart for S_H then "zeros out" at the next subgroup. From this point on the FIR CUSUM would be the same as a basic CUSUM procedure, and the ARLs for the latter would then apply.

TABLE 8.5 Additional Data Generated from $N(0, 1)$ Illustrating Effect of FIR CUSUM When the Process Is in Control ($k = 0.5$, $h = 5$)

			With FIR		Without FIR	
Subgroup	$\bar{x}$	z	S_H	S_L	S_H	S_L
(Reset)	—	—	2.5	−2.5	0	0
21	−0.28	−0.56	1.44	−2.56	0	−0.06
22	0.07	0.14	1.08	−1.92	0	0
23	0.21	0.42	1.00	−1.00	0	0
24	0.46	0.92	1.42	0	0.42	0
25	0.55	1.10	2.02	0	1.02	0
26	0.77	1.54	3.06	0	2.06	0
27	−0.30	−0.60	1.96	−0.10	0.96	−0.10
28	0.09	0.18	1.64	0	0.64	0
29	0.69	1.38	2.52	0	1.52	0
30	0.44	0.88	2.90	0	1.90	0
31	−0.26	−0.52	1.88	−0.02	0.88	−0.02
32	−0.34	−0.68	0.70	−0.20	0	−0.20
33	−0.28	−0.56	0	−0.26	0	−0.26

[a] Indicates out-of-control signal.

TABLE 8.6 ARL Values for Various CUSUM Schemes Using $h = 5$ and $k = 0.5$[a]

Mean Shift (in Standard Units)	Basic CUSUM (1)	Shewhart–CUSUM ($z = 3.5$) (2)	FIR CUSUM (3)	Shewhart–FIR CUSUM ($z = 3.5$) (4)
0	465	391	430	359.7
0.25	139	130.9	122	113.9
0.50	38.0	37.15	28.7	28.09
0.75	17.0	16.80	11.2	11.15
1.00	10.4	10.21	6.35	6.32
1.50	5.75	5.58	3.37	3.37
2.00	4.01	3.77	2.36	2.36
2.50	3.11	2.77	1.86	1.86
3.00	2.57	2.10	1.54	1.54
4.00	2.01	1.34	1.16	1.16
5.00	1.69	1.07	1.02	1.02

[a] Columns 1 and 3 were adapted from Lucas and Crosier (1982a); columns 2 and 4 were adapted from Lucas (1982). The last five values in column 4 were obtained using the one-sided CUSUM scheme ARL values in Table 2 as (more accurate) approximations to the two-sided values.

By this time it should be apparent that the only way an $\overline{X}$ chart could ever be competitive with the CUSUM procedures described here (with and without FIR) for detecting small mean shifts would be for runs criteria to be designed in such a way as to approximate (or improve upon) the ARL properties of a CUSUM scheme. This would not be possible, however, since Moustakides (1986) proved that a CUSUM scheme is the optimal scheme for detecting a shift of a certain magnitude among all procedures that have the same in-control ARL.

The CUSUM procedures presented in this chapter are easy to use and should not be difficult for production personnel to employ. [See, e.g., Box and Luceño (1997, p. 83) for a similar view.] Therefore, CUSUM procedures should be used to a greater extent than has been the case.

8.2.3 Combined Shewhart–CUSUM Scheme

Either of the two CUSUM procedures discussed so far can be improved and made competitive with an $\overline{X}$ chart (or an X chart) for detecting large mean shifts.

A CUSUM procedure is often compared to a Shewhart chart—sometimes erroneously. Hawkins and Olwell (1998, p. 43) point out that a basic CUSUM with $k = 3$ and $h = 0$ is equivalent to a Shewhart chart. That is, the CUSUM decision interval is exceeded when $|z| > 3$, just as it is with a Shewhart chart. Because of this equivalence, it follows that an X chart is the best procedure for detecting a shift of six standard deviations. Of course, we are generally interested in detecting much smaller shifts, but the authors' point is that a Shewhart chart is optimal for detecting a very large shift. [Note that this does not contradict the result of Moustakides (1986) because of the equivalence of a CUSUM procedure with $h = 0$, $k = 3$, and a Shewhart chart.] Therefore, it is desirable to modify a basic CUSUM scheme so that it will be competitive with a Shewhart chart for detecting a large mean shift. That is, since a CUSUM procedure is the optimal procedure for detecting a shift of a specified size, it will not be optimal for shifts of other sizes. If we tune a CUSUM to detect a small shift, it will not be optimal for detecting any large shift. Accordingly, a modification is necessary if we want a CUSUM procedure to have good properties for detecting both small and large shifts.

This modification entails using $\overline{X}$-chart limits in conjunction with a CUSUM scheme. The result is called a *combined Shewhart–CUSUM scheme*, as introduced by Lucas (1982). The difference between this scheme and the other two CUSUM schemes is that an out-of-control signal can be received not just from S_H or S_L but also from the z value. We *could* select $z = 3$ as the threshold value so that an out-of-control signal would be received if either the absolute value of z exceeds 3 or either S_H or S_L exceeds h. This would then be, in essence, a modified $\overline{X}$-chart procedure since signals are also received from an $\overline{X}$ chart when z exceeds 3. Accordingly, the in-control ARL for the combined Shewhart–CUSUM scheme must be less than the in-control ARL for an $\overline{X}$ chart, which, we recall, is approximately 370.

It can be seen (from Lucas, 1982) that the in-control ARL for a CUSUM scheme with $h = 5$, $k = 0.5$, $z = 3$ and without FIR is 223.4, whereas the in-control ARL for the same scheme *with* FIR is 206.5. Since these numbers are considerably less than 370, a z-value greater than 3 is preferable. Table 8.6 gives the ARL values for (1) a basic CUSUM scheme (the first type introduced), (2) a combined Shewhart–CUSUM scheme with $z = 3.5$, (3) an FIR CUSUM scheme, and (4) a combined Shewhart–FIR CUSUM scheme with $z = 3.5$. By comparing columns 1 and 2, we can see the effect of the Shewhart limits, especially in detecting shifts of at least 3-sigma. This improvement, however, is offset to some extent by a considerable reduction in the in-control ARL. By comparing columns 3 and 4, however, we can see that adding Shewhart limits to a FIR CUSUM has virtually no effect for mean shifts of any magnitude. Lucas (1982) shows the beneficial effects of adding Shewhart limits to a basic CUSUM and of adding the FIR option to a combined Shewhart–CUSUM scheme. (Here this can be seen by comparing columns 2 and 4.) Thus, adding Shewhart limits can be beneficial if added to a basic CUSUM, but not if they are added to a FIR CUSUM. In particular, we can see that the addition of Shewhart limits to a FIR CUSUM has no effect on the ARL for shifts greater than 1-sigma and decreases the in-control ARL considerably. Thus, the actual effect of the Shewhart limits is just the opposite of the intended effect.

To this point only integral values of h have been discussed. Lucas and Crosier (1982a) provide some insight into the effect that increasing h by 0.1 will have on the ARL values. For example, increasing h from 5.0 to 5.1 increases the in-control ARL for the FIR CUSUM from 430 to 478, but the ARL values for the various mean shifts increase very little, and they are virtually the same for shifts of at least 1-sigma.

One important point should be made in regard to ARL values for a FIR CUSUM scheme. These ARL values are valid only when one of the indicated shifts takes place at the time the sums are reset and the headstart value is applied. We saw in Table 5.8 that the effect of the headstart value can dissipate rather rapidly when the process is, in fact, in control. Thus, if a process is in control when the sums are reset but there is a mean shift quite some time later, these tabular ARL values do not apply. Instead, the appropriate ARL values are the ones that are given for the corresponding scheme without the FIR option. For example, if a combined Shewhart–FIR CUSUM is being used, the appropriate ARL values will then be the ones for the combined Shewhart–CUSUM scheme. Those values, however, are strictly applicable only when both S_H and S_L are zero at the time of the shift. This requirement will not always be met but the expected value of S_H and S_L is zero when the process is in control at the estimated value of μ.

In summary, if we knew that shifts occur only when the sums are reset, our best bet would be to use a FIR CUSUM without Shewhart limits. That would be very unrealistic, however, since it would mean that the process is constantly out of control. A better strategy would be to use a combined Shewhart–FIR CUSUM with either h slightly larger than 5 (e.g., 5.2) or z larger than 3.5 (e.g., 3.7) or both, the objective being to increase the in-control ARL somewhat without causing much of an increase in the ARL values for the various shifts. Again, such decisions must include considerations of nonstatistical factors such as the cost of searching for nonexistent assignable causes and the cost of not detecting shifts of various magnitudes. These factors should be considered in conjunction with the tables in Lucas (1982) and Lucas and Crosier (1982a).

The use of a combined Shewhart–CUSUM scheme in clinical chemistry is described by Westgard, Groth, Aronsson, and de Verdier (1977), and another (nonmanufacturing) application is described in Blacksell, Glefson and Chamnpood (1994). Another methodological reference not previously mentioned is Lucas and Crosier (1982b).

It should be noted that the superiority of the CUSUM procedures discussed in this section is for sudden jumps in the process mean. The advantage of these procedures over an $\overline{X}$ chart is somewhat less, but still significant, when the change is a slow drift (see Bissell 1984, 1986).

8.2.4 Computation of CUSUM ARLs

Unlike an $\overline{X}$ chart, for which ARLs can be computed by taking the reciprocal of the sum of the tail areas for the standard normal distribution, ARLs for CUSUM procedures cannot be as easily computed. This is because the chance of obtaining a signal at time i depends upon the value of the cumulative sums at time $i-1$. The work on computing ARLs for CUSUM procedures dates from Brook and Evans

(1972), who used a Markov chain approach. Such an approach gives exact ARLs only when the assumed distribution is discrete, however, since a finite number of states is used. For a discrete distribution there is a finite number of possible values for the cumulative sums at the next time period but an infinite number if a continuous distribution is assumed. Therefore, for a normal distribution the ARLs will be approximate, with the quality of the approximation depending on the number of states. Generally more than 10 states are used, and some researchers have used trial and error to determine the number of states at which the ARLs seem to begin to converge.

A Markov transition matrix is constructed, with the elements of the matrix being the transitional probabilities for the various states. See Brook and Evans (1972) for details. Woodall (1984) provided a procedure that reduces the computing time since all states cannot be occupied at a given point in time, and Hawkins (1992) gave a method for computing the ARLs for an arbitrary distribution.

An integral equation approach can also be used for computing the ARLs. Champ and Rigdon (1991) compared the Markov chain approach and the integral equation approach and, in particular, stated that "the Markov chain approach begins by approximating the problem, and then by obtaining the exact solution to the approximate problem, whereas the integral equation approach begins with the exact problem and finds an approximate solution to it." They show that the two approaches give the same approximation if the product midpoint rule is used in conjunction with the integral equations. They conclude that the integral equation approach is preferable since there are better methods than the product midpoint rule.

It is not practical to attempt to use the integral equation approach to obtain the ARL for each type of CUSUM procedure, however. In particular, Lucas and Crosier (1982a) were unable to use this approach to obtain all the ARLs in their tables, and had to use a Markov chain approach to generate many of their tabular values. The authors stated that a particular integral equation approach used by earlier researchers "does not allow the properties of the FIR feature to be evaluated easily".

8.2.5 Robustness of CUSUM Procedures

Very little attention has been given to determining the robustness of CUSUM schemes when the assumptions of normality and independence are violated. Kemp (1967) showed that nonnormality can be a serious problem, although his note contained very few numerical results. More recently, Hawkins and Olwell (1998) gave a few numerical results for a CUSUM applied to individual observations, using two symmetric distributions and two asymmetric distributions; details are given later. Faddy (1996) showed that nonnormality does affect the in-control ARLs of CUSUM schemes, especially the Shewhart–CUSUM scheme. (See also Ryan and Faddy, 2000.) A Shewhart–CUSUM scheme is sensitive to nonnormality because it inherits the sensitivity to nonnormality of a Shewhart chart. The other CUSUM procedures are also quite sensitive to normality, but this is not reflected by the in-control ARLs for a two-sided scheme because the errors relative to each side tend to be somewhat offsetting for a skewed distribution. (We may view a two-sided CUSUM as a combination of two one-sided CUSUMs run simultaneously.) This means, however, that a mean-shift ARL could be greatly different from what is assumed under normality, but for such distributions, such as a chi-square, the variance is also changing.

Therefore, we will consider only the in-control ARLs. Table 8.7 contains the in-control ARLs for a basic CUSUM scheme, an FIR CUSUM scheme, and a Shewhart–CUSUM scheme, respectively, with a subgroup size of 5 and normality assumed but the actual distribution being chi-square with the degrees of freedom (d.f.) given by r. The values of h and k are 5.0 and 0.5, respectively. This table is produced from separate tables given by Faddy (1996). The latter used a Markov chain approach for obtaining the ARLs, so there is a small percentage error.

Since the in-control ARL for the basic CUSUM scheme is 465 under normality, we can observe that there is not a major problem unless the degrees of freedom for the chi-square distribution is extremely small. Similarly, with a normal-theory value of 430.4 for the FIR CUSUM, there is also obviously not a problem unless the degrees of freedom is quite small. Notice that the situation is quite different for the Shewhart–CUSUM scheme, however, as the degrees of freedom does not have to be quite small for there to be a difference between the normal-theory ARL and the actual ARL that exceeds 100. In particular, since the normal-theory value is 391, the difference is approximately 100 at $r = 10$.

We gain a better picture of the effect of non-normality under the assumption of a chi-square distribution by looking at the results for a one-sided CUSUM scheme. This is apparent from the results given in Table 8.8. Notice for the lower one-sided scheme that the ARL is more than double the normal-theory value of 930 when the d.f. is less than 10.

TABLE 8.7 Estimated In-Control ARLs for Normality-Based Two-Sided CUSUM Schemes When the Actual Distribution is Chi-Square with r Degrees of Freedom ($n = 5$)

Basic CUSUM		FIR CUSUM		Shewhart–CUSUM	
r	ARL	r	ARL	r	ARL
2	330.0	2	310.7	2	167.8
3	363.4	3	341.0	3	199.0
4	383.6	4	359.4	4	222.0
6	406.9	6	380.5	6	254.4
8	419.9	8	392.2	8	276.3
10	428.2	10	400.0	10	292.3
25	450.0	25	419.5	25	344.7
50	457.8	50	426.5	50	368.9
100	462.2	100	430.4	100	383.1
500	466.0	500	434.7	500	395.6

TABLE 8.8 Estimated In-Control ARLs for Normality-Based One-Sided CUSUM Schemes When the Actual Distribution Is Chi-Square with r Degrees of Freedom ($n = 5$)

Lower		Upper	
r	ARL	r	ARL
4	2963.5	4	440.3
6	2298.2	6	493.9
8	1995.2	8	531.2
10	1818.8	10	559.4
25	1390.7	25	664.1
50	1227.4	50	728.8
100	1127.8	100	780.4
500	1011.8	500	858.6

The following is a rough explanation of the magnitude of the ARLs relative to the normal-theory values. Unlike the case for a normal distribution, it is not possible to have any very short run lengths for a lower-sided CUSUM when the distribution has considerable right skewness. To see this, consider the case of $n = 1$ for a chi-square distribution with 4 d.f. Since the variance is 8, $K = 0.5(\sqrt{8})$ when the CUSUM is *not* computed using a standard score approach (i.e., not dividing by the appropriate standard deviation). We then have $H = -5(\sqrt{8})$, so that the value of the chi-square random variable would have to be -11.554 in order for the CUSUM threshold to be exceeded on the first observation. Clearly this is impossible since negative values cannot occur. Since the largest possible contribution to the cumulative sum must be less than -2.586 since $0 - 4 + 0.5(\sqrt{8}) = -2.586$, it is obviously impossible for an in-control run length to be less than 5.

Another factor is that since the distribution is skewed to the right, large values will occur much more frequently than with the normal distribution. The occurrence of such values will of course cause the lower-sided CUSUM to go to zero.

There is thus a problem with one-sided CUSUM schemes, but not as great a problem for two-sided schemes since the effects of nonnormality for the lower and upper one-sided schemes are somewhat offsetting. We may quantify the latter using the following result [originally given by Kemp (1962); see also Van Dobben de Bruyn (1968)]:

$$\frac{1}{\text{ARL(two sided)}} = \frac{1}{\text{ARL(upper)}} + \frac{1}{\text{ARL(lower)}} \tag{8.4}$$

It follows that the two-sided ARL is equal to the product of the two one-sided ARLs divided by their sum. It is also apparent from Eq. (8.4) that the two-sided

ARL will be influenced most by the smaller of the two one-sided ARLs, with the influence of the other ARL dependent upon its "smallness."

To illustrate, consider the upper-sided and lower-sided estimated ARLs in Table 8.8 for $r = 4$ and the corresponding two-sided figure in Table 8.7. Using Eq. (8.4), we obtain

$$\begin{aligned}\frac{1}{\text{ARL(two-sided)}} &= \frac{1}{440.294} + \frac{1}{2963.50}\\ &= 383.340\end{aligned}$$

The latter is obviously very close to the corresponding entry (383.608) in Table 8.7. (Of course, we cannot expect exact agreement since the entries in Tables 8.7 and 8.8 are estimated numbers.)

Hawkins and Olwell (1998, pp. 74–76) gave a few numerical results to show the influence of nonnormality on a one-sided basic CUSUM scheme applied to individual observations, with the scheme designed for an in-control ARL of 1000. They showed that nonnormality can be a major problem, with the in-control ARL being quite small for certain distributions and certain values of k, but equal to 24,789 for the standard uniform distribution [i.e, uniform on (0, 1)]. Thus, nonnormality can be a very serious problem.

It is traditional to use a (k, h) combination such as (0.5, 5.0), with k chosen to be half of the shift, in standard deviation units, that one wishes to detect. Hawkins and Olwell (1998) suggest, however, that a way to combat the nonnormality problem is to use a larger value of h. This means that k would have to be adjusted so that the new (k, h) combination has approximately the same ARL properties as the old combination.

Specifically, if h is increased, then k will have to be decreased, as increasing h will increase the ARLs but decreasing k will decrease the ARLs. One difficulty in implementing such an approach, however, is that there are no available tables that could be used to facilitate the selection of any combination of k and h, such as $k = 0.23$ and $h = 6.4$. There is, however, a CUSUM web site of the School of Statistics at the University of Minnesota (www.stat.umn.edu) that was designed to be a companion to Hawkins and Olwell (1998). The software available at that site, which uses the code of Hawkins (1992), can be used to, for example, find the value of h, for a given value of k while searching for desirable (k, h) combinations. It is also possible to evaluate ARLs for nonnormal distributions.

A potentially serious problem that a small value of k could cause is that parameter estimation errors could cause a CUSUM to be unstable. Specifically, a small value of k is for detecting a small mean shift, but for the purposes of computing ARLs, estimation errors can be disguised as mean shifts and cause the ARLs to be highly sensitive to estimation errors. [This is demonstrated in Ryan and Faddy (2000).] For example, assume that h and k are selected so that a CUSUM procedure will have good properties if the actual distribution is a chi-square distribution with eight degrees of freedom. But the mean of the historical data almost certainly will not equal the mean of the assumed distribution, and for the purpose of computing ARLs, the error in estimating the mean will have the same effect as a mean shift.

This can be a very serious problem when only a small amount of historical data is available for estimating the parameter(s), as shown by Hawkins and Olwell (1998, p. 160), and can even be a serious problem when a very large amount of historical data is used (Ryan and Faddy, 2000).

A reasonable approach would be to use a (k, h) combination that works well under normality but also produces reasonable results for departures from normality that might be expected and try to ensure that a very large amount of historical data is used for parameter estimation.

Of course, if departures from normality are expected to be extreme, and to be the rule rather than the exception, then the objective should be to design a CUSUM scheme for the anticipated degree of nonnormality and to use a combination of (k, h) values that provide reasonable ARL properties under alternative degrees of nonnormality.

Another possibility would of course be to try to transform to normality and apply a CUSUM scheme to the transformed values.

8.2.6 CUSUM Procedures for Individual Observations

We first discussed CUSUM schemes for subgroup averages so as to facilitate a comparison of CUSUM schemes with an $\overline{X}$ chart, since the latter is the most frequently used control chart. However, Hawkins and Olwell (1998, p. 24) make the point that it is more efficient to use a CUSUM procedure for individual observations, while recognizing that CUSUM procedures applied to subgroups will sometimes be appropriate due to economies of scale. Specifically, if we are going to "accumulate" data, as is done when CUSUM procedures are used, it might seem as though we should be computing cumulative sums as the data are arriving and not necessarily wait until each subgroup is formed. But subgroups are generally formed by observations that are all obtained at almost the same point in time, and a CUSUM scheme using $n > 1$ has more power than a CUSUM scheme based on $n = 1$. Therefore, the argument made by Hawkins and Olwell in favor of CUSUM schemes for individual observations seems somewhat overstated.

The CUSUM methods of Lucas (1982) and Lucas and Crosier (1982a), which were illustrated in Sections 8.2.1–8.2.3, are also applicable to individual observations. Specifically, one would compute

$$Z = \frac{X - \overline{X}}{\hat{\sigma}_x}$$

and then proceed as in Section 8.2.1 once the z-values are obtained. As with the X chart discussed in Chapter 5, the standard deviation of an individual observation, σ_x, could be estimated using either $\overline{\text{MR}}/d_2$ or s/c_4. The use of CUSUM schemes for individual observations might parallel the suggested use of an X chart in Chapter 5. Specifically, both estimators could be used in Stage 1, with s/c_4 used in Stage 2. Using s/c_4 in Stage 2 will make a CUSUM procedure more stable (i.e., less influenced by sampling variability), just as the use of the estimator causes the control limits of an X chart to have less sampling variability than it does when the moving range estimator is used.

We will not illustrate CUSUM procedures for individual observations since the mechanics are the same as for CUSUM procedures once the z-statistics are computed.

8.3 CUSUM PROCEDURES FOR CONTROLLING PROCESS VARIABILITY

Although much attention has been given to CUSUM procedures for controlling a process mean, CUSUM procedures for controlling process variability have received considerably less attention.

A CUSUM procedure for controlling σ using individual observations was proposed by Hawkins (1981). The approach is as follows. Let

$$W_i = \frac{|X_i/\sigma|^{1/2} - 0.82218}{0.34914}$$

where $E(|X_i/\sigma|^{1/2}) = 0.82218$ and $\text{Var}(|X_i/\sigma|^{1/2}) = (0.34914)^2$ when $X \sim N(0, \sigma^2)$. If X has a nonzero mean, successive differences between the X values (i.e., $X_i - X_{i-1}$) would be used in place of X_i.

The reason for using $|X/\sigma|^{1/2}$ is that this will have approximately a normal distribution when $X \sim N(0, \sigma^2)$. Therefore, W will be approximately $N(0, 1)$. Consequently, the general CUSUM procedure that has been used in the previous section and elsewhere can also be used here, but Hawkins (1981) presented the procedure in which the cumulative sum of deviations from the mean (0.82218) was used.

Hawkins and Olwell (1998, pp. 67–70) discuss this procedure, although presenting it somewhat differently with $(X_i - \mu)/\sigma$ replacing X_i/σ. They point out that this CUSUM is effective for detecting increases in σ^2 but not for detecting small decreases, stating that the ARL for detecting a small variance decrease can be *greater* than the in-control ARL. Of course, this is caused by the fact that the statistic that is being CUSUMed does not have a symmetric distribution, although Hawkins and Olwell (1998) state that $\sqrt{|(X - \mu)/\sigma|}$ is very close to being a normal random variable. Of course, the way to achieve process improvement is to reduce variability, so the inability to quickly detect small variance decreases is indeed a shortcoming of the procedure, as the authors point out.

Whenever individual observations are used, a change in the mean can cause a signal on a chart for monitoring variability, and conversely. Here the expected value of $\sqrt{|(X - \mu)/\sigma|}$ will of course change when μ changes, regardless of whether σ also changes. So a signal may not necessarily mean a change in σ. (Recall the discussion in Section 5.3 for an X chart.)

This general problem of parameter-change ARLs being greater than the in-control ARL was discussed in Section 6.1.4.4; the problem is certainly not unique to this CUSUM procedure. This does remind us, however, of the desirability of having ARL-unbiased charts as we obviously want to be able to quickly detect small but significant reductions in variances, nonconformity rates, and percentages of nonconforming units.

Hawkins and Olwell (1998, p. 70) state that a CUSUM of the sum of squared deviations from a subgroup mean is the optimal diagnostic for detecting a step change in the variance, assuming normality, with nothing assumed about the mean. Thus, this is a CUSUM of $(n-1)s^2$. Since the statistic that is being CUSUMed is distributed as a constant times a chi-square random variable, this approach could also be undermined by the asymmetry in the distribution of the statistic that is being CUSUMed.

Therefore, it is preferable to design a CUSUM scheme that takes into consideration the distribution of the statistic that is being used, rather than trying to transform to approximate normality. Hawkins and Olwell (1998, p. 144) derive the form for k, using the gamma distribution, with k a function of the in-control variance and a step change increase in the variance that one wishes to detect. Then the form of the CUSUM statistic is $S_{H_i} = \max[0, (s_i^2 - k) + S_{H_{i-1}})$, with s_i^2 denoting the ith sample variance. Letting σ_0^2 and σ_1^2 denote the former and the latter, respectively,

$$k = \frac{\ln(\sigma_0^2/\sigma_1^2)\sigma_0^2\sigma_1^2}{\sigma_0^2 - \sigma_1^2} \tag{8.5}$$

Assume that $\sigma_0^2 = 1$ and $\sigma_1^2 = 4$. Then $k = 1.848$. If we wished to have the in-control ARL set to 930, then $h = 4.1213$. The ARL for detecting a shift to $\sigma_1^2 = 4$ is 3.2.

It is of interest at this point to determine how long we would expect to have to wait before the shift is detected on an s chart with .001 probability limits. Recall from Section 4.7.4 that

$$\frac{(n-1)S^2}{\sigma^2} \sim \chi^2_{n-1}$$

so that

$$P\left[\frac{(n-1)S^2}{\sigma^2} > \chi^2_{.999}\right] = .001$$

It then follows that

$$P\left(S^2 > \frac{\sigma^2}{n-1}\chi^2_{.999}\right) = .001 \qquad P\left(S > \frac{\sigma}{\sqrt{n-1}}\chi_{.999}\right) = .001$$

In this instance $\sigma = 2$ and $n = 4$, so

$$P\left(S > \frac{2}{\sqrt{3}}\chi_{.999}\right) = .001$$

We need to determine $P[S > (1/\sqrt{3})\chi_{.999}]$. This can be done by using the more familiar chi-square distribution. We know from Table 4.3 that $\chi^2_{.999}$ for $n = 4$ is 16.2660. Thus,

$$P\left[S^2 > \left(\frac{2}{\sqrt{3}}\right)^2 \chi^2_{.999}\right] = .001$$

so that

$$P\left[S^2 > \tfrac{4}{3}(16.2660)\right] = .001$$

where $S^2 \sim \frac{4}{3}\chi^2_3$. What we need is $P[S^2 > 1/3(16.2660)]$ since σ is assumed to equal 1. It can be determined that

$$\begin{aligned} P\left[S^2 > \frac{1}{3}(16.2660)\right] &= P\left[S^2 > (4)\left(\frac{1}{3}\right)\left(\frac{16.2660}{4}\right)\right] \\ &= P\left[S^2 > \frac{4}{3}(4.0665)\right] \\ &= .25437 \end{aligned}$$

Accordingly, the ARL for the UCL is $(1/.25437) = 3.93$. Therefore, using an s chart (or an s^2 chart) with .001 probability limits, we would expect the shift from $\sigma = 1$ to $\sigma = 2$ to be detected with the fourth subgroup.

8.4 CUSUM PROCEDURES FOR NONCONFORMING UNITS

The improved control limits for a p chart and an np chart given by Ryan and Schwertman (1997) could not be used directly to produce an "improved" CUSUM procedure since a statistic was not used. That is, regression equations were used, but there was no random variable directly involved that could be CUSUMed.

In order to have a good CUSUM procedure for nonconforming unit data that is analogous to the CUSUM scheme for measurement data that was presented in Section 8.2.1, it would be necessary to use a transformation that produces a (transformed) statistic that is approximately normally distributed. Thus, one possibility would be to define

$$z_i = \frac{\sin^{-1}\sqrt{(x_i + 3/8)/(n + 3/4)} - \sin^{-1}\sqrt{\bar{p}}}{\sqrt{1/4n}} \tag{8.6}$$

and then use the z_i values the same way they were used in Section 8.2.1.

The utility of such a CUSUM procedure can be illustrated as follows. It was stated in Section 6.1.1 that for $n = 400$ a reduction in p from 0.10 to 0.08 would not be detected, on the average, until after 46 samples, when a p chart or an np chart is used. In Table 8.9 values for X are randomly generated for two binomial

TABLE 8.9 Illustration of the Proposed CUSUM Procedure[a] ($n = 400, k = 0.5, h = 5$)

Sample Number	x^a	z_A	z_{NA}	Arcsin Transformation S_H	Arcsin Transformation S_L	Normal Approximation S_H	Normal Approximation S_L
			$p_{-} = .10$, LCL = 22, UCL = 58				
1	47	1.17	1.17	0.67	0	0.67	0
2	38	−0.29	−0.33	0	0	0	0
3	39	−0.12	−0.17	0	0	0	0
4	46	1.01	1.00	0.51	0	0.50	0
5	42	0.38	0.33	0.39	0	0.33	0
6	36	−0.63	−0.67	0	0.13	0	0.17
7	46	1.01	1.00	0.51	0	0.50	0
8	37	−0.46	−0.50	0	0	0	0
9	40	0.05	0	0	0	0	0
10	35	−0.80	−0.83	0	0.30	0	0.33
			$p_{\sim} = .08$, LCL = 22, UCL = 58				
11	34	−0.98	−1.00	0	−0.78	0	−0.83
12	31	−1.53	−1.50	0	−1.81	0	−1.83
13	33	−1.16	−1.17	0	−2.47	0	−2.50
14	29	−1.90	−1.83	0	−3.87	0	−3.83
15	33	−1.16	−1.17	0	−4.53	0	−4.50
16	39	−0.12	−0.17	0	−4.15	0	−4.17
17	29	−1.90	−1.83	0	−5.55[b]	0	−5.50[b]
18	39	−0.12	−0.17				
19	34	−0.98	−1.00				

[a] Values of x are simulated; the LCL and UCL are 3-sigma np chart limits; z_A represents the z value using the arcsin transformation and z_{NA} designates the z value from the normal approximation.
[b] Value exceeds h.

distributions with $n = 400$: first for $p = 0.10$ and then for $p = 0.08$. The two sets of values for z, S_H, and S_L are for the arcsin method given in Eq. (8.7) and for the normal approximation given by

$$z_i = \frac{x_i - np}{\sqrt{np(1-p)}} \tag{8.7}$$

Notice that the two sets of z values differ somewhat, and the S_H and S_L values differ by a similar amount. (We should expect the differences to be much greater, however, when p is considerably less than .10.) Thus, although the arcsin method will be generally preferable, the use of the normal approximation z in Eq. (8.5) gives the same result in this instance. The specific result is that sample 17 provides evidence that there has been a decrease in p, as indicated by the fact that $S_L > h$.

Since the change actually occurred with the 11th sample, 6 additional samples were required to detect the change. Since

$$\sigma_{\overline{p}} = \sqrt{\frac{p(1-p)}{n}}$$

which equals .015 when $n = 400$ and $p = .10$, the decrease in p from .10 to .08 is equal to $1.33\sigma_{\overline{p}}$. Interpolating in Table 8.3 produces 7.331 as the expected number of samples needed to detect a mean shift equal to 1.33σ. Thus, the result for this illustrative example is in general agreement with the theory. Notice that the z_{NA} values for samples 11–17 are all within $(-3, 3)$, so a signal would not be received from a p chart or an np chart.

As indicated earlier in this chapter, runs criteria can also be used to detect such a shift, but it should be remembered that the probability of a point lying outside of 1-, 2-, or 3-sigma limits can differ considerably from the assumed normal distribution probabilities. Consequently, this should be kept in mind if runs criteria are to be used with the x values.

In spite of the fact that the runs criteria probabilities can be off considerably, we would certainly suspect that a decrease in p had occurred by sample 17 since there are eight consecutive points below the midline (10–17). The "assumed" probability of this occurring due to chance is $(0.5)^8 = .0039$, whereas the actual probability is .0026

Can a better approach be devised? Hawkins and Olwell (1998) prefer CUSUM techniques that directly incorporate the binomial distribution, emphasizing (p. 75) that whereas a CUSUM approach is the optimal method for detecting a step change, a CUSUM applied to transformed data is not optimal. Specifically, the arcsin transformation approach given by Eq. (8.5) will not transform data to exact normality, so a normality-based CUSUM procedure cannot be the optimal procedure. From a practical standpoint, however, normality does not exist in applications, as has been emphasized previously, and since "all models are wrong," we cannot expect to design an optimal CUSUM in any application because we do not know the true distribution.

Hawkins and Olwell (1998, p. 123) gave a CUSUM procedure for binomial data that is based on the binomial distribution. The general form is the same as that given in Eq. (8.3), except that z_i is replaced by X_i, the value of the, binomial random variable, and a minus sign precedes the reference value in S_L. Specifically, with p_0 denoting the in-control value of p and p_1 denoting the out-of-control value that a CUSUM procedure is to be designed to detect, the reference value for detecting an increase in p is

$$k^+ = -n \frac{\ln\left(\dfrac{1-p_1}{1-p_0}\right)}{\ln\left(\dfrac{p_1(1-p_0)}{p_0(1-p_1)}\right)}$$

and $k^- = -k^+$ for detecting a decrease.

For the current example with $p_0 = .10$ and $p_1 = .08$, $k^- = 35.866$. In general, we will not be able to specify the in-control ARL exactly because we are working with discrete data. If we use $k^- = 36$ and desired to have the in-control ARL be as close to 930 as possible, we would set $h = 23$. This would produce an in-control ARL of 1029.49 and an ARL for detecting the shift to $p_1 = .08$ of 6.31.

This has some advantages over the CUSUM of the arcsin-transformed values that was illustrated earlier in this section. The properties of that procedure are not known exactly because it is possible to transform only to approximate normality with an arcsin transformation. Therefore, the properties of that procedure are unknown, whereas the properties of the binomial CUSUM are known if we assume that the binomial model is the correct model and that p is known. Of course, p is generally unknown and autocorrelation might make the binomial model assumption implausible. Nevertheless, the binomial CUSUM is more appealing, even though it may not be as intuitive as the CUSUM of the transformed values. In particular, most control chart users would probably prefer to use the same general approach for both measurement and attribute data. Certainly this would be easier to understand and to explain to management.

The ARL properties of the two procedures will probably not differ greatly. The ARL for detecting the shift was estimated as 7.3 with the transformation approach, whereas with the binomial CUSUM it is 6.3. Of course, there is also granulation when a transformation is used, so the 7.3 is just an estimate.

Reynolds and Stoumbos (1998, 1999a,b, 2000) have also considered the problem of monitoring nonconforming units using either a CUSUM approach or a sequential probability ratio test (SPRT). Reynolds and Stoumbos (1999a) consider the problem of monitoring nonconforming units when inspection is performed continuously, and Reynolds and Stoumbos (1999b) consider the problem when not all units are inspected.

Reynolds and Stoumbos (1999b) found that a CUSUM based on the Bernoulli distribution (i.e., the binomial distribution with $n = 1$) performs similar to a CUSUM based on the binomial distribution for detecting small-to-moderate changes in p, but the former is superior to the latter for detecting large changes in p. This is fairly intuitive because with a Bernoulli CUSUM a signal can be received after a given observation, whereas with a binomial CUSUM a signal can be received only after n observations. If a large change in p has occurred, we would expect this to be apparent after inspecting only a small number of units.

Both CUSUM procedures are superior to a p chart for detecting both small and large shifts. See also Gan (1993).

8.5 CUSUM PROCEDURES FOR NONCONFORMITY DATA

Cumulative sum procedures can also be used with nonconformity data. Several such procedures are discussed in this section. *If* the normal approximation to the Poisson distribution could be assumed to be adequate, a CUSUM procedure could be used with

$$z_{NA} = \frac{c - \bar{c}}{\sqrt{\bar{c}}}$$

where NA designates normal approximation. The superiority of the transformation approach was demonstrated in the preceding section, however, so it would be logical to use one of those transformations in a CUSUM procedure. For example, if the transformation $y = \sqrt{c} + \sqrt{c+1}$ is used, this would lead to

$$\begin{aligned} z_T &= \frac{(\sqrt{c} + \sqrt{c+1}) - \sqrt{4\bar{c}+1}}{1} \\ &= \sqrt{c} + \sqrt{c+1} - \sqrt{4\bar{c}+1} \end{aligned}$$

since the mean of $(\sqrt{c} + \sqrt{c+1})$ is estimated by $\sqrt{4\bar{c}+1}$.

The results for the standard approach and the transformation approach are given in Table 8.10, using the data in Table 6.5. Recall that this same type of comparison was made for a p-chart example, and the results did not differ greatly since p was not extremely small. There is obviously a big difference with this nonconformity data, however. In particular, the z values differ considerably at the two extremes; namely, $c \geq 15$ and $c \leq 2$. This causes the cumulative sums to differ greatly and results in the two out-of-control messages being received at different points. (*Note:* $\sqrt{4\bar{c}+1}$ was used in computing z_T because $\mu_y = \sqrt{4\lambda+1}$, where $y = \sqrt{c} + \sqrt{c+1}$ and $\bar{c}$ is the estimator of the Poisson mean, λ. Alternatively, $\bar{y}$ could have been used. The two values will not differ much, particularly when the c values do not differ greatly.)

Lucas (1985) presents a CUSUM procedure for Poisson data that does not utilize a normal approximation and differs somewhat from the procedure given in Lucas and Crosier (1982a) for variables data. Specifically, the cumulative sums are obtained from the formulas

$$S_{Hi} = \max[0, (c_i - k_H) + S_{Hi-1}]$$

and

$$S_{Li} = \max[0, -(c_i - k_L) + S_{Li-1}]$$

where the first formula is for detecting an increase in the number of nonconformities and the second is for detecting a decrease. It is stated that the reference value should be chosen close to

$$k_H = \frac{\mu_d - \mu_a}{[\ln(\mu_d) - \ln(\mu_a)]} \tag{8.8}$$

where μ_a is the acceptable level (number of nonconformities) and μ_d is the level such that a shift to this level is to be detected quickly. This is actually for a one-sided CUSUM in which an upward shift is to be detected. For a two-sided

TABLE 8.10 CUSUM Calculations ($k = 0.5$, $h = 5$)

c	$z_{\text{NA}} = (c - \bar{c})/\sqrt{\bar{c}}$	$z_T = \sqrt{c} + \sqrt{c+1} - \sqrt{4\bar{c}+1}$	$S_H(z_{\text{NA}})$	$S_H(z_{\text{T}})$	$S_L(z_{\text{NA}})$	$S_L(z_{\text{T}})$
9	0.52	0.57	0.02	0.07	0	0
15	2.71	2.28	2.23	1.85	0	0
11	1.25	1.19	2.98	2.54	0	0
8	0.16	0.24	2.64	2.28	0	0
17	3.43	2.78	5.57[a]	4.56	0	0
11	1.25	1.19	6.32	5.25[a]	0	0
5	−0.93	−0.90	4.89	3.85	−0.43	−0.40
11	1.25	1.19	5.64	4.54	0	0
13	1.98	1.76	7.12	5.80	0	0
7	−0.20	−0.12	6.42	5.18	0	0
10	0.89	0.89	6.81	5.57	0	0
12	1.61	1.48	7.92	6.55	0	0
4	−1.29	−1.35	6.13	4.70	0.79	−0.85
3	−1.66	−1.86	3.97	2.34	1.95	−2.21
7	−0.20	−0.12	3.27	1.72	1.65	−1.83
2	−2.02	−2.44	0.75	0	3.17	−3.77
3	−1.66	−1.86	0	0	4.33	−5.13[a]
3	−1.66	−1.86	0	0	5.49[a]	−6.49
6	−0.57	−0.49	0	0	5.56	−6.48
2	−2.02	−2.44	0	0	7.08	−8.42
7	−0.20	−0.12	0	0	6.78	−8.04
9	0.52	0.57	0.02	0.07	−5.76	−6.97
1	−2.39	−3.18	0	0	−7.65	−9.65
5	−0.93	−0.90	0	0	−8.08	−10.05
8	0.16	0.24	0	0	−7.42	−9.31

[a] Value exceeds $h = 5$.

CUSUM, which is generally recommended, a second value of k, k_L, would be calculated using the level that corresponds to a shift in the opposite direction. Equation (8.8) could also represent the general form of k_L, although Hawkins and Olwell (1998, p. 113) write the numerator and denominator in such a way that each is positive. That is, for the lower one-sided CUSUM the numerator would be written as $\mu_a - \mu_d$, and similarly for the denominator.

Applying this method to the nonconformity data in Table 6.5, we might set μ_d at 10.31 and compare the results with the results for the other CUSUM procedures. This would provide a fair comparison with the other CUSUM procedures, as $\mu_d = 10.31$ is one standard deviation above $\mu_a = 7.56$, and the use of $k = 0.5$ with the other (normal approximation) procedures is appropriate when a one standard deviation shift is to be detected quickly. For a downward shift, $\mu_d = 4.81$

is one standard deviation below the mean. These two values of μ_d will produce different k values. Using $\mu_d = 10.31$,

$$\begin{aligned} k_H &= \frac{10.31 - 7.56}{\ln(10.31) - \ln(7.56)} \\ &= 8.86 \end{aligned}$$

whereas $\mu_d = 4.81$ produces

$$\begin{aligned} k_L &= \frac{4.81 - 7.56}{\ln(4.81) - \ln(7.56)} \\ &= 6.08 \end{aligned}$$

Thus, different k values will be used in the two cumulative sums. Following Lucas's (1985) recommendation, we round these to the nearest integer, so that the two sets of cumulative sums will be obtained using

$$S_{Hi} = \max[0, (c_i - 9) + S_{Hi-1}]$$

and

$$S_{Li} = \max[0, -(c_i - 6) + S_{Li-1}]$$

This will also require the use of two different values, denoted by Lucas as h_H and h_L. As with measurement data, h is determined from a table of ARL values. The general form of the reference value given by Hawkins and Olwell (1998, p. 113) is the same as that given by Lucas (1985) and was originally derived by Johnson and Leone (1962). Using the Poisson CUSUM program available from the web site www.stat.umn.edu, we find that for $k_H = 9$ and $h_H = 15$ the in-control ARL is 908.5 and the ARL for detecting the shift from $\mu_a = 7.56$ to $\mu_d = 10.31$ is 11.2. For the lower one-sided scheme, with $k_L = 6$ and $h_L = 11$, the in-control ARL is 827.5 and the ARL for detecting the shift to $\mu_d = 4.81$ is 9.2.

The results are given in Table 8.11. Notice that the upward-shift signal from S_H is received one observation after a change occurred (see Section 6.2.4), and the downward-shift signal from S_L is received five observations sooner than the signal from the c chart with the regression-based limits and the charts that incorporate a square root transformation.

Comments made at the end of Section 8.4 also apply here, as it is generally desirable to use a Poisson-based CUSUM rather than a normal approximation approach, although the latter is more intuitive. We should note, however, that the ARLs of CUSUMs of variables that are assumed to be Poisson are very sensitive to departures from the assumed Poisson distribution, as discussed by Hawkins and Olwell (1998, p. 119). Since the Poisson mean and variance are the same, a test for overdispersion could be performed, as described in Section 6.2.6.

TABLE 8.11 CUSUM Calculations for Poisson CUSUM ($k_H = 9, k_L = 6, h_H = 15, h_L = 11$)

c	S_H	S_L
9	0	0
15	6	0
11	8	0
8	7	0
17	15	0
11	17[a]	0
5	13	1
11	15	0
13	19	0
7	17	0
10	18	0
12	21	0
4	16	2
3	10	5
7	8	4
2	1	8
3	0	11
3	0	14[a]
6	0	14
2	0	18
7	0	17
9	0	14
1	0	19
5	0	18
8	0	16

[a] Threshold value is reached or exceeded.

The binomial and Poisson reference values had been previously given by Johnson and Leone (1962).

8.6 EXPONENTIALLY WEIGHTED MOVING AVERAGE CHARTS

This type of control chart is similar to a CUSUM procedure in that it has utility in detecting small shifts in the process mean. It is due to Roberts (1959), who labeled the chart a *geometric moving average chart.* It recent years, however, the term *exponentially weighted moving average (EWMA) chart* has been used, and this term will be used for the rest of this chapter.

8.6.1 EWMA Chart for Subgroup Averages

The general idea is to plot the value of a statistic that we will designate as w_t:

$$w_t = \lambda \overline{x}_t + (1 - \lambda) w_{t-1} \tag{8.9}$$

The expression given by Eq. (8.9) is essentially a weighted average in which the subgroup average at time t, $\overline{x}_t$, is given a weight λ $(0 < \lambda \leq 1)$ and the value of the expression at time $t - 1$ is given a weight $1 - \lambda$.

The iterative calculations begin with $w_0 = \overline{\overline{x}}$. The customary approach is to obtain the control limits from

$$\overline{\overline{x}} \pm 3 \frac{\hat{\sigma}}{\sqrt{n}} \sqrt{\left(\frac{\lambda}{2 - \lambda}\right) [1 - (1 - \lambda)^{2t}]} \tag{8.10}$$

which follows from the fact that

$$\sigma_{w_t}^2 = \frac{\sigma^2}{n} \left(\frac{\lambda}{2 - \lambda}\right) [1 - (1 - \lambda)^{2t}] \tag{8.10a}$$

The use of the exact expression for σ_{w_t} in Eq. (8.10) will thus produce variable control limits, which will become wider over time. If $\lambda \geq 0.02$, however, $(1 - \lambda)^{2t}$ will be close to zero if $t \geq 5$, so $\sigma_{w_t}^2$ could then be approximated by $\sigma^2/n[\lambda/(2 - \lambda)]$. Constant control limits can then be obtained using

$$\overline{\overline{x}} \pm 3 \frac{\hat{\sigma}}{\sqrt{n}} \sqrt{\frac{\lambda}{2 - \lambda}} \tag{8.11}$$

Accordingly, the exact variance can be used in producing the control limits for the first four or five time periods [i.e., using Eq. (8.10)] and then Eq. (8.11) used thereafter to produce constant limits.

We should note, however, that the control limits given by Eq. (8.10) are obtained acting as if $\overline{\overline{x}}$ is not a random variable. That is, the variance expression in Eq. (8.10a) is a conditional variance, conditioned on the value of $\overline{\overline{x}}$. If desired, one might produce (wider) control limits that additionally incorporate the variance of $\overline{\overline{x}}$, although this is not the standard approach.

The value of λ would have to be determined. As with any control chart or CUSUM procedure, the general idea is to detect a shift that is of any consequence as quickly as possible without having an unacceptably high probability of receiving false alarms.

We consider the general case of L-sigma limits, where $L = 3$ in Eqs. (8.10) and (8.11). Similar in spirit to the tables that are available for designing a CUSUM procedure, Lucas and Saccucci (1990) give tables of ARL values that provide some guidance in the selection of L, λ, and n. (See also Crowder, 1989.)

We will assume that n has been determined and use the data in Table 8.1 to illustrate this technique. For detecting a 1-sigma shift, the expanded tables relative to Lucas and Saccucci (1990) but not given therein suggest that one reasonable combination of L and λ would be $L = 3.00$ and $\lambda = 0.25$. This combination would provide an in-control ARL of 492.95 and an ARL of 10.95 for detecting the 1-sigma shift. (Note that these are very close to the corresponding values for a CUSUM procedure with $h = 5$ and $k = 0.5$ that are 465 and 10.4, respectively.)

We can see from Table 8.12 that with this EWMA scheme the shift would be detected with subgroup 20, which is also when the shift would be detected using a CUSUM procedure with $k = 0.5$ and $h = 5$ (as can be seen from Table 8.2). This should come as no surprise in light of the ARL values for the two procedures. [The calculations were started using $w_0 = 0$ since the data were initially generated from an $N(0, 1)$ distribution and recall that the shift occurred with subgroup 13.]

Thus, an EWMA procedure with good choices for L and λ should provide results that closely parallel the results obtained using a CUSUM procedure with good choices for h and k. So which one should be used? The amount of required computation is about the same for each, but the CUSUM procedure has a few advantages. First, the S_H, and S_L values are not scale dependent, so the values can be compared for different products and for different processes. The exponentially weighted moving averages are, of course, scale dependent, although that problem could be easily remedied by standardizing the averages, which would then require

TABLE 8.12 EWMA Calculations with $\lambda = 0.25$

Subgroup Number	Subgroup Average	w_t	Exact Control Limits	Approximate Control Limits
1	0.41	0.1025	±0.3750	
2	1.11	0.3544	±0.4688	
3	−0.04	0.2558	±0.5140	
4	−1.04	−0.0682	±0.5378	
5	−0.04	−0.0612	±0.5508	
6	−0.68	−0.2159	±0.5579	
7	−0.85	−0.3744	±0.5619	
8	0.81	−0.0783		±0.5669
9	0.62	0.0963		↓
10	−0.38	−0.0228		
11	0.54	0.1179		
12	0.18	0.1134		
13	0.33	0.1826		
14	0.32	0.2170		
15	0.44	0.2728		
16	0.22	0.2596		
17	0.88	0.4147		
18	0.16	0.3510		
19	0.95	0.5008		
20	1.09	0.6481		

changing the general form of the control limits. The CUSUM procedure also incorporates z-values that indicate how many standard deviations a subgroup average is from the midline, whereas z-values are not an inherent part of the EWMA calculations (but could be easily added). Another problem with EWMA procedures vis-à-vis CUSUM procedures is the "inertia problem," as it has been referred to in the literature. This refers to the fact that if the EWMA has a small(large) value and there is an increase(decrease) in the mean, the EWMA can be slow to detect the change, as compared to a CUSUM chart. How often this problem is likely to occur in practice is arguable, however.

8.6.2 EWMA Misconceptions

There has been some confusion in the literature regarding how λ should be chosen, in addition to other misconceptions. Specifically, the use of an EWMA for forecasting and the use of an EWMA control chart for process control have been somewhat confused. An EWMA provides the optimal forecast for a process that can be fit by a first-order integrated moving average model [an IMA(1,1)]. An EWMA control chart is used for *independent* data, however. Consequently, it would not make any sense to choose the value of λ in such a way as to minimize the sum of squares of the forecast errors. Since the mean is assumed to be constant when λ is selected, we would expect such an approach to produce a value of λ that is essentially zero. In fact, w_t would be a biased estimator of the process mean for each value of t if $\lambda \neq 0$, since $E(\bar{\bar{x}}) = \mu$, by assumption, so we can see intuitively that we would expect λ to be approximately zero. Of course, when a change in the process mean occurs, we want to detect it as quickly as possible. This would suggest that λ not be close to zero, but it should be apparent that the larger the value of λ, the higher the false alarm rate will be.

There are other misconceptions surrounding the use of an EWMA chart. Some writers have claimed that the EWMA chart with 3-sigma limits and $\lambda = .4$ performs approximately the same as a Shewhart chart with the Western Electric runs rules. This matter was addressed by Woodall (1991), who pointed out that the tables in Lucas and Saccucci (1990) show that an EWMA chart with this value of λ and 3.054-sigma limits is 500, whereas the ARL for a Shewhart chart with the runs rules is only 91.75, as shown by Champ and Woodall (1987).

Another mistake is the claim that the EWMA chart begins to resemble a CUSUM chart when λ approaches zero. That is obviously false since the EWMA would approach a constant ($\bar{\bar{x}}$) as λ approached zero.

8.6.3 EWMA Chart for Individual Observations

If data are not collected in subgroups, an EWMA would be applied to the individual observations. The same general approach is used; the equations are just appropriately modified. Specifically, $\bar{x}_t$ in Eq. (8.9) would be replaced by x_t, $\bar{\bar{x}}$ in Eqs. (8.10) and (8.11) would be replaced by $\bar{x}$, and there would be no $\sqrt{n}$ term in Eqs. (8.10) and (8.11) since n would equal 1.

8.6.4 Shewhart–EWMA Chart

Since an EWMA procedure is good for detecting small shifts but is inferior to a Shewhart chart for detecting large shifts, it is desirable to combine the two, as was done with a CUSUM chart (see Section 8.2.3). As with a Shewhart–CUSUM scheme, the general idea is to use Shewhart limits that are larger than 3-sigma limits. Lucas and Saccucci (1990) state that it is possible to design a Shewhart–EWMA scheme with ARL properties similar to that of a good Shewhart–CUSUM scheme. They recommend using k-sigma limits for the Shewhart portion with $4.0 \leq k \leq 4.5$. Unfortunately, there are apparently no tables of Shewhart–EWMA ARLs in the literature to aid in designing a Shewhart–EWMA scheme, although tables were given in Lucas and Saccucci (1987).

8.6.5 FIR–EWMA

Lucas and Saccucci (1990) discuss adding an FIR feature to an EWMA scheme. This requires using two one-sided EWMA schemes with different starting values. Van Gilder (1994) discusses a powertrain application at General Motors in which an FIR–EWMA has been used.

8.6.6 EWMA Chart with Variable Sampling Intervals

As with the other types of control charts, a variable sampling scheme can be used for an EWMA chart. Such an approach is discussed by Saccucci, Amin, and Lucas (1992) and Reynolds (1996).

8.6.7 EWMA Chart for Grouped Data

Steiner (1998) gives an EWMA approach for grouped data. Grouped data occur in industry in various ways; perhaps the simplest is when a pass–fail gauge is used. But gauges that divide the data into more than two groups are also used. For k groups, one approach is to use the midpoint of each interval; another approach is to use a representative value for each interval such that when the process is in control the sample mean and variance using these values is equal to the process mean and variance. Steiner (1998) favors the latter approach.

8.6.8 EWMA Chart for Variances

Chang and Gan (1994) considered the use of an EWMA chart for detecting a change in the process variance. MacGregor and Harris (1993) state that their exponentially weighted moving variance chart and exponentially weighted mean-squared deviation chart are particularly useful for monitoring changes in variability with individual observations.

8.6.9 EWMA for Attribute Data

The CUSUM procedures for nonconforming units data and for nonconformity data were discussed in Sections 8.4 and 8.5, respectively. Similarly, an EWMA chart can also be used for attribute data. Comparatively little work has been done in this area, however. See Gan (1990a,b).

8.7 SUMMARY

We have seen that CUSUM procedures, in particular, have better overall properties than the properties of Shewhart charts. Another advantage of CUSUM procedures is that they are not prone to become ARL biased (see Section 6.1.4.4), even when applied to attribute data (Acosta–Mejia, 1998). This is another advantage of CUSUM procedures since attribute data come from skewed distributions and Shewhart attribute charts can be seriously ARL biased. The EWMA charts can be easily ARL biased, however, but the parameters can generally be selected in such a way so as to make the chart ARL unbiased (Acosta–Mejia, 1998).

REFERENCES

Acosta–Mejia, C. A. (1998). Personal communication.

Bissell, A. F. (1984). The performance of control charts and Cusums under linear trend. *Applied Statistics 33*(2): 145–151.

Bissell, A. F. (1986). Corrigendum. *Applied Statistics 35*(2): 214.

Blacksell, S .D., L. J. Glefson, R. A. Lunt, and C. Chamnpood (1994). Use of combined Shewhart-CUSUM control charts in internal quality control of enzyme-linked immunosorbent assays for the typing of foot and mouth disease virus-antigen. *Revue Scientifique et Technique de L'Office International des Epizooties 13*(3): 687–699.

Box, G. E. P. and A. Luceño (1997). *Statistical Control by Monitoring and Feedback Adjustment.* New York: Wiley.

Brook, D. and D. A. Evans (1972). An approach to the probability distribution of CUSUM run length. *Biometrika 59*(3): 539–549.

Champ, C. W. and S. E. Rigdon (1991). A comparison of the Markov chain and the integral equation approaches for evaluating the run length distribution of quality control charts. *Communications in Statistics—Simulation and Computation 20*(1): 191–204.

Champ, C. W. and W. H. Woodall (1987). Exact results for Shewhart control charts with supplementary runs rules. *Technometrics 29*(4): 393–399.

Chang, T. C. and F. F. Gan (1994). Optimal designs of one-sided EWMA charts for monitoring a process variance. *Journal of Statistical Computation and Simulation 49*(1/2): 33–48.

Crowder, S. V. (1989). Design of exponentially weighted moving average schemes. *Journal of Quality Technology 21:* 155–162.

Deming, W. E. (1993). *The New Economics for Industry, Government, and Education.* Cambridge, MA: Center for Advanced Engineering Study, Massachusetts Institute of Technology.

Faddy, B. J. (1996). The effect of non-normality on the performance of cumulative sum quality control schemes. Honours thesis, Department of Statistics, University of Newcastle, Australia.

Gan, F. F. (1990a). Monitoring observations generated from a binomial distribution using modified exponentially weighted moving average control chart. *Journal of Statistical Computation and Simulation 37:* 45–60.

Gan, F. F. (1990b). Monitoring Poisson observations using modified exponentially weighted moving average control charts. *Communications in Statistics—Simulation and Computation 19:* 103–124.

Gan, F. F. (1991). An optimal design of CUSUM quality control charts. *Journal of Quality Technology 23*(4): 279–286.

Gan, F. F. (1993). An optimal design of CUSUM control charts for binomial counts. *Journal of Applied Statistics 20*(4): 445–460.

Hawkins, D. M. (1981). A CUSUM for a scale parameter. *Journal of Quality Technology 13*(4): 228–231.

Hawkins, D. M. (1992). Evaluation of average run lengths of cumulative sum charts for an arbitrary data distribution. *Communications in Statistics B21*(4): 1001–1020.

Hawkins, D. M. and D. H. Olwell (1998). *Cumulative Sum Charts and Charting for Quality Improvement.* New York: Springer-Verlag.

Johnson, N. L. and F. C. Leone (1962). Cumulative sum control charts: Mathematical principles applied to their construction and use. Part III. *Industrial Quality Control 19*(2): 22–28.

Kemp, K. W. (1962). The use of cumulative sums for sampling inspection schemes. *Applied Statistics 11:* 16–31.

Kemp, K. W. (1967). An example of errors incurred by erroneously assuming normality for CUSUM schemes. *Technometrics 9*(3): 457–464.

Lucas, J. M. (1982). Combined Shewhart-CUSUM quality control schemes. *Journal of Quality Technology 14*(2): 51–59.

Lucas, J. M. (1985). Counted data CUSUMs. *Technometrics 27*(2): 129–144.

Lucas, J. M. and R. B. Crosier (1982a). Fast initial response for CUSUM quality control schemes: Give your CUSUM a head start. *Technometrics 24*(3): 199–205.

Lucas, J. M. and R. B. Crosier (1982b). Robust CUSUM. *Communications in Statistics—Theory and Methods 11*(23): 2669–2687.

Lucas, J. M. and M. S. Saccucci (1987). Exponentially weighted moving average control schemes: Properties and enhancements. Faculty Working Series Paper 87-5, Department of Quantitative Methods, Drexel University.

Lucas, J. M. and M. S. Saccucci (1990). Exponentially weighted moving average control schemes: Properties and enhancements (with discussion). *Technometrics 32:* 1–29.

MacGregor, J. E. and T. J. Harris (1993). The exponentially weighted moving variance. *Journal of Quality Technology 25*(2): 106–118.

Montgomery, D. C. (1996). *Introduction to Statistical Quality Control*, 3rd ed. New York: Wiley.

Montgomery, D. C. and W. H. Woodall (1997). Concluding remarks. *Journal of Quality Technology 29*(2): 157.

Moustakides, G. V. (1986). Optimal stopping times for detecting changes in distributions. *Annals of Statistics 14:* 1379–1387.

Page, E. S. (1954). Continuous inspection schemes. *Biometrika 41:* 100–115.

Reynolds, M. R., Jr. (1996). Shewhart and EWMA variables sampling interval control charts with sampling at fixed times. *Journal of Quality Technology 28*(2): 199–212.

Reynolds, M. R., Jr. and Z. G. Stoumbos (1998). The SPRT chart for monitoring a proportion. *IIE Transactions 30:* 545–561.

Reynolds, M. R., Jr. and Z. G. Stoumbos (1999a). A CUSUM chart for monitoring a proportion when inspecting continuously. *Journal of Quality Technology 31*(1): 87–108.

Reynolds, M. R., Jr. and Z. G. Stoumbos (1999b). A general approach to modeling CUSUM charts for a proportion. *IIE Transactions* (to appear).

Reynolds, M. R., Jr. and Z. G. Stoumbos (2000). Monitoring a proportion using CUSUM and SPRT control charts. *Frontiers in Statistical Quality Control 6* (to appear).

Roberts, S. W. (1959). Control chart tests based on geometric moving averages. *Technometrics 1*(3): 239–250.

Ryan, T. P. and B. J. Faddy (2000). The effect of non-normality on CUSUM procedures. *Frontiers in Statistical Quality Control 6* (to appear).

Ryan, T. P. and N. C. Schwertman (1997). Optimal limits for attributes control charts. *Journal of Quality Technology 29*(1):86–98.

Saccucci, M. S., R. W. Amin, and J. M. Lucas (1992). Exponentially weighted moving average schemes with variable sampling intervals. *Communications in Statistics—Simulation and Computation 21*(3): 627–657.

Steiner, S. H. (1998). Grouped data exponentially weighted moving average charts. *Applied Statistics 47*(2): 203–216.

Van Dobben de Bruyn, C. S. (1968). *Cumulative Sum Tests: Theory and Practice.* London: Griffin.

Van Gilder, J. F. (1994). Application of EWMA to automotive onboard diagnostics. *In ASQC Quality Congress Proceedings*, pp. 43–51. Milwaukee, WI: American Society for Quality Control.

Westgard, J. O., T. Groth, T. Aronsson, and C. de Verdier (1977). Combined Shewhart–CUSUM control chart for improved quality control in clinical chemistry. *Clinical Chemistry 23*(10): 1881–1887.

Woodall, W. H. (1984). On the Markov-chain approach to the two-sided CUSUM procedure. *Technometrics 26*(1): 41–46.

Woodall, W. H. (1986). The design of CUSUM quality control charts. *Journal of Quality Technology 18:* 99–102.

Woodall, W. H. (1991). Letter to the editor. *Quality Engineering 3*(3): vii–viii.

Woodall, W. H. and D. C. Montgomery (1999). Research issues and ideas in statistical process control. *Journal of Quality Technology* (to appear).

EXERCISES

1. What are the guiding factors for determining the values of h and k for a CUSUM scheme?

2. Assume that a CUSUM scheme is being used to monitor the process mean, and a $1.5\sigma_{\overline{x}}$ increase in the process mean suddenly occurs.
   - **(a)** If $h = 5$ and $k = 0.5$ are being used, how many subgroups would we expect to observe before a signal is received?
   - **(b)** How many subgroup averages would we expect to have to plot with an $\overline{X}$ chart before observing an average above the UCL?
   - **(c)** Explain why your answers to (a) and (b) do not depend on the subgroup size.
   - **(d)** Now assume that $n = 4$ and a $0.75\sigma_x$ shift occurs. (Note that this is the same as a $1.5\sigma_{\overline{x}}$ shift.) Compare the ARLs of a CUSUM based on subgroup averages of $n = 4$ with a CUSUM that uses the individual observations. Comment.
3. Indicate when it might be inadvisable to reset S_H and S_L to zero after an out-of-control signal has been received.
4. Fill in the blanks below where a CUSUM procedure is being used to control the process mean, with $k = .05$.

z	S_H	S_L
0.48	0	0
	3.0	
		1.3
2.13		

5. Explain why it is better to look at the two one-sided in-control ARLs for a two-sided CUSUM scheme than to look only at the single (two-sided) in-control ARL when examining the effects of nonnormality.
6. Describe when a Shewhart–CUSUM scheme should be used, and explain why $z = 3.5$ is used for the Shewhart criterion instead of $z = 3.0$.
7. Explain the difference between a Bernoulli CUSUM and a binomial CUSUM.
8. Assume that individual observations can be modeled by a chi-square distribution with 20 degrees of freedom. Would you use a normality-based CUSUM approach or not? If not, how would you proceed?
9. What will be the effect of using a relatively large value of λ (say, at least 0.4) in an EWMA scheme?
10. Use an EWMA scheme with $L = 3.00$ and $\lambda = 0.20$ for the data in Table 8.1 and compare your results with the results that were obtained using the CUSUM procedure with $k = 0.5$ and $h = 5$. Also compare with the results shown in Table 8.12 using $L = 3.00$ and $\lambda = 0.25$. Comment.
11. Explain the advantages and disadvantages of applying a normalizing transformation to attribute data and then applying a CUSUM procedure to the transformed values.

CHAPTER 9

Multivariate Control Charts for Measurement Data

We will assume that p variables $X_1, X_2, \ldots, X_p$ are to be simultaneously monitored, and we will further assume (initially) $\mathbf{X} \sim N_p(\boldsymbol{\mu}, \boldsymbol{\Sigma})$, with $\mathbf{X} = (X_1, X_2, \ldots, X_p)'$. That is, $\mathbf{X}$ is assumed to have a multivariate normal distribution. The general idea is to monitor $\boldsymbol{\mu}$, the correlations between the X_i, and the $\text{Var}(X_i)$. A change in at least one mean, correlation (or covariance), or variance constitutes an out-of-control process.

The expressions "when the quality of a product is defined by more than one property, all the properties should be studied collectively" (Kourti and MacGregor, 1996) and "the world is multivariate" suggest that multivariate procedures should be routinely used. Although not used as often as univariate control chart procedures, multivariate procedures have nevertheless been used in many different types of applications. For example, Nijhius, deJong, and Vandeginste (1997) discuss the use of multivariate control charts in chromatography and Colon and GonzalezBarreto (1997) discuss an application to printed circuit (PC) boards that has 128 variables.

The number of variables used can be quite large. In applications in the chemical industry even thousands of variables may be involved (MacGregor, 1998). Dimension reduction methods are obviously needed when there is a large number of variables. Such methods are covered briefly in Section 9.10.

The usual practice when multiple quality variables are to be controlled simultaneously, however, has been to maintain a separate (univariate) chart for each variable. When many charts are maintained, there is a not-so-small probability that at least one chart will emit an out-of-control message due to chance alone.

For example, Meltzer and Storer (1993) described a scenario in which approximately 200 control charts for individual variables were used, and false signals occurred so frequently that both the production shop and the engineering personnel lost confidence in the charts. If 200 separate charts were to be used and points were plotted one at a time, the probability of having a false signal from at least one chart at a particular point in time is $1 - (1 - .0027)^{200} = .42$, *if* the quality characteristics were independent, normality was assumed, and 3-sigma limits were used. Therefore, this might be used as an approximation if the correlations between the quality characteristics were quite small, since the actual probability cannot be determined analytically. Obviously company personnel will lose faith in control charts if false signals are received about 50% of the time.

What would happen if a much smaller number of charts were used? Assume that an $\overline{X}$ chart is being maintained for each of eight quality variables. Again assuming normality and independence, the probability that at least one of the charts will signal an out-of-control condition when there is no change in any of the means or standard deviations is, assuming 3-sigma limits, $1 - (1 - .0027)^8 = .0214$. Thus, a false signal will occur on at least one of the charts approximately 2% of the time. This is analogous to the simultaneous inference problem in statistics, where an adjustment is necessary to account for the fact that multiple hypothesis tests are being performed simultaneously. Similarly, there is a need for some type of "correction" when multiple control charts are used simultaneously.

There is a stronger need for a multivariate chart when quality variables are correlated, as there is then a greater potential for misleading results when a set of univariate charts is used. Here the problems are the likelihood of receiving a false alarm and, more importantly, the likelihood of *not* receiving a signal when the multivariate process is out of control. This is illustrated in Figure 9.1, which shows a typical elliptical control region for the case of two variables that are positively correlated, relative to the box which would represent the rectangular control region for the pair of Shewhart charts.

The ellipse might be used for controlling the *bivariate process*. A point that falls outside of the ellipse would indicate that the bivariate process may be out of control. The bivariate control ellipse has a major weakness, however, in that time order is not indicated; here the ellipse will be used simply to make a point.

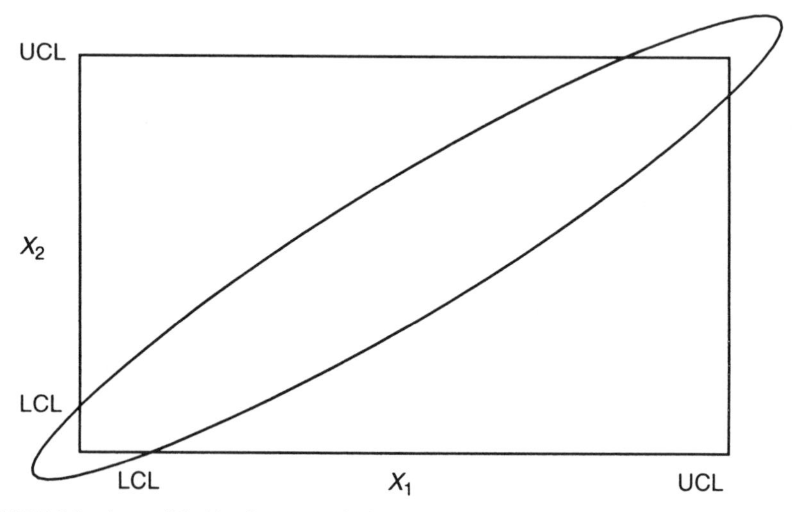

FIGURE 9.1 A possible bivariate control ellipse relative to the rectangular control region for a pair of Shewhart charts.

Readers interested in the use of control ellipses for subgroup averages are referred to Montgomery (1996, p. 363).

Assume that the two variables are highly correlated when the bivariate process is in control. If points were to start plotting outside the ellipse, this would suggest that the correlation has been disturbed (i.e., the process is out of control). Notice, however, that there is plenty of room for points to be outside the ellipse but still be inside the box. Conversely, Figure 9.1 also shows that there is room for points to plot inside the ellipse but outside the box. Thus, the two individual charts could easily either fail to signal when they should signal (the first case) or signal when they should not signal (the second case). See also Alt, Smith and Jain (1998) for additional illustration/explanation of this relationship.

The narrowness of the ellipse will depend upon the degree of correlation between the variables—the higher the correlation in absolute value, the narrower the ellipse. Thus, for two highly correlated variables, the univariate charts would frequently signal that the two characteristics are in control when the correlation structure is out of control.

False signals will occur at an even higher rate if the quality characteristics are correlated. This should be intuitively apparent. Assume, for example, that there are two quality characteristics that have a very high positive correlation. If one of the characteristics plots above its upper control limit, then the other characteristic will likely plot close to, if not above, its upper control limit simply because of the high correlation. If this did not occur, then the correlation between the two variables would be out of control.

Thus, the use of a control chart for each of many quality characteristics can cause problems, and it is better to use a single multivariate chart whenever possible.

We will first consider procedures when there is a moderate number of quality variables (say, 10 or so) and will then consider scenarios where some type of dimension-reduction approach will be desirable. There is frequently a large number of variables used in certain applications, such as PC board applications, where the variables could be the positions on the board, and in certain applications in the chemical industry.

For example, MacGregor (1995) states that in the chemical industry it is not uncommon for measurements to be made on hundreds or even thousands of process variables and on 10 or more quality variables. It would be impractical to try to incorporate a very large number of variables into a single multivariate chart, as this would stretch the data too thin since a very large number of parameters (means, variances, and covariances) would have to be estimated. Therefore, dimension reduction would first be necessary before process control could be applied. Dimension reduction techniques are discussed briefly in Section 9.10.

It should be noted that the control charts presented in this chapter are quite different from the Multi-Vari chart proposed by Seder (1950). The latter is actually not a type of control chart. Rather, it is primarily a graphical tool for displaying variability due to different factors and is covered in Section 11.7.

As with the univariate charts, multivariate charts can be used when there is subgrouping or when individual (multivariate) observations are to be used. Multivariate charts are not as easy to understand and use as are univariate charts, however. In particular, when a signal is received on a univariate chart, the operator knows to look for an assignable cause for the variable that is being charted. For a multivariate chart, however, a signal could be caused by a mean shift in one or more variables, or a variance shift, or a change in the correlation structure. So the user will have to determine what type of change has triggered the signal and also which variables are involved. The survey results of Saniga and Shirland (1977) suggested that only about 2% of companies were using multivariate charts. Two decades later the percentage probably has not changed very much.

9.1 HOTELLING'S T^2 DISTRIBUTION

Some of the multivariate procedures for control charts are based heavily on Hotelling's T^2 distribution, which was introduced by Hotelling (1947). This is the multivariate analogue of the univariate t distribution that was covered in Chapter 3. Recall that

$$t = \frac{\overline{x} - \mu}{s/\sqrt{n}}$$

has a t distribution. If we wanted to test the hypothesis that $\mu = \mu_0$, we would then have

$$t = \frac{\overline{x} - \mu_0}{s/\sqrt{n}}$$

so that

$$\begin{aligned} t^2 &= \frac{(\overline{x} - \mu_0)^2}{s^2/n} \\ &= n(\overline{x} - \mu_0)(s^2)^{-1}(\overline{x} - \mu_0) \end{aligned} \tag{9.1}$$

When Eq. (9.1) is generalized to p variables, it becomes

$$T^2 = n(\overline{\mathbf{x}} - \boldsymbol{\mu}_0)'\mathbf{S}^{-1}(\overline{\mathbf{x}} - \boldsymbol{\mu}_0)$$

where

$$\overline{\mathbf{x}} = \begin{bmatrix} \overline{x}_1 \\ \overline{x}_2 \\ \vdots \\ \overline{x}_p \end{bmatrix} \qquad \boldsymbol{\mu}_0 = \begin{bmatrix} \mu_1^0 \\ \mu_2^0 \\ \vdots \\ \mu_p^0 \end{bmatrix}$$

$\mathbf{S}^{-1}$ is the inverse of the sample variance–covariance matrix, $\mathbf{S}$, and n is the sample size upon which each $\overline{x}_i$, $i = 1, 2, \ldots, p$, is based. (The diagonal elements of $\mathbf{S}$ are the variances and the off-diagonal elements are the covariances for the p variables. See Section 3.4 for the mathematical definition of a sample covariance.)

It is well known that when $\boldsymbol{\mu} = \boldsymbol{\mu}_0$

$$T^2 \sim \frac{p(n-1)}{n-p} F_{(p,n-p)}$$

where $F_{(p,n-p)}$ refers to the F distribution (covered in Chapter 3) with p degrees of freedom for the numerator and $n - p$ for the denominator.

Thus, if $\boldsymbol{\mu}$ were specified to be $\boldsymbol{\mu}_0$, this could be tested by taking a single p-variate sample of size n, then computing T^2 and comparing it with

$$\frac{p(n-1)}{n-p} F_{\alpha(p,n-p)}$$

for a suitably chosen value of α. (Here α denotes the upper tail area for the F distribution.) One suggested approach for determining α is to select α so that $\alpha/2p = .00135$—the 3-sigma value for a univariate chart. Accordingly, this approach would lead to $\alpha = .0054$ when $p = 2$. If $T^2 > \mathrm{F}_{\alpha(p,n-p)}$, we would then conclude that the multivariate mean is no longer at $\boldsymbol{\mu}_0$. We consider the selection of α further in Section 9.2.1.

9.2 A T^2 CONTROL CHART

In practice, $\boldsymbol{\mu}$ is generally unknown, so it is necessary to estimate $\boldsymbol{\mu}$ analogous to the way that μ is estimated when an $\overline{X}$ chart or X chart is used. Specifically, when there are rational subgroups, $\boldsymbol{\mu}$ is estimated by $\overline{\overline{\mathbf{x}}}$, where

$$\overline{\overline{\mathbf{x}}} = \begin{bmatrix} \overline{\overline{x}}_1 \\ \overline{\overline{x}}_2 \\ \vdots \\ \overline{\overline{x}}_p \end{bmatrix}$$

Each $\overline{\overline{x}}_i$, $i = 1, 2, \ldots, p$, is obtained the same way as with an $\overline{X}$ chart; namely, by taking k subgroups of size n and computing $\overline{\overline{x}}_i = (1/k)\sum_{j=1}^{k} \overline{x}_{ij}$. (Here $\overline{x}_{ij}$ is used to denote the average for the jth subgroup of the ith variable.) As with an $\overline{X}$ chart (or any other chart) the k subgroups would be tested for control by computing k values of T^2 [Eq (9.2)] and comparing each against the UCL. If any T^2 value falls above the UCL (there is no LCL), the corresponding subgroup would be investigated.

Thus, one would plot

$$T_j^2 = n(\overline{\mathbf{x}}^{(j)} - \overline{\overline{\mathbf{x}}})' \, \mathbf{S}_p^{-1} (\overline{\mathbf{x}}^{(j)} - \overline{\overline{\mathbf{x}}}) \tag{9.2}$$

for the jth subgroup ($j = 1, 2, \ldots, k$), where $\overline{\mathbf{x}}$ denotes a vector with p elements that contains the subgroup averages for each of the p characteristics for the jth subgroup. (Here, $\mathbf{S}_p^{-1}$ is the inverse of the "pooled" variance–covariance matrix,

$\mathbf{S}_p$, which is obtained by averaging the subgroup variance–covariance matrices over the k subgroups.)

Each of the k values of Eq. (9.2) would be compared with

$$\text{UCL} = \left(\frac{knp - kp - np + p}{kn - k - p + 1} \right) F_{\alpha(p, kn-k-p+1)} \tag{9.3}$$

[See, e.g., Alt (1982a, p. 890), Alt (1985), or Jackson (1985).] If any of the T_j^2 values exceed the UCL from Eq. (9.3), the corresponding subgroup(s) would be investigated.

We should note that the selected value of α is almost certainly not going to be the actual probability that at least one point in the Stage 1 analysis plots above the UCL when all of the historical data do not come from the same multivariate normal distribution. This probability will depend in part on the number of plotted points. Furthermore, the fact that the parameters are estimated causes the random variables that correspond to the plotted points to not be independent. Simulation would have to be used to determine (estimate) the true α value, as discussed by Sullivan and Woodall (1996). Similarly, ARLs for Stage 2 would also have to be determined by simulation (see the discussion in Section 9.9). The reader should keep this in mind while reading the following sections.

9.2.1 Identifying the Source of the Signal

The investigation would proceed somewhat differently, however, than it would for, say, an $\overline{X}$ chart in which only one quality characteristic is involved. Specifically, it is necessary to determine which quality characteristic(s) is causing the out-of-control signal to be received. There are a number of possibilities, even when $p = 2$. Assume for the moment that we have two quality characteristics that have a very high positive correlation (i.e., ρ close to 1.0). We would then expect their average in each subgroup to almost always have the same relationship in regard to their respective averages in $\overline{\overline{\mathbf{x}}}$. For example, if $\overline{x}_1^{(j)}$ exceeds $\overline{\overline{x}}_1$, then $\overline{x}_2^{(j)}$ will probably exceed $\overline{\overline{x}}_2$. Similarly, if $\overline{x}_1^{(j)} < \overline{\overline{x}}_1$, we could expect to observe $\overline{x}_2^{(j)} < \overline{\overline{x}}_2$. That is, we would not expect them to move in opposite directions relative to their respective averages since the two characteristics have a very high *positive* correlation (assuming that ρ has not changed). If they do start moving in opposite directions, this would indicate that the "bivariate" process is probably out of control. This out-of-control state could result in a value of Eq. (9.2) that far exceeds the value of Eq. (9.3). Unless one of the two deviations, $\overline{x}_1^{(j)} - \overline{\overline{x}}_1$ or $\overline{x}_2^{(j)} - \overline{\overline{x}}_2$, in absolute value exceeds $\overline{\overline{x}}_1 + 3(\overline{R}_1/d_2\sqrt{n})$ or $\overline{\overline{x}}_2 + 3(\overline{R}_2/d_2\sqrt{n})$, respectively, the individual $\overline{X}$ charts will not give an out-of-control message, however. This will be illustrated later with a numerical example.

There are a variety of other out-of-control conditions that *could* be quickly detected using an $\overline{X}$ chart or CUSUM procedure, however. For example, if the two deviations are in the same direction and one of the deviations is more than three standard deviations from the average ($\bar{\bar{x}}$) for that characteristic, the deviation will show up on the corresponding $\overline{X}$ chart. It is *not* true, however, that a deviation that causes an out-of-control signal to be received on at least one of the $\overline{X}$ charts will also cause a signal to be received on the multivariate chart. (This will also be illustrated later.) Thus, it is advisable to use a univariate procedure (such as an $\overline{X}$ chart or Bonferroni intervals) in conjunction with the multivariate procedure.

Bonferroni intervals are described in various sources, including Alt (1982b). They essentially serve as a substitute for individual $\overline{X}$ charts (or X charts) and will usually be as effective as $\overline{X}$ charts in identifying the quality characteristic(s) that cause the out-of-control message to be emitted by the multivariate chart.

The general idea is to construct p intervals (one for each quality characteristic) for each subgroup that produces an out-of-control message on the multivariate chart. Thus, for the jth subgroup the interval for the ith characteristic would be

$$\bar{\bar{x}}_i - t_{\alpha/2p,k(n-1)}\, s_{p_i}\sqrt{\frac{k-1}{kn}} \le \bar{x}_i^{(j)} \le \bar{\bar{x}}_i + t_{\alpha/2p,k(n-1)}\, s_{p_i}\sqrt{\frac{k-1}{kn}} \tag{9.4}$$

where s_{p_i} designates the square root of the pooled sample variance for the ith characteristic and the other components are as previously defined. If Eq. (9.4) is not satisfied for the ith characteristic, the values of that characteristic would then be investigated for the jth subgroup. If an assignable cause is detected and removed, the entire subgroup would be deleted (for all p characteristics) and the UCL recomputed.

Although the Bonferroni approach is frequently used, it is also flawed. It is well known that the Bonferroni interval approach is quite conservative, with the significance level for the set of intervals being much less than α, especially when there are correlations that are large in absolute value. This is shown later in this section.

As discussed by Hayter and Tsui (1994), it is difficult to determine the significance level to be used for the Bonferroni intervals. Even if the variables were independent, the significance level would still be less than α. For example, assume $p = 2$. The probability of not receiving a signal when there is no process change is $(1 - \alpha/2)^2$, so the probability of receiving a signal is $1 - (1 - \alpha/2)^2 = \alpha - \alpha^2/4$. The difference between α and the actual overall significance level will be much greater when the two variables are highly correlated. Hayter and Tsui (1994) point out that, under the assumption of normality, the Dunn–Sidak inequality (Dunn, 1958) leads to the selection of $1 - (1 - \alpha)^{1/p}$ as the significance level for each Bonferroni interval. The use of this significance level would result in a less conservative procedure than the use of α/p, although the difference between the two may be small.

Hayter and Tsui (1994) discuss the construction of confidence intervals for the means of the variables such that the overall significance level will be α. Their approach to multivariate quality control consists primarily of the set of confidence intervals, however, and such intervals cannot be used to detect a change in the correlation structure. Thus, if such an out-of-control condition is a distinct possibility in a given application, the Hayter and Tsui (1994) simultaneous confidence interval approach will have to be supplemented with a multivariate procedure. (Note that the Bonferroni interval approach has the same shortcoming, as do all methods that do not utilize the covariances in the variance–covariance matrix.)

The construction of the confidence intervals is straightforward for $p = 2$, as the tables of Odeh (1982) can be used, assuming normality, to determine the significance level for each variable, so that the overall significance level will be α. Hayter and Tsui (1994) illustrate the use of the tables for bivariate individual observations with $\alpha = .05$. Using their notation, the intervals are of the form $\overline{X}_i \pm \sigma_{\overline{X}_i} C_{R,\alpha}$. We would question the use of $\alpha = .05$, however, as this would mean that we would expect to receive a false signal on every 20th multivariate observation. Therefore, a much smaller value of α seems desirable.

When $p \geq 3$, those tables can be used only when the variables have an equicorrelation structure, however. The correlations will usually be unequal, however, so there is a need to determine the individual significance levels in a different manner. Hayter and Tsui (1994) state that numerical integration techniques could be used when $p = 3, 4$ but indicate that such an approach will be infeasible for $p \geq 5$. Consequently, simulation apparently must be used.

A better method is therefore needed—one that provides practitioners with the capability to produce a set of intervals with a desired overall α.

When $p = 2$, an exact (under normality) pair of confidence intervals may be constructed using the tables of Odeh (1982), as illustrated by Hayter and Tsui (1994). (Those tables were originally designed for use with order statistics but are applicable here.).

When $p > 2$, exact intervals may be constructed (e.g., using Odeh's tables) only in the case of an equicorrelation structure, but such a configuration of correlations is not likely to exist in any application. Nevertheless, we can compare the exact intervals under the assumption of an equicorrelation structure with Bonferroni intervals to show how conservative the latter are.

Since the values in Odeh's tables do not incorporate a subgroup size or a sample size, to provide a relevant comparison we will assume that there is a very large number of subgroups in the Stage 1 analysis. Then the appropriate value from the t-table would be essentially the same as $z_{\alpha/2}$ and the square root term in expression (9.4) would be approximately 1. We will let ρ denote the common correlation between the variables. With these assumptions, it is a matter of comparing $z_{\alpha/2p}$ with $C_{R,\alpha}$ from Odeh's tables, for different values of ρ and p. We will use $\alpha = .01$, which is roughly in line with Alt's suggestion for α if there is a small-to-moderate number of variables.

Table 9.1 shows that there is very little difference between the two values when both the correlation and the number of variables are low, but at the other extreme, there is a large percentage difference when $\rho = 0.90$ and $p = 10$.

A critical decision is the choice of α. Some researchers have contended that we should not be concerned, however, with making the type I error rate "right" for each univariate confidence interval or hypothesis test. For example, Kourti and MacGregor (1996), in discussing the use of a multivariate chart followed by some

TABLE 9.1 Comparison of Bonferroni and Exact Values for $\alpha = .01$

ρ	p	$z_{\alpha/2p}$	$C_{R,\alpha}$	$\dfrac{z_{\alpha/2p} - C_{R,\alpha}}{C_{R,\alpha}}$
.10	2	2.8070	2.8059	0.04%
	5	3.0902	3.0884	0.06
	10	3.2905	3.2883	0.07
.50	2	2.8070	2.7943	0.45
	5	3.0902	3.0603	0.98
	10	3.2905	3.2465	1.36
.90	2	2.8070	2.7154	3.37
	5	3.0902	2.8744	7.51
	10	3.2905	2.9794	10.44

univariate procedure, state the following. "However, once a multivariate chart has detected an event, there is no longer a need to be concerned with precise control limits which control the type I error (α) on the univariate charts. A deviation at the chosen level of significance has already been detected. Diagnostic univariate plots are only used to help decide on the variables which are causing the deviation."

While it is certainly true that the probability of the multivariate signal is controlled by the selected type I error rate for the selected multivariate procedure, the type I error rate for the univariate test is nevertheless important. If a control chartist used univariate charts with 3-sigma limits after a multivariate signal is received, as Kourti and MacGregor (1996) imply would be acceptable, the type I error rate for the set of univariate charts could far exceed the type I error rate for the multivariate procedure, depending on the correlation structure and the number of variables.

Consider the following scenario. Assume that 20 variables are being monitored with some type of multivariate control chart, and we will initially assume that the variables are independent. Assume further that the historical data suggest that there is generally a mean shift in at most two variables whenever the multivariate process is out of control. For the purpose of illustration we will assume that $\alpha = .0027$ is used for the multivariate procedure and there is a sizable mean shift in exactly one variable. If the same α were used for each univariate procedure, the probability would be $1 - (1 - .0027)^{19} = .05$ of receiving a false signal from at least one of the other variables. This would be considered much too large for a *single* univariate control chart procedure, since it would produce an in-control ARL of only 20, so it should probably also be considered too large in a multivariate setting. Of course, if the variables are correlated, the false-alarm probability would be much less, but a probability of, say, .02 would correspond to an in-control ARL of 50, so this would also be somewhat undesirable.

The point is that there are *two* false-alarm rates, not one, and the second one cannot be ignored. The control chart user would have to decide what would be a tolerable false-alarm rate for the univariate procedure. Since a process is not stopped, it is reasonable to assume that a higher false-alarm rate could be tolerated than for the multivariate procedure. But how much higher? If the rate is too high, users could become discouraged when they repeatedly cannot find a problem.

Arguments can be made for and against having the two error rates approximately equal. Assume, for example, that a T^2 chart is being used, there are 10 quality variables, and there has been a mean shift for two of the variables. It would seem reasonable to have the significance level for whatever univariate procedure is used be the same as that which results from the use of 3-sigma limits for normally distributed data. This would give a user the same chance of detecting the shift as would exist if univariate charts had been constructed for each of the two variables. But we also have to think about what the false-alarm probability is when the univariate procedure is applied to the set of eight variables for which there has been no parameter change. For simplicity, assume that all pairwise correlations are .8 and eight X charts with 3-sigma limits are used. Then, assuming known parameter values and using the Odeh (1982) tables in reverse, we find that the probability of a false alarm is approximately .01. The user would have to decide whether or not such a false-alarm probability is too high, recognizing that if the false-alarm probability is .01 each time a signal is received from the multivariate procedure, then a needless search for assignable causes can be expected to happen once for every 100 multivariate signals.

9.2.2 Regression Adjustment

Another approach that can be useful in detecting the cause of the signal is regression adjustment (Hawkins, 1991, 1993; Hawkins and Olwell, 1998, pp. 198–207). Such an approach can be especially useful when a multivariate process consists of a sequence of steps that are performed in a specified sequence, such as a "cascade process" (Hawkins, 1991). A cause-selecting chart, presented in Section 12.8, is one type of regression adjustment chart.

If, for example, raw material were poor, then product characteristics for units of production that utilized this raw material would also probably be poor. Obviously it would be desirable to identify the stage at which the problem occurred. Clearly there is no point in looking for a process problem when the real problem is the raw material.

9.2.3 Recomputing the UCL

Recomputing the UCL that is to be subsequently applied to *future* subgroups entails recomputing S_p and $\bar{\bar{\mathbf{x}}}$ *and* using a constant and an F-value that are different from the form given in Eq. (9.3). The latter results from the fact that different distribution theory is involved since future subgroups are assumed to be independent of the "current" set of subgroups that is used in calculating S_p and $\bar{\bar{\mathbf{x}}}$. (The same thing happens with $\overline{X}$ charts; the problem is simply ignored through the use of 3-sigma limits.)

For example, assume that a subgroups had been discarded so that $k - a$ subgroups are used in obtaining S_p and $\bar{\bar{\mathbf{x}}}$. We shall let these two values be represented by S_p^* and $\bar{\bar{\mathbf{x}}}^*$ to distinguish them from the original values, S_p and $\bar{\bar{\mathbf{x}}}$, before any subgroups are deleted. Future values to be plotted on the T^2 chart would then be obtained from

$$n(\bar{\mathbf{x}}^{(\text{future})} - \bar{\bar{\mathbf{x}}}^*)'(S_p^*)^{-1}(\bar{\mathbf{x}}^{(\text{future})} - \bar{\bar{\mathbf{x}}}^*) \qquad (9.5)$$

where $\bar{\mathbf{x}}^{(\text{future})}$ denotes an arbitrary vector containing the averages for the p characteristics for a single subgroup obtained in the future. Each of these future values would be plotted on the multivariate chart and compared with

$$\text{UCL} = \left(\frac{p(k-a+1)(n-1)}{(k-a)n-k+a-p+1}\right) F_{\alpha[p,(k-a)n-k+a-p+1]} \qquad (9.6)$$

where a is the number of the original subgroups that is deleted before computing S_p^* and $\bar{\bar{\mathbf{x}}}^*$. Notice that Eq. (9.6) does *not* reduce to Eq. (9.3) when $a = 0$, nor should we expect it to since Eq. (9.3) is used when testing for control of the entire *set* of subgroups that is used in computing S_p and $\bar{\bar{\mathbf{x}}}$. [*Note*: Eqs. (9.5) and (9.6) are variations of a result given in Alt (1982a).]

9.2.4 Characteristics of Control Charts Based on T^2

Although plotting T^2 values on a chart with the single (upper) control limit given in the preceding section is straightforward (provided that computer software is available), this approach has limitations in addition to the previously cited limitation of not being able to identify the reason for a signal without performing additional computations.

In the retrospective analysis stage, the estimated variances and covariance(s) are pooled over the subgroups. Therefore, possible instability in the subgroup variances or in the correlations between the variables *within* each subgroup is lost because of the pooling. Assume that there are two process characteristics and let the pooled estimated variances for X_1 and X_2 be given by a and b, respectively. Since these are actually average values, the numbers in a given data set could be altered considerably without affecting the T^2 values (such as making all of the subgroup variances equal to a and b, respectively, or causing them to vary greatly, while retaining the original subgroup means).

This problem is not unique to a T^2 chart, however, as the same problem can occur when an $\overline{X}$ chart is used. It is simply necessary to use a chart for monitoring multivariate variability in addition to using a T^2 chart, as is necessary in the univariate case.

A change in the correlation structure, occurring *between* subgroups, can be detected in Stage 1, however, as will be illustrated in the next section.

In the process-monitoring stage (Stage 2), the only statistic computed from the current subgroup and used in the computation of T^2 is $\bar{\mathbf{x}}^{(\text{future})}$ since neither $\bar{\bar{\mathbf{x}}}$ nor $\mathbf{S}_p$ involves any statistics computed using real-time data. This is analogous to the Stage 2 use of an $\overline{X}$ chart. Thus, there is no measure of current variability that is used in computing the T^2 values that are plotted in real time. Therefore, whereas a variability change could occur at time i, it will not be detected by the T^2 values unless the change is so great that it (accidentally) causes a large change in the vector of subgroup means. A value of T^2 that plots above the control limit could signify either that the multivariate mean or the correlation structure is out of control or that they are both out of control. Therefore, in Stage 2 the multivariate chart functions in essentially the same general way as a univariate $\overline{X}$ chart in regard to mean and variance changes, as the T^2 chart is not suitable for detecting variance changes. Charts for detecting shifts in the variances are discussed in Section 9.4.

Thus, it is inappropriate to say (as several writers have said) that the T^2 chart confounds mean shifts and variance shifts since a variance shift generally cannot be detected in either the retrospective analysis stage or the process-monitoring stage. [Sullivan and Woodall (1998a) do, however, present a method for separate identification of a variance shift or a mean shift in Stage 1 when individual observations are plotted.]

Hawkins and Olwell (1998, p. 192) point out that the T^2 test statistic is the most powerful affine invariant test statistic for testing $H_0 : \boldsymbol{\mu} = \boldsymbol{\mu}_0$ against $H_1 : \boldsymbol{\mu} \neq \boldsymbol{\mu}_0$. [An affine invariant test statistic is one whose value is unaffected by a full-rank linear transformation of $\mathbf{x}$ (for the case of individual observations).] From a practical standpoint, this means that a T^2 chart makes sense, relative to the suggested alternatives, if it is not possible to anticipate the direction of the shift in $\boldsymbol{\mu}$. Specifically, we have the same general relationships in the multivariate setting that exist in the univariate case. That is, if we fix the sample size, the T^2 statistic will be the optimal test statistic for detecting a change in $\boldsymbol{\mu}$, just as $\overline{X}$ is the optimal test statistic for detecting a change in μ in the univariate case. If we move beyond a single sample, however (as we should surely do since a control chart is a sequence of plotted points, not a single point), the optimality of T^2 is lost just as the optimality of $\overline{X}$ is lost in the univariate case. Then the CUSUM of Healy (1987), which reduces to a univariate CUSUM, will be optimal.

However, if the direction of the shift in $\boldsymbol{\mu}$ when the process goes out of control *can* be anticipated, then we can do better than we could using T^2. The optimal non-CUSUM approach is the test statistic that Healy (1987) "CUSUMed," whereas in the general case, the CUSUM approach of Crosien (1988) seems to work well. Healy's test statistic is given by

$$Z = (\mathbf{X} - \boldsymbol{\mu}_0)\boldsymbol{\Sigma}^{-1}(\boldsymbol{\mu}_1 - \boldsymbol{\mu}_0) \tag{9.7}$$

with $\boldsymbol{\mu}_0$ denoting the in-control multivariate mean (which will generally have to be estimated), and $\boldsymbol{\mu}_1$ representing the out-of-control mean that one wishes to detect as quickly as possible.

Although the criticism of the T^2 chart in the literature seems overly harsh, it would obviously be much better to use a multivariate control procedure that allows a shift in the multivariate mean to be detected apart from a shift in the correlation structure. We will look at some proposed methods that make this possible in Section 9.5.

As with the univariate charts discussed in the preceding chapters, a practitioner may wish to consider the discussion of Section 4.2 and select α for the retrospective analysis such that the probability of at least one T_j^2 value plotting above the UCL is approximately .00135. Unfortunately, it is not easy to determine how to select α to accomplish this.

9.2.5 Determination of a Change in the Correlation Structure

If a signal is received from the T^2 chart but not from at least one of the univariate tests, such as a univariate control chart, this would cause us to suspect that the correlation structure may be out of control rather than one or more process characteristics being out of control. This is illustrated in the next section.

9.2.6 Illustrative Example

An example will now be given for $p = 2$ that illustrates how an out-of-control condition can be detected with the multivariate chart but would not be detected with the two $\overline{X}$ charts. The data set to be used is given in Table 9.2. We will not assume that a shift in a particular direction can be anticipated, so the T^2 chart will be the optimal approach. Assume that each pair of values represents an observation on each of the two variables. Thus, there are 20 subgroups (represented by the 20 rows of the table), with four observations in each subgroup for each variable.

TABLE 9.2 Data for Multivariate Example[a] ($p = 2, k = 20, n = 4$)

Subgroup Number	First Variable				Second Variable			
1	72	84	79	49	23	30	28	10
2	56	87	33	42	14	31	8	9
3	55	73	22	60	13	22	6	16
4	44	80	54	74	9	28	15	25
5	97	26	48	58	36	10	14	15
6	83	89	91	62	30	35	36	18
7	47	66	53	58	12	18	14	16
8	88	50	84	69	31	11	30	19
9	57	47	41	46	14	10	8	10
10	26	39	52	48	7	11	35	30
11	46	27	63	34	10	8	19	9
12	49	62	78	87	11	20	27	31
13	71	63	82	55	22	16	31	15
14	71	58	69	70	21	19	17	20
15	67	69	70	94	18	19	18	35
16	55	63	72	49	15	16	20	12
17	49	51	55	76	13	14	16	26
18	72	80	61	59	22	28	18	17
19	61	74	62	57	19	20	16	14
20	35	38	41	46	10	11	13	16

[a]The multivariate observations are (72, 23), (84, 30), ..., (46, 16).

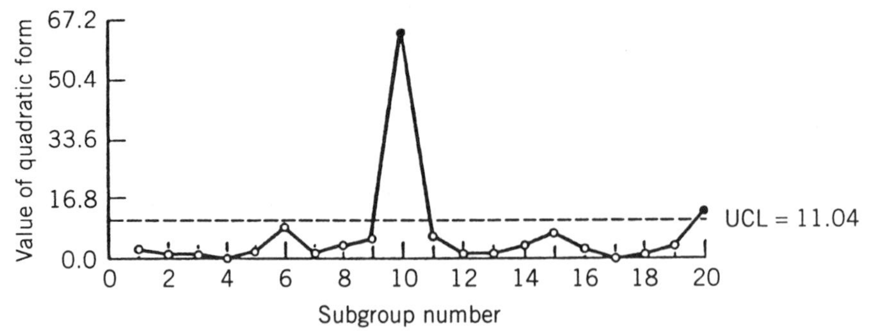

FIGURE 9.2 Multivariate chart for Table 9.2 data.

TABLE 9.3 T^2 Values in Figure 9.2

Subgroup	Subgroup Averages		
Number	First Variable	Second Variable	T^2
1	71.00	22.75	2.24
2	54.50	15.50	0.65
3	52.50	14.25	1.27
4	63.00	19.25	0.22
5	57.25	18.75	1.53
6	81.25	29.75	8.98
7	56.00	15.00	1.32
8	72.75	22.75	3.77
9	47.75	10.50	4.95
10	41.25	20.75	63.76
11	42.50	11.50	6.55
12	69.00	22.25	1.37
13	67.75	21.00	1.36
14	67.00	19.25	3.26
15	75.00	22.50	7.41
16	59.75	15.75	2.76
17	57.75	17.25	0.12
18	68.00	21.25	1.33
19	63.50	17.25	3.50
20	40.00	12.50	13.04
$\bar{\bar{x}}$	60.38	18.49	

When, for this data set, the values obtained using Eq. (9.2) are plotted against the UCL given by Eq. (9.3), with $\alpha = .0054$, the result is the chart given in Figure 9.2. There are two points that exceed the UCL; in particular, the value for subgroup 10 far exceeds the UCL. (The T^2 values are given in Table 9.3.)

Before explaining why the value for subgroup 10 is so much larger than the other values, it is of interest to view the individual $\overline{X}$ charts. These are given in Figures 9.3*a*, *b*.

It should be observed that at subgroup 10 the average that is plotted on each chart is inside the control limits. Thus, the use of two separate $\overline{X}$ charts would not have detected the out-of-control condition at that point. But how can the process be out of control at that point when the $\overline{X}$ charts indicate control? The answer is that the bivariate process is out of control. At subgroup 10 the positions of the two averages relative to their respective midlines differ greatly; one is well below its midline whereas the other is somewhat above its midline. If two variables have a high positive correlation (as they do in this case), we would expect their relative positions to be roughly the same over time. That is, they should either be both above their respective midlines or below their midlines, and the distances to the midline should be about the same on each chart. Notice that this holds true for virtually every other subgroup except 10.

Thus, if these were actual data values and the points were plotted in real time, there would be reason to suspect that something was wrong with the process when the data in subgroup 10 were obtained.

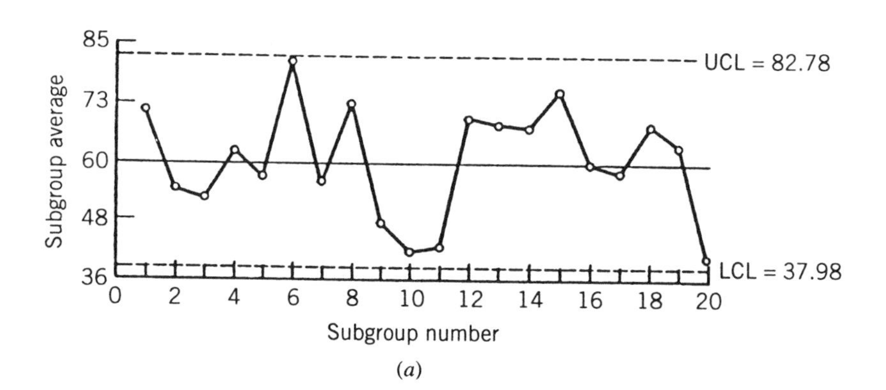

(*a*)

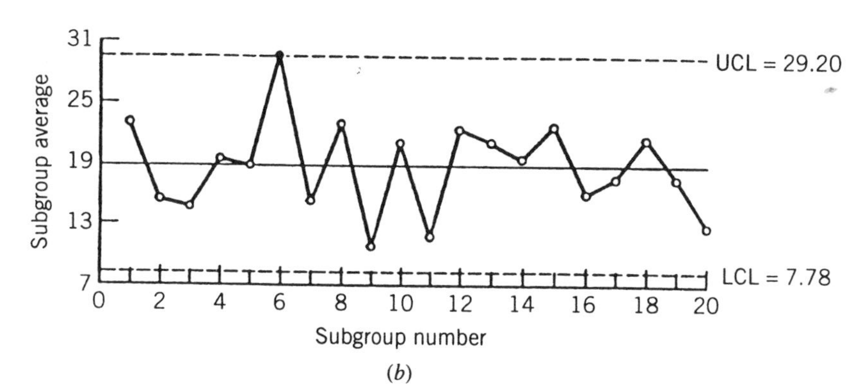

(*b*)

FIGURE 9.3 $\overline{X}$ chart for Table 9.2 data: (*a*) first variable; (*b*) second variable.

Conversely, no out-of-control message is received at subgroup 6 on the multivariate chart, but Figure 9.3*b* shows that the subgroup average for the second variable is above the UCL. Does this mean that a search for an assignable cause should not be initiated since no message was emitted by the multivariate chart at subgroup 6? The main advantage of a multivariate chart is that it can be used as a substitute for separate $\overline{X}$ charts, but if for some reason the latter were also constructed, then they should be used in the same way as they would be used if a multivariate chart were not constructed. This example simply shows that a multivariate chart is not a perfect substitute for separate $\overline{X}$ charts. Obviously we should not expect this to be the case since we cannot easily set the false-alarm rate for the set of univariate charts equal to the false-alarm rate for the multivariate chart.

9.3 MULTIVARIATE CHART VERSUS INDIVIDUAL $\overline{X}$ CHARTS

The preceding example illustrates one of the primary advantages of a single multivariate chart over p separate (univariate) X charts and relates to an important general advantage that can be explained as follows. For a multivariate chart with p characteristics, the probability that the chart indicates control when the process is actually in control at the multivariate average $\overline{\overline{\mathbf{x}}}$ is $1 - \alpha$, which equals $1 - .0027p$ when α is chosen in accordance with the suggested procedure of selecting α so that $\alpha/2p = .00135$. As Alt (1982a) demonstrates, the use of Bonferroni's inequality leads to the result that with p separate $\overline{X}$ charts the probability that each of the p averages (in a given subgroup) will fall within the control limits is at least $1 - .0027p$ when the multivariate process is in control. Thus, the p separate charts could conceivably indicate that the process is in control more often than it actually is in control. The difference between the two probabilities is virtually zero when the p variables are independent. Specifically, for the p separate charts the probability is $(1 - .0027)^p$ compared to $1 - .0027p$ for the multivariate chart. Recall that in Chapter 4 it was demonstrated that $.0027n$ gives a reasonably good approximation of $1 - (1 - .0027)^n$ when $n \leq 50$. That result applies directly here since if $.0027n$ is used to estimate $1 - (1 - .0027)^n$, then $1 - .0027n$ would be used to estimate $(1 - .0027)^n$, and the estimate would be with the same accuracy.

What this implies is that when the p quality characteristics are virtually unrelated (uncorrelated) and $\alpha/2p = .00135$, it will not make much difference whether a single multivariate chart is used or p separate $\overline{X}$ charts are used. However, when the characteristics are highly correlated, it can make a considerable difference; the illustrative example showed how the difference can occur.

The (somewhat unjust) criticisms of the T^2 chart would also apply to any of the various schemes that have been proposed that are functions of both μ and Σ. This includes the schemes given by Pignatiello and Runger (1990), Crosier (1988), and Alwan (1986).

9.4 CHARTS FOR DETECTING VARIABILITY AND CORRELATION SHIFTS

Recent research on multivariate charts has been oriented toward detecting changes in the multivariate mean, with less attention given to detecting variance changes or changes in the correlations between the process characteristics.

Some multivariate charts for controlling the process variability have been developed by Alt (1973, Chapter 7) and are also described in Alt (1986) and Alt et al. (1998). These charts are based on $|\mathbf{S}|$, the determinant of $\mathbf{S}$. This was termed the *generalized variance* by Wilks (1932). A multivariate range procedure was given by Siotani (1959a,b).

As discussed by Alt et al. (1998), the multivariate analogue of the univariate s chart is to plot $|\mathbf{S}|^{1/2}$ against control limits given by

$$\text{UCL} = \left(b_3 + 3\sqrt{b_1 - b_3^2}\right)|\mathbf{\Sigma}_0|^{1/2} \qquad \text{LCL} = \left(b_3 - 3\sqrt{b_1 - b_3^2}\right)|\mathbf{\Sigma}_0|^{1/2}$$

with the centerline given by $b_3|\mathbf{\Sigma}_0|^{1/2}$ and

$$b_1 = (n-1)^{-p}\prod_{i=1}^{p}(n-i)$$

$$b_3 = \left(\frac{2}{n-1}\right)^{p/2}\frac{\Gamma(n/2}{\Gamma[(n-p)/2]}$$

This type of chart can be useful in detecting an increase (or decrease) in one or more of the individual variances or covariances, but it would be desirable to supplement the chart with some univariate procedure.

This is because in using $|\mathbf{S}|^{1/2}$ we face difficulties that are similar to the difficulties that we face when we use a T^2 chart. Specifically, a signal could be caused by a variance change or a correlation change, and we also do not know which variable(s) to investigate. Even worse, we can have changes in the variances and correlations and not have a change in $|\mathbf{S}|$, although these changes may be unlikely to occur in practice. To illustrate, $|\mathbf{\Sigma}| = \sigma_1^2\sigma_2^2(1-\rho_{12}^2)$ when $p = 2$. Clearly certain changes in the variances and correlation could occur that would leave $|\mathbf{\Sigma}|$ unchanged, such as an increase in one variance that is offset by a decrease in the other variance, or changes in all three. If there are multiple assignable causes, this could easily happen.

For example, assume that the variability of two quality characteristics is being controlled by a multivariate chart and both variances are initially 2.0. We will assume that the covariance (covered in Section 3.5.1) is 1.0 so that the correlation coefficient is 0.50 and $|\mathbf{S}| = 3.0$. Now assume that an assignable cause increases each variance to $2\sqrt{1.25}$ and the covariance increases to $\sqrt{13}$. The correlation coefficient will increase to $\sqrt{13}/4 = 0.90$ but $|\mathbf{S}|$ will remain at 3.0.

As another example, assume that there are two process characteristics and an assignable cause doubles the subgroup variance for each characteristic but the correlation between the two characteristics remains unchanged. The value of $|\mathbf{S}|$ for the next subgroup could have almost the same value that it had at the last subgroup before the assignable cause occurred. To illustrate, assume that the values of one of the characteristics change from (2, 4, 6, 8) to (4, 8, 12, 16) and the values of the other characteristic change from (1, 11/3, 19/3, 9) to (2, 22/3, 38/3, 18) in going from subgroup k to subgroup $k + 1$; $|\mathbf{S}|$ will remain unchanged. Notice also that $|\mathbf{S}| = 0$ since the characteristics have a perfect positive correlation. This shows that values of $|\mathbf{S}|$ could be somewhat misleading for two characteristics that have a high positive or negative correlation, as the strength of the correlation could cause the value of $|\mathbf{S}|$ to remain fairly small when there has been a sizable increase in either or both variances.

Another problem in using $|\mathbf{S}|$ or $|\mathbf{S}|^{1/2}$ is there is not a closed-form expression for the distribution of either (see, e.g., Bagai, 1965), although normal and chi-square approximations have been given. Guerrerocusumano (1995) also argued against the use of the generalized variance and proposed a conditional entropy approach. See also Hawkins (1992).

Tang and Barnett (1996a) proposed a decomposition scheme that is similar in spirit to the decomposition schemes presented in Section 9.5.2 that can be used in conjunction with methods for controlling the multivariate mean. Tang and Barnett (1996b) showed by simulation that the methods proposed in Tang and Barnett (1996a) are superior to other methods. Unfortunately, however, these methods are moderately complicated and cannot be implemented without appropriate software.

9.4.1 Application to Table 9.2 Data

Which, if any, of these methods can be applied to the illustrative example in Section 9.2.6? First, since the data were not simulated, there was not a "built-in shift" in the means, the variances, or the correlation structure. Nevertheless, since the two subgroup averages are within their respective chart limits in Figures 9.3*a*, *b*, we might assume that there has not been a mean shift.

It was suggested in Section 9.2.6 that it is the correlation between the two process characteristics that is out of control. This seems apparent for this example, but an apparent change in the correlation will not always be as obvious as it is in this example.

When there are only two process characteristics, there are various ways that the correlation could be monitored. One simple approach would be to construct a scatter plot of the subgroup means. If there is a high correlation between the process characteristics (and hence between the subgroup averages), most of the points should practically form a line, with points that are well off the line representing (perhaps) a change in the relationship between the subgroup averages. A less subjective approach would be to regress $\overline{X}_1$ against $\overline{X}_2$, using historical data, and construct a control chart of the residuals. That is, the subgroup average for one process characteristic would be regressed against the subgroup average for the other characteristic. The control limits would be given by $0 \pm 3\hat{\sigma}_e$, where $\hat{\sigma}_e$ denotes the estimated standard deviation of a residual. The estimated standard deviation of the ith residual, assuming $\overline{X}_1$ to be the dependent variable, is given by $\hat{\sigma}_e\sqrt{1 - (1/nk) - (\bar{x}_{2i} - \bar{\bar{x}}_2)/S_{\bar{x}_2\bar{x}_2}}$, with the denominator in the second fraction denoting the corrected sum of products for $\overline{X}_2$, and the historical data are assumed to consist of k subgroups of size n for each of the two variables.

Such a chart could be viewed as a chart of the constancy of the relationship between the two sets of subgroup averages. A somewhat similar chart, a cause-selecting chart, is presented in Section 12.8.

For Stage 2, process monitoring, the regression equation developed in Stage 1 would be used in addition to the control limits for the residuals. Thus, for the next pair of subgroup means, the residual would be computed by substituting the means into the regression equation, and the residual would be plotted against the control limits. This is similar to the use of any Shewhart chart.

One problem with this approach is that the results are not invariant to the choice of the dependent variable. That is, the relationship between each residual and the control limits depends on whether $\overline{X}_1$ or $\overline{X}_2$ is the dependent variable. For the Table 9.2 data, if we select the first variable as the dependent variable, then at subgroup 10 the standardized residual will be large since the subgroup mean for the second variable is too large relative to the mean of the first variable. Conversely, if the second variable is selected as the dependent variable, the value of the subgroup mean for the first variable is too small, so the standardized residual will be a large negative number. It can be shown that the absolute values of these standardized residuals will not be the same, in general. (In this example, we have 3.49 and −3.60, respectively, for these two cases.)

If we select the first option for this example, we obtain the chart shown in Figure 9.4. We observe that this chart resembles Figure 9.2, although it will not necessarily resemble the T^2 chart.

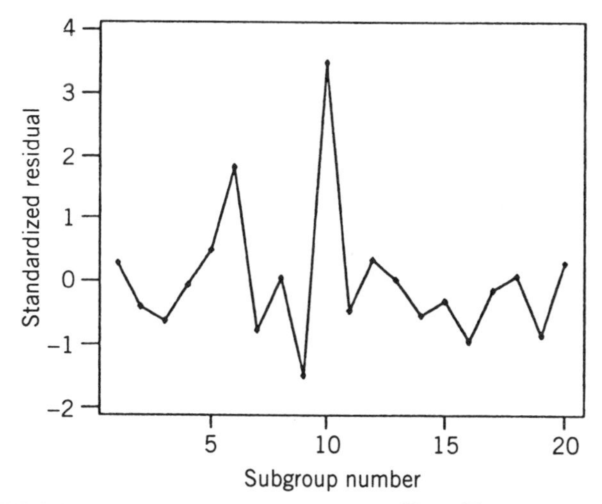

FIGURE 9.4 Residual chart for the regression of $\overline{X}_1$ on $\overline{X}_2$ for Table 9.2 data.

9.5 CHARTS CONSTRUCTED USING INDIVIDUAL OBSERVATIONS

The multivariate charts presented to this point in the chapter are applicable when subgrouping is used. A multivariate analogue of the X chart was presented by Jackson (1956, 1959) and Jackson and Morris (1957). The UCL that was given is an approximation, however, which should be used only for a large number of observations. The exact UCL is given by Alt (1982a, p. 892). It was originally derived by Alt (1973, p. 114), and is also derived in Tracy, Young, and Mason (1992).

Using Alt's notation, if m individual multivariate observations are to be used for estimating the mean vector and variance–covariance matrix (the estimates denoted by $\overline{\mathbf{x}}_m$ and $\mathbf{S}_m$, respectively), each *future* observation vector, $\mathbf{x}$, would, in turn, be used in computing

$$Q = (\mathbf{x} - \overline{\mathbf{x}}_m)'\mathbf{S}_m^{-1}(\mathbf{x} - \overline{\mathbf{x}}_m) \tag{9.8}$$

and comparing it against the UCL. The latter is

$$\frac{p(m+1)(m-1)}{m^2 - mp} F_{\alpha(p,m-p)} \tag{9.9}$$

where, as before, p denotes the number of characteristics.

Tracy et al. (1992) derived the form of the control limits for the *retrospective analysis* (Stage 1). They stated that Q multiplied times a constant has a beta distribution. Specifically, $Q \sim ((m-1)^2/m)B(p/2, (m-p-1)/2)$. (See Section 3.6.11 for the beta distribution.) Therefore, the control limits are given by

$$\begin{aligned} \text{LCL} &= \left(\frac{(m-1)^2}{m}\right) B\left(1 - \frac{\alpha}{2}; \frac{p}{2}; \frac{m-p-1}{2}\right) \\ \text{UCL} &= \left(\frac{(m-1)^2}{m}\right) B\left(\frac{\alpha}{2}; \frac{p}{2}; \frac{m-p-1}{2}\right) \end{aligned} \tag{9.10}$$

Alternatively, the control limits could be written in terms of the percentiles of the corresponding F-distribution, using the relationship between the percentiles of a beta distribution and the percentiles of the corresponding F-distribution that was given in Section 3.6.11. Those control limit expressions are given by Tracy et al. (1992).

We should note, however, that these limits are exact only when applied to a single point in Stage 1. Since a Stage 1 analysis is an analysis of all of the observations in Stage 1, the value of α that is used will not apply to the set of points. There are two problems: (1) multiple points are being compared against the UCL, and (2) the deviations from the UCL are correlated since the deviations are functions of the common observations that were used in computing the control limit. Therefore, the UCL to give a specified α for a set of observations in Stage 1 would have to be determined by simulation.

9.5.1 Retrospective (Stage 1) Analysis

Unfortunately, a T^2 chart for individual observations will often perform poorly when used in Stage 1, as illustrated by Sullivan and Woodall (1996). The problem is the manner in which the variance for each variable is estimated. The general approach given in the literature is to compute the sample variances and covariances in the usual way. But if a mean shift in, say, a single variable has occurred in the historical data, the estimate of the variance of that variable will be inflated by the shift. Therefore, using sample variances in Stage 1 is unwise in the multivariate case, just as it is unwise to rely on the sample variance (only) in the univariate case.

Accordingly, Holmes and Mergen (1993) suggested a logical alternative: Estimate the variances and covariances using the multivariate analogue of the moving range approach for a univariate X chart. Specifically, the estimator of $\mathbf{\Sigma}$ is $\hat{\mathbf{\Sigma}} = \mathbf{V}'\mathbf{V}/[2(n-1)]$, with $\mathbf{v}_i = \mathbf{x}_{i+1} - \mathbf{x}_i$, $i = 1, 2, \ldots, n$. Sullivan and Woodall (1996) compared this approach with several other methods and found that it performed well. We will denote the estimator obtained using the moving range approach as $\hat{\mathbf{\Sigma}}_2$; the estimator obtained using the sample variances and covariances will be denoted as $\hat{\mathbf{\Sigma}}_1$.

We will consider this approach using part of a data set that was originally given by Holmes and Mergen (1993) and also has been analyzed by other authors, including Sullivan and Woodall (1996, 1998a). The data, which are given in Table 9.4, consist of 56 multivariate individual observations from a European plant that produces gravel. The two variables that are to be controlled are the two sizes of the gravel: large and medium. (The percentage of small gravel was also part of the original data set but is not used here since the percentages would then add to 100 for each observation.)

Sullivan and Woodall (1996) gave the control charts constructed using the standard approach for estimating the variances and covariances and using the approach of Holmes and Mergen (1993). All of the plotted points were below the UCL using the standard approach, whereas with the latter approach two of the points were well above the UCL. Sullivan and Woodall (1996) concluded that there is a shift in the mean vector after observation 24. Since this shift is near the middle of the data set, the shift will of course affect $\hat{\mathbf{\Sigma}}_1$ more than $\hat{\mathbf{\Sigma}}_2$, analogous to what happens in the univariate case. The two limits can thus be expected to differ considerably.

Since the sum of the two percentages in Table 9.4 is close to 100 for each multivariate observation, there must necessarily be a high negative correlation between the two variables. In fact, the correlation is -0.769. Because of this relatively high correlation, a multivariate chart will be far superior to the use of two univariate charts. Sullivan and Woodall (1996) determined the control limits for the T^2 chart by simulation.

A multivariate analogue to a moving range chart can be developed using $\hat{\mathbf{\Sigma}}_2$. This estimator might be used for assessing the stability of $\mathbf{\Sigma}$ in Stage 1, although the exact distribution of the estimator is unknown. [An approximation to the distribution is given by Scholz and Tosch (1994).]

TABLE 9.4 Multivariate Individual Observations, $p = 2$ (%)

Large	Medium	Large	Medium	Large	Medium
5.4	93.6	2.5	90.2	5.8	86.9
3.2	92.6	3.8	92.7	7.2	83.8
5.2	91.7	2.8	91.5	5.6	89.2
3.5	86.9	2.9	91.8	6.9	84.5
2.9	90.4	3.3	90.6	7.4	84.4
4.6	92.1	7.2	87.3	8.9	84.3
4.4	91.5	7.3	79.0	10.9	82.2
5.0	90.3	7.0	82.6	8.2	89.8
8.4	85.1	6.0	83.5	6.7	90.4
4.2	89.7	7.4	83.6	5.9	90.1
3.8	92.5	6.8	84.8	8.7	83.6
4.3	91.8	6.3	87.1	6.4	88.0
3.7	91.7	6.1	87.2	8.4	84.7
3.8	90.3	6.6	87.3	9.6	80.6
2.6	94.5	6.2	84.8	5.1	93.0
2.7	94.5	6.5	87.4	5.0	91.4
7.9	88.7	6.0	86.8	5.0	86.2
6.6	84.6	4.8	88.8	5.9	87.2
4.0	90.7	4.9	89.8		

It would be desirable to assess the "added value" of such a chart, however, since it is known that a moving range chart adds very little in the univariate case, as discussed in Chapter 5.

9.5.2 Stage 2 Analysis: Methods for Decomposing Q

Various ways of partitioning or decomposing the Q-statistic for the purpose of trying to identify the variable(s) causing the signal have been proposed, including those given by Hawkins (1991, 1993) and Mason, Tracy, and Young (1995). (The latter shall be referred to as the MTY decomposition.) For the sake of consistency with the notation used in the literature, the Q-statistic in Section 9.5 will be represented by T^2 in this section.

Assume for simplicity that we have two process variables. As shown by Mason, Tracy, and Young (1997), we could decompose T^2 as $T^2 = T_1^2 + T_{2\cdot 1}^2 = T_2^2 + T_{1\cdot 2}^2$. The unconditional statistics T_1^2 and T_2^2 are each given by $T_j^2 = (x_j - \overline{x}_j)^2/s_j^2$, $j = 1, 2$, and the conditional statistics are given by $T_{1\cdot 2} = r_{1\cdot 2}/\left(s_1\sqrt{1 - R_{1\cdot 2}^2}\right)$ and $T_{2\cdot 1} = r_{2\cdot 1}/\left(s_2\sqrt{1 - R_{2\cdot 1}^2}\right)$. As pointed out by Mason et al. (1997), the use of each T_j^2 statistic is equivalent to using a Shewhart chart for the jth variable. The standard deviations of the two variables are denoted by s_1 and s_2, with $r_{1\cdot 2}$ and $r_{2\cdot 1}$ denoting a residual resulting from the regression of the first variable on the second and the second variable on the first, respectively, and similarly for R^2, the square of the (first-order) correlation between the two variables.

If the correlation between the two variables is disturbed at a particular point in time, we would expect at least one of the two residuals to be large, with the regression equations obtained using the historical data. If the correlation structure is not disturbed but there is a mean shift for at least one of the variables, this should be detected by the unconditional statistics, and perhaps also by the conditional statistics.

For only two variables we could easily compute both decompositions, but the number of unique terms in the different decompositions increases (greatly) with p—from 12 for $p = 3$ to 5120 for $p = 10$. Software is thus essential, and a program for obtaining the significant components of the decomposition is given by Langley, Young, Tracy, and Mason (1995). Mason et al. (1997) give a computational procedure that might further reduce the number of necessary computations. Their suggested approach entails first computing all of the (unconditional) T_i^2 and then computing as many of the conditional statistics as appears necessary.

9.5.2.1 *Illustrative Example*

We will illustrate the MTY decomposition using an example with $p = 3$ so as to keep the necessary computations from becoming unwieldy without software. Simulation will be used so that the state of nature will be known. The data were simulated from a bivariate normal distribution with mean vector $\mathbf{0}$ and variance–covariance matrix $\boldsymbol{\Sigma}$ with unit variances and covariances equal to 0.9 for the first 50 points, which shall constitute the historical data. Of course, with unit variances this implies that the correlations are also 0.9. At observation 51 the mean of X_1 was generated to be 4, and the correlations were altered to be 0.1 at observation 52.

This was accomplished by generating the first 50 values for X_1 and X_2 as, using a modification of the approach of Dunnett and Sobel (1955), $X_{ij} = (0.1)^{1/2}Z_{ij} + (0.9)^{1/2}Z_{0j}$, $i = 1, 2$, $j = 1, 2, \ldots, 50$. The values for observation 51 were generated as $X_i = (0.1)^{1/2}Z_i + (0.9)^{1/2}Z_0 + k_i$, $i = 1, 2$, with $k_1 = 4$, $k_2 = 0$. The values for observation 52 were generated as $X_i = (0.9)^{1/2}Z_i + (0.1)^{1/2}Z_0$, $i = 1, 2$. (Here $Z \sim V(0, 1)$.)

Using the MTY approach, the unconditional T_1^2 statistic should detect the mean shift, and the correlation shift should be detected by either $T_{1.2}$ or $T_{2.1}$. The first 50 observations will of course be used for the historical data, with the means and standard deviations computed using these observations. The expected value of T_1 is approximately 4, so the expected value of T_1^2 is approximately 16. For the simulated data $T_1 = 4.55$, so $T_1^2 = 20.67$. Since $T_1^2 > 3^2 = 9$, this statistic detects the mean shift. As shown by Mason, et al. (1995), the conditional statistics $T_{1\cdot 2}$ and $T_{2\cdot 1}$ are each distributed as $[(n+1)/n]F_{1,n-1}$. Mason et al. (1997) use $\alpha = .05$ in their example. We might question using a significance level this large, but even using such a value here does not allow us to detect the correlation change immediately (i.e., with the 52nd point). However, when 50 additional points are generated, a signal is received on 27 of those points. If

$\alpha = .00135$ had been used, the signal would have been received on 16 of the 50 points, so for this example the selection of α does not have a large effect, and this is undoubtedly due to the large change in the correlation between X_1 and X_2.

9.6 WHEN TO USE EACH CHART

In summary, multivariate charts were presented for controlling the multivariate process mean using either subgroups or individual observations. Multivariate dispersion charts were mentioned only briefly; the details are given in Alt (1973, 1986). When subgroups are being used, the value of Eq. (9.2) is plotted against the UCL of Eq. (9.3) when testing for control of the "current" set of subgroups. For control using future subgroups obtained individually, Eq. (9.5) is computed for each future subgroup and compared with the UCL given by Eq. (9.6). When individual observations are used rather than subgroups, control using future observations is checked by comparing the value of Eq. (9.8) with that of Eq. (9.9). When a current set of individual observations is being tested for control, the value of Eq. (9.8) would be compared with the control limits given in Eqs. (9.10). Analogous to the case with subgrouping, vectors would be discarded if removable assignable causes are detected so that, in general, one would actually compute

$$(\mathbf{x} - \overline{\mathbf{x}}_{m-a})'\mathbf{S}_{m-a}^{-1}(\mathbf{x} - \overline{\mathbf{x}}_{m-a}) \tag{9.11}$$

where $m - a$ denotes the number from the original m observations that are retained. [Equation (9.8) assumes that they are all retained.] If a vectors are discarded, the value of Eq. (9.11) would then be compared with

$$\frac{p(m-a+1)(m-a-1)}{(m-a)^2-(m-a)p} F_{\alpha(p,m-a-p)}$$

for testing future multivariate observations.

As indicated previously in the material on multivariate procedures for subgroups, a bivariate control ellipse could be used when $p = 2$. Such an ellipse could also be constructed when individual observations are plotted, as is illustrated in Jackson (1959). But it has the same shortcoming as the ellipse for subgroups; that is, the time order of the individual observations is lost when they are plotted.

9.7 ACTUAL ALPHA LEVELS FOR MULTIPLE POINTS

The remarks made concerning the α value for an $\overline{X}$ chart when one point is plotted versus the plotting of multiple points also applies here. Specifically, the α given in this chapter as a subscript for F and χ^2 applies, strictly speaking, when one (individual) observation or one subgroup is being used. When current control is being tested for a set of individual observations or a set of subgroups and the process is in control, the probability that at least one point exceeds the UCL far exceeds α. For example, if $\alpha = .0054$ for $p = 2$ (in accordance with the suggested approach of having $\alpha/2p = .00135$), the probability that at least one value of Eq. (9.8) exceeds Eq. (9.10), assuming $\bar{\mathbf{x}}_{75} \doteq \boldsymbol{\mu}$, is approximately 0.334 when the process is in control. (The probability cannot be determined exactly since the deviations from the control limits are correlated, as previously discussed.) If the parameters were known, to make $\alpha = .0054$ for the entire set of 75 observations would require using $\alpha = .000072$ in Eq. (9.10).

9.8 REQUISITE ASSUMPTIONS

The multivariate procedures presented in this chapter are also based upon the assumptions of normality and independence of observations. Everitt (1979) examined the robustness of Hotelling's T^2 to multivariate nonnormality and concluded that the (one-sample) T^2 statistic is quite sensitive to skewness (but not kurtosis) and the problem becomes worse as p increases. Consequently, it is reasonable to conclude that multivariate control chart procedures that are similar to Hotelling's T^2 will also be sensitive to multivariate nonnormality.

A number of methods have been proposed for assessing multivariate nonnormality, including those given by Royston (1983), Koziol (1982), and Small (1978, 1985).

9.9 EFFECTS OF PARAMETER ESTIMATION ON ARLs

The effects of parameter estimation on the ARLs for univariate $\overline{X}$ and X charts were discussed in Section 4.14. The general result is that all of the ARLs are increased when parameters are estimated rather than assumed to be known. In the multivariate case, the problem is more involved because the means, variances, and covariances (or correlations) must all be estimated, and the effect of parameter estimation cannot be so easily described. Bodden and Rigdon (1999) explain that extremely large sample sizes will be necessary in order for the in-control ARL of a multivariate individual observation (T^2) chart to be close to the nominal, parameters-known value, unless p is small. This is due to the fact that overestimating a correlation reduces the in-control ARL by more than underestimating a correlation increases the in-control ARL. Thus, the larger the value of p, the greater the departure from the in-control ARL when the parameters are assumed to be known. There are only 3 correlations to estimate when $p = 3$, but 190 correlations must be estimated when $p = 20$. It is not uncommon for multivariate charts to be used when p is much larger than 20.

Bodden and Rigdon (1999) illustrate a Bayesian approach that shrinks the estimates of the correlations toward zero and thus moves the in-control ARL closer to the parameters-known ARL. The variances would be estimated in the usual way.

Although not considered by Bodden and Rigdon (1999), this same general problem with the in-control ARL undoubtedly exists with any multivariate control chart procedure, as the variance–covariance matrix must also be estimated with the multivariate CUSUM and EWMA procedures discussed in Sections 9.11 and 9.12.

9.10 DIMENSION REDUCTION TECHNIQUES

When there is a very large number of variables that seem to be important, an alternative to attempting to control all of the relevant variables is to reduce the set in some manner. There are various ways of doing this, but with most approaches we lose the actual variables and are left with transformed variables that will generally not have physical meaning. Nevertheless, MacGregor (1998) reports success with such approaches.

9.11 MULTIVARIATE CUSUM CHARTS

Several multivariate CUSUM charts have been given in the literature, including two by Crosier (1988). The first of these is simply a CUSUM of the square root of the T^2 statistic. The choice of this rather than a CUSUM of the T^2 statistic itself is based on the author's preference for forming a CUSUM of distance rather than a CUSUM of squared distance. Crosier (1988) provided values of h for $p = 2$, 5, 10, 20 so as to produce an in-control ARL of 200 or 500 and found that $k = \sqrt{p}$ worked well for detecting a shift from $\mu_T = 0$ to $\mu_T = 1$. (The ARL values were determined using a Markov chain approach.)

The other approach given by Crosier (1988) was to use a direct multivariate generalization of a univariate CUSUM. Recall from Section 8.2.1 that the two univariate CUSUM statistics, S_H and S_L, are each obtained by subtracting a constant from the z-score and accumulating the sum of this quantity and the previous sum. To connect this univariate CUSUM to the Crosier scheme, we need to first write the univariate CUSUM for S_H as $S_{H_i} = \max(0, (X_i - \hat{\mu}) - k\hat{\sigma} + S_{H_{i-1}})$, which we could regard as the "unstandardized" version of an upper CUSUM for individual observations. If each of these scalar quantities (except $\hat{\sigma}$) were replaced by vectors, there would be the problem of, in particular, determining the maximum of a vector and the null vector, in addition to the selection of $\mathbf{k}$. These problems led Crosier (1988) to consider a multivariate CUSUM of the general form

$$Y_n = (\mathbf{S}_n \hat{\mathbf{\Sigma}}^{-1} \mathbf{S}_n)^{1/2}$$

with

$$\mathbf{S}_n = \begin{cases} \mathbf{0} & \text{if } C_n \le k \\ (\mathbf{S}_{n-1} + \mathbf{X}_n - \hat{\boldsymbol{\mu}})\left(1 - \dfrac{k}{C_n}\right) & \text{if } C_n > k \end{cases}$$

and

$$C_n = [(\mathbf{S}_{n-1} + \mathbf{X}_n - \hat{\boldsymbol{\mu}})' \hat{\mathbf{\Sigma}}^{-1} (\mathbf{S}_{n-1} + \mathbf{X}_n - \hat{\boldsymbol{\mu}})]^{1/2}$$

A good choice is $k = 0.5$, which of course is the value of k that is generally used in univariate CUSUM schemes. Crosier (1988) gave ARL curves for various values of h.

Both of these CUSUM procedures permit the use of the FIR feature and the Shewhart feature. Crosier (1988) indicates that the multivariate CUSUM scheme may be preferable because it permits faster detection of shifts than the first CUSUM procedure and because the CUSUM vector can give an indication of the direction that the multivariate mean has shifted.

Clearly the most important use of CUSUM procedures is for Stage 2, but they can also be used for Stage 1. Sullivan and Woodall (1998b) suggest setting $k = 0$ when a multivariate CUSUM procedure is used in Stage 1, so as to make the procedure sensitive to small shifts. This is quite reasonable because the presence of special causes in Stage 1 can create problems with parameter estimation. They also suggest using the multivariate analogue of the moving range estimator that was presented in Section 9.5.1. Sullivan and Woodall (1998b) also provide an exact Stage 1 procedure that has a specified probability of detecting either a step change or a trend, even though the parameters are estimated.

See also Pignatiello and Runger (1990) for additional information on multivariate CUSUM procedures.

9.12 MULTIVARIATE EWMA CHARTS

Lowry, Woodall, Champ and Rigdon (1992) introduced a multivariate exponentially weighted moving average (MEWMA) chart for use with multivariate individual observations. As discussed in Section 8.6.3, a univariate EWMA chart can be used for either individual observations or subgroup means, and we have the same option in a multivariate setting.

The MEWMA chart of Lowry et al. (1992) is the natural multivariate extension of the univariate EWMA chart. The statistic that is charted is

$$\mathbf{Z}_t = \mathbf{R}\mathbf{X}_t + (\mathbf{I} - \mathbf{R})\mathbf{Z}_{t-1} \tag{9.12}$$

where $\mathbf{X}_t$ denotes either the vector of individual observations or the vector of subgroup averages at time t and $\mathbf{R}$ is a diagonal matrix that would be of the form $r\mathbf{I}$ unless there is some a priori reason for not using the same value of r for each variable.

As in the univariate case, we could use a MEWMA chart for Stage 1, although the literature considers only Stage 2. It is questionable how useful a Stage 1 MEWMA approach might be when compared to other available methods, however, so we will be concerned only with Stage 2.

If the MEWMA were used for Stage 1 and no out-of-control signal were received, we might think of $\mathbf{Z}_{t-1}$ as being the value of the MEWMA at the end of Stage 1. If some other approach were used in Stage 1, then we could let $\mathbf{Z}_{t-1} = \bar{\mathbf{x}}$, where the latter would be the vector of subgroup averages (one average for each variable), computed using only the multivariate individual observations that were not discarded in the Stage 1 analysis.

Analogous to the T^2 control chart, the value of a quadratic form is computed for the MEWMA. Specifically, the computed statistic is

$$W_t^2 = \mathbf{Z}_t' \mathbf{\Sigma}_Z^{-1} \mathbf{Z}_t$$

with a signal being received if W_t^2 exceeds a threshold value, h, that is determined so as to produce a desired in-control ARL. With the univariate EWMA chart, the value of r and the value of L for L-sigma control limits can be determined so as to provide an acceptable in-control ARL and a parameter-change ARL for a parameter change that one wishes to detect as quickly as possible. Similarly, extensive tables could be produced for the MEMWA chart, but this has not been done as yet.

It would be somewhat impractical to try to produce such tables, however, since the tables would necessarily be quite voluminous due to the fact that a four-way table would have to be constructed—for h, r, p, and λ. The latter is the noncentrality parameter, which would be $\lambda = (\boldsymbol{\mu} - \boldsymbol{\mu}_0)' \mathbf{\Sigma}_x^{-1} (\boldsymbol{\mu} - \boldsymbol{\mu}_0)$, assuming that $\boldsymbol{\mu}_0$ denotes the in-control mean vector and p continues to denote the number of variables.

Another problem is that a particular value of λ does not have intuitive meaning. We might attempt to view λ in terms of an equal change in all of the parameters, but λ will still depend upon $\mathbf{\Sigma}_x^{-1}$. Some tables are given by Prabhu and Runger (1997), but the choice of λ may, in general, be difficult.

Evaluating ARLs for MEWMA charts entails some additional complexities, however. Lowry et al. (1992) showed that the parameter(s) change ARL depends only on the value of λ, and obtained the ARLs by simulation.

Runger and Prabhu (1996) defined the MEWMA somewhat differently, using $\mathbf{X}_t - \boldsymbol{\mu}_0$ in place of $\mathbf{X}_t$ in Eq. (9.12). This form of the MEWMA allows the ARLs to be conveniently computed using a two-dimensional Markov chain such that only one element of the vector of means of $\mathbf{Z}_t$ is considered to be nonzero. See Runger and Prabhu (1996) for details. Alternatively, the ARLs could be determined using the program of Bodden and Rigdon (1998), which uses an integral equation approach. The user either specifies the desired in-control ARL with the program producing the necessary control limit or specifies the control limit and the program computes the in-control ARL.

9.12.1 Design of a MEWMA Chart

As stated previously, there are no published tables that are sufficiently extensive to permit the selection of h and r so as to produce a desired in-control ARL and a parameter-change ARL for one or more parameters whose change a user would want to detect as quickly as possible. For a given (h, r) combination, the ARL will depend upon p and λ, so four-way tables would have to be constructed, as stated previously. The program described by Bodden and Rigdon (1998) could be used to produce the in-control ARL for a given (h, p, r) combination, and one might in this way hone in on a reasonable (h, r) combination for a given value of p. For a moderate value of p (say, $p \doteq 5$), a reasonable starting point would be $r = 0.15$ and $h = 15$. Alternatively, the user could specify a desired value of the in-control ARL and the program would provide h.

We should also note that the specification of a parameter change that is to be detected is nontrivial. For example, is a user interested in detecting a shift of the same order of magnitude (in standard units) for each process characteristic, or should shifts of possibly different amounts across the variables be considered of equal importance? Since the ARL performance of the chart depends only on the value of a noncentrality parameter (see Section 9.12), it seems desirable to use variables in an MEWMA scheme where parameter changes in standard units are such that equal parameter changes are viewed as being equally important. For example, assume that it is critical to detect a 1-sigma change in one process characteristic but a 1.5-sigma change in the other characteristic will not cause any serious problems. Then we could frequently receive out-of-control signals for a set of parameter changes that are not considered to be of critical importance.

Sullivan and Woodall (1998b) recommend that a multivariate EWMA be applied in both reverse time order and in regular time order when used in Stage 1. See Prabhu and Runger (1997) for additional information on designing MEWMA charts.

9.12.2 Searching for Assignable Causes

As with any multivariate chart, a major problem is not automatically knowing "where to look" when a signal is received. Lowry et al. (1992) recommend monitoring the principal components if they are interpretable. [See, e.g., Morrison (1990) for a discussion of principal components.] Principal components are generally not readily interpretable, however. When the principal components are not interpretable, they recommend using the univariate EWMA values. This is a reasonable thing to do since the individual values are the components of the $\mathbf{Z}_t$ vector. The general idea is to use this information simply to determine which process characteristic(s) to check, as opposed to performing a formal test with each EWMA value compared against a threshold value. If the latter were done, then there would be some simultaneous inference problems that would have to be addressed, with a Bonferroni-type adjustment being required. Instead, one might simply use the individual EWMA values to rank the variables in terms of the strength of the evidence that the mean of a particular variable has changed, as is discussed generally by Doganaksoy, Faltin, and Tucker (1991).

9.13 APPLICATIONS OF MULTIVARIATE CHARTS

Multivariate control charts are undoubtedly used much less frequently than univariate charts. In view of the various options that practitioners have for attempting to detect assignable causes, it is interesting to study how multivariate charts have been used in various fields of application.

In addition to the applications mentioned at the beginning of the chapter, other applications of multivariate control charts and other multivariate methods in quality improvement work are described in Hatton (1983), Andrade, Muniategui, and Prada (1997), and Majcen, Rius, and Zupan (1997).

9.14 MULTIVARIATE PROCESS CAPABILITY INDICES

Univariate process capability indices were presented in Chapter 7. Several multivariate process capability indices have also been proposed, including those given by Wierda (1993) and Kotz and Johnson (1993).

Wierda (1993) defines a multivariate capability index that is a function of the estimate of the probability of the product being declared a good product, where "good" means that the specifications on each characteristic are satisfied. Such an index does not have the intuitive appeal as univariate capability indices, however, so the index is defined as $\mathrm{MC}_{pk} = \frac{1}{3}\Phi^{-1}(\theta)$, where θ, the probability of a good product, would have to be estimated. The intent is to make MC_{pk} comparable to C_{pk}.

Kotz and Johnson (1993) present several types of multivariate process capability indices, including multivariate generalizations of C_p and C_{pm}.

9.15 SUMMARY

Since the world is multivariate rather than univariate, there is a strong need for multivariate process control procedures. There are difficulties when one attempts to capture the appropriate dimensionality in process control procedures, however, as was discussed in some of the preceding sections. Because of the complexities involved, research is likely to continue on new and better procedures.

Mason, Champ, Tracy, Wierda, and Young (1997) gave a recent assessment and comparison of multivariate control chart procedures. See also Lowry and Montgomery (1995) and Alt et al. (1998).

APPENDIX

For both stages, what is calculated and plotted on a chart when monitoring the process mean is the value of a *quadratic form*. In general, a quadratic form can be written as

$$\mathbf{x}'\mathbf{A}\mathbf{x}$$

where $\mathbf{A}$ is a matrix (a rectangular array of numbers arranged in rows and columns), $\mathbf{x}'$ is a row vector, and $\mathbf{x}$ is a column vector that contains the elements of $\mathbf{x}'$ written in a column. Numerically, if we let

$$\mathbf{A} = \begin{bmatrix} 6 & 4 \\ 4 & 3 \end{bmatrix}$$

and

$$\mathbf{x} = \begin{bmatrix} 2 \\ 5 \end{bmatrix}$$

it can be shown that

$$\begin{aligned} \mathbf{x}'\mathbf{A}\mathbf{x} &= [2 \quad 5] \begin{bmatrix} 6 & 4 \\ 4 & 3 \end{bmatrix} \begin{bmatrix} 2 \\ 5 \end{bmatrix} \\ &= [(2 \times 6) + (5 \times 4)] \times 2 + [(2 \times 4) + (5 \times 3)] \times 5 \\ &= 179 \end{aligned}$$

Readers unfamiliar with matrix algebra are referred to books such as Searle (1982).

REFERENCES

Alt, F. B. (1973). *Aspects of Multivariate Control Charts.* M. S. Thesis, Georgia Institute of Technology, Atlanta GA.

Alt, F. B. (1982a). Multivariate quality control: State of the art. *ASQC Quality Congress Transactions*, 886–893.

Alt, F. B. (1982b). Bonferroni inequalities and intervals. In S. Kotz and N. Johnson, eds. *Encyclopedia of Statistical Sciences*, Vol. 1, pp. 294–300. New York: Wiley.

Alt, F. B. (1985). Multivariate quality control. In S. Kotz and N. Johnson, eds. *Encyclopedia of Statistical Sciences*, Vol. 6, pp. 110–122. New York: Wiley.

Alt, F. B. (1986). SPC of dispersion for multivariate data. *ASQC Quality Congress Transactions*, 248–254.

Alt, F. B., N. D. Smith, and K. Jain (1998). Multivariate quality control. In H. M. Wadsworth, ed. *Handbook of Statistical Methods for Engineers and Scientists*, 2nd ed., Chapter 21. New York: McGraw-Hill.

Alwan, L. C. (1986). CUMSUM quality control: multivariate approach. *Communications in Statistics—Theory and Methods 15*: 3531–3543.

Andrade, J. M., S. Muniategui, and D. Prada (1997). Use of multivariate techniques in quality control of kerosene production. *Fuel 76*(1): 51–59.

Bagai, O. P. (1965). The distribution of the generalized variance. *Annals of Mathematical Statistics 36*: 120–129.

Bodden, K. M. and S. E. Rigdon (1998). A program for approximating the in-control ARL for the MEWMA chart. *Journal of Quality Technology 31*: 120–123.

Bodden, K. M. and S. E. Rigdon (1999). A shrinkage estimator of the covariance for use with multivariate control charts. Manuscript.

Chan, L. K., S. W. Cheng, and F. A. Spiring (1991). A multivariate measure of process capability. *International Journal of Modelling and Simulation 11*: 1–6.

Colon, H. I. E. and D. R. GonzalezBarreto (1997). Component registration diagnosis for printed circuit boards using process-oriented basis elements. *Computers and Industrial Engineering 33*(1/2), 389–392.

Crosier, R. B. (1988). Multivariate generalizations of cumulative sum quality control schemes. *Technometrics 30*: 291–303.

Doganaksoy, N., F. W. Faltin, and W. T. Tucker (1991). Identification of out-of-control quality characteristics in a multivariate manufacturing environment. *Communications in Statistics—Theory and Methods 20*: 2775–2790.

Dunn, O. J. (1958). Estimation of the means of dependent variables. *Annals of Mathematical Statistics 29*: 1095–1111.

Dunnett, C. W. and M. Sobel (1955). Approximations to the probability integral and certain percentage points of a multivariate analogue of Student's t-distribution. *Biometrika 42*: 258–260.

Everitt, B. S. (1979). A Monte Carlo investigation of the robustness of Hotelling's one and two-sample T^2 tests. *Journal of the American Statistical Association 74*(365): 48–51.

Guerrerocusumano, J. L. (1995). Testing variability in multivariate quality control—a conditional entropy measure approach. *Information Sciences 86*(1/3): 179–202.

Hatton, M. B. (1983). Effective use of multivariate control charts. Research publication GMR-4513, General Motors Research Lab, Warren, MI.

Hawkins, D. L. (1992). Detecting shifts in functions of multivariate location and covariance parameters. *Journal of Statistical Planning and Inference 33*(2): 233–244.

Hawkins, D. M. (1991). Multivariate quality control based on regression-adjusted variables. *Technometrics 33*(1): 61–75.

Hawkins, D. M. (1993). Regression adjustment for variables in multivariate quality control. *Journal of Quality Technology 25*(3): 170–182.

Hawkins, D. M. and D. H. Olwell (1998). *Cumulative Sum Charts and Charting for Quality Improvement*. New York: Springer-Verlag.

Hayter, A. and K. Tsui (1994). Identification and quantification in multivariate quality control problems. *Journal of Quality Technology 26*(3): 197–207.

Healy, J. D. (1987). A note on multivariate CUSUM procedures. *Technometrics 29*(4): 409–412.

Holmes, D. S. and A. E. Mergen (1993). Improving the performance of the T^2 chart. *Quality Engineering 5*: 619–625.

Hotelling, H. (1947). Multivariate quality control. In C. Eisenhart, M. W. Hastay, and W. A. Wallis, eds. *Techniques of Statistical Analysis*. New York: McGraw-Hill.

Jackson, J. E. (1956). Quality control methods for two related variables. *Industrial Quality Control 12*(7): 4–8.

Jackson, J. E. (1959). Quality control methods for several related variables. *Technometrics 1*(4): 359–377.

Jackson, J. E. (1985). Multivariate quality control. *Communications in Statistics—Theory and Methods 14*: 2657–2688.

Jackson, J. E. and R. H. Morris (1957). An application of multivariate quality control to photographic processing. *Journal of the American Statistical Association 52*(278): 186–189.

Johnson, N. L. and S. Kotz (1970). *Distributions in Statistics, Continuous Univariate Distributions* Vol. 2. New York: Wiley.

Kotz, S. and N. L. Johnson (1993). *Process Capability Indices*. New York: Wiley.

Koziol, J. A. (1982). A class of invariant procedures for assessing multivariate normality. *Biometrika 69*: 423–427.

Kourti, T. and J. F. MacGregor (1996). Multivariate SPC methods for process and product monitoring. *Journal of Quality Technology 28*(4): 409–428.

Langley, M. P., J. C. Young, N. D. Tracy, and R. L. Mason (1995). A computer program for monitoring multivariate process performance. In *Proceedings of the Section on*

Quality and Productivity, pp. 122–123. Alexandria, VA: American Statistical Association.

Lowry, C. A. and D. C. Montgomery (1995). A review of multivariate control charts. *IIE Transactions 27*(6): 800–810.

Lowry, C. A., W. H. Woodall, C. W. Champ, and S. E. Rigdon (1992). A multivariate exponentially weighted moving average chart. *Technometrics 34*: 46–53.

MacGregor, J. F. (1995). Using on-line data to improve quality. W. J. Youden Memorial Address given at the 39th Annual Fall Technical Conference, St. Louis, MO.

MacGregor, J. F. (1998). Interrogating large industrial databases. Invited talk given at the 1998 Joint Statistical Meetings, Dallas, TX.

Majcen, N., F. X. Rius, and J. Zupan (1997). Linear and non-linear multivariate analysis in the quality control of industrial titanium dioxide white pigment. *Analytica Chimica Acta 348*: 87–100.

Mason, R. L., N. D. Tracy, and J. C. Young (1995). Decomposition of T^2 for multivariate control chart interpretation. *Journal of Quality Technology 27*: 99–108.

Mason, R. L., N. D. Tracy, and J. C. Young (1997). A practical approach for interpreting multivariate T^2 control chart signals. *Journal of Quality Technology 29*: 396–406.

Mason, R. L., C. W. Champ, N. D. Tracy, S. J. Wierda, and J. C. Young (1997). Assessment of multivariate process control techniques. *Journal of Quality Technology 29*(2): 140–143.

Meltzer, J. S. and R. H. Storer (1993). An application of multivariate control charts to a fiber optic communication subsystem testing process. Manuscript.

Mestek, O., J. Pavlik, and M. Suchanek (1994). Multivariate control charts: Control charts for calibration curves. *Fresenius' Journal of Analytical Chemistry 350*(6): 344–351.

Montgomery, D. C. (1996). *Introduction to Statistical Quality Control*, 3rd ed. New York: Wiley.

Morrison, D. F. (1990). *Multivariate Statistical Methods*, 3rd ed. New York: McGraw-Hill.

Nijhius, A., S. deJong, and B. G. M. Vandeginste (1997). Multivariate statistical process control in chromatography. *Chemometrics and Intelligent Laboratory Systems 38*(1): 51–62.

Odeh, R. E. (1982). Tables of percentage points of the maximum absolute value of equally correlated normal random variables. *Communications in Statistics—Simulation and Computation 11*: 65–87.

Pignatiello, J. J. and G. C. Runger (1990). Comparison of multivariate CUSUM charts. *Journal of Quality Technology 22*: 173–186.

Prabhu, S. S. and G. C. Runger (1997). Designing a multivariate EWMA control chart. *Journal of Quality Technology 29*: 8–15.

Royston, J. P. (1983). Some techniques for assessing multivariate normality based on the Shapiro-Wilk *W*. *Applied Statistics 32*(2): 121–133.

Runger, G. C. and S. S. Prabhu (1996). A Markov chain model for the multivariate exponentially weighted moving averages control chart. *Journal of the American Statistical Association 91*: 1701–1706.

Saniga, E. M. and L. E. Shirland (1977). Quality control in practice ... a survey. *Quality Progress 10*(5): 30–33.

Scholz, F. W. and T. J. Tosch (1994). Small sample uni- and multivariate control charts for means. *Proceedings of the American Statistical Association, Quality and Productivity Section.*

Searle, S. R. (1982). *Matrix Algebra Useful for Statistics.* New York: Wiley.

Seder, L. A. (1950). Diagnosis with diagrams—Part I. *Industrial Quality Control 6*(4): 11–19.

Siotani, M. (1959a). On the range in the multivariate case. *Proceedings of the Institute of Statistical Mathematics 6*: 155–165 (in Japanese).

Siotani, M. (1959b). The extreme value of the generalized distances of the individual points in the multivariate normal sample. *Annals of the Institute of Statistical Mathematics 10*: 183–203.

Small, N. J. H. (1978). Plotting squared radii. *Biometrika 65*: 657–658.

Small, N. J. H. (1985). Testing for multivariate normality. In S. Kotz and N. Johnson, eds. *Encyclopedia of Statistical Sciences*, pp. 95–100, New York: Wiley.

Sullivan, J. H. and W. H. Woodall (1996). A comparison of multivariate control charts for individual observations. *Journal of Quality Technology 28*(4): 398–408.

Sullivan, J. H. and W. H. Woodall (1998a). Change-point detection of mean vector or covariance matrix shifts using multivariate individual observations. Paper presented at the 1998 Joint Statistical Meetings, Dallas, TX.

Sullivan, J. H. and W. H. Woodall (1998b). Adapting control charts for the preliminary analysis of multivariate observations. *Communications in Statistics—Simulation and Computation 27*(4): 953–979.

Tang, P. F. and N. S. Barnett (1996a). Dispersion control for multivariate processes. *Australian Journal of Statistics 38*(3): 235–251.

Tang, P. F. and N. S. Barnett (1996b). Dispersion control for multivariate processes—some comparisons. *Australian Journal of Statistics 38*(3): 253–273.

Tracy, N. D., J. C. Young, and R. L. Mason (1992). Multivariate control charts for individual observations. *Journal of Quality Technology 24*(2): 88–95.

Wierda, S. J. (1993). A multivariate process capability index. *ASQC Annual Quality Transactions*, 342–348.

Wierda, S. J. (1994). *Multivariate Statistical Process Control.* Groningen Theses in Economics, Management and Organization, Walters-Noordhoff, Groningen, The Netherlands.

Wilks, S. S. (1932). Certain generalizations in the analysis of variance. *Biometrika 24*: 471–494.

EXERCISES

1. Assume that there are two correlated process characteristics, and the following values have been computed:

$$\bar{\bar{x}}_1 = 47 \qquad \bar{\bar{x}}_2 = 53 \qquad S_p^{-1} = \begin{bmatrix} 1.5 & -1 \\ -1 & 1 \end{bmatrix}$$

(a) What would the UCL for *future* control of the bivariate process be if $k = 25$, $n = 5$, $\alpha = .0054$, and no subgroups were deleted from the original set (i.e., $a = 0$)? (Use $F_{.0054(2,99)} = 5.51$ in calculating the UCL).

(b) Which of the following (future) subgroup averages would cause an out-of-control signal: (1) $\bar{x}_1 = 48$, $\bar{x}_2 = 54$; (2) $\bar{x}_1 = 46$, $\bar{x}_2 = 54$; (3) $\bar{x}_1 = 47$, $\bar{x}_2 = 54$; and (4) $\bar{x}_1 = 50$, $\bar{x}_2 = 56$?

(c) It can be shown (using S_p) that the sample correlation coefficient is 0.816 (i.e., the two characteristics have a high positive correlation). Use this fact to explain why one of the subgroups in part (b) caused an out-of-control signal even though $\bar{x}_1 - \bar{\bar{x}}_1$ and $\bar{x}_2 - \bar{\bar{x}}_2$ were both comparatively small.

2. What assumptions must be made (and should be verified) before a multivariate chart can be used for controlling the multivariate process mean (using either individual observations or subgroups)?

3. An experimenter wishes to test a *set* of (past) individual observations for control. What approach would you recommend?

4. It was stated that the use of Bonferroni intervals in conjunction with a multivariate chart for subgroups is essentially a substitute for individual $\overline{X}$ charts. The two approaches will not necessarily produce equivalent results, however, due in part to the fact that standard deviations are used in computing Bonferroni intervals, whereas ranges are used (typically) with $\overline{X}$ charts. For the data given in Table 9.2, construct the two intervals for subgroups 6 and 10. Notice that there is agreement for the latter but that the results differ slightly for the former. (Use $s_{p_1} = 14.90$, $s_{p_2} = 7.52$, and $t_{.00135,60} = 3.13$ in constructing the Bonferroni intervals.)

5. When a signal is received from a multivariate chart such as a T^2 chart, would you recommend the use of (a) Bonferroni intervals, (b) a Shewhart chart for each variable, or (c) neither? Explain.

6. The "standards given" approach for an $\overline{X}$ chart was discussed in Chapter 4. Could a T^2 value be compared against a UCL in the form of $p(n-1)/(n-p)F_{p,n-p}$ when a target value of μ is to be used? Explain.

7. An experimenter wishes to construct a multivariate chart for individual observations with three process characteristics.

(a) What value of α should be used?

(b) What is the numerical value of the constant that would be multiplied times $F_\alpha(p, m-p)$ in determining the UCL if $m = 60$?

(c) To which observations should this UCL be applied, past or future?

8. A bivariate control ellipse could be used with either subgroups or individual observations, but what is one major shortcoming of that approach?

9. Discuss the advantages and disadvantages of selecting α such that $\alpha/2p = .00135$.

10. Construct a multivariate CUSUM chart for the data in Table 9.2 and compare with the results obtained using the T^2 chart in Figure 9.2.

11. Using the data in Table 9.4, construct the two multivariate charts using $\hat{\boldsymbol{\Sigma}}_1$ and $\hat{\boldsymbol{\Sigma}}_2$ as given in Section 9.5.1.

12. Since a shift in the mean vector was believed to have occurred after observation 24 for the data in Table 9.4, use the approach of Mason, Tracy, and Young for each observation after 24, treating the first 24 multivariate observations as the base set. What do you conclude?

CHAPTER 10

Miscellaneous Control Chart Topics

In this chapter we discuss several topics that, although important, would generally not be presented in individual chapters. These include pre-control, short-run SPC, charting autocorrelated data, nonparametric control chart approaches, Bayesian control chart methods, the economic design of control charts, administrative applications of control charts, software for control charting, and applications of control charts in specific industries and for specific company functions.

10.1 PRE-CONTROL

This topic is presented first in this chapter, not because of its importance but rather because many companies seem to have used it. Bhote (1991), Shainin (1984), Shainin and Shainin (1989), and others have claimed that pre-control is a superior alternative to control charts, and there is evidence (Bhote, 1988) that Motorola has used pre-control extensively. We will examine these claims and also discuss critical assessments of pre-control that have appeared in the literature (Ledolter and Swersey, 1997), as well as consider the properties of suggested modifications to pre-control. The properties of these modified procedures have been studied by Steiner (1997).

The first paper on pre-control was a technical report by Satterthwaite (1954). Pre-control was originally advocated for machining operations, for which an operator has to set up a machine and then determine whether or not the setup was performed properly. The contention has been that it is almost impossible to maintain statistical control over such processes. Consequently, the operator uses pre-control to determine when process adjustments are necessary.

The technique has been popularized by Dorian Shainin, a well-known quality consultant who has contended that control charts are of no value if one wishes to control a process by the specification limits. What seems to have been overlooked in some of these papers, however, is that an acceptance control chart has control limits that are *determined from the specification limits* (see Section 4.9.1), so it would be more appropriate to compare pre-control with an acceptance chart than an $\overline{X}$ chart.

Pre-control has been advocated as a (superior) replacement for $\overline{X}$ and R charts, and indeed this is evident from the titles of the papers by Traver (1985), Shainin (1984), and Shainin and Shainin (1989). But pre-control is quite different from an $\overline{X}$ chart, since the control limits of the latter are not determined from specification limits. Thus, although the comparison of pre-control with an $\overline{X}$ chart is appropriate since pre-control has been advocated as a replacement for an $\overline{X}$ chart, this is almost like comparing apples and oranges.

Pre-control is similar to a zone chart (Jaehn, 1987) in that zones are used and decisions are made in accordance with the zone in which a unit of production falls. Unlike a zone chart, however, the zones in pre-control are determined from the specification limits. Specifically, the midline on the chart is the target value for a process characteristic, with a line also being drawn halfway between each specification limit and the midline of the chart. These lines are called the pre-control lines. The lines thus define four zones between the specification limits and, of course, two zones outside the limits. The two zones between the pre-control lines are called the "green zones," the other two zones within the specification limits are called the "yellow zones," and the zones outside the specification limits are termed the "red zones."

Process capability is initially assessed, which would correspond to Stage 1 for control chart usage. Once a process is declared capable, which occurs when five consecutive observations fall in the green zones, the counterpart to real-time process control begins. Two consecutive units are sampled and the process is stopped if at least one of the units is in a red zone or both are in one of the yellow zones or in different yellow zones.

One obvious problem with this technique is that only five observations are used to determine whether or not the process is a "capable process." Clearly that is far too few. Ledolter and Swersey (1997) showed that because of this small number there is a substantial probability that a process with poor capability is erroneously declared capable.

This can be explained as follows, using a somewhat different approach. As in Ledolter and Swersey (1997), we will assume that the appropriate distribution is a normal distribution, that the process mean is centered between the specification limits, that the mean and variance of the normal distribution are known and the observations are independent, and we will let "c" denote the value of C_{pk} (equals C_p since the process is centered.) As in Chapter 7 and in other chapters, we will let USL denote the upper specification limit, and LSL will denote the lower specification limit. For simplicity we will think in terms of C_p. Since $C_p = c$, it follows that USL $= \mu + 3\sigma c$ and LSL $= \mu - 3\sigma c$. Since the process is centered and the pre-control lines are halfway between the mean and the specification limits, the pre-control lines must be at $\mu + 3\sigma c/2$ and $\mu - 3\sigma c/2$.

We can easily determine the probability that five consecutive observations fall within these pre-control lines. Specifically, we want

$$P\left(\mu - \frac{3}{2}\sigma c \leq X \leq \mu + \frac{3}{2}\sigma c\right) = P\left(-\frac{3}{2}c \leq \frac{X - \mu}{\sigma} \leq \frac{3}{2}c\right)$$

and then raise this probability to the fifth power. Since the middle term in the double inequality defines a standard normal random variable, we may use Table B in the Appendix to the book to find the probability for a given value of c. For example, for $c = 1$ the probability of five consecutive observations falling between the pre-control lines is $(.8664)^2 = .4882$. A value of $c = 2$ corresponds to six-sigma quality (see Section 17.3), and the probability is .9886. In many American and Japanese industries $c = 1.33$ is considered to be the smallest acceptable value, and the probability that corresponds to this value is .7900.

What do these numbers mean? If the process capability is extremely good, such as six-sigma capability, it is unlikely that the process would be declared unacceptable. Hardly any company is going to have such process capability, however, so there will be a substantial probability that a process that would otherwise be declared acceptable will in fact be declared unacceptable by the criterion for five consecutive observations. Satterthwaite (1954) recognized that 5 observations are not sufficient and recommended that 12 observations be used. Obviously this would reduce the probability of declaring a capable process unacceptable, but there are other problems with pre-control.

Specifically, Table 2 of Ledolter and Swersey (1997) shows that the false-alarm rate of pre-control is extremely high when C_p is much less than 1.30. For example, when $C_p = 1.00$, the in-control ARL is only 44.4. This would be totally unacceptable for virtually any industrial application, so pre-control should not be used as a substitute for a control chart for such a C_p value. When $C_p = 1.30$, the in-control ARL for an $\overline{X}$ chart with a subgroup size of 2 is virtually the same as the in-control ARL for pre-control, but the $\overline{X}$ chart has much better shift detection capability. This superiority increases as C_p increases from 1.30.

Steiner (1997) compares pre-control with an acceptance chart, while claiming that comparing pre-control with an $\overline{X}$-chart is inappropriate since the charts are used for different purposes. Obviously there is some validity to such a claim, as noted earlier in this section, but an argument can be made for comparing pre-control with an $\overline{X}$ chart since some writers have argued that pre-control should be used instead of an $\overline{X}$ chart. The author also examines two-stage pre-control (Salvia, 1988), modified pre-control (Gruska and Heaphy, 1991), and three other variations.

Steiner (1997) concludes: "Classical Pre-control and Two-stage Pre-control are good methods when the process standard deviation σ lies in the range $T/10 \le \sigma \le 11T/75$, where T is the tolerance range." The author recommends that modified pre-control not be used. If we assume a process with the process mean centered between the specification limits, this translates to $1.14 \le C_p \le 1.67$. Clearly the false alarm rate for classical pre-control will be quite high when C_p is close to 1.14, and the mean shift detection capability will be poor when C_p is close to 1.67. Whether the comparison is made with an $\overline{X}$ chart or an acceptance control chart, we cannot expect *any* form of pre-control to compare favorably because, as noted by Ledolter and Swersey (1997), the Neyman–Pearson Lemma guarantees that a procedure based on the sample mean will have mean shift detection properties that are superior to pre-control when the two types of procedures have the same false alarm rate. [The Neyman–Pearson Lemma is discussed in virtually all mathematical statistics books. See, e.g., Hogg and Craig (1978).] Thus, an $\overline{X}$ chart and an acceptance chart must be superior to any form of pre-control.

This result is also suggested by common sense since the use of the actual value of an observation should produce results that are superior to the results that are obtained when only a region in which an observation falls is recorded.

In summary, pre-control has been used by many companies despite the fact that very little has been written about it in the technical literature. Its touted advantages have essentially been based upon folklore, but those claims have essentially been refuted by Ledolter and Swersey (1997).

10.2 SHORT-RUN SPC

Not all processes are continuing processes. There are many "one-time-only" processes in industry, including but not limited to job shop operations. There are also short production runs that are motivated by the adoption of just-in-time (JIT) manufacturing. In their review of short-run SPC methods, Del Castillo, Grayson, Montgomery, and Runger (1996) distinguish between short runs where entirely different setups are needed and short runs where the same general setup is used, with relatively minor modifications needed to accommodate different parts. They refer to the latter as repetitive production and to the former as nonrepetitive production.

How is a process to be controlled with very little data? The obvious answer is that it cannot be controlled very well, in general, when the limits are constructed solely from historical data. In particular, recall the discussion in Section 4.14 about the amount of historical data that is needed for a Shewhart chart to have ARL properties that are close to those for the parameters-known case. In order to accomplish the latter with the parameters estimated from a small amount of data, we would have to use k-sigma limits with $k < 3$ since the ARLs increase (decrease) when the sample size decreases (increases) when 3-sigma limits are used. Since the value of k that would be needed to produce a desired in-control ARL cannot be determined analytically, simulation would have to be used, somewhat analogous to the simulations that were performed by Quesenberry (1993) for $k = 3$.

One point to keep in mind is there is not a Stage 2 in short-run SPC. Recognizing that parameter changes will not be detected immediately unless the change is relatively large, we might want to create an artificial Stage 2, provided that a production run is not extremely short. As stated above, the value of k could be adjusted so as to provide the desired ARL values *for the rest of the production run.* The control limits could also be applied retrospectively, although if the cutoff is made in such a way that Stage 1 contains very few observations, running a CUSUM with different starting points, as suggested by Sullivan and Woodall (1996), would be expected to work better.

The approach most often advocated for the short-run SPC problem is to chart deviations from target values [see, e.g., Bothe (1987a,b, 1989), Farnum (1992), and Montgomery (1996)]. This has been referred to as a "deviations from nominal (DNOM) chart". The general idea is to take advantage of the fact that even though there may be short runs for individual parts, different types of parts are often produced using the same equipment. This approach does have some shortcomings, however. One obvious problem is that control is relative to nominal values rather than in terms of estimates of process parameters. As discussed in Section 4.7.11, this is undesirable, in general. Another problem with the standard DNOM approach is that one must assume that the variance of the process characteristic is the same for the different types of parts.

Farnum (1992) presents a good introduction to the short-run problem and provides some solutions. In particular, he points out that whereas some writers have recommended using a *standardized DNOM chart* [i.e., plotting $\sqrt{n}(\overline{X}_i - T_i)/\sigma_i$, where the subscript indexes the particular part type], this approach does have some shortcomings. One obvious problem with this is that an estimate of σ_i is needed, and since we would generally expect to have individual observations, not subgroups, in a short-run situation, a user of this type of chart might not have enough observations to obtain a reasonably good estimate of σ (i.e., an estimate whose corresponding estimator would have a small variance) until the end of a production. Unless the part was manufactured in later runs, this would not solve the short-run problem.

Farnum (1992) considers various models and makes the "mild assumption" (p. 139) that the mean of a process variable is equal to the target value. Such an assumption could have serious consequences, however, since controlling to target values is not the same as statistical control of a process, although the former could have some value when very little data are available. We must point out, however, that the in-control ARL will be less than the nominal ARL value if the target value differs from the process mean, assuming for the moment that σ is known. Furthermore, the difference in the two ARL values may be unacceptably large. For example, assume that the target value for the mean is 50, the actual mean is 48, and the standard deviation is 3. The control limits for individual observations should thus be at 39 and 57, but instead are at 41 and 59. This "slight" difference will cause the in-control ARL to be only 100.62, assuming normality, even though the mean is less than one standard deviation from the target value. This results from the fact that the LCL is too large. The area under the standard normal curve increases greatly as one moves toward zero from either $+3$ or -3, and the gain far exceeds the loss relative to the other control limit. Practitioners who would consider plotting deviations from target values should be aware of these potential problems.

Obviously similar problems will occur if σ is misspecified (or misestimated). It is possible to estimate σ without using data, however, and the following approach can be useful in the absence of any data or a sufficient amount of data. Assume that a company has a convincing amount of evidence suggesting that process averages have not deviated greatly from target values for a certain repetitive process that has minor periodical changes to accommodate different types of parts. Assume further that the specification limits define an interval outside of which values of a process characteristic seldom occur and that the individual values seem to be approximately normally distributed. Then σ could be estimated as (USL − LSL)/6. This results from assuming that the upper specification limit (USL) is approximately equal to $\mu + 3\sigma$ and the lower specification limit (LSL) is approximately equal to $\mu - 3\sigma$.

The estimate of σ obtained in this manner could then be combined with the target value to produce the control limits. Alternatively, with different minor setups, different random variables will be involved, but the user may want to use the same chart throughout. The plotted statistic might then be the standardized variable $Z_t = (X_t - \mu_t)/\sigma_t$, with t denoting the tth setup and the random variable X having the same general representation, differing only over the setups.

One aspect of the problem that seems to be overlooked by most researchers is that in the short-run case control limits are needed at the beginning of the run. Even if one subscribes to the use of a DNOM chart, process variability must still be estimated. Very little is accomplished if the estimate of (common) variability is not made until after data are available from short runs on similar parts. Obviously the control limits that are computed after a set of short runs could be applied in the future, but this means that we would have to forego some real-time process monitoring so as to accumulate a data base for a Stage 1 analysis. This is obviously somewhat unappealing but might be the best that could be done.

From Section 4.13 we know that there will be considerable control chart variability when parameters are estimated from small amounts of data. Consequently, estimating parameters as data arrive will produce results with *extremely* high variability. As in statistics, in general, there is simply not much that can be done when there is very little data. The problem is especially acute in control chart applications since the variability in control chart performance, even when parameters are known or estimated with a very large amount of data, results from the fact that very extreme percentiles of a distribution are being estimated.

Hillier (1969) and Yang and Hillier (1970) were apparently the first to investigate how one should proceed when a small amount of data (they considered only subgroup data) are available, their work being motivated by the work of King (1954). Hillier (1969) provided a table of constants to use in constructing the control limits for $\overline{X}$ and R charts when a small number of subgroups is used.

Wheeler (1991) and Griffith (1996) are publications on constructing control charts for short runs. Wasserman (1993) presents a Bayesian approach to the problem and Wasserman (1995) suggests an adaptation of the EWMA chart for short runs.

One possible approach to the short-run problem would be to try bootstrapping. With bootstrapping one attempts to essentially "enlarge" the amount of information in the sample. Strictly speaking, this cannot be done, however, and there is no clear evidence that bootstrapping works in control chart applications. This is discussed further in Section 10.6.

Short-run SPC methods have been highly controversial. See, for example, the spirited discussion in Maguire (1999).

10.3 CHARTS FOR AUTOCORRELATED DATA

Autocorrelated data are quite prevalent, especially in the process industries. Various authors (e.g., Alwan and Roberts, 1988) have recommended that a time series model be fit to the data, with the model residuals subsequently used to control the process. Since the model *errors* would be independently and identically distributed with a mean of zero and a constant variance if the true model were fit (and also normally distributed if normality were assumed), it might seem as though an X chart of the residuals would have the same properties as when an X chart is applied to individual observations.

Unfortunately, this is not true. The ability of the chart to detect a mean shift depends on the model that is assumed to be the true model. (Of course, we recognize that the "true model" will generally be unknown.) It can be shown that a residuals chart will have poor properties for detecting mean shifts for time series models that are frequently assumed for autocorrelated data, and for likely parameter values within these models. This was first demonstrated in print by Longnecker and Ryan (1991) and Ryan (1991), where an AR(1) model was assumed, and also illustrated for additional models in Longnecker and Ryan (1992). Some of this ground has been covered more recently by Zhang (1997). For example, Table 2 of Zhang (1997) shows that the ARL for detecting a 1-sigma mean shift for an AR(1) process with $\phi = 0.5$ is 123.47. Thus, a residuals chart will have poor properties for detecting mean shifts even when the autocorrelation is only moderate. For autocorrelations in the neighborhood of 0.8–0.9, the ARLs are extremely large.

Wardell, Moskowitz, and Plante (1994) suggested that k-sigma limits on a standard chart be used, with k slightly different from 3 so as to adjust for the autocorrelation. They did not, however, provide any guidance as to how the adjustment should be made. Michelson (1994) and Zhang (1997) used simulation to determine the best choice of k for certain time series models so as to make the in-control ARL as close as possible to the normal-theory value of 370. The tables given by Michelson (1994) are more extensive than the tables given by Zhang (1997), and although the tabular entries are different for the two sets of tables, this is probably due to the fact that a very large number of simulations must be used in order to estimate the ARLs with any degree of precision since the run lengths have very large standard deviations. Therefore, since the ARLs and the standard deviation of the run lengths will not differ greatly, in general, the estimated ARLs of Zhang (1997) will have a standard deviation that is about $\frac{1}{32}$ times the run length standard deviation since Zhang estimated ARLs from 1000 simulations ($\doteq 32^2$).

The tables of Michelson (1994) can be used in the following way. Assume an AR(1) with $\phi = 0.7$, and the in-control ARL is to be 370. This would require the use of $k = 2.94$. The ARL for detecting a 1-sigma shift would be approximately 55, as compared to 44 when the data are independent. Thus, the loss in detection power is not severe.

If ϕ were less than 0.7, then k would be closer to 3. If ϕ is much larger than 0.7, then it is not possible to select a value of k such that an X chart would have approximately the properties that it would have for independent data. Consequently, for extreme correlation, such as $\phi \doteq 0.9$, it is difficult to overcome the deleterious effects of the autocorrelation. Using the tables of Michelson (1994), we would set $k = 2.80$ and would have an ARL of approximately 68 for detecting a 1-sigma shift. Although data from an AR(1) process with $\phi > 0.9$ would not be encountered very often, Zhang (1997) shows that if $\phi = 0.95$, the appropriate value of k is 2.54, and the ARL for detecting a 1-sigma shift is about 116.

Thus, the control limits of an X chart can be suitably modified and the properties will not differ greatly from the properties of an X chart applied to independent data, provided that the autocorrelation is not extreme. If the autocorrelation is quite high, this means that there is far less information in a set of observations than there would be if the correlation was much lower.

Negatively autocorrelated process data occur much less frequently. For such data, either a residuals chart or a Shewhart chart can be used, although the former has much better properties. For example, Zhang (1997) shows that if $\phi = -0.50$, the ARL for detecting a 1-sigma shift with a residuals chart is only 10.45, whereas the ARL is approximately 42 for an X chart with 2.97-sigma limits.

Box and Jenkins (1976, p. 32) give an example of 70 consecutive yields from a batch chemical process for which the estimated first-order autocorrelation is -0.79. We may show that the ARL for detecting a 1-sigma mean shift is only 2.95, and the ARLs for other mean shifts are also quite small.

Why does a residuals chart have such exceptional properties for detecting mean shifts of negatively autocorrelated data but poor properties when the data are positively autocorrelated? This may be understood by looking at the appropriate expected values. The model for an AR(1) process may be written as

$$X_t = \mu + \phi(X_{t-1} - \mu) + \varepsilon_t$$

Parameters are assumed known in the literature on residuals charts (e.g., Zhang, 1997). This greatly simplifies the theoretical development, which would be practically impossible without this assumption. With the assumption the residuals are the same as the errors. Therefore, we may write the residuals, e_t, as

$$e_t = X_t - \mu - \phi(X_{t-1} - \mu)$$

Assuming a mean shift from μ to μ^* at time $t = T$, we have

$$E(e_T) = \mu^* - \mu \tag{10.1}$$

but

$$E(e_{T+k}) = (\mu^* - \mu)(1 - \phi) \qquad k \geq 1 \tag{10.2}$$

Thus, the entire shift is captured by the first residual, but subsequent residuals capture only a fraction of the shift, with the fraction being small when ϕ is large.

It is somewhat better to look at the standardized residuals (errors), since these relate more directly to a residuals control chart. It is well known [see, e.g., Box and Jenkins (1976, p. 58) or Cryer (1986, p. 62)] that for an AR(1) model $\sigma_\varepsilon^2 = \sigma_x^2(1-\phi^2)$. Consider

$$Z_T = \frac{e_t - 0}{\sigma_e}$$

with $E(e_t) = 0$. Using Eqs. (10.1) and (10.2) we obtain

$$E(Z_T) = \frac{\mu^* - \mu}{\sigma_x\sqrt{1-\phi^2}} = \frac{1}{\sqrt{1-\phi^2}}\left(\frac{\mu^* - \mu}{\sigma_x}\right) \tag{10.3}$$

and

$$E(Z_{T+k}) = \sqrt{\frac{1-\phi}{1+\phi}}\left(\frac{\mu^* - \mu}{\sigma_x}\right) \tag{10.4}$$

We can see from Eq. (10.3) that when ϕ is positive, the first standardized residual actually captures more than the shift (in standard units). Subsequent residuals, however, capture only a fraction of the shift, as is apparent from Eq. (10.4). For example, when $\phi = 0.7$, 140% of the shift is captured by the first residual, but only 42% by subsequent residuals. Compare this to an X chart in which 100% of the shift is captured by each plotted point. The problem is that 140% is not enough to ensure that the shift will be detected with high probability by the first residual. If it is not detected by the first residual, then it will take much longer to detect the shift with the subsequent residuals than it would to detect the same shift with independent data. The situation is much different when ϕ is negative, however, as then the subsequent residuals have much greater power. For example, when $\phi = -0.7$, the first residual has the same power as with $\phi = 0.7$ but the subsequent residuals capture 238% of the shift. Thus, the residuals chart will have far greater power than an X chart for independent data for this model and this value of ϕ.

Runger (1998) considers the "ideal stirred tank" in illustrating how various procedures can be expected to perform similarly. With the stirred tank, the autocorrelated observations can be modeled by an EWMA. The primary result is that a control chart of the raw data can thus be expected to have approximately the ARL properties of an EWMA, designed to detect a step change in white noise, applied to the residuals, after an appropriate time series model has been fitted.

Some people have suggested that the sampling interval be widened sufficiently so that the *sampled* observations are not highly autocorrelated. For an AR(1), if the autocorrelation between units *produced* is ϕ, then if every kth unit were sampled, the correlation between the *sampled units* would be ϕ^k (see, e.g., Box and Luceño, 1997, p. 107). Obviously we would have to obtain a sequence of consecutively produced units in order to have any idea of the autocorrelation between consecutive units of production. Of course, we need a rather large number of observations to determine an appropriate model to fit to the data. This suggests that at least 100 consecutive observations (and preferably more) be obtained. The question of sampling frequency should be addressed *before* looking at the data. For example, a practitioner should determine what would be a reasonable sampling rate if the data were independent. Then the autocorrelation for a normal sampling frequency could be determined, and a decision made as to how to proceed. It is quite possible that the autocorrelation could be high, but the autocorrelation for a normal sampling frequency could be such that adjusted control limits (as in Michelson, 1994) could have reasonable properties.

If not, the sampling frequency might be decreased slightly. That is, the interval between samples would be increased. It should be kept in mind, however, that tight control of the process will not be possible if the normal sampling frequency is decreased considerably. So it is largely a question of economics: What is the cost of detecting an out-of-control condition later than would occur with more frequent sampling?

10.3.1 Autocorrelated Attribute Data

Woodall (1997) recommended that the effect of autocorrelation on the performance of attribute control charts be considered. Very little attention has been given to the charting of correlated attribute data, however. Wisnowski and Keats (1999) briefly consider the monitoring of autocorrelated proportion data and state that EWMA-based monitoring schemes hold promise. This assertion is based on the work of Harvey and Fernandes (1989), who claim that correlated count data can generally be modeled with an EWMA approach. See also the discussion of autocorrelated count data in Section 6.1.7.

10.4 CHARTS FOR BATCH PROCESSES

In certain industries, such as the semiconductor industry, batch processing is used. For example, in a semiconductor application, a diffusion furnace may process 150 units at a time, five units with a polisher, and only a single unit with an etcher. Thus, there is neither continuous processing nor a constant subgroup size. This creates some complexities, and the use of standard control charting techniques can result in a large percentage of the plotted points falling outside the control limits. Joshi and Sprague (1997) explain how this results from autocorrelation and give an "effective sample size" in an effort to adjust for the different subgroup sizes.

10.5 CHARTS FOR MULTIPLE-STREAM PROCESSES

There are many processes that have multiple streams. One example, as described by Mortell and Runger (1995), is where a machine involved in the production of shampoo bottles has multiple heads. Other examples are processes that involve different machines, different operators, multiple suppliers, and so on.

Such scenarios are amenable to the use of a *group control chart*. The group control chart concept is traceable at least as far back as Boyd (1950) and is described in detail in Burr (1976), Nelson (1986), and Montgomery (1996). The output from the different streams might be highly correlated or perhaps almost uncorrelated.

The objectives when a group chart is used are (1) to detect when all of the streams are out of control for the plotted statistic, and (2) to detect when only a single stream is out of control. All of the streams being out of control would suggest a common assignable cause, whereas only one stream being out of control would suggest a problem with one apparatus.

Assume that an $\overline{X}$ chart and R chart will be used. If each stream has the same target value and approximately equal variation, then control limits for the $\overline{X}$ chart would be determined by averaging the subgroup averages over both the subgroups and the streams. Nelson (1986) uses $\overline{\overline{X}}$ to denote this average, although different notation, such as $\overline{\overline{\overline{X}}}$, seems to be needed. Similarly, the average range could be represented as $\overline{\overline{R}}$. The control limits for the $\overline{X}$ chart would be given by $\overline{\overline{\overline{X}}} \pm A_2\overline{\overline{R}}$, where A_2 is determined by the subgroup size only and thus does not depend on the number of streams. Nelson (1986) states that the points that are plotted on the group chart would be the largest and smallest subgroup averages. For example, if there are six streams, the stream with the largest subgroup average and the stream with the smallest subgroup average would be plotted at each point in time. This is not quite right, however, because the control limits are not for testing extreme values. The control limits would have to be determined using extreme value theory .

Mortell and Runger (1995) review the literature on the subject and propose new methods.

10.6 NONPARAMETRIC CONTROL CHARTS

If there is strong evidence of non-normality and the user wishes to take a nonparametric approach, there are several options. Bootstrapping, mentioned briefly in Section 10.2, is a popular nonparametric approach that has been used extensively in various areas of statistics. The general idea is to resample a sample that has been obtained, acting as if the sample is representative of the population. An empirical sampling distribution is then obtained for whatever statistic is desired. Unfortunately, however, bootstrapping fails under many scenarios, and standard bootstrapping methods perform poorly in estimating extreme percentiles of a distribution. Control chart construction of course requires that extreme percentiles be estimated. Some of the bootstrap control chart papers that have appeared in the literature have been flawed. The poor performance of standard bootstrapping techniques in control chart applications is illustrated by Jones and Woodall

(1998). Can any bootstrap approach produce satisfactory control limits? This seems highly unlikely.

Consider the following. One rule-of-thumb (Runger, Willemain, Grayson, and Messina, 1997) is that at least $3/(0.01p)$ observations are needed to estimate the pth or $100(1-p)$th percentile of a distribution. Thus, in order to estimate the 0.135 percentage point of a distribution (i.e., the tail area is .00135), at least $3/0.00135 = 2222$ observations would be needed. Assume that an X-chart is to be constructed for an unknown distribution. If bootstrap control limits were attempted, we would be trying to reproduce the control limits that could be produced reasonably well using thousands of observations, when we are resampling from a sample that is probably an order of magnitude smaller. Thus, it is no surprise that bootstrap control limits are inadequate.

Other nonparametric control chart approaches that are not bootstrap-based have been given by Amin, Reynolds, and Bakir (1995) and Hackl and Ledolter (1991). A review of nonparametric control chart approaches was given by Bakir (1998).

Hackl and Ledolter (1991) proposed an EWMA chart for individual observations that was based on the ranks of the observations. They showed that with such a procedure there is very little loss in efficiency when normality holds and a considerable gain when there is non-normality. Of course, there is no such thing as a normal distribution in practice (see, e.g., Geary, 1947), so using such a nonparametric procedure would be a logical alternative to fitting a distribution to historical data when the historical data suggest that the corresponding population may be more than slightly nonnormal.

10.7 BAYESIAN CONTROL CHART METHODS

Bayesian techniques incorporate prior information in addition to information from a sample. There is, of course, a substantial amount of prior information that is produced whenever control charts are used, as all past data are, in essence, prior information relative to the present point in time. Consequently, Bayesian techniques do not have quite the same opportunity to be superior to standard control chart approaches that such techniques often have over one-shot frequentist approaches when prior information is available. So when can Bayesian methods be useful in control charting?

Bayesian methods have been applied to various types of control charts, including multivariate charts. Mason et al. (1997) state: "All multivariate SPC procedures rely heavily on the covariance structure, whether it be known or estimated from historical data". These authors also indicate that it is preferable for the subgroup size to exceed the number of process characteristics and additionally point out that missing data in the multivariate case is problematic in the absence of information about the correlation structure.

Nandram and Cantell (1999) consider the case of multivariate individual observations when there is prior information on the covariance structure and multivariate normality is assumed, although the latter is not required for their approach. They show that incorporating prior information in a semiconductor manufacturing application produces better results than the Tracy et al. (1992) frequentist approach.

10.8 CONTROL CHARTS FOR VARIANCE COMPONENTS

Control charts for variance components have been considered by various authors, including Hahn and Cockrum (1987), Woodall and Thomas (1995), Laubscher (1996), and Park (1998). Obviously it is easier to control, and preferably reduce, variability when the components of the total variability can be identified. See also Roes and Does (1995).

Recall from Section 4.17 that the implicit model when an $\overline{X}$ chart for measurement data is used is $Y_{ij} = \mu + \varepsilon_{ij}$, with i denoting the ith subgroup and j denoting the jth observation within the ith subgroup. With such a model there is no attempt to identify and monitor specific sources of variation.

10.9 NEURAL NETWORKS

Neural networks have become very prominent. Literally thousands of papers have been written on neural networks and their applications, and many of these papers can be found in the journal *Neural Networks: The Official Journal of the International Neural Network Society*. Neural networks have been advocated for use in many different areas, including process control. Hamburg, Booth, and Weinroth (1996) describe the use of neural networks for monitoring nuclear material balances, pointing out that these are often autocorrelated, with autocorrelation signaling an out-of-control process. Hwarng (1991) used a neural network approach to pattern recognition for an $\overline{X}$ chart used with either an R chart or an s chart, and Nieckula and Hryniewicz (1997) also discuss the use of neural networks in conjunction with an $\overline{X}$ chart.

Despite the popularity of neural networks, their superiority over other techniques has not been clearly established. The approach is nonparametric, with neural networks intended to simulate learning that is performed by the human brain. This is the lure of the approach, but neural networks have some negative features, including the fact that different answers will generally be produced when neural networks are repeatedly run *on the same data set*. Readers interested in neural networks are advised to proceed with caution until all of the nuances are well understood.

10.10 ECONOMIC DESIGN OF CONTROL CHARTS

The control charts that have been presented in the preceding chapters were not based directly on economic considerations. Specifically, the sample size and 3-sigma control limits were presented without the sample size and width of the limits being determined from cost criteria. These factors, in addition to the sampling interval, could just as well be determined from cost considerations.

The general idea is to minimize some function of the costs of producing items that are not within specification limits, the costs of detecting and eliminating assignable causes, and the costs of false alarms. One obvious shortcoming of economic designs is that target values of process characteristics are not considered, nor is process control part of the general scheme of things. This is discussed further in Section 10.10.1.

The design of control charts using cost criteria has received very little attention in industry, although it has received some attention in the literature. Keats, Del Castillo, von Collani, and Saniga (1997) review recent developments. Another fairly recent review was given by Ho and Case (1994). One reason that is given for not using economically designed control charts is that actual costs are difficult to obtain. This is not really a valid reason, however, as the economic models that have been proposed will work fairly well when only estimates of the actual costs are available.

The costs that need to be determined or estimated include the cost of sampling, the cost of searching for an assignable cause when none exists (i.e., a false signal), the cost of detecting and removing an assignable cause, and the cost of producing nonconforming units.

Admittedly, many of the economic models that have been proposed are somewhat complicated (as compared to, say, an $\overline{X}$ chart), but computer programs can prevent this from being an impediment to their use.

Most of the models that have been proposed have been for an $\overline{X}$ chart, and the work dates from Girshick and Rubin (1952), although the economic model proposed by Duncan (1956) is better known among the earliest economic models. Chiu and Wetherill (1974) presented a simple model as an alternative to the model proposed by Duncan (1956). The latter assumed a model in which the process is presumed to start in control, with a subsequent mean shift due to a single assignable cause. The time until the assignable cause occurs is assumed to have an exponential distribution with parameter λ, so the process is assumed to operate in control for λ^{-1} hours. Samples of size n are drawn h hours apart, and the control limits are "k-sigma" limits. As usual, a search for the assignable cause is initiated when a point falls outside the control limits, and the process continues in operation until the cause is detected. The repair cost is not charged against the income from the process. Duncan (1956) has shown that the expected loss (cost) per hour of operation is

$$L = \frac{\lambda BM + \alpha T/h + \lambda w}{1 + \lambda B} + \frac{b + cn}{h}$$

where

$$B = \left(\frac{1}{p} - \frac{1}{2} + \frac{\lambda h}{12}\right) h + en + D$$

and the other symbols are defined as follows:

α	probability of a point falling outside the control limits when the process is in control
M	increased net loss per hour attributable to a greater percentage of unacceptable items when the assignable cause occurs
$b + cn$	cost of taking a sample of size n and maintaining the control chart
T	average cost of looking for an assignable cause when none exists
w	average cost of looking for an assignable cause when one exists
en	time required to take a sample and compute the results
D	average time required to discover the assignable cause after a signal has been received

The general idea is to determine the values of h, k, and n that will minimize the expected loss. The minimization procedure proposed by Duncan is somewhat involved, but Chiu and Wetherill (1974) gave a simple procedure for approximating the minimum of Duncan's loss function.

More recently, von Collani (1986, 1988, 1989) provided a model that does not require the use of a computer and involves the estimation of only three economic parameters.

Although researchers have concentrated on the economic design of standard charts, economic design principles can also be applied to more advanced charts. Linderman and Love (1999) discuss economic design of a multivariate EWMA chart.

10.10.1 Economic-Statistical Design

Woodall (1986, 1987) pointed out that economic models are not consistent with the idea of maintaining tight control of a process. This is apparent from the *np*-chart examples of Williams, Looney, and Peters (1985), where it is economically optimal to allow 64% of the items produced to be defective, and in an example given by Lorenzen and Vance (1986), in which an economic model is formulated for an out-of-control state with 11.3% nonconforming items whereas the in-control state has 1.36% nonconforming items. Furthermore, the emphasis is on process deterioration, not process improvement, as the latter cannot be detected with a purely economic approach.

Consequently, it is necessary to use statistical constraints in conjunction with an economic model. Specifically, the maximum value for the false-alarm probability and the average time to signal (ATS) or ARL for a parameter change that one wishes to detect could be specified. (The ATS would be preferred over the ARL if the sample size and/or sampling frequency were not constant.) Saniga (1989) proposes such a design for a joint $\overline{X}$ and R chart. Although the cost of an economic-statistical design will be greater than or equal to the cost for an economic design, as noted by Saniga (1989) the cost for an economic-statistical design could be less than the cost for an economic design if the parameter change that actually occurs is not the change for which the design was constructed.

Saniga, Davis, and McWilliams (1995) provided a computer program that could be used for the construction of economic, statistical, or economic-statistical design of attributes charts.

10.11 CHARTS WITH VARIABLE SAMPLE SIZE AND/OR VARIABLE SAMPLING INTERVAL

With economic designs the sample size and sampling interval are determined using economic considerations. Once determined, these are fixed. Another possibility is to use either a variable sample size or a variable sampling interval, or both. The general idea is that the sample size can be decreased or the sampling interval increased whenever a process exhibits good control. If a point plots very close to a control limit, it would be reasonable to suspect that the next plotted point may fall outside of the limit. Therefore, the sampling interval could be decreased, or the sample size increased, or both. Conversely, if a point plots very close to the midline of a chart, there is no obvious reason for concern, so the sampling interval might be increased or the sample size decreased.

This is the general idea behind variable sampling interval (VSI) and variable sample size (VSS) charts. A moderate number of papers have been written on VSI and VSS control chart schemes. In particular, a VSI $\overline{X}$ chart was presented by Reynolds, Amin, Arnold, and Nachals (1988), and properties of such charts were studied by Reynolds (1995). The properties of a VSS $\overline{X}$ chart were studied by Prabhu, Runger, and Keats (1993) and Costa (1994). Prabhu, Montgomery, and Runger (1994) and Costa (1997) studied an $\overline{X}$ chart when both the sample size and the sampling interval were allowed to vary. Cumulative sum charts with variable sampling intervals were presented by Reynolds, Amin, and Arnold (1990).

10.12 USERS OF CONTROL CHARTS

If a business is to run efficiently, more than just the manufactured products need to be "in control." Other factors such as clerical errors, accounts receivable, salesmen's expenses, and a host of other factors can be charted with the purpose of keeping them in a state of statistical control.

Although the use of control charts for administrative applications has not received very much attention, some actual applications are discussed in this section with the intention of providing the reader with some insight as to their (relatively untapped) potential value.

It is important to remember, however, that statistical quality improvement consists of more than just control charts, so general statistical quality improvement principles can still be applied in situations in which control charts cannot be used without modification. For example, if clerical errors are being charted with the objective of keeping them "in control," would a c chart with 3-sigma limits be suitable, or would management prefer to establish an upper limit based on what it considers to be an excessive number of errors, or perhaps not even use an upper limit? Whichever the case may be, the mere maintenance of some type of control chart will certainly indicate the variability of clerical errors over time and should lead to the detection of assignable causes when the variability is judged to be excessive.

One such assignable cause could be inadequate training. Lobsinger (1954) describes the use of *np*-type charts in the airline industry that uncovered the need for additional training for certain individuals. The end result was that basic training time was cut in half, and this was accomplished by using the feedback from the control charts.

Various other interesting applications of control charts were described in journals decades ago. For example, a description of the use of an *np* chart for clerical errors (using 2σ limits) in a mail order plant can be found in Ballowe (1950), where the application is described in considerable detail. The use of an $\overline{X}$ chart for cost control, by controlling the number of work hours per thousand units produced, is described in Schiesel (1956). Pringle (1962) gives a detailed account of the successful application of $\overline{X}$ and R charts for controlling the volume of speech transmitted on a switched telephone call. Bicking (1955) describes the application of X and R charts to (1) inventory control, (2) rating of technical personnel, and (3) analysis of indirect expense. Enrick (1973) also discusses the use of statistical control limits for inventory control. Latzko (1977) discusses the use of control charts for controlling the encoding of bank checks. See also Latzko (1986). There have been many applications of SQC principles to accounting data, although the method used has often been some acceptance sampling procedure. [See, e.g., Buhl (1955) and Dalleck (1954).]

10.12.1 Recent Control Chart Nonmanufacturing Applications

The papers that were mentioned in the preceding section were published decades ago. In this section we discuss recent applications of control charts in various types of organizations. The list of articles given is intended to be representative, not exhaustive. Nevertheless, the list should provide a feel for the wide variety of possible control chart applications. Simply stated, control charts can be applied to almost everything.

10.12.1.1 Health Care

The use of control charts in health care is extensive. Cinimera and Lease (1992) describe the use of control charts for monitoring medication errors. Schramm and Freund (1993) detail the use of control charts to solve planning problems in a hospital. Monitoring intensive care unit performance with control charts is described by Chamberlin, Lane, and Kennedy (1993). Hand, Piontek, and Klemka-Walden (1994) discuss the use of control charts in assessing the outcomes of Medicare patients with pneumonia. Finison, Spencer, and Finison (1993) describe the use of individual observations charts for monitoring days between infections. See also Rodriguez (1996).

10.12.1.2 Financial

Herath, Park, and Prueitt (1995) describe how projects can be monitored using cash flow control charts. Bruch and Lewis (1994) show how control charts can be used to manage accounts receivable in a health care organization.

10.12.1.3 Environmental

Berthouex (1989) describes the use of control charts for wastewater treatment.

10.12.1.4 Clinical Laboratories

A historical article on the use of control charts in clinical laboratories is given by Levey and Jennings (1992).

10.12.1.5 Analytical Laboratories

Mullins (1994) and Howarth (1995) discuss the use of control charts in analytical laboratories in tutorial-type articles.

10.12.1.6 Civil Engineering

Gebler (1990) describes the interpretation of control charts in concrete production.

10.12.1.7 Education

Melvin (1993) describes the application of control charts to educational systems.

10.12.1.8 Law Enforcement/Investigative Work

Charnes and Gitlow (1995) describe the use of $\overline{X}$ and R charts in corroborating bribery in jai alai. An interesting feature of this application is that the charts were used *backward* in time. That is, the control limits were projected backward rather than forward.

10.12.1.9 Lumber

Maki and Milota (1993) describe the use of control charts in sawmill operations. Since "fractions of an inch or a few degrees can mean millions of dollars in revenue," it is desirable to use control charts in an attempt to minimize sawing variation. With trees not being as plentiful as in the past, it is desirable to try to maximize the quantity and quality of wood from each tree. One can attempt to minimize sawing variation by identifying assignable causes such as dull saw blades, misplacement of the log, or feeding the log to the saw too fast (Brown, 1982).

10.12.1.10 Athletic Performance

Clark and Clark (1997) illustrate the role that control charts played in the younger Clark's quest to improve his free-throw shooting.

10.13 SOFTWARE FOR CONTROL CHARTING

Many of the control charts that were presented in the previous chapters could be easily maintained by hand and/or hand calculator. For example, the points to be plotted on an $\overline{X}$, R, or c chart can be easily determined. At the other extreme, determining the points to be plotted on a multivariate chart for averages could not easily be performed without a computer, and determining the trial control limit for the chart would be extremely laborious without a computer.

In general, the more sophisticated (and superior) control chart procedures could not be used very efficiently without a computer. Even when the simpler control charts are used, it is easier to view and analyze historical data with a computer than when the charts are maintained by hand.

Developers of SPC software have concentrated on the microcomputer, and on the IBM-PC in particular. In the early 1980s there were very few such packages on the market, but by the middle of the decade the number had increased to several dozen, and the number exceeded 100 during the 1990s. On the whole, the software has been easy to use and could thus be easily used by plant personnel. The developers generally concentrated on the standard control charts for measurement and attribute data. Although user friendly, many packages were somewhat limited in terms of data input flexibility as well as the way in which the data could be handled.

Several software directories have been published to aid people who are searching for SPC software. One such directory has been published in the March issue of *Quality Progress* each year, starting in 1984. Other directories have appeared in *Quality* and *Industrial Engineering*.

These packages are generally in the price range of \$500–\$1000, with a few packages above \$1000, and software for computers larger than micros is also above \$1000.

BIBLIOGRAPHY

Alwan, L. C and H. V. Roberts (1988). Time series modeling for statistical process control. *Journal of Business and Economic Statistics* 6(1): 87–95.

Amin, R. W., M. R. Reynolds Jr., and S. Bakir (1995). Nonparametric control charts based upon the sign statistic. *Communications in Statistics—Theory and Methods 24:* 1597–1623.

Armour, N., R. Morey, R. Kleppinger, and K. Pitts (1985). Statistical process control for LSI manufacturing: What the handbooks don't tell you. *RCA Engineer 30*(3): 44–53.

Bakir, S. T (1998). Bibliography and review of distribution-free quality control charts. Paper presented at the 1999 Joint Statistical Meetings, Dallas, TX.

Ballowe, J. M (1950). Results obtained during five years of operation in a mail order plant. ASQC Annual Quality Control Conference Papers, Paper 10.

Beall, G. (1956). Control of basis weight in the machine direction. *Tappi 39:* 26–29.

Beaudry, J. P (1956). Statistical quality control methods in the chemical industry. *ASQC National Convention Transactions*, 626–627.

Berthouex, P. M (1989). Constructing control charts for wastewater treatment plant operation. *Research Journal of the Water Pollution Control Federation 61*(9/10): 1534.

Bhote, K. (1988). *Strategic Supply Management*. New York: American Management Association.

Bhote, K. (1991). *World Class Quality*. New York: American Management Association.

Bicking, C. A (1955). Quality control as an administrative aid. *ASQC National Convention Transactions*, 347–357.

Bingham, R. S (1957). Control charts in multi-stage batch processes. *Industrial Quality Control 13*(12): 21–26.

Bothe, D. (1987a). J.I.T. calls for S.P.C. *Job Shop Technology* January, pp. 21–23.

Bothe, D. (1987b). *SPC for Short Production Runs*. Northville, MI: International Quality Institute.

Bothe, D. (1989). A powerful new control chart for job shops. In *ASQC Annual Quality Congress Transactions*, pp. 265–270. Milwaukee, WI: American Society for Quality Control.

Box, G. E. P. and G. M. Jenkins (1976). *Time Series Analysis: Forecasting and Control*, rev. ed. San Francisco: Holden-Day.

Box, G. E. P. and A. Luceño (1997). *Statistical Control by Monitoring and Feedback Adjustment*. New York: Wiley.

Boyd, D. F. (1950). Applying the group chart for $\overline{X}$ and R. *Industrial Quality Control 7:* 22–25.

Breunig, H. L. (1964). Statistical control charts in pharmaceutical industry. *Industrial Quality Control 21*(2): 79–86.

Bruch, N. M. and L. L. Lewis (1994). Using control charts to help manage accounts receivable. *Healthcare Financial Management 48*(7): 44.

Brown, T. D. (1982). *Quality Control in Lumber Manufacturing*. San Francisco: Miller-Freeman.

Buhl, W. F. (1955). Statistical controls applied to clerical and accounting procedures. *ASQC National Convention Transactions*, 9–25.

Burr, I. W. (1976). *Statistical Quality Control Methods*. New York: Dekker.

Chamberlin, W. H., K. A. Lane, and J. N. Kennedy (1993). Monitoring intensive care unit performance using statistical quality control charts. *International Journal of Clinical Monitoring and Computing 10*(3): 155–161.

Charnes, J. M. and H. S. Gitlow (1995). Using control charts to corroborate bribery in jai alai. *The American Statistician 49:* 386–389.

Chiu, W. K. and G. B. Wetherill (1974). A simplified scheme for the economic design of $\overline{X}$-charts. *Journal of Quality Technology 6*(2): 63–69.

Ciminera, J. L. and M. P. Lease (1992). Developing control charts to review and monitor medication errors. *Hospital Pharmacy 27*(3): 192.

Clark, T. and A. Clark (1997). Continuous improvement on the free throw line. *Quality Progress 30*(10): 78–80.

Costa, A. F. B. (1994). $\overline{X}$ charts with variable sample size. *Journal of Quality Technology 26:* 155–163.

Costa, A. F. B. (1997). $\overline{X}$ chart with variable sample size and sampling intervals. *Journal of Quality Technology 29*(2): 197–204.

Cryer, J. D. (1986). *Time Series Analysis*. Boston: Duxbury.

Cyffers, B. (1957). Setting snuff packing machines under control. *Revue de Statistique Appliquee 5*(1): 67–76.

Dalleck, W. C. (1954). Quality control at work in airline accounting. *ASQC Quality Control Convention Papers*, 489–498.

Deile, A. J. (1956). How General Foods prevents accidents—by predicting them. *Management Methods 10*(4): 45–48.

Del Castillo, E., J. M. Grayson, D. C. Montgomery, and G. C. Runger (1996). A review of statistical process control techniques for short run manufacturing systems. *Communications in Statistics—Theory and Methods 25*(11): 2723–2737.

Desmond, D. J. (1961). The testing of ceiling fans. *Quality Engineering 25*(3): 77–80.

Doornbos, R. (1959). Efficient weight control of margarine packets. *Statistica Neerlandica 13*(3): 323–328.

Duffy, D. J. (1960). A control chart approach to manufacturing expense. *Journal of Industrial Engineering 11*(6): 451–458.

Duncan, A. J. (1956). The economic design of $\overline{X}$-charts used to maintain current control of a process. *Journal of the American Statistical Association 51*(274): 228–242.

Duncan, J. M. (1957). Statistical quality control in a petroleum control laboratory. *ASTM Bulletin No. 219*, 40–43 (January).

Enrick, N. L. (1956). Survey of control chart applications in textile processing. *Textile Research Journal 26:* 313–316.

Enrick, N. L. (1973). Control charts in operations research. *ASQC Annual Quality Conference Transactions*, 17–26.

Erdman, E. J. and L. E. Bailey (1969). The production line SQC helped. *Quality Progress* 2(8): 20–22.

Farnum, N. (1992). Control charts for short runs: Nonconstant process and measurement error. *Journal of Quality Technology 24:* 138–144.

Field, E. G. (1957). How to determine yarn strength control limits. *Textile Industry 121:* 109–112.

Finison, L. J., M. Spencer, and K. S. Finison (1993). Total quality measurement in health care: Using individuals charts in infection control. In *ASQC Quality Congress Transactions*, pp. 349–359. Milwaukee, WI: American Society for Quality Control.

Gadzinski, C. and R. W. Hooley (1957). The control of magnesium alloy castings. *Industrial Quality Control 14*(5): 14–19.

Geary, R. C. (1947). Testing for normality. *Biometrika 34:* 209–242.

Gebler, S. H. (1990). Interpretation of quality-control charts for concrete production. *ACI Materials Journal 87*(4): 319–326.

Girshick, M. A. and H. Rubin (1952). A Bayes' approach to a quality control model. *Annals of Mathematical Statistics 23:* 114–125.

Griffith, G. K. (1996). *Statistical Process Control for Long and Short Runs*, 2nd ed. Milwaukee, WI: Quality.

Gruska, G. F. and M. S. Heaphy (1991). Stop light control—revisited. *American Society for Quality Control Statistics Division Newsletter 11:* 11–12.

Hackl, P. and J. Ledolter (1991). A control chart based on ranks. *Journal of Quality Technology 23:* 117–124.

Hahn, G. J. and M. B. Cockrum (1987). Adapting control charts to meet practical needs: A chemical processing application. *Journal of Applied Statistics 14:* 33–50.

Hamburg, J. H., D. E. Booth, and G. J. Weinroth (1996). A neural network approach to the detection of nuclear material losses. *Journal of Chemical Information and Computer Sciences 36*(3): 544–553.

Hance, L. H. (1956). Statistical quality control, a modern management tool. *Modern Textiles 37:* 61–64.

Hand, R., F. Piontek, and L. Klemka-Walden (1994). Use of statistical control charts to assess outcomes of medical care: Pneumonia in Medicare patients. *American Journal of the Medical Sciences 307*(5): 329–334.

Harrison, H. B. (1956). Statistical quality control will work on short-run jobs. *Industrial Quality Control 13*(3): 8–11.

Harrison, P. J. (1964). The use of cumulative sum (Cusum) techniques for the control of routine forecasts of product demand. *Operations Research 12*(2): 325–333.

Hart, R. F. (1984). Steel by Shewhart. *Quality 23*(6): 66–69.

Harvey, A. C. and C. Fernandes (1989). Time series modeling for count or correlated observations. *Journal of Business and Economic Statistics 7:* 407–422.

Herath, H. S. B., C. S. Park, and G. C. Prueitt (1995). Monitoring projects using cash flow control charts. *The Engineering Economist 41*(1): 27.

Hillier, F. S. (1969). $\overline{X}$- and R-chart control limits based on a small number of subgroups. *Journal of Quality Technology 1:* 17–26.

Ho, C. and K. E. Case (1994). Economic design of control charts, a literature review for 1981–1991. *Journal of Quality Technology 26*(1): 39–53.

Hogg, R. V. and A. T. Craig (1978). *Introduction to Mathematical Statistics*, 4th ed. New York: Macmillan.

Hoogendijk, J. (1962). Quality care of a graphic product. *Sigma 8*(1): 10–12.

Howarth, R. J. (1995). Quality control charting for the analytical laboratory: Part 1. Univariate methods. *Analyst 120:* 1851–1873.

Hull, A. M. (1956). Optimizing maintenance costs through quality control. *Industrial Quality Control 12*(2): 4–8.

Hwarng, H. B. (1991). Pattern recognition on Shewhart control charts using a neural approach. Ph.D. thesis, Arizona State University.

Jaehn, A. H. (1987). Zone control charts—SPC made easy. *Quality 26:* 51–53.

Jones, L. A. and W. H. Woodall (1998). The performance of bootstrap control charts. *Journal of Quality Technology 30*(4): 362–375.

Joshi, M. and K. Sprague (1997). Obtaining and using statistical process control limits in the semiconductor industry. In V. Czitrom and P. D. Spagon, eds. *Statistical Case Studies for Industrial Process Improvement*, Chapter 24. Philadelphia: American Statistical Association and Society for Industrial and Applied Mathematics.

Keats, J. B., E. Del Castillo, E. von Collani, and E. M. Saniga (1997). Economic modeling for statistical process control. *Journal of Quality Technology 29*(2): 144–147.

King, E. P. (1954). Probability limits for the average chart when process standards are unspecified. *Industrial Quality Control 10*(6): 62–64.

Kornetsky, A. and A. Kramer (1957). Quality control program for the processing of sweet corn. *Food Technology 11:* 188–192.

Kreft, I. J. (1980). Control charts help set firm's energy management goals. *Industrial Engineering 12*(12): 56–58.

Kroll, F. W. (1957). Effective quality control program for the industrial control laboratory. *Statistical Methods in the Chemical Industry, ASQC*, pp. 1–14 (January 12).

Laubscher, N. F. (1996). A variance components model for statistical process control. *South African Statistical Journal 30:* 27–47.

Latzko, W. J. (1977). Statistical quality control of MICR documents. *ASQC Annual Quality Conference Transactions*, 117–123.

Latzko, W. J. (1986). *Quality and Productivity for Bankers and Financial Managers*. New York: Marcel Dekker.

Ledolter, J. and A. Swersey (1997). An evaluation of pre-control. *Journal of Quality Technology 29:* 163–171.

Lee, E. P. (1956). A statistical quality control approach to weights and measures. *ASQC National Convention Transactions*, 543–551.

Levey, S. and E. R. Jennings (1992). Historical perspectives: The use of control charts in the clinical laboratory. *Archives of Pathology and Laboratory Medicine 116*(7): 791–798.

Linderman, K. and T. E. Love (2000). Economic and economic-statistical designs for MEWMA control charts. *Journal of Quality Technology* (to appear).

Littauer, S. B. (1956). The application of statistical control techniques to the study of industrial and military accidents. *Transactions of the New York Academy of Sciences, Series II 18*(3): 272–277.

Llewellyn, R. W. (1960). Control charts for queuing applications. *Journal of Industrial Engineering 11*(4): 332–335.

Lobsinger, D. L. (1954). Some administrative attributes of SQC. *Industrial Quality Control 10*(6): 20–24.

Lobsinger, D. L. (1957). Application of statistical quality control in administrative areas of a company. *TAPPI 40:* 209A–211A.

Longnecker, M. T. and T. P. Ryan (1991). A deficiency for residuals charts for correlated data. Technical Report No. 131, Department of Statistics, Texas A&M University.

Longnecker, M. T. and T. P. Ryan (1992). Charting correlated process data. Technical Report No. 166, Department of Statistics, Texas A&M University.

Lorenzen, T. J. and L. C. Vance (1986). The economic design of control charts: A unified approach. *Technometrics 28*(1): 3–10.

Maguire, M., ed. (1999). Statistical gymnastics revisited: A debate on one approach to short-run control charts. *Quality Progress 32*(2): 84–94.

Maki, R. G. and M. R. Milota (1993). Statistical quality control applied to lumber drying. *Quality Progress 26*(12): 75–79.

Mandel, B. J. (1975). C-chart sets mail fines. *Quality Progress 8*(10): 12–13.

Mansfield, E. and H. H. Wein (1958). A regression control chart for costs. *Applied Statistics 7*(2): 48–57.

Mason, R. L., C. W. Champ, N. D. Tracy, S. Wierda, and J. C. Young (1997). Assessment of multivariate process control techniques. *Journal of Quality Technology 29*(2): 140–143.

Meddaugh, E. J. (1975). The bias of cost control charts toward type II errors. *Decision Sciences 6*(2): 367–382.

Melvin, C. A. (1993). Application of control charts to educational systems. *Performance Improvement Quarterly 6*(3): 74.

Meyer, J. J. (1963). Statistical sampling and control and safety. *Industrial Quality Control 19*(12): 14–17.

Meyer, T. R., J. H. Zambone, and F. L. Curcio (1957). Application of statistical literature survey. *Industrial Quality Control 19*(2): 21–24.

Michelson, D. K. (1994). *Statistical Process Control for Correlated Data*. Ph.D. dissertation, Department of Statistics, Texas A&M University, College Station, TX.

Montgomery, D. C. (1980). The economic design of control charts: A review and literature survey. *Journal of Quality Technology 12*(2): 75–87.

Montgomery, D. C. (1982). Economic design of an $\overline{X}$ control chart. *Journal of Quality Technology 14*(1): 40–43.

Montgomery, D. C. (1996). *Introduction to Statistical Quality Control*, 3rd ed. New York: Wiley.

Mortell, R. R. and G. C. Runger (1995). Statistical process control of multiple stream processes. *Journal of Quality Technology 27*(1): 1–12.

Mullins, E. (1994). Introduction to control charts in the analytical laboratory. *Analyst 119:* 369–375.

Nandram, B. and B. Cantell (1999). A Bayesian multivariate control chart with a parsimonious covariance. Manuscript.

Nelson, L. S. (1986). Control chart for multiple stream processes. *Journal of Quality Technology 18*(4): 255–256.

Newchurch, E. J., J. S. Anderson and E. H. Spencer (1956). Quality control in petroleum research laboratory. *Analytical Chemistry 28:* 154–157.

Nieckula, J. and O. Hryniewicz (1997). Neural network support for Shewhart X-bar control chart. *Systems Science 23*(1): 61–75.

Osinski, R. V. (1962). Use of median control charts in the rubber industry. *Industrial Quality Control 19*(2): 5–8.

Oxenham, J. P. (1957). An application of statistical control techniques to a post office. *Industrial Quality Control 14*(3): 5–10.

Park, C. (1998). Design of $\overline{X}$ and EWMA charts in a variance components model. *Communications in Statistics—Theory and Methods, 27*(3): 659–672.

Patte, W. E. (1956). General techniques in pulp and paper mills. *ASQC National Convention Transactions*, 185–193.

Prabhu, S. S., G. C. Runger, and J. B. Keats (1993). An adaptive sample size $\overline{X}$ chart. *International Journal of Production Research 31:* 2895–2909.

Prabhu, S. S., D. C. Montgomery, and G. C. Runger (1994). A combined adaptive sample size and sampling interval $\overline{X}$ control scheme. *Journal of Quality Technology 26:* 164–176.

Preston, F. W. (1956). A quality control chart for the weather. *Industrial Quality Control 12*(10): 4–6.

Pringle, J. B. (1962). SQC methods in telephone transmission maintenance. *Industrial Quality Control 13*(2): 6–11.

Quesenberry, C. P. (1993). The effect of sample size on estimated limits for $\overline{X}$ and X charts. *Journal of Quality Technology 25*(4): 237–247.

Reynolds, M. R., Jr. (1995). Evaluating properties of variable sampling interval control charts. *Sequential Analysis 14:* 59–97.

Reynolds, M. R., Jr., R. W. Amin, J. C. Arnold, and J. A. Nachlas (1988). $\overline{X}$ charts with variable sampling intervals. *Technometrics 30:* 181–192.

Reynolds, M. R., Jr., R. W. Amin, and J. C. Arnold (1990). CUSUM charts with variable sampling intervals. *Technometrics 32:* 371–384 (discussion: 385–396).

Rhodes, W. L. and J. F. Petrycki (1960). Experience with control charts on ink-film thickness and sharpness. *TAPPI 43:* 429–433.

Rodriguez, R. N. (1996). Health care applications of statistical process control: Examples using the SAS System. In *Proceedings of the 21st Annual Conference*, pp. 1381–1396. SAS Users Group International (SUGI). Cary, NC: SAS.

Roes, K. C. B. and R. J. M. M. Does (1995). Shewhart-type charts in nonstandard situations. *Technometrics 37*(1): 15–24 (discussion: 24–40).

Runger, G. C. (1998). Control charts for autocorrelated data: Observations or residuals? Manuscript.

Runger, G. C., T. R. Willemain, J. M. Grayson, and W. W. Messina (1997). Statistical process control for massive data sets. Paper presented at the 41st Annual Fall Technical Conference, Baltimore, MD.

Ryan, T. P. (1991). Discussion (of "Some statistical process control methods for autocorrelated data" by D. C. Montgomery and C. M. Mastrangelo). *Journal of Quality Technology 23*(3): 200–202.

Salvia, A. A. (1988). Stoplight control. *Quality Progress 21:* 39–42.

Sandon, F. (1956). A regression control chart for use in personnel selection. *Applied Statistics 5*(1): 20–31.

Saniga, E. M. (1989). Economic statistical control-chart designs with application to $\overline{X}$ and R charts. *Technometrics 31:* 313–320.

Saniga, E. M., D. J. Davis, and T. P. McWilliams (1995). Economic, statistical, and economic statistical design of attributes charts. *Journal of Quality Technology 27:* 56–73.

Satterthwaite, F. E. (1954). A simple, effective, process control method. Report 54-1, Rath & Strong, Boston.

Schiesel, E. E. (1956). Statistical cost control and analysis. *ASQC National Convention Transactions*, 553–558.

Schramm, W. R. and L. E. Freund (1993). Application of economic control charts by a nursing modeling team. Henry Ford Hospital applies techniques to solve aggregate planning problems. *Industrial Engineering 25*(4): 27–31.

Schreiber, R. J. (1956). The development of engineering techniques for the evaluation of safety programs. *Transactions of the New York Academy of Sciences, Series II 18*(3): 266–271.

Schumacher, R. B. F. (1976). Quality control in a calibration laboratory, Part II. *Quality Progress 9*(2): 16–20.

Shainin, D. (1984). Better than good old $\overline{X}$ and R charts asked by vendees. In *ASQC Quality Congress Transactions*, 302–307. Milwaukee, WI: American Society for Quality Control.

Shainin, D. and P. Shainin (1989). Pre-control vs. $\overline{X}$ and R charting: Continuous or immediate quality improvement? *Quality Engineering 1:* 419–429.

Steiner, S. H. (1997). Pre-control and some simple alternatives. *Quality Engineering 10*(1): 65–74.

Sullivan, J. and W. H. Woodall (1996). A control chart for preliminary analysis of individual observations. *Journal of Quality Technology 28*(3): 265–278.

Tracy, N. D., J. C. Young, and R. L. Mason (1992). Multivariate control charts for individual observations. *Journal of Quality Technology 24:* 88–95.

Traver, R. W. (1985). Pre-control: A good alternative to $\overline{X}$ and R charts. *Quality Progress 17:* 11–14.

Vance, L. C. (1983). A bibliography of statistical quality control chart techniques, 1970–1980. *Journal of Quality Technology 15*(2): 59–62.

Vardeman, S. and J. A. Cornell (1987). A partial inventory of statistical literature on quality and productivity through 1985. *Journal of Quality Technology 19*(2): 90–97.

von Collani, E. (1986). A simple procedure to determine the economic design of an $\overline{X}$ control chart. *Journal of Quality Technology 18:* 145–151.

von Collani, E. (1988). A unified approach to optimal process control. *Metrika 35:* 145–159.

von Collani, E. (1989). *The Economic Design of Control Charts.* Stuttgart: Teubner.

Walter, J. T. (1956). How reliable are lab analyses? *Petroleum Refiner 35:* 106–108.

Wardell, D. G., H. Moskowitz, and R. D. Plante (1994). Run-length distributions of special-cause charts for correlated processes. *Technometrics 36*(1): 3–17 (discussion: 17–27).

Wasserman, G. S. (1993). Short run SPC based upon the second order dynamic linear model for trend detection. *Communications in Statistics: Simulation and Computation 22*(4): 1011–1036.

Wasserman, G. S. (1995). An adaptation of the EWMA chart for short run SPC. *International Journal of Production Research 33*(10): 2821–2833.

Way, C. B. (1961). Statistical quality control applications in the food industry. *Industrial Quality Control 17*(11): 30–34.

Wheeler, D. J. (1991). *Short Run SPC.* Knoxville, TN: SPC Press.

Williams, W. W., S. W. Looney, and M. H. Peters (1985). Use of curtailed sampling plans in the economic design of *np*-control charts. *Technometrics 27:* 57–63.

Wisnowski, J. W. and J. B. Keats (1999). Monitoring the availability of assets with binomial and correlated observations. *Quality Engineering 11*(3): 387–393.

Woodall, W. H. (1986). Letter to the editor. *Technometrics 28:* 408–409.

Woodall, W. H. (1987). Conflicts between Deming's philosophy and the economic design of control charts. In H.-J. Lenz, G. B. Wetherill, and P.-Th. Wilrich, eds. *Frontiers in Statistical Quality Control,* Vol. 3, pp. 242–248. Würzburg, Germany: Physica-Verlag.

Woodall, W. H. (1997). Control charts based on attribute data: Bibliography and review. *Journal of Quality Technology 29*(2): 172–183.

Woodall, W. H. and E. V. Thomas (1995). Statistical process control with several components of common cause variability. *IIE Transactions 27:* 757–764.

Yang, C.-H. and F. S. Hillier (1970). Mean and variance control chart limits based on a small number of subgroups. *Journal of Quality Technology 2:* 9–16.

Yegulalp, T. M. (1975). Control charts for exponentially distributed product life. *Naval Research Logistics Quarterly 22*(4): 697–712.

Zhang, N. F. (1997). Detection capability of residual control chart for stationary process data. *Journal of Applied Statistics 24*(4): 475–492.

EXERCISES

1. Assume normality, known parameter values, and that the process mean is halfway between the specification limits. When pre-control is used to determine if a process is capable, what is the probability of concluding that the process is capable if $C_p = 1.50$?
2. What is the probability of detecting a shift from μ to $\mu + \sigma$ if $C_p = 1.67$? (Again assume normality, a centered process, and known parameters.) What would be the corresponding probability if an $\overline{X}$ were used with $n = 2$? Comment.
3. Explain why no short-run SPC approach could be expected to have good properties in the absence of assumptions such as that parameter values are known.
4. Explain why it is natural to consider Bayesian methods when control charts are used.
5. List four application areas of control charts for which it might be practical to use k-sigma limits with, say, k values of 2 or 2.5 rather than 3.
6. Explain why a control chart of residuals will have poor properties when applied to data from an AR(1) process with a positive autocorrelation.

PART III

Beyond Control Charts: Graphical & Statistical Methods

CHAPTER 11

Other Graphical Methods

The cliché "a picture is worth a thousand words" could perhaps be modified to "a picture is often better than several numerical analyses" when adapted to the field of statistics. In this chapter we present some additional graphical tools for displaying data, including some powerful methods that have gained acceptance within the past 25 years. Recall that histograms and scatter plots were presented in Chapter 2 and control charts have been covered in other chapters.

The book *How to Lie with Statistics* (Huff, 1954) is replete with examples of misleading graphical displays that can serve to remind us of the much-quoted statement of Disraeli concerning different types of lies and statistics. The problem, of course, is not with statistics but rather in the way in which they are used. Football experts have been prone to say that statistics are for losers, but *all* of the game statistics generally indicate why a team was the loser. One issue that must be faced when using graphical displays is that there is generally not going to be one right way to proceed in terms of choosing appropriate graphical methods or in terms of how a particular method is used.

The graphical tools discussed in this chapter all have value as stand-alone procedures but can and should be used in conjunction with the statistical techniques discussed in the succeeding chapters, as well as with control charts.

The numerous control charts illustrated in Part II are graphical displays of a two-dimensional type, in which whatever is being charted is generally graphed against the sample number (i.e., against time).

In general, we would want a graphical display to provide us with answers to questions such as the following: What is the general shape of the distribution of the data? Is it close to the shape of a normal distribution or is it markedly nonnormal? Are there any numbers that are noticeably larger or smaller than the rest of the numbers?

11.1 STEM-AND-LEAF DISPLAY

A stem-and-leaf display is one of the newer graphical techniques alluded to in the first paragraph of this chapter. It is one of many techniques that are generally referred to as exploratory data analysis (EDA) methods, as popularized in the book on EDA by Tukey (1977).

We want our graphical displays to be reasonably compact but not to sacrifice important details about the data while providing a summarization of the data. With a histogram we lose information about the individual values after we put them into classes and display the histogram. A stem-and-leaf display, however, provides us with essentially the same information as a histogram *without* losing the individual values.

There are many different ways to create a stem-and-leaf display, depending upon the type of data that are to be displayed and what the user wishes to show. We shall illustrate a basic display using the data in Table 2.1 and compare the result with the histogram that was given in Figure 2.2. The basic idea is to display "leaves" that correspond to a common "stem" and to do this in such a way that the display does not require much space. With a two-digit number, the choice for the stem is the first digit, and the second digit is for the leaf. This leads to the stem-and-leaf display in Figure 11.1.

In Figure 11.1 the numerals to the left of the vertical line represent 10's. The numerals to the right of the vertical line represent, when combined with the 10's digit, the actual numbers that are distributed within that range (e.g., 20–29). Thus if we look at the "2" line, we see 1 (representing 21), 3 (representing 23), 4 (representing 24), and so on. These are taken from the actual data values in Table 2.1. If a data value occurred more than once, the numeral is repeated the number of times that it occurred in the sample. For example, in Table 2.1 there are two data values of 32; thus on the "3" line of the stem-and-leaf chart there are two numeral 2's to the right of the vertical line—each representing 32.

We can observe that this particular stem-and-leaf display requires roughly the same amount of space as the corresponding histogram, and when turned on its side has exactly the same shape as the histogram. Thus, it provides the same information as a histogram, but with the advantage that it also allows us to see the individual values. Obviously this provides us with better insight into the data and the data collection procedures. For example, if we look at the stem "6" and the corresponding leaves, we can see that all of the digits are listed twice except for the 4 and the 7. If this had been actual data, we would certainly want to question those numbers, as we would expect to see more variation in the digit counts.

```
2 | 1 3 4 5 7 8
3 | 0 1 2 2 3 3 4 5 6 7 9
4 | 0 0 1 1 2 2 3 3 4 5 5 6 6 7 8 9 9
5 | 0 0 1 1 1 1 2 2 2 2 3 3 4 4 5 5 5 6 6 6 7 7 8 8 8 8 9 9 9
6 | 0 0 1 1 2 2 3 3 4 4 4 5 5 6 6 7 8 8 9 9
7 | 0 1 2 3 4 5 6 6 8 9
8 | 1 2 3 4 5 7
```

FIGURE 11.1 Stem-and-leaf display of the data in Table 2.1.

Velleman and Hoaglin (1981, p. 14) discuss an example in which the pulse rates of 39 Peruvian Indians were displayed in a histogram and in a stem-and-leaf display. The latter revealed that all of the values except one were divisible by 4, thus leading to the conjecture that 38 of the values were obtained by taking 15-second readings and multiplying the results by 4, with the other value obtained by doubling a 30-second reading resulting (perhaps) from missing the 15-second mark. Thus, the stem-and-leaf display provided some insight into how the data were obtained, whereas this information was not provided by the histogram.

An interesting industrial use of stem-and-leaf displays is described by Godfrey (1985), who relates that the Japanese train schedules are actually stem-and-leaf displays in which the hours are the stems and the leaves are the minutes, and the display runs in two directions—one for arrivals and the other for departures.

Another example is the Atlanta phone directory, which took the general form of a stem-and-leaf display starting with the December 1985 issue. (Of course, the phone directories for various other cities are now constructed the same way.) Instead of each last name being listed as many times as there are phone numbers of people with that last name, the last name was listed only once. Thus, the last name was the stem, and the leaves were the various combinations of the first two names and initials.

Readers interested in further reading on stem-and-leaf displays are referred to Velleman and Hoaglin (1981) and Emerson and Hoaglin (1983).

11.2 DOT DIAGRAMS

Another way to display one-dimensional data is through the use of dot diagrams, which have also been termed one-dimensional scatter plots [Chambers, Cleveland, Kleiner, and Tukey (1983)]. The first label seems slightly better; it will be used in Chapter 13 and is also used in Box et al. (1978).

A dot diagram is simply a one-dimensional display in which a dot is used to represent each point. For example, Figure 11.2 is a dot diagram of the numbers 60, 20, 24, 10, 12, 17, 26, 35, 42, 50, and 87. The dot diagram portrays the relationship between the numbers and, in this instance, allows us to see the separation between the number 87 and the rest of the numbers. This might cause us to question whether or not that number was recorded correctly. If so, the number might be classified as an *outlier*; an observation that is far removed from the main body of the data. (The classification of outliers is somewhat subjective. If the distribution of 100 or so numbers is reasonably symmetric, we might classify a number as an outlier if it differs from $\bar{x}$ by more than $3s$, where $\bar{x}$ and s are the average and standard deviation, respectively, of the numbers.)

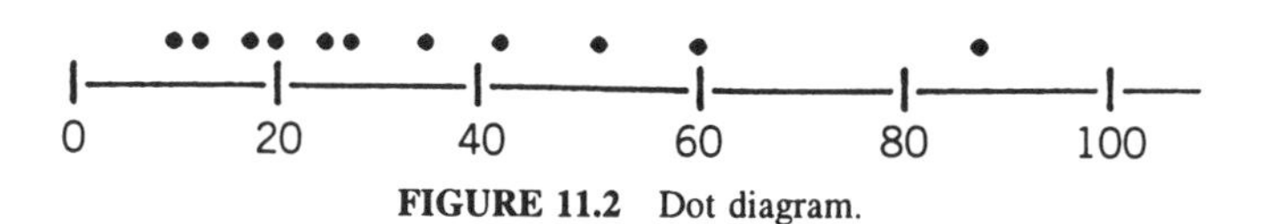

FIGURE 11.2 Dot diagram.

A dot diagram does have some limitations, however. In particular, we would not want to use it if we have a large number of observations. Box et al., (1978, p. 25) suggest at most 20 observations, as the dots would tend to run together and give rather poor resolution. We also cannot accurately determine the individual values from the diagram. It simply shows us the relationship between the numbers for a small set of numbers.

11.2.1 Digidot Plot

This plot, due to Hunter (1988), is a combination of a time sequence plot and a stem-and-leaf display. If a data set contains a time element, it is lost when a stem-and-leaf display is constructed. Since it is almost always useful to look at time sequence plots, it is desirable to add a time sequence feature to a stem-and-leaf display. Figure 11.3 is a digidot plot for the numbers 15.5, 16.6, 16.7, 17.2, 17.3, 18.9, 18.1, 18.4, 18.6, 19.2, 19.3, 19.2, and 15.0. Notice that the ordering in the stem-and-leaf display is determined by the time sequence, not by numerical order.

11.3 BOXPLOT

We should perhaps start by pointing out that this type of display was not named for G. E. P. Box. Rather, it is another EDA tool that was introduced and popularized by Tukey (1977). It derives its name from the fact that the middle half of a set of data is depicted by the area between the top and bottom of a box (rectangle). It is an easily constructed display that is routinely used.

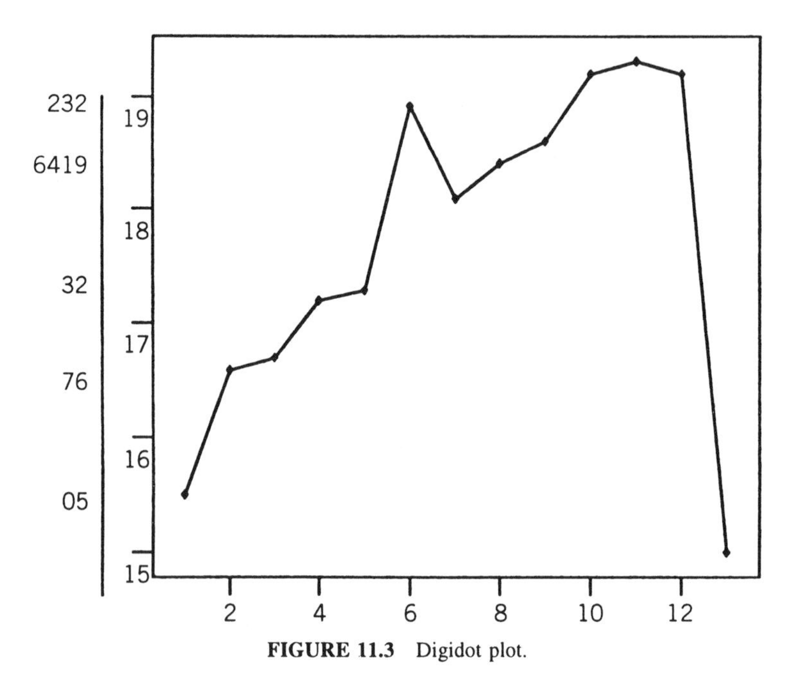

FIGURE 11.3 Digidot plot.

There are several ways to construct a boxplot; we shall begin with what has been termed a *skeletal boxplot* by Velleman and Hoaglin (1981, p. 66). Assume that we have a sample of 16 numbers as follows: 18, 19, 23, 24, 26, 29, 31, 33, 35, 37, 39, 40, 42, 45, 47, and 49. Notice that these numbers have been placed in ascending order and that there are no large gaps between any of the numbers.

The first step in constructing the box is to calculate the median, which was defined in Chapter 3 as the middle value when there is an odd number of observations and the average of the two middle values when there is an even number of observations. Here we have an even number so the median is 34 (the average of 33 and 35). The next step is to compute the two *hinges*, which are equivalent to the two medians for each half of the data when there is an even number of observations but cannot be explained quite so simply when there is an odd number of observations. This is illustrated later.

We can observe that the median of the first eight numbers is $25[(24+26)/2]$, and the median for the second eight numbers is $41[(40+42)/2]$. These two "hinges" determine the top and bottom of the box, and the overall median is indicated by a horizontal line running across the box. A vertical line is then drawn from the top of the box to a point representing the largest value, and a line is similarly drawn down from the bottom of the box to reach the smallest value. The skeletal boxplot for these 16 values is shown in Figure 11.4.

The fact that the two vertical lines are short relative to the length of the box indicates that there are no values that are either much larger or much smaller than the other values. Also, the midline being slightly closer to the top than to the bottom of the box indicates that the numbers above the median are, as a group, closer to the median than the group of numbers below the median. The box will generally contain roughly half of the observations, so the display also indicates how the middle half of the data are dispersed. We can see from Table B in the Appendix to the book that for a normal distribution roughly 50% of the data are contained within $\mu \pm 0.67\sigma$. Thus, for a normal distribution the length of the box will be close to 1.34σ. Therefore, the display could also

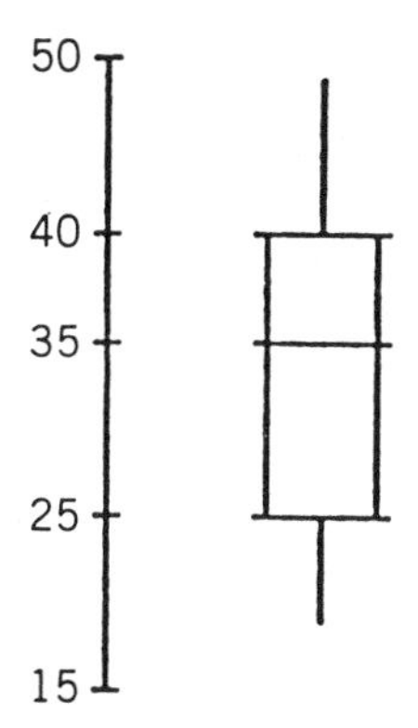

FIGURE 11.4 Skeletal boxplot for 16 numbers.

provide us with a rough estimate of σ for data that are approximately normally distributed.

The formal name for the plot in Figure 11.4 is a "box-and-whiskers plot," with the vertical lines being thought of as representing whiskers. The shortened name is what is generally used, however.

When there is an odd number of observations, the two hinges cannot be explained quite as simply and intuitively as when there is an even number. If we think about assigning the median to both halves of the data (only for the purpose of explanation), we would then have an even number of observations in each half. The hinges would then be the averages of the two middle values in each half. To illustrate, if we were to modify the previous example by deleting the 49 so as to produce 15 observations, the hinges would then be 25 and 39.5. Notice that only the upper hinge changes since the largest value is the one that is being deleted.

Some authors have presented a boxplot in which the bottom and the top of the box are the first and third quartiles, respectively (i.e., the 25th and 75th percentiles). [See, e.g., Chambers et al. (1983, p. 21).] The endpoints of the box should be about the same when they are obtained from the quartiles (for the 15 observations the quartiles are 24.5 and 39.75), but the quartiles require more work to obtain. Thus, if the necessary calculations are being performed with pencil and paper (which is what EDA tools were originally intended for), the use of hinges is preferable, whereas with a computer it really does not make much difference. (Of course, if there are more than 20 or so numbers, the use of a computer becomes essential since the data have to be sorted into ascending order before the plot can be constructed.)

There are other variations of boxplots that are more sophisticated and thus provide more information than skeletal boxplots. Two uses of these other types of boxplots deserve special mention: (1) the determination of outliers, and (2) the comparison of groups. The latter will be illustrated in the chapter on design of experiments (Chapter 13); the former can be illustrated as follows.

Following Velleman and Hoaglin (1981, p. 68), we will call the difference between the two hinges the *H-spread*. An observation is then considered to be an outlier if it either exceeds the upper hinge plus (1.5 × H-spread) or is less than the lower hinge minus (1.5 × H-spread). Remembering that for a normal distribution (1) mean = median, (2) H-spread $\doteq 1.34\sigma$, and (3) the distance from each hinge to the mean is approximately equal to 0.67σ, it then follows that this rule for classifying outliers is roughly equivalent to classifying a value as an outlier if it is outside the interval $\mu \pm 2.68\sigma$. (Notice that this is in general agreement with the statement made earlier in the chapter.)

To illustrate, we shall modify the original example with 16 observations by adding one very small number (2) and one very large number (75). The 18 numbers are thus 2, 18, 19, 23, 24, 26, 29, 31, 33, 35, 37, 39, 40, 42, 45, 47, 49, and 75. The question to be addressed at this point is whether or not these two additional values would be classified as outliers. If we let UH = upper hinge,

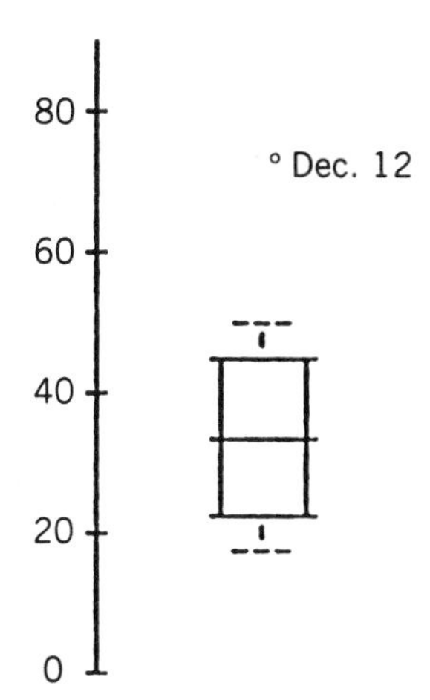

FIGURE 11.5 Boxplot for 18 numbers.

LH = lower hinge, and H-S = H-spread, we thus have

$$\begin{aligned} &\text{UH} + (1.5 \times \text{H-S}) \\ &\quad = 42 + (1.5 \times 18) \\ &\quad = 69 \end{aligned} \qquad \begin{aligned} &\text{LH} - (1.5) \times \text{H-S}) \\ &\quad = 24 - (1.5 \times 18) \\ &\quad = -3 \end{aligned}$$

Since 75 exceeds 69, we would classify the former as an outlier and try to determine whether or not it is a valid data point. These two boundary values, −3 and 69, are termed *inner fences* by Velleman and Hoaglin (1981, p. 68). Their *outer fences* are obtained by using 3.0 rather than 1.5 as the multiplier of H-S, which is equivalent to using boundaries of approximately $\mu \pm 4.69\sigma$ for data that are roughly normally distributed.

The boxplot for the 18 observations is shown in Figure 11.5. The dashed horizontal lines represent the two values 18 and 49, which are the most extreme values of those that lie within the inner fences. (The latter are not shown on the plot.) The value 75 that is outside the inner fences is designated by a circle or some other symbol, and identifying information is provided so that the validity of the value can be investigated.

11.4 NORMAL PROBABILITY PLOT

Most of the statistical procedures used in quality improvement work are based on the assumption that the population of data values from which a sample is obtained is approximately normally distributed. There are various ways to check this assumption; one such method is to construct a *normal probability plot*, which was discussed briefly in Chapter 2. This is easy to do with appropriate computer software but a bit laborious to do by hand. Nevertheless, we shall discuss and illustrate hand-generated normal plots so that readers who will be using software will have some understanding of how the plots are generated.

Normal probability paper is generally used for hand-constructed normal plots, although such paper is by no means essential. We shall not illustrate the construction of normal paper since such paper is readily available. The best source for obtaining normal probability paper and probability paper for other distributions is probably TEAM (Technical Engineering Aids for Management).* Readers interested in the construction of normal probability paper are referred to Nelson (1976).

We shall concentrate on explaining the meaning of the points that are plotted and then plot a set of such points without using normal paper. We would generally want to have at least 20 observations before constructing a normal probability plot, but for the sake of simplicity, we will use a sample consisting of the following 10 observations: 8, 10, 12, 14, 15, 15, 16, 18, 19, and 20. Could these data have come from a normal distribution with some (unknown) values of μ and σ? This is the question that we wish to address. We are not interested in estimating μ and σ from a normal plot, although that can be done. Remembering from Chapter 3 that we can always transform a random variable X where $X \sim N(\mu, \sigma^2)$ into a random variable Z where $Z \sim N(0, 1)$, we wish to determine what the z-values would be if the sample data were *exactly* normally distributed. We should not expect the sample data to be exactly normally distributed, however, as there will be sampling variability even if the population were normally distributed. Thus, even when the latter is true, there will generally not be an exact linear relationship between the sample values and the theoretical z-values, but when they are plotted against each other, the result should be something close to a straight line.

We will define a plotting position p_i for the ith ordered observation as $p_i = (i - 0.5)/n$, where n is the number of observations. (It might seem more logical to use i/n, which would then give the exact fraction of the sample values that are less than or equal to the ith ordered value, but when $i = n$, the fraction would be 1.0, and the z-value that corresponds to an area of 1.0 under the standard normal curve is plus infinity, i.e., not a real number.) We then compute the z-value that would correspond to a cumulative area of p_i under the standard normal curve. These z-values are sometimes called "inverse (standard) normal values" and could be obtained (approximately) by interpolating in Table B in the Appendix to the book. For example, $p_i = (1 - 0.5)/10 = 0.05$ corresponds to $z = -1.645$. The other values are given in Table 11.3. (It should be noted that other expressions for p_i are also used, but they differ very little from what is given here.)

We can observe that the differences between the z-values are not always equal for constant differences between the sample values (e.g., the difference between 8 and 10 is the same as the difference between 10 and 12, but the

---

* Catalogue and price list are available from TEAM, Box 25, Tamworth, NH 03886

TABLE 11.1 Values for a Normal Probability Plot

Sample Value	p	z
8	0.05	−1.645
10	0.15	−1.036
12	0.25	−0.674
14	0.35	−0.385
15	0.45	−0.126
15	0.55	0.126
16	0.65	0.385
18	0.75	0.674
19	0.85	1.036
20	0.95	1.645

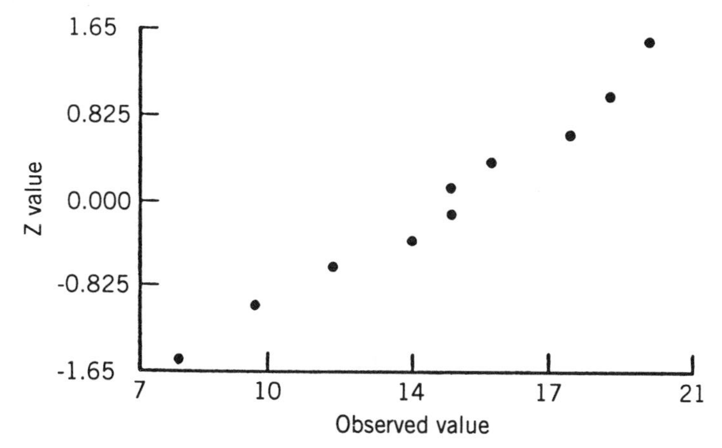

FIGURE 11.6 Normal probability plot of the data in Table 11.1.

differences in the corresponding z-values are unequal.) Thus, we will not obtain a straight line when we plot the sample values against the z-values, but closer observation of the differences between adjacent pairs for all 10 numbers reveals that it should be fairly close to a straight line. The actual plot is shown in Figure 11.6.

It is worth noting that there are various forms of a normal probability plot, with the variations resulting from different quantities for the ordinate of the plot. For example, Figure 11.7 is a normal probability plot of the same data produced by Minitab, Release 12. Notice that "cumulative probability" serves as the ordinate, with the unequal distances for equal differences in cumulative probabilities reflecting the shape of a normal distribution. Notice also that it is useful to have

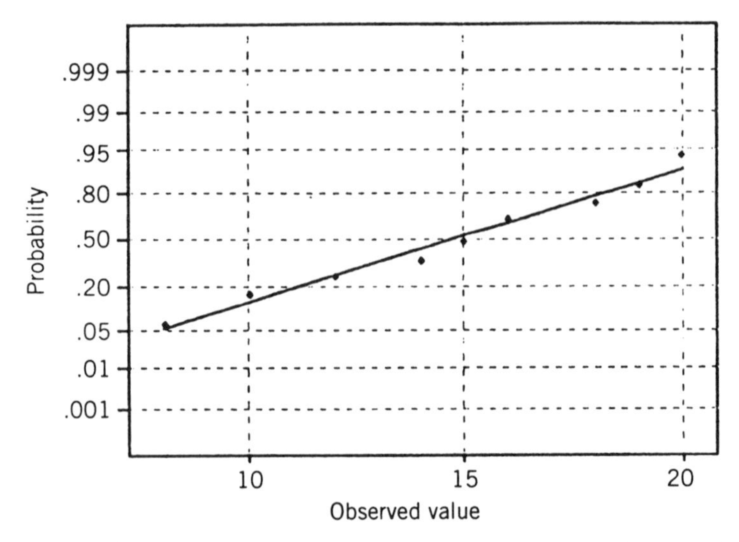

Average: 14.7
StDev: 3.86005
N: 10

Anderson–Darling normality test
A^2: 0.185
p-Value: 0.878

FIGURE 11.7 Computer-generated normal probability plot.

the results of a test of non-normality, in this case the Anderson–Darling test, printed with the normal probability plot.

A variation of the normal plot is the half-normal plot (Daniel, 1959), which utilizes the half-normal (also called *folded normal*) distribution, with only positive z-values being used. See also Zahn (1975a,b).

11.5 PLOTTING THREE VARIABLES

One technique for displaying three-dimensional data is a *casement display*, introduced by Tukey and Tukey (1983) and discussed by Chambers et al. (1983). A casement display is a set of two-variable scatter plots. If the third variable is discrete (and possibly categorical), a scatter plot is produced for each value of that variable, and the plots assembled in logical order. If the third variable is continuous, intervals for that variable would be constructed and the scatter plots then produced.

A similar technique introduced by Tukey and Tukey (1981) is a *draftsman's display*, which is the set of the three two-variable scatter plots arranged in a particular manner.

11.6 DISPLAYING MORE THAN THREE VARIABLES

Much of the work that has been done on the display of four or more variables has been motivated by Chernoff (1973), who introduced the concept of faces for displaying multivariate data. For example, the following two faces would represent multivariate data that differ only on the variable represented by the mouth:

Other methods for displaying many variables include a *star plot*, in which the length of each ray of a star would represent the value of each of the variables. This is essentially the same idea as a *glyph* (Anderson, 1960), which has also been called a metroglyph. There are various other techniques for displaying multivariate data, including the use of *weathervanes* and *trees*. These various methods are discussed in some detail in Chambers et al. (1983) and in Chapter 9 of Wadsworth et al. (1986). Boardman (1985) discussed variations of some of these methods that can be used in plotting hourly data.

11.7 MULTI-VARI CHART

A multi-vari chart is a graphical device that is helpful in assessing variability due to three or more factors. (This should not be confused with the multivariate control charts that were discussed in Chapter 9.)

Seder (1950) presented the idea of a chart for looking at variability due to different sources, although the chart is more frequently associated with Dorian Shainin. The general idea seems to predate Seder (1950), however, as Juran (1974) points out that "the concept of the basic vertical line ... had previously been employed by J. M. Juran (and possibly others), who derived it from the method long used by financial editors for showing stock market prices".

The "basic vertical line" to which Juran refers connects the largest and smallest observations in each group. Juran (1974) gives an example of three charts that show different problems—excessive variability on a single piece, excessive piece-to-piece variability, and excessive time-to-time variability. To illustrate the latter, assume that the widths of five consecutive roller bearings that come off a production line are recorded every 20 minutes and vertical lines are used to connect the most extreme of each set of five observations. The chart appears as in Figure 11.8.

We may observe that not only is there a shift in the mean level during the first two hours, but there is also an increase in variability at the last three times relative to the variability during the first hour and 40 minutes, in addition to a mean shift. In fact, the variability has become so great that the smallest observation at 11:00 is below the lower specification limit (LSL).

Since ranges are plotted in a multi-vari chart, we might compare such a chart with an R chart. Subgroup ranges are, of course, also plotted with the latter, but

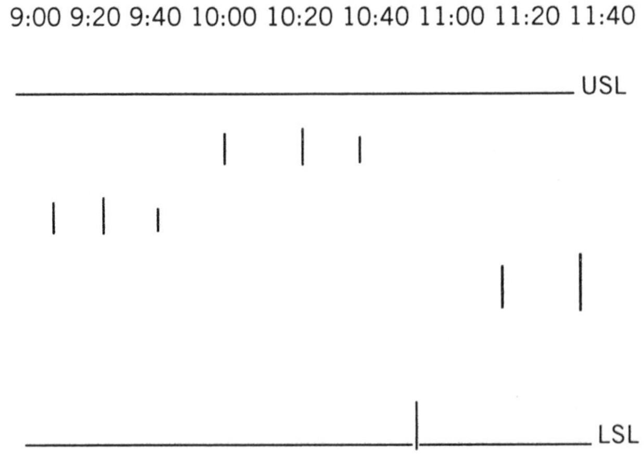

FIGURE 11.8 Multi-vari chart showing variation over time.

the ranges are plotted against control limits for ranges. The ranges in Figure 11.8 are plotted against specification limits. This does not allow us to see whether or not the ranges have become excessive relative to their inherent variability, but it does allow us to see the variability relative to the specification limits.

More than one variable can be used in a multi-vari chart, as illustrated in Zaciewski and Nemeth (1995). See also Delott and Gupta (1990) for illustration of a multi-vari chart.

Unfortunately, there are very few articles in the literature that describe applications of multi-vari charts, which leads one to believe that this is not a frequently used tool. Such an impression is also evident from the title of the paper by Zaciewski and Nemeth (1995), "The Multi-Vari Chart: An Underutilized Quality Tool."

11.8 PLOTS TO AID IN TRANSFORMING DATA

One important function of graphical methods is to provide the user of statistical methods with some insight into how data might be transformed so as to simplify the analysis. Assume that we are provided with data as in Table 11.2. If we graph these data in a scatter plot, we obtain the plot shown in Figure 11.9. Notice that only a small fraction of the horizontal axis is used with this scaling convention. Assume that our objective is to transform Y and/or X so that we have approximately a linear relationship between the new variables that is of the form $Y^* = mX^* + b$.

We can see from Figure 11.9 that Y increases sharply as X approaches 10. How do you think the plot would have looked if the 10 on the horizontal axis

TABLE 11.2
Sample Data

Y	X
9	3
15	4
24	5
3	2
36	6
65	8

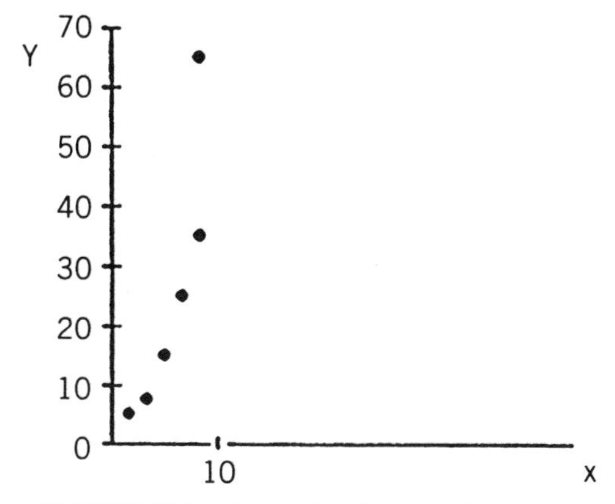

FIGURE 11.9 Scatter plot of data in Table 11.2.

had been placed at the end of the line that represents that axis? The result would be that the curve would not only be bent down considerably, but it would also appear to be straightened somewhat. The reason is that one unit on the horizontal scale would equal roughly six units on the vertical scale, and the average ratio of Y to X is 4.7 : 1 as X varies from 2 to 8. The scatter plot with this new scaling is shown in Figure 11.10.

This plot gives the false impression that there is approximately a linear relationship between Y and X, but we have not transformed either Y or X so the relationship has not changed. If we let $Y^* = \sqrt{Y}$ and $X^* = X$ (i.e., no transformation of X), we will bend the curve down by virtue of shortening the vertical axis. This will produce the data given in Table 11.3 and the scatter plot given in Figure 11.11.

Although there is not much difference between the scatter plots in Figures 11.10 and 11.11, there is obviously a considerable difference in the values of the variables plotted on the vertical axis. The scale of the vertical axis in Figure 11.11 is appropriate since the values of Y^* are virtually the same

TABLE 11.3 Transformed Data

$Y^* = \sqrt{Y}$	$X^* = X$
3	3
3.87	4
4.90	5
1.73	2
6	6
8.06	8

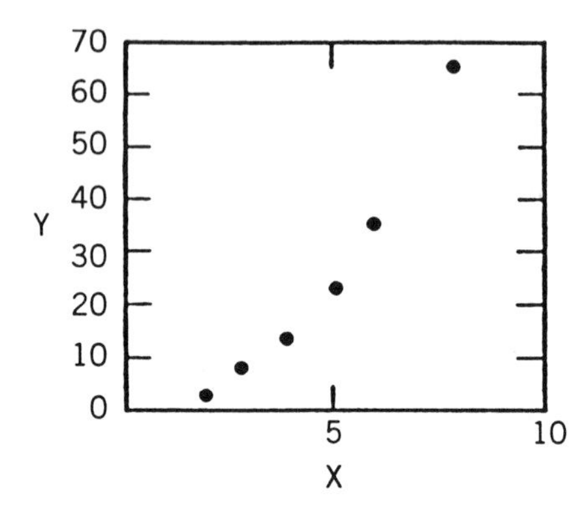

FIGURE 11.10 Scatter plot with increased horizontal scale.

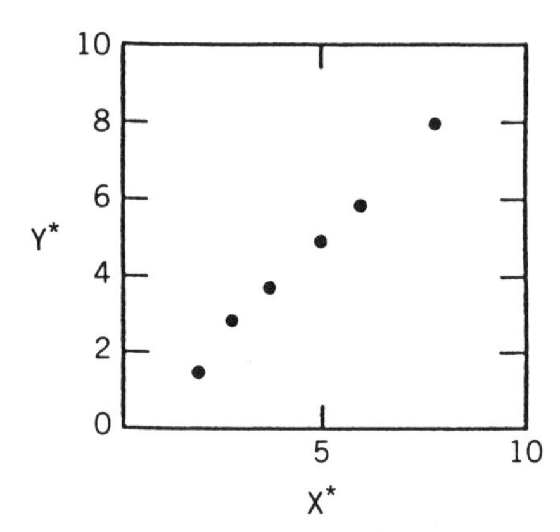

FIGURE 11.11 Scatter plot with transformed data.

as the corresponding values of X, so if the line were extended to the origin, it would form approximately a 45° line, which is what results when two variables have identical values. This, of course, is not to suggest that equal scales should always be used for the two axes, as with many sets of data this will not be possible. Rather, it should be done whenever possible and practical.

The use of transformations to straighten out curves and for other purposes is discussed in detail by Mosteller and Tukey (1977), Velleman and Hoaglin (1981), Box and Cox (1964), Emerson and Stoto (1983), and Emerson (1983).

11.9 SUMMARY

Within the past few decades many new and powerful graphical methods have been developed. Some of these are simple paper-and-pencil techniques, although others essentially require the use of a computer. With the increasing sophistication of computing devices, there is no reason why these superior techniques should not be used.

Old standbys, such as histograms and scatter plots, still have value, although stem-and-leaf displays are more informative than histograms and can be used for small to moderate amounts of data. Similarly, a conventional scatter plot can be modified in accordance with the suggestions of Tufte (1983) so as to make it more informative. The latter is an excellent treatise on the construction of good graphical displays, including the type of displays often found in newspapers and other types of "nonstatistical" publications. The text by Chambers et al. (1983) is another excellent reference that, on the other hand, emphasizes "statistical graphics" (e.g., probability plotting) rather than general principles in constructing graphical displays. The reader interested in additional reading is referred to these two books, primarily, as well as to the other references cited in this chapter.

REFERENCES

Anderson, E. (1960). A semigraphical method for the analysis of complex problems. *Technometrics* 2(3): 387–391. [Originally published in *Proceedings of the National Academy of Sciences* (1957), *13*: 923–927.]

Boardman, T. J. (1985). The use of simple graphics to study hourly data with several variables. In R. D. Snee, L. B. Hare, and J. R. Trout, eds. *Experiments in Industry: Design, Analysis, and Interpretation of Results*. Milwaukee, WI: Quality.

Box, G. E. P. and D. R. Cox (1964). An analysis of transformations. *Journal of the Royal Statistical Society, Series B 26*(2): 211–243 discussion: pp. 244–252.

Box, G. E. P., W. G. Hunter, and J. S. Hunter (1978). *Statistics for Experimenters*. New York: Wiley.

Chambers, J. M., W. S. Cleveland, B. Kleiner, and P. A. Tukey (1983). *Graphical Methods for Data Analysis*. Boston: Duxbury.

Chernoff, H. (1973). The use of faces to represent points in K-dimensional space graphically. *Journal of the American Statistical Association 68*(342): 361–368.

Daniel, C. (1959). Use of half-normal plots in interpreting factorial two-level experiments. *Technometrics 1*(4): 311–341.

Delott, C. and P. Gupta (1990). Characterization of copperplating process for ceramic substrates. *Quality Engineering 2*(3): 269–284.

Emerson, J. D. (1983). Mathematical aspects of transformation. In D. C. Hoaglin, F. Mosteller, and J. W. Tukey, eds. *Understanding Robust and Exploratory Data Analysis*, Chapter 8. New York: Wiley.

Emerson, J. D. and D. C. Hoaglin (1983). Stem-and-leaf displays. In D. C. Hoaglin, F. Mosteller, and J. W. Tukey, eds. *Understanding Robust and Exploratory Data Analysis*, Chapter 1. New York: Wiley.

Emerson, J. D. and M. A. Stoto (1983). Transforming data. In D. C. Hoaglin, F. Mosteller, and J. W. Tukey, eds. *Understanding Robust and Exploratory Data Analysis*, Chapter 4. New York: Wiley.

Godfrey, A. B. (1985). Training and education in quality and reliability—A modern approach. *Communications in Statistics, Part A (Theory and Methods) 14*(11): 2621–2638.

Huff, D. (1954). *How to Lie with Statistics*. New York: Norton.

Hunter, J. S. (1988). The digidot plot. *The American Statistician 42*(1): 54.

Juran, J. M. (1974). Quality improvement. In J. M. Juran, F. M. Gryna, Jr., and R. S. Bingham, Jr., eds. *Quality Control Handbook*, Chapter 16. New York: McGraw-Hill.

Mosteller, F. and J. W. Tukey (1977). *Data Analysis and Regression*. Reading, MA: Addison-Wesley.

Nelson, L. S. (1976). Constructing normal probability paper. *Journal of Quality Technology 8*(1): 56–57.

Seder, L. A. (1950). Diagnosis with diagrams—Part I. *Industrial Quality Control 6*(4): 11–19.

Tufte, E. R. (1983). *The Visual Display of Quantitative Information*. Cheshire, CT: Graphics Press.

Tukey, J. W. (1977). *Exploratory Data Analysis*. Reading, MA: Addison-Wesley.

Tukey, J. W. and P. A. Tukey (1983). Some graphics for studying four-dimensional data. In *Computer Science and Statistics: Proceedings of the 14th Symposium on the Interface*, pp. 60–66. New York: Springer-Verlag.

Tukey, P. A. and J. W. Tukey (1981). Graphical display of data sets in three or more dimensions. In V. Barnett, ed. *Interpreting Multivariate Data*, Chapters 10, 11, and 12. Chichester, UK: Wiley.

Velleman, P. V. and D. C. Hoaglin (1981). *ABC of EDA*. Boston: Duxbury.

Wadsworth, H. M., K. S. Stephens, and A. B. Godfrey (1986). *Modern Methods for Quality Control and Improvement*. New York: Wiley.

Zaciewski, R. D. and L. Nemeth (1995). The multi-vari chart: An underutilized quality tool. *Quality Progress 28*(10): 81–83.

Zahn, D. A. (1975a). Modifications of and revised critical values for the half-normal plot. *Technometrics 17*(2): 189–200.

Zahn, D. A. (1975b). An empirical study of the half-normal plot. *Technometrics 17*(2): 201–211.

EXERCISES

1. To see how a normal probability plot would look for nonnormal data, adding to what was shown in Section 3.5.8.1, consider the following data sets:

(a) Uniform (0, 1) ($n = 20$)	(b) t_{30} ($n = 20$)	(c) χ^2_5 ($n = 20$)
0.2639	−0.0674	8.1278
0.6985	0.2307	1.0846
0.1529	−0.0616	6.2232
0.5760	−0.7760	2.8507
0.2816	−0.5425	2.4717
0.5663	−1.0136	2.7880
0.9196	−0.8606	6.8144
0.7045	−0.2912	18.2239
0.6802	−0.6809	3.6573
0.2446	0.1260	2.4020
0.7704	0.3174	3.1241
0.9092	1.1917	6.6827
0.2409	1.6762	7.3324
0.1398	1.4463	3.5686
0.2363	−2.3747	3.9961
0.3464	0.3117	9.1297
0.2449	0.7305	7.0888
0.1580	−0.7936	3.3413
0.3487	−1.4877	5.4827
0.6722	−0.0027	3.3195

Construct a normal probability plot for each of these distributions, preferably by computer or using normal probability paper. [*Note*: A uniform (0, 1) distribution is flat and graphs as a unit square, the t distribution was covered in Chapter 3, and a chi-square distribution with five degrees of freedom is skewed (i.e., has a long tail) to the right.]

2. The following 20 numbers were generated from a normal distribution with $\mu = 20$ and $\sigma = 4$:

11.5360	22.0665	16.0396	18.8414
11.4074	16.7529	22.8638	18.3333
18.2744	23.4409	12.9792	17.8925
27.2872	26.7256	20.9274	18.3116
21.1735	18.8092	20.5048	18.1904

(a) Construct a boxplot and identify any outliers.

(b) Estimate σ using the values corresponding to the top and bottom of the box. How does your estimate compare with the known value of σ?

3. Construct a stem-and-leaf display using each of the data sets in exercise 1. Does the general shape of each display conform to what you would expect considering the nature of the distributions in 1? (Use only the integer part of the number and the first decimal place.)
4. Construct a dot diagram for the data in (b) of exercise 1. What does the diagram tell you relative to what you might have expected?
5. Construct a digidot plot for the data given at the beginning of Section 11.3, using the order in which the data were listed as the time sequence.
6. For the simulated data in exercise 1, use appropriate software (if available), and construct a half-normal plot for each of the three simulated data sets. Compare the results with the results obtained using a full normal plot.
7. Construct a boxplot for the chi-square data listed in exercise 1. Consider the plot relative to outliers. In general, what would we expect, relative to outlier highlighting, if we construct a boxplot for data from a highly skewed distribution such as a χ^2_5?

CHAPTER 12

Linear Regression

Regression analysis is one of the two most widely used statistical procedures; the other is analysis of variance, which is covered in Chapter 13.

There are various procedures within the broad area of linear regression that have direct application in quality improvement work, and a regression approach is the standard way of analyzing data from designed experiments.

The word *regression* has a much different meaning outside the realm of statistics than it does within it; literally it means to revert back to a previous state or form. In the field of statistics, the word was coined by Sir Francis Galton (1822–1911) who observed that children's heights regressed toward the average height of the population rather than digressing from it. (This is essentially unrelated to the present-day use of regression, however.)

In this chapter we present regression as a statistical tool that can be used for (1) description, (2) prediction, and (3) estimation. A regression control chart and a cause–selecting control chart are also illustrated.

12.1 SIMPLE LINEAR REGRESSION

In (univariate) regression there is always a single "dependent" variable and one or more "independent" variables. For example, we might think of the number of nonconforming units produced within a particular company each month as being dependent upon the amount of time that is devoted to maintaining control charts and using other statistical tools. In this case the amount of time (in minutes, say) would be the single independent variable. *Simple* is used to denote the fact that a single independent variable is being used.

Linear does not have quite the meaning that one would expect. Specifically, it does not necessarily mean that the relationship between the dependent variable and the independent variable is a straight-line relationship. The equation

$$Y = \beta_0 + \beta_1 X + \varepsilon \tag{12.1}$$

is a linear regression equation, but so is the equation

$$Y = \beta_0' + \beta_1' X + \beta_{11} X^2 + \varepsilon \tag{12.2}$$

A regression equation is linear if it is linear in the parameters (the betas), and both of these equations satisfy that condition.

If the ε in Eq. (12.1) were absent, the equation would then be in the general form of the equation for a straight line that is given in algebra books. Its presence indicates that there is not an exact linear relationship between X and Y.

Regression analysis is not used for variables that have an exact linear relationship, as there would be no need for it. For example, the equation

$$F = \tfrac{9}{5}C + 32 \tag{12.3}$$

expresses temperature in Fahrenheit as a function of temperature measured on the Celsius scale. Thus, F can be determined exactly for any given value of C. This is not the case for Y in Eq. (12.1), however, as β_0 and β_1 are generally unknown and therefore must be estimated. Even if they were known [as is the case with the constants $\frac{9}{5}$ and 32 in Eq. (12.3)], the presence of ε in Eq. (12.1) would still prevent Y from being determined exactly for a given value of X.

The ε is generally thought of as an error term. This does not mean that a mistake is being made, however; it is merely a symbol used to indicate the lack of an exact relationship between X and Y.

The first step in any regression analysis with a single X is to plot the data. This would be done to see if there is evidence of a linear relationship between X and Y as well as for other purposes such as checking for outlying observations (outliers).

Assume that Y denotes the number of nonconforming units produced each month and X represents the amount of time, in hours, devoted to using control charts and other statistical procedures each month. Assume that data from a company's records for the preceding 12 months are given in Table 12.1.

We can see from Table 12.1 that there is apparently a moderately strong relationship between X and Y. Specifically, as X increases Y decreases, although we can see that the relationship is far from being exact. We can obtain a better idea of the strength of the relationship by simply plotting the data in the form of a scatter plot, which was illustrated in Chapter 2. The scatter plot is given in Figure 12.1.

As noted in Chapter 11, there are a number of ways to make a scatter plot, and some attention needs to be given to the scaling of the axes. If, for example, the months had been numbered and the numbers used in place of the dots, the plot would then show that the amount of time devoted to quality improvement is generally increasing from month to month, and that the number of nonconforming units produced each month is simultaneously declining. The intent with Figure 12.1 is to simply show the strength of the linear relationship, however.

TABLE 12.1 Quality Improvement Data

Month	X, Time Devoted to Quality Improvement (hours)	Y, Number of Nonconforming Units
January	56	20
February	58	19
March	55	20
April	62	16
May	63	15
June	68	14
July	66	15
August	68	13
September	70	10
October	67	13
November	72	9
December	74	8

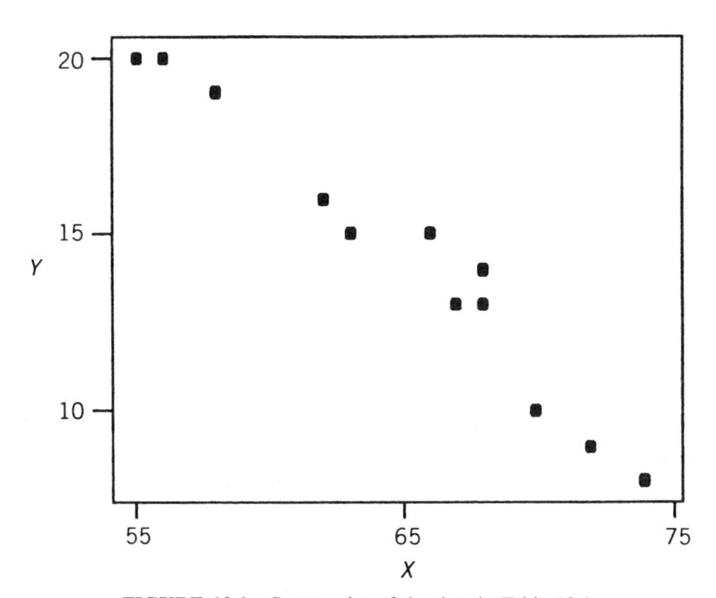

FIGURE 12.1 Scatter plot of the data in Table 12.1.

Since there is evidence of a fairly strong relationship between X and Y, we could develop a regression equation for the purpose of predicting Y from X. In essence, we will be fitting a line through the center of the points in Figure 12.1. But then what do we mean by "center?" Minimizing the sum of the (signed) deviations from each point to the line will not uniquely determine the line, but minimizing the sum of the *squared* distances will do so. This is known as the *method of least squares*, which is the most commonly used method for determining a regression equation, although there are several other methods. Thus, $\sum_{i=1}^{n} \varepsilon_i^2$ is to be minimized where $\varepsilon_i = Y_i - \beta_0 - \beta_1 X_i$.

It can be shown using simple methods of calculus that for a sample of size n, minimizing $\sum_{i=1}^{n} \varepsilon_i^2$ produces two equations, which when solved simultaneously provide estimators for β_0 and β_1. These least squares estimators are of the general form

$$\hat{\beta}_1 = \frac{\sum XY - (\sum X)(\sum Y)/n}{\sum X^2 - (\sum X)^2/n}$$

and

$$\hat{\beta}_0 = \overline{Y} - \hat{\beta}_1 \overline{X}$$

Obviously $\hat{\beta}_1$ would have to be computed first since it is used in computing $\hat{\beta}_0$.

Using the values in Table 12.1, we obtain

$$\begin{aligned}\hat{\beta}_1 &= \frac{10896 - (779)(172)/12}{50991 - (779)^2/12} \\ &= \frac{-269.67}{420.92} \\ &= -0.64\end{aligned}$$

and

$$\begin{aligned}\hat{\beta}_0 &= 14.33 - (-0.64)(64.92) \\ &= 55.92\end{aligned}$$

(*Note:* Round-off error can be a troublesome problem in regression analysis. The calculated values are displayed here using two decimal places, but the intermediate calculations are not rounded off to two places. In general, such rounding should not be performed for any statistical analysis but particularly not for regression analysis in which serious problems can ensue. Of course, regression computations are generally performed by a computer, not by hand computation.)

The *prediction equation* is of the general form

$$\hat{Y} = \hat{\beta}_0 + \hat{\beta}_1 X$$

So with these data

$$\hat{Y} = 55.92 - 0.64X$$

This prediction equation could then be used to predict future values of Y (say, for next year) for given values of X. The general idea would be to predict what is impossible or impractical to observe. For example, if an admissions committee is having a difficult time trying to decide whether or not to admit a student to a college, it would be helpful to be able to predict reasonably well what his or her grade point average would be at the end of 4 years. The student's Y would thus not be observable, but it could be predicted by a prediction equation obtained by using the records of previous students. (Many colleges and universities actually use regression analysis for this purpose.)

For the data in Table 12.1, Y could be observed, but that would require 100% inspection. Thus, it would be preferable to predict Y if that could be done with accuracy.

The prediction equation could also be used for descriptive purposes. With $\hat{\beta}_1 = -0.64$ we can state that, on the average, there is approximately a decrease of 0.64 nonconforming units for every additional hour devoted to quality improvement.

We might attempt to interpret $\hat{\beta}_0$ in a similar manner, that is, we could claim that there should be about 56 nonconforming units if no time was devoted to quality improvement. The problem with this assertion is that X ranges from 55 to 74, so the interval does not even come close to containing zero.

In general, a prediction equation with a single X should be used only for values of X that either fall within the interval of values used in producing the equation or else are only slightly outside the interval. The user who employs a prediction equation for values well outside that interval is guilty of *extrapolation*. See Hahn (1977) for a discussion of the potential hazards of extrapolation.

12.2 WORTH OF THE PREDICTION EQUATION

It is easy to obtain a prediction equation, but its worth needs to be assessed before it is used. If it can be used to predict reasonably well the values of Y used in developing the equation, then it stands to reason that future, unobservable values of Y should also be well predicted.

Using the 12 values of X to obtain the corresponding predicted values of Y, the predicted and observed values are given, along with their differences in Table 12.2.

Remembering that Y must be an integer so that $\hat{Y}$ would be rounded off to the nearest integer, we can see that the predicted values are quite close to the observed values. We can also observe that the prediction is poorest in the middle of the year. This is because there is "pure error" in the middle of the year due to the fact that the X values are the same in June and August, but the Y values are different. This causes the two points to plot vertically on a scatter plot. A regression line cannot be vertical (since the slope would be undefined), so points that plot vertically are said to constitute pure error since they cannot be accommodated by a regression line.

In spite of the presence of pure error, the predicted values are obviously close to the observed values. A comparison of the two sets of values is, of course, subjective so what is needed is an objective means of assessing the worth of

TABLE 12.2 Predicted Y Values

Y	$\hat{Y}$	$Y - \hat{Y}$
20	20.04	−0.04
19	18.76	0.24
20	20.68	−0.68
16	16.20	−0.20
15	15.56	−0.56
14	12.35	1.65
15	13.64	1.36
13	12.35	0.65
10	11.07	−1.07
13	13.00	0.00
9	9.79	−0.79
8	8.51	−0.51

a prediction equation. The most commonly used measure is R^2, which can be written as

$$R^2 = 1 - \frac{\sum(Y - \hat{Y})^2}{\sum(Y - \overline{Y})^2} \tag{12.4}$$

We want the numerator of the fraction to be as small as possible. If the numerator were zero, then R^2 would equal 1.0. This would mean that the observed and predicted values were all the same. This will not happen, of course, but we would like to see R^2 as close to 1.0 as possible. There is no dividing line between good and bad R^2 values, although values in excess of 0.90 will generally indicate that the equation has good predictive ability.

The smallest possible value of R^2 is zero. This can be seen as follows. Since $\hat{\beta}_0 = \overline{Y} - \hat{\beta}_1\overline{X}$, $\hat{Y}$ can be written as

$$\begin{aligned} \hat{Y} &= \overline{Y} - \hat{\beta}_1\overline{X} + \hat{\beta}_1 X \\ &= \overline{Y} + \hat{\beta}_1(X - \overline{X}) \end{aligned}$$

If there were no relationship between X and Y, we would expect $\hat{\beta}_1$ to be close to zero. If $\hat{\beta}_1 = 0$ then $\hat{Y} = \overline{Y}$ and substituting $\overline{Y}$ for $\hat{Y}$ in the equation for R^2 produces $R^2 = 0$. Furthermore, substituting $\overline{Y} + \hat{\beta}_1(X - \overline{X})$ for $\hat{Y}$ in Eq. (12.4) and expanding the expression would show that the numerator could never be larger than the denominator.

For the present example

$$R^2 = 1 - \frac{7.90}{180.67} = 0.96$$

This means that 96% of the variation in Y is explained (accounted for) by using X to predict Y.

12.3 ASSUMPTIONS

To this point the only assumption that needed to be made is that the true (unknown) relationship between X and Y can be adequately represented by the model given in Eq. (12.1). That in itself is an important assumption, however, as Box and Newbold (1971) showed what can happen when that model is inappropriately used. Specifically, Coen, Gomme, and Kendall (1969) demonstrated a strong relationship between stock prices and car sales seven quarters earlier. Unfortunately, the $Y - \hat{Y}$ values were correlated over time, and when Box and Newbold fit an appropriate model that accounted for the error structure, they showed that no significant relationship existed between the two variables. This example is also discussed in Box, Hunter, and Hunter (1978, p. 496).

Plotting the residuals (the $Y - \hat{Y}$ values) is thus an important part of what is referred to as *model criticism*, that is, checking on whether or not the postulated model is an appropriate model. If the model is appropriate, the residuals should have no discernible pattern when plotted against time, $\hat{Y}$, or any other variable. This applies not just to regression models but to statistical models in general, including time-series models and analysis of variance models. For example, the use of residual analysis in analyzing data from designed experiments is illustrated in Chapter 13.

The use of an appropriate model thus implies that the errors should be independent. (The residuals are, strictly speaking, not independent since they must sum to zero when β_0 is included in the model, but they are used to check the assumption of independent errors.)

If hypothesis tests and confidence intervals are to be used, it is also necessary to assume that the errors are approximately normally distributed. The mean of the errors should be zero and the variance should be constant so that it does not depend upon X.

These assumptions can be conveniently represented by

$$\varepsilon \sim \text{NID}\ (0, \sigma^2)$$

This means that the errors are independent (ID) and (approximately) normally (N) distributed with a mean of zero and a constant variance of σ^2.

12.4 CHECKING ASSUMPTIONS THROUGH RESIDUAL PLOTS

In simple linear regression the residuals should be plotted against (1) X or $\hat{Y}$, (2) time (if the time order of the data has been preserved), and (3) any other variable that an experimenter decides might be important. For the first type of

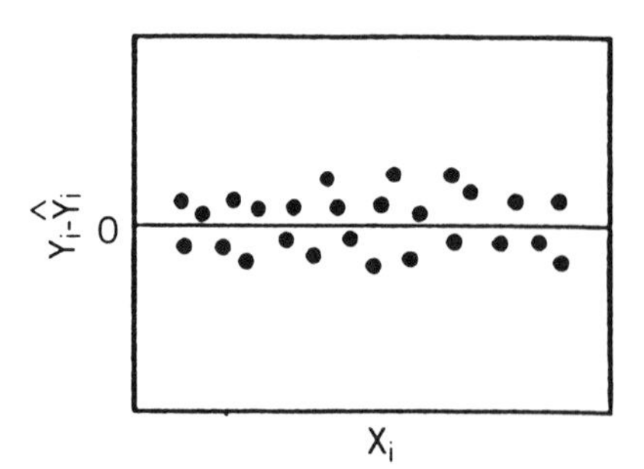

FIGURE 12.2 Residual plot.

plot, the desired configuration is along the lines of what is shown in Figure 12.2. Specifically, we would like to see all of the points close to the midline (which would cause R^2 to be close to 1.0), and form a tight cluster that can be enclosed in a rectangle. A rectangular configuration would provide evidence that the error variance is constant.

If there were any points that were far removed from the other points, such points might be labeled (residual) outliers, or they could represent data that were recorded incorrectly.

If Eq. (12.1) is used when a time-series model should have been used, a plot of the residuals against time would reveal a nonrandom pattern such as a sequence of several positive residuals followed by several negative residuals.

If the error variance either increases or decreases as X increases, this could also be detected by plotting the residuals against X. Such a problem can often be remedied by a transformation of X. If transforming X does not solve the problem, then *weighted least squares* would have to be used [see Ryan (1997, Chapter 2)].

Residual plots can also be used for checking the assumption that the postulated model is an appropriate model. If the plot of the residuals against X was in the form of a parabola, then an X^2 term would probably be needed in the model as in Eq. (12.2). Similarly, if the plot of the residuals against an omitted variable exhibited a linear trend, then a linear term in that variable should be added to the model.

12.5 CONFIDENCE INTERVALS AND HYPOTHESIS TESTS

As discussed in Chapter 3, confidence intervals are generally more informative than hypothesis tests. Both can be used in simple linear regression as long as the assumption of approximate normality of the error terms is plausible. If not, an experimenter can choose between, say, robust regression and nonparametric regression. These are covered in Ryan (1997). Iteratively reweighted least squares is another possibility. It is an alternative to robust regression when the error term has a distribution that is believed to be markedly nonnormal. See Box and Draper (1986).

Confidence intervals on the regression parameters are of the general form

$$\hat{\beta}_i \pm ts_{\hat{\beta}_i}$$

A confidence interval for β_0 is of the form

$$\hat{\beta}_0 \pm ts\sqrt{\sum X^2 / \left(n\sum(X - \overline{X})^2\right)}$$

and for β_1 the general form is

$$\hat{\beta}_1 \pm t\frac{s}{\sqrt{\sum(X - \overline{X})^2}}$$

where, for both expressions, $s = \sqrt{\sum(Y - \hat{Y})^2/(n-2)}$ and $t = t_{\alpha/2,n-2}$.

We certainly want the confidence interval for β_1 to not include zero. If it did include zero, the prediction equation would be of little value, and a test of the hypothesis that $\beta_1 = 0$ would not be rejected.

For the data in Table 12.1 we should be inclined to assume that the interval does not come close to including zero since the prediction equation has high predictive ability. (The reader is asked to show this in exercise 1.)

A confidence interval for β_0 is sometimes of value even though it is just the constant term. It is generally not desirable to test the hypothesis that $\beta_0 = 0$ and use the outcome of that test to determine whether or not the constant term would be used in the model. Only rarely would we want to use a regression equation that does not have a constant term. If we knew that $Y = 0$ when $X = 0$ *and* we were interested in predicting Y when X is close to zero, then we might omit the constant term and use what is known as *regression through the origin.*

The hypothesis test for $\beta_1 = 0$ is of the form

$$t = \frac{\hat{\beta}_1}{s_{\hat{\beta}_1}}$$

where, as previously stated, $s_{\hat{\beta}_1} = s/\sqrt{\sum(X - \overline{X})^2}$.

12.6 PREDICTION INTERVAL FOR Y

In addition to obtaining $\hat{Y}$ for a given value of X (to be denoted in this and the next section as X_0), an experimenter could construct a prediction interval around $\hat{Y}$. The general form is

$$\hat{Y} \pm ts\sqrt{1 + \frac{1}{n} + \frac{(X_0 - \overline{X})^2}{\sum(X - \overline{X})^2}}$$

where s is as previously defined, and t is $t_{\alpha/2,n-2}$ for a $100(1-\alpha)\%$ confidence interval.

To illustrate, for the data in Table 12.1, $s = \sqrt{7.90/10} = 0.89$ and for a 95% prediction interval $t_{0.025,10} = 2.228$. Therefore, for $X_0 = 65$ the interval is obtained as

$$14.28 \pm 2.228(0.89)\sqrt{1 + \frac{1}{12} + \frac{(65 - 64.92)^2}{420.92}}$$

$$= 14.28 \pm 2.06$$

so that the lower limit is 12.22 and the upper limit is 16.34.

It should be noted that this is not a confidence interval since confidence intervals are constructed only for parameters and Y is not a parameter. The general interpretation for a prediction interval is the same as for a confidence interval, however. That is, we are $100(1-\alpha)\%$ confident that the prediction interval that is to be constructed will contain Y.

A prediction interval for Y is related to a regression control chart, which is discussed in the next section.

12.7 REGRESSION CONTROL CHART

Mandel (1969) presented a regression control chart and provided a number of administrative applications. The general concept differs only slightly from a prediction interval for Y.

In fact, the assumptions for a regression control chart are exactly the same as those for simple linear regression. The general idea is to monitor the dependent variable using a control chart approach.

The centerline on the chart is at $\hat{Y} = \hat{\beta}_0 + \hat{\beta}_1 X$. Mandel (1969) provided an argument for obtaining the control limits from

$$\hat{Y} \pm 2s \tag{12.5}$$

rather than from

$$\hat{Y} \pm 2s\sqrt{1 + \frac{1}{n} + \frac{(X_0 - \overline{X})^2}{\sum(X - \overline{X})^2}} \tag{12.6}$$

which would seem to be the natural choice for obtaining the control limits. The argument given is that the use of control limits obtained from $\hat{Y} \pm 2s$ will provide tighter control for extreme values of X_0, which is where tight control is generally desired.

Control lines obtained from Eq. (12.6) will be curved since the value of the expression under the radical depends upon the value of X_0. For values of X_0 that

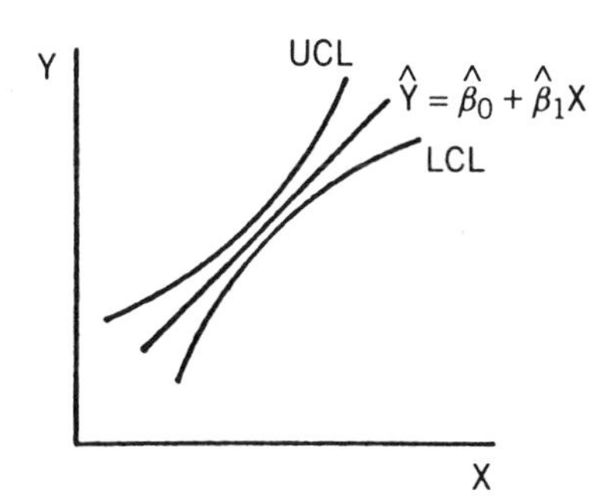

FIGURE 12.3 Regression control chart with limits obtained from Eq. (12.6).

are equidistant from $\overline{X}$, the control lines will be the same distance apart as in Figure 12.3.

The control lines in Figure 12.3 will be the closest when $X_0 = \overline{X}$. Even at that point, however, the lines will be further apart than the lines obtained using Eq. (12.5).

For the data in Table 12.1, the most extreme value of X relative to $\overline{X} = 64.92$ is 55. Using Eq. (12.6) with $X_0 = 55$, the values on the two control lines would be obtained from 20.72 ± 2.04, whereas using Eq. (12.5) the values would be 20.72 ± 1.78.

In addition to the administrative applications described in Mandel (1969), regression control charts have also been used for tool wear, as have acceptance control charts (described in Chapter 4).

12.8 CAUSE-SELECTING CHARTS

A control chart that is very similar to a regression control chart is a "cause-selecting chart." The name and the concept are due to Zhang (1984). See also Zhang (1985, 1989). An example of the use of this type of chart in the auto industry is given by Constable, Cleary, Tickel, and Zhang (1988). Wade and Woodall (1993) provided a review and analysis of cause-selecting control charts.

The general idea is to try to distinguish between quality problems that occur at one stage in a process from problems that occur at a previous processing step. For example, a problem at a particular point in a process could actually be caused by poor raw materials rather than a problem at the location where the chart is being maintained.

A modification of the original idea of the chart is necessary, due in part to the fact that Zhang (1984, 1985) assumed that the bivariate process was initially in control. That is, there was the implicit assumption that Stage 1 could be bypassed. A better idea would be to use a more customary Stage 1 approach. Let Y be the output from the second step and let X denote the output from the first step. The relationship between Y and X would be modeled—this might be a simple linear regression model or perhaps a more complex model.

Assume that a model is fit from available historical data such that the model errors can be assumed to be independent and approximately normally distributed. Then a chart of the standardized residuals could be constructed; this would be simply an individual observations chart of the standardized residuals. Zhang called the standardized residuals the cause-selecting values. Zhang estimated the standard deviation of the residuals by using the average moving range approach, but such an approach overlooks the fact that the residuals have different standard deviations. A more reasonable approach would be to estimate the standard deviation of each residual using a typical regression approach. If each residual were divided by its standard deviation (given in Section 9.4.1), the appropriate control limits would be -3 and $+3$. This would be appropriate for Stage 1. Another approach would be to use *standardized deletion residuals* (Ryan, 1997, p. 53), where n regression equations would have to be computed, with the ith observation ($i = 1, 2, \ldots n$) being left out of each equation. Clearly the ith observation should not be used if it is bad. Standardized deletion residuals are generally available (under a different name) from statistical software. Since a standardized deletion residual has a t-distribution (Cook and Weisberg, 1982, p. 20), whereas a standardized residual does not, each standardized deletion residual could be compared against control limits of $0 \pm t_{.00135,n-3}$. Such an approach could work well if there were a temporary process upset in Stage 1.

If a process shift occurred during Stage 1 such that a substantial fraction of the data is not from the in-control process distribution, a different approach may be needed. The least trimmed sum of squares (LTS) approach (Rousseeuw, 1984) can be used in identifying aberrant data points when used sequentially (see Ryan, 1997), and the algorithm of Hawkins (1994) might be used in implementing the approach. Similarly, other robust regression approaches might be used.

If a point is outside the control limits, this could be caused by Y being out of control or by X being out of control. Therefore, a chart on X must also be constructed so that a signal from the residuals chart can be interpreted unambiguously. The chart, which here is an X chart in more ways than one, could be constructed using a traditional approach with the standard deviation estimated using the moving range approach, although the likelihood ratio test method of Sullivan and Woodall (1996), described in Section 5.3, would be preferable.

A different approach is necessary for Stage 2. For the X chart, the observations that are declared to be good at the end of Stage 1 should be used to compute the sample standard deviation (as in Section 3.4), with the control limits then obtained as $\overline{x} \pm 3s_x/c_4$. (The standard deviation should not be estimated using moving ranges here for the reason given in Section 5.1.1.)

A residuals chart would not be appropriate for Stage 2, at least not without modification, since the variance of $Y_i - \hat{Y}_i = e_i$ is different for the two stages. Specifically, in Stage 1 Y_i is used in computing $\hat{Y}_i$, whereas this is not the case in Stage 2, since the prediction equation would be obtained using the good data that remain at the end of Stage 1. Note that if a Y-value is declared spurious at the end of Stage 1, the corresponding X-value would also have to be deleted and the regression equation recomputed. A deleted X-value could still be used in computing the control limits for the X chart, however, as it would be illogical to delete a valid observation and thus unnecessarily increase the variability of the performance of the X chart in Stage 2.

The appropriate variance for Stage 2 is the variance that is used in constructing a prediction interval for Y, and Wade and Woodall (1993) recommend that this approach be used in Stage 2. Since this approach produces nonconstant control limits, constant control limits could be produced simply by plotting each residual in real time divided by its appropriate standard deviation. It would be logical to set the control limits at $0 \pm t_{.00135,n-2}$, where n would be the number of good observations at the end of Stage 1 that are used in determining the (possibly revised) prediction equation for Stage 2.

Consider what should happen on each chart for the four combinations of each of the two processes being in control or out of control. First consider Stage 1. If the mean of the first process is out of control by more than a small amount, the X chart should provide a signal. How would this affect the residuals chart? There would not likely be a signal from the chart if the value of Y was affected in an additive manner, with the change in Y in line with the change in X. In other words, there would be no signal if the worsening in quality of Y was directly related to the worsening in quality of X, and the deterioration in the quality of Y was approximately the amount that would be predicted from the prediction equation. Thus, a non-signal from the residuals chart and a signal from the X chart would allow the user to properly focus attention on the first process.

Conversely, a signal from the residuals chart but not from the X chart would suggest that the second process was out of control.

For the other two possibilities, no signal on either chart would obviously suggest that both processes are in control, whereas signals from both charts would suggest that both processes are out of control. There could be problems in properly interpreting the residuals chart if a sizable percentage of observations from the second process were out of control, as the regression equation could follow the bad observations with the consequence that the good observations could be falsely identified as being bad. This could happen even if there were very few bad observations (e.g., see Stefanski, 1991).

There are a few problems with interpretation that could occur in Stage 2. The use of the X chart would be straightforward, but there could be problems with the residuals chart. For example, if the first process is badly out of control, a signal could be received from the residuals chart that could be caused by the relationship between Y and X for an unusually large or small value of X not being well represented by the regression equation.

12.9 INVERSE REGRESSION

Another important application of simple linear regression for quality improvement is in the area of calibration. Assume that two measuring tools are available — one is quite accurate but expensive to use and the other is not as expensive but also not as accurate. If the measurements obtained from the two devices are highly correlated (perhaps each pair of measurements differs by almost a constant amount), then the measurement that would have been made using the expensive measuring device could be predicted fairly well from the measurement that is actually obtained using the less expensive device.

Let Y = measurement from the less expensive device and X = measurement from the accurate device. The general idea is to first regress Y on X in the usual manner to obtain the prediction equation

$$\hat{Y} = \hat{\beta}_0 + \hat{\beta}_1 X$$

If we then "solve" for X, we obtain

$$X = \frac{\hat{Y} - \hat{\beta}_0}{\hat{\beta}_1}$$

What is being estimated (predicted) is X instead of Y, however, so for a known value of Y, Y_c, the equation is

$$\hat{X}_c = \frac{Y_c - \hat{\beta}_0}{\hat{\beta}_1}$$

Thus, the measurement that would have been obtained using the accurate measuring device is estimated (predicted) using a known measurement from the other device.

This approach has been termed the *classical estimation approach* to calibration. Krutchkoff (1967) reintroduced the concept of "inverse" regression to estimate X where X is regressed on Y. Thus,

$$\hat{X}_c^* = \hat{\beta}_0^* + \hat{\beta}_1^* Y_c$$

where the asterisks are used to signify that the estimated value of X_c and the parameter estimates will be different from the estimates obtained using the classical approach.

The only way that $\hat{X}_c$ and $\hat{X}_c^*$ could have the same value would be if X and Y were perfectly correlated. Thus, when there is a strong relationship between X and Y, not only will the difference between $\hat{X}_c$ and $\hat{X}_c^*$ be essentially inconsequential, but we should also be able to do a good job of estimating the actual measurement.

To illustrate, assume that we have data as in Table 12.3. We can see that there is almost an exact relationship between X and Y. Specifically, $X = Y + 0.1$ for the first nine values, and $X = Y$ for the tenth value. If the true relationship between X and Y were well represented by these data, and if the measurements given for X differ hardly at all from the actual measurements, then it should be possible to accurately estimate X with Y, and thus accurately estimate the actual measurement.

TABLE 12.3 Data for Illustrating Inverse Regression

Y	X
2.3	2.4
2.5	2.6
2.4	2.5
2.8	2.9
2.9	3.0
2.6	2.7
2.4	2.5
2.2	2.3
2.1	2.2
2.7	2.7

For the regression of Y on X we obtain

$$\hat{Y} = -0.14375 + 1.021X$$

so that

$$\hat{X}_c = \frac{Y_c + 0.14375}{1.021}$$

Regressing X on Y produces

$$\hat{X}_c^* = 0.176 + 0.9655Y_c$$

where the coefficients are obtained by simply reversing the roles of X and Y in the computational formulas. At $Y_c = 2.2$, we obtain $\hat{X}_c = 2.296$ and $\hat{X}_c^* = 2.300$, so, as expected, there is hardly any difference in the two values. It should also be noted that $X = 2.3$ when $Y = 2.2$ in Table 12.3.

For additional reading on inverse regression and the classical estimation approach to calibration in quality improvement, the reader is referred to Crocker (1985). See also Montgomery and Peck (1992) and Ryan (1997). The classical and inverse methods are compared by Chow and Shao (1990).

12.10 MULTIPLE LINEAR REGRESSION

Multiple regression conjures up thoughts of a disease by one author who prefers to discuss "fitting equations to data," which is the title of a well-known text (Daniel and Wood, 1980). *Multiple* is used in this section to mean that there is more than one X, that is, more than one independent variable.

The model for multiple linear regression is just an extension of the model for simple linear regression. Specifically, the model is

$$Y = \beta_0 + \beta_1 X_1 + \beta_2 X_2 + \cdots + \beta_k X_k + \varepsilon$$

for k independent variables, and the assumptions for the error term are the same as for simple regression.

Whereas the necessary calculations for simple regression can be easily performed by hand calculator, this is not true for multiple regression. For this reason and also because a detailed discussion of multiple regression is not feasible due to space limitations, the computational formulas will not be illustrated here. Rather, the reader will be referred to other texts for the details as well as to software for handling multiple regression analyses.

12.11 ISSUES IN MULTIPLE REGRESSION

The user of multiple regression must ponder more questions than one faces when using simple regression. How many regressors (i.e., independent variables) should be used and which ones should be used? How does the user know when he is extrapolating beyond the experimental region when the region is multidimensional and cannot be easily pictured? Should alternatives to least squares be used, and, if so, when? These and other questions will be addressed in this section.

In practice, there is generally more than one regressor that is highly correlated with a particular dependent variable. For example, a student's college grade point average (GPA) could be predicted using his/her high school GPA, but there are other variables such as aptitude test scores that should be considered for inclusion in the prediction equation.

12.11.1 Variable Selection

It might seem as though the objective should be to include as many regressors in the prediction equation as are thought to be related to the dependent variable. In fact, this is the strategy that would be followed if an experimenter wanted to maximize R^2, as R^2 will virtually always increase (and can never decrease) when additional variables are added to a prediction equation.

A price would be paid for such a strategy, however, since $\text{Var}(\hat{Y})$ cannot decrease (and will almost certainly increase) when new regressors are added, as was originally demonstrated by Walls and Weeks (1969).

Thus, the objective should be to essentially identify a point of diminishing returns such that adding additional regressors will create more harm [in terms of increasing $\text{Var}(\hat{Y})$] than good (in terms of increasing R^2).

A commonly used statistic for making this determination is the C_p statistic proposed by Mallows (1973). It is defined as

$$C_p = \frac{\text{SSE}_p}{\hat{\sigma}^2_{\text{full}}} - n + 2p$$

where p is the number of parameters in the model, SSE_p is the residual sum of squares [equal to $\sum(Y - \hat{Y})^2$], $\hat{\sigma}^2_{\text{full}}$ is the estimate of the error variance using all of the available regressors, and n is the sample size.

With k potential regressors there are $2^k - 1$ possible models (1023 when $k = 10$), and the C_p statistic can be used to allow the experimenter to focus attention on a small number of these possible models. Specifically, the idea is to look hard at those prediction equations for which C_p is small and is close to p. The final choice of a model from this group can then be made using other statistical considerations as well as possibly nonstatistical factors such as the cost of acquiring the data and personal preference. We hope that the model we select is a good proxy for the true (unknown) model, which is almost certainly nonlinear.

Other methods that can be used for selecting regressors are discussed in regression books such as Ryan (1997).

12.11.2 Extrapolation

It is easy to determine whether or not extrapolation has occurred when there is only one regressor. This is not true when there are, say, five or six regressors. Weisberg (1985) discusses the problem and presents two methods that can be used to determine if extrapolation has occurred.

12.11.3 Multicollinear Data

Problems occur when at least two of the regressors are related in some manner. For example, two of the regressors may be highly correlated (say, $r = .95$). This has the effect of inflating the variances of the least squares estimators, but it will generally have little effect on $\text{Var}(\hat{Y})$. Therefore, multicollinear data should be of no great concern if prediction is the objective, but it is another matter if the objective is estimation.

Solutions to the problem of multicollinear data include discarding one or more of the variables causing the multicollinearity. Another solution is to use ridge regression, which allows an experimenter to use all of the regressors that he may want to use, but in a manner different from least squares. [See, e.g., Hoerl and Kennard (1981).]

12.11.4 Residual Plots

Residual plots are used extensively in multiple regression, and for the same general purposes as in simple regression; that is, for checking on the model assumptions, and for determining if the model could be improved.

The number of necessary residual plots is, of course, much greater in multiple regression. The residuals should generally be plotted against $\hat{Y}$, each of the regressors, time (if possible), and any potential regressor that might later seem important. [Partial residual plots are often more valuable than regular residual plots, however, for determining appropriate transformations of the regressors. Examples of this can be found in Gunst and Mason (1980, p. 251) and Ryan (1997, p. 153). Partial residuals to be plotted versus the jth regressor $(j = 1, 2, \ldots, k)$ are defined as $e_i + \hat{\beta}_j X_{ij}$, where e_i is the ith residual and X_{ij} is the ith observation on regressor X_j.]

Residual plots virtually require the use of a computer, and Henderson and Velleman (1981) illustrate how a user can interact with a computer in developing a regression model.

12.11.5 Regression Diagnostics

It is desirable to be able to detect extreme observations that may represent data that have been recorded incorrectly and also to detect observations that are much more influential than the other observations. Ideally, we would want each of the n observations to have equal influence in determining the coefficients of the regressors in a prediction equation, but this does not happen, in general. Methods for detecting extreme observations and influential observations can be found in Belsley, Kuh, and Welsch (1980), Cook and Weisberg (1982), and Ryan (1997).

Invalid points can, of course, be discarded. Points of high influence can either be discarded or given less influence as in Krasker and Welsch (1982).

12.11.6 Transformations

A regression model can often be improved by transforming one or more of the regressors, and possibly the dependent variable as well. Logarithmic and other simple transformations can often improve the value of a prediction equation.

Transformations can also often be used to transform a nonlinear regression model into a linear one. For example, the nonlinear model

$$Y = \beta_0 \beta_1^X \varepsilon$$

can be transformed into a linear model by taking the logarithm of both sides of the equation so as to produce

$$\ln Y = \ln \beta_0 + X \ln \beta_1 + \ln \varepsilon$$

Simple linear regression could then be applied to the latter.

The reader is referred to Chapter 8 of Box and Draper (1986), Chapter 16 and the Appendix of Mosteller and Tukey (1977), and Chapter 6 of Ryan (1997) for detailed information concerning the use of transformations.

12.12 SOFTWARE FOR REGRESSION

As with control charts, the user of regression analysis has a multitude of software from which to choose. Mainframe packages such as the SAS* System and Minitab (both of which also run on a PC) have excellent regression capabilities, and the same can be said of a number of other general-purpose statistical software packages. The numerical accuracy for regression data of a variety of microcomputer statistical packages is assessed in a study conducted by Lesage and Simon (1985).

12.13 SUMMARY

The rudiments of regression analysis have been presented in this chapter, with an emphasis on regression tools that can be used in quality improvement, such as, a regression control chart.

Multiple regression was treated somewhat briefly. This is because multiple regression analyses are generally performed on a computer, and also because a detailed exposition of multiple regression would require a considerable amount of space.

There is a wealth of reading matter on multiple regression. Readers seeking a capsuled but more detailed account of multiple regression than provided here are referred to Ryan (1998). Readers seeking a complete account of almost all aspects of multiple regression (including nonlinear regression) are referred to Ryan (1997). See also the regression books of Draper and Smith (1998), Montgomery and Peck (1992), Weisberg (1985), and Neter, Kutner, Nachtsheim, and Wasserman (1996). Additional information on the use of multiple regression in quality improvement can be found in Hinchen (1968), Hotard and Jordan (1981), and Crocker (1985).

REFERENCES

Belsley, D. A., E. Kuh, and R. Welsch (1980). *Regression Diagnostics: Identifying Influential Data and Sources of Collinearity.* New York: Wiley.

Box, G. E. P. and N. R. Draper (1986). *Empirical Model Building and Response Surfaces.* New York: Wiley.

Box, G. E. P. and P. Newbold (1971). Some comments on a paper by Coen, Gomme, and Kendall. *Journal of the Royal Statistical Society, Series A 134*(2): 229–240.

Box, G. E. P., W. G. Hunter, and J. S. Hunter (1978). *Statistics for Experimenters.* New York: Wiley.

Chow, S.-C. and J. Shao (1990). On the difference between the classical and inverse methods of calibration. *Applied Statistics 39*: 219–228.

* SAS is a registered trademark of SAS Institute, Inc., Cary, NC, USA.

Coen, P. J., E. E. Gomme, and M. G. Kendall (1969). Lagged relationships in economic forecasting. *Journal of the Royal Statistical Society, Series A 132*(2): 133–152.

Conover, W. S. (1980). *Practical Nonparametric Statistics*, 2nd ed. New York: Wiley.

Constable, G. K., M. J. Cleary, C. Tickel, and G. X. Zhang (1988). Use of cause-selecting charts in the auto industry. In *ASQC Quality Congress Transactions*, pp. 597–602. Milwaukee, WI: American Society for Quality Control.

Cook, R. D. and S. Weisberg (1982). *Residuals and Influence in Regression*. New York: Chapman and Hall.

Crocker, D. C. (1985). How to use regression analysis in quality control. In *Basic References in Quality Control: Statistical Techniques*, Vol. 9. Milwaukee, WI: American Society for Quality Control.

Daniel, C. and F. S. Wood (1980). *Fitting Equations to Data*, 2nd ed. New York: Wiley.

Draper, N. R. and H. Smith (1998). *Applied Regression Analysis*, 3rd ed. New York: Wiley.

Gunst, R. F. and R. L. Mason (1980). *Regression Analysis and Its Application*. New York: Dekker.

Hahn, G. J. (1977). The hazards of extrapolation in regression analysis. *Journal of Quality Technology 9*(4): 159–165.

Hawkins, D. M. (1994). The feasible solution algorithm for least trimmed squares regression. *Computational Statistics and Data Analysis 17*: 185–196.

Henderson, H. V. and P. F. Velleman (1981). Building multiple regression models interactively. *Biometrics 37*(2): 391–411.

Hinchen, J. D. (1968). Multiple regression in process development. *Technometrics 10*(2): 257–269.

Hoerl, A. E. and R. W. Kennard (1981). Ridge regression–1980—advances, algorithms, and applications. *Journal of Mathematical and Management Sciences 1*(1): 5–83.

Hotard, D. G. and J. D. Jordan (1981). Regression analysis is applied to improve product quality. *Industrial Engineering 13*(3): 68–75.

Krasker, W. S. and R. E. Welsch (1982). Efficient bounded influence regression estimation. *Journal of the American Statistical Association 77*(379): 595–604.

Krutchkoff, R. G. (1967). Classical and inverse regression methods of calibration. *Technometrics 9*(3): 425–439.

Lesage, J. P. and S. D. Simon (1985). Numerical accuracy of statistical algorithms for microcomputers. *Computational Statistics and Data Analysis 3*(1): 47–57.

Mallows, C. L. (1973). Some comments on C_p. *Technometrics 15*(4): 661–675.

Mandel, B. J. (1969). The regression control chart. *Journal of Quality Technology 1*(1): 1–9.

Montgomery, D. C. and E. A. Peck (1992). *Introduction to Linear Regression Analysis*, 2nd ed. New York: Wiley.

Mosteller, F. and J. W. Tukey (1977). *Data Analysis and Regression*. Reading, MA: Addison-Wesley.

Neter, J., M. H. Kutner, C. Nachtsheim, and W. Wasserman (1996). *Applied Linear Regression Models*, 3rd ed. Homewood, IL: Irwin.

Rousseeuw, P. J. (1984). Least median of squares. *Journal of the American Statistical Association 79*: 871–880.

Ryan, T. P. (1997). *Modern Regression Methods*. New York: Wiley.

Ryan, T. P. (1998). Linear regression. In H. M. Wadsworth, ed. *Handbook of Statistical Methods for Engineers and Physical Scientists*, 2nd ed., Chapter 14. New York: McGraw-Hill.

Stefanski, L. A. (1991). A note on high-breakdown estimators. *Statistics and Probability Letters 11*: 353–358.

Sullivan, J. H. and W. H. Woodall (1996). A control chart for preliminary analysis of individual observations. *Journal of Quality Technology 28*: 265–278.

Wade, M. R. and W. H. Woodall (1993). A review and analysis of cause-selecting control charts. *Journal of Quality Technology 25*(3): 161–169.

Walls, R. C. and D. L. Weeks (1969). A note on the variance of a predicted response in regression. *American Statistician 23*(3): 24–26.

Weisberg, S. (1985). *Applied Linear Regression*, 2nd ed. New York: Wiley.

Zhang, G. X. (1984). A new type of control charts and a theory of diagnosis with control charts. In *ASQC Annual Quality Congress Transactions*, pp. 175–185. Milwaukee, WI: American Society for Quality Control.

Zhang, G. X. (1985). Cause-selecting control charts—A new type of quality control charts. *The QR Journal 12*: 221–225.

Zhang, G. X. (1989). A new diagnosis theory with two kinds of quality. In *ASQC Annual Quality Congress Transactions*, pp. 594–599. Madison, WI: American Society for Quality Control.

EXERCISES

1. Obtain a 95% confidence interval for β_1 using the data in Table 12.1. Notice that the interval does not come close to including zero. This means that we would reject the hypothesis that $\beta_1 = 0$ for any reasonable significance level, but it does not necessarily mean, in general, that the prediction equation has any value (i.e., R^2 could be small). Show this by writing R^2 as a function of t^2, with t denoting the t-statistic for testing $\beta_1 = 0$.

2. Assume that $\hat{\beta}_1 = 1.38$ in a simple linear regression prediction equation. What does this number mean relative to X and Y?

3. Regression data are often coded by subtracting the mean of each variable from the values of the respective variables. Consider the following coded data:

X	−3	−2	−1	0	1	2	3
Y	9	4	1	2	1	4	7

(a) Construct a scatter plot of the data.

(b) Does the scatter plot suggest a linear relationship?

(c) Fit a linear regression equation and then plot the residuals against $\hat{Y}$.

(d) Why does that plot have a parabolic shape?

(e) What does this suggest about how the prediction equation should be modified?

4. Consider the following calibration data in which X represents a measurement made on a measuring instrument that is highly accurate, and Y represents the corresponding measurement (on the same physical unit) made with a less accurate instrument.

X	3.6	3.5	3.7	3.7	3.4	3.8	3.7	3.5	3.6	3.5
Y	3.3	3.2	3.4	3.3	3.1	3.5	3.4	3.2	3.2	3.2

Obtain the estimates of X using the classical estimation approach to calibration and comment on the apparent worth of the approach for these data.

5. What assumption(s) does an experimenter need to make if he or she is going to develop a prediction equation solely for the purpose of obtaining predicted values for Y?

6. Consider the data in Table 12.1. What would be wrong with using the prediction equation to predict Y when $X = 45$?

7. For the following data:

X	23	31	26	25	28	30	34	38	29	32	24
Y	41	50	45	46	48	50	58	60	47	52	43

(a) Plot the data using a scatter plot.

(b) Determine $\hat{\beta}_0$ and $\hat{\beta}_1$ for the model $Y = \beta_0 + \beta_1 X + \varepsilon$.

(c) Determine R^2.

(d) Construct the 95% confidence interval for β_1.

(e) Construct the 95% prediction interval for Y using $X_0 = 25$.

8. There are many applications of simple linear regression for which we know that Y will be zero when X is zero. Would this suggest that β_0 should not be used in the model? Provide a counterargument.

CHAPTER 13

Design of Experiments

In the early 1980s, when industrial personnel began to apply statistical quality control methods extensively in industry in the United States, the emphasis was on control charts. This prompted a number of prominent statisticians to recommend the use of "conversational" or active statistical tools to supplement the use of "listening" or passive statistical tools. Specifically, a control chart is a listening tool in that data are obtained from a process but no attempt is made to see what happens when the process is changed.

Assume that our objective is to control the diameter of a machined part within fixed limits. Our chances of successfully doing so will be increased if we can identify the factors that affect the diameter (e.g., temperature, humidity, pressure, machine setting) and the extent to which the diameter is dependent upon each factor. If there is a strong dependency, the levels of these factors could be set to, ideally, produce a desired diameter, and *if* there were no other factors that had to be controlled, there would just be random variation about the target value.

In this chapter we present experimental design procedures that can be used to identify the factors that affect the quality of products and services. The intent is to present a capsule account of the statistical principles of experimental design. We emphasize the planning of experiments and the fact that statistical experiments should generally be viewed as being part of a sequence of such experiments, as opposed to a "one-shot-only" approach. Many books have been written on experimental design in which the emphasis is on the analysis of standard designs. Such analysis is de-emphasized in this chapter, as are the computational formulas and tables that are inherent in such analyses. Instead, the reader is referred to such books (and journal articles) for additional reading.

13.1 A SIMPLE EXAMPLE OF EXPERIMENTAL DESIGN PRINCIPLES

We begin with a practical design problem that has been discussed in a number of sources, including Hahn (1977) and Hicks (1982). The objective is to compare 4 different brands of tires for tread wear using 16 tires (4 of each brand) and 4 cars in an experiment. We can often learn more about experimental design by looking at how *not* to design an experiment than we can by reading about the analyses of data from standard designs.

What would be an illogical way to design this experiment? One simple approach would be to assign the tires to the cars in such a way that each car would have all four tires of a given brand. This would be a poor design; even if the cars were identical, the drivers might have different driving habits, and the driving conditions might also be expected to vary.

With such a design, the differences in tire wear among the four brands would be, in statistical terminology, "confounded" (i.e., confused) with differences between cars, drivers, and driving conditions.

A simple way to alleviate this problem would be to assign the tires to the cars in such a way that each car will have one tire of each brand. The design layout might then appear as in Figure 13.1, where the letters, A, B, C, and D denote the four brands. This would also be a poor design, however, because brands A and B would be used only on the front of each car, and brands C and D. would be used only on the rear positions. Thus, the brand effect would be confounded with the position effect.

The way to remedy this problem would be to construct the design layout in such a way that each brand is used once at each position as well as once with each car. One possible configuration is given in Figure 13.2. With this design the brand differences would not be confounded with wheel position differences

	Car			
Wheel Position	1	2	3	4
LF	A	B	A	B
RF	B	A	B	A
LR	D	C	D	C
RR	C	D	C	D

FIGURE 13.1 Tire brand layout — unacceptable design.

	Car			
Wheel Position	1	2	3	4
LF	A	B	C	D
RF	B	A	D	C
LR	C	D	A	B
RR	D	C	B	A

FIGURE 13.2 Tire brand layout — acceptable design.

or with differences in cars, drivers, or driving conditions. (Of course, there still might be a driver-by-position effect since tires do not wear evenly at each wheel position, and an overly aggressive driver might thus create problems.)

We have thus looked at two unacceptable designs and one acceptable design. (Another unacceptable design would be to randomly assign the 16 tires to the 4 cars, as this would likely create an imbalance similar to what was seen for the first two designs.)

The statistical analysis of these types of designs will be discussed later in the chapter; the intent at this point is to introduce the reader to the types of considerations that need to be made in designing an experiment.

13.2 PRINCIPLES OF EXPERIMENTAL DESIGN

One point that cannot be overemphasized is the need to have processes in a state of statistical control, if possible, when designed experiments are carried out. If control is not achieved, the influence of factors that might affect the mean of a process characteristic is confounded (i.e, confused) with changes in the mean of the characteristic caused by the process being out of control. For example, consider Figure 13.3.

Assume that mixing speed has been too low and there is a desire to see if mixing speed can be improved. The process engineer believes that the problem can be corrected and so takes what he or she believes is the appropriate action. An experiment is performed with measurements taken before and after the changes to see if the action has had the desired effect. For the sake of simplicity, assume that sampling was performed such that every fourth unit was sampled and the same sampling was performed after the adjustment. Assume that A and B in

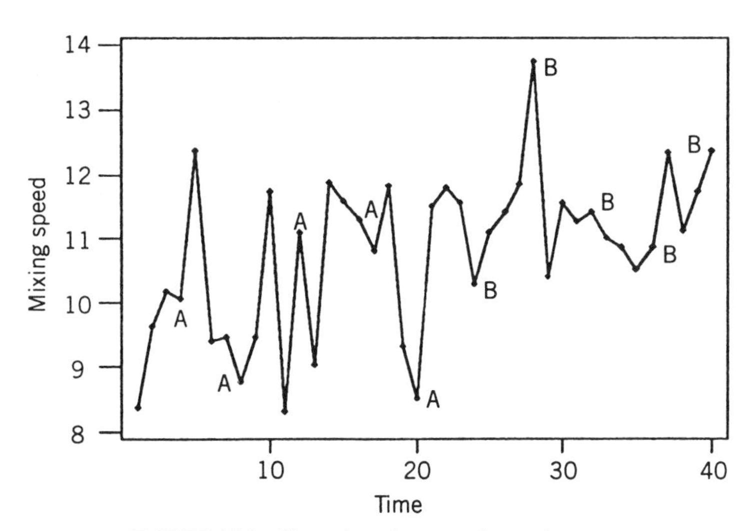

FIGURE 13.3 Illustration of an out-of-control process.

Figure 13.3 denote measurements that are taken before and after the change, respectively, and that are used in the comparison. The 40 points that are plotted are all of the mixing speed values. Notice that there appears to be a mean shift. In fact, a 1.5-sigma mean shift was built into the simulated data, with the shift occurring with observation 21. If a t-test (see Section 13.4) were performed and a 5% significance level were used, the conclusion would be that the means differ and thus the process changes have had the desired effect. But the change in this example is due to the process being out of control.

Thus, the wrong conclusion is drawn, and this has occurred because experimentation was performed on a process that was not in a state of statistical control. In an actual experiment an experimenter would not likely be so unfortunate as to have a process go out of control at the same point at which process changes, unrelated to the assignable cause(s), were instituted, but if this did occur, there would be no way to separate the two effects. At the very least, a small amount of contamination could easily occur that could cause erroneous conclusions to be drawn.

Box, Bisgaard, and Fung (1990, p. 190) recommend that processes be brought into the best state of control possible and experimentation be performed in such a way that blocks are formed to represent intervals of time when the mean appears to be constant. One potential impediment to the implementation of such an approach, however, is that it may be difficult to determine when the mean changes have occurred unless the changes are large.

General guidelines on the statistical design of experiments are given by Coleman and Montgomery (1993). They list seven steps that should be performed sequentially: (1) recognition of and statement of the problem, (2) choice of factors and levels, (3) selection of the response variable(s), (4) choice of experimental design, (5) conduction of the experiment, (6) data analysis, and (7) conclusions and recommendations. Montgomery (1996) points out that the problem statement is often too broad. The factors that are studied in the initial stages of sequential experimentation are those that are believed to be important. The levels of a factor should be far enough apart so as to show that the factor has a significant effect if in fact it does so. The choice of levels will be discussed in more detail later in the chapter. In pointing out that all (or nearly all) real-world experiments have multiple responses, Montgomery (1996) states that "the classical single-response case occurs only in textbooks and journal articles". The multiple-response case is naturally more complex than the single-response case, however, so the former is covered briefly in Section 13.16.9 after the latter is presented.

13.3 STATISTICAL CONCEPTS IN EXPERIMENTAL DESIGN

Assume that the objective is to determine the effect of two different levels of temperature on process yield, where the current temperature is 250°F and the experimental setting is 300°F. We shall also assume for the purpose of illustration that temperature is the only factor that is to be varied. Later, it will be explained why varying one factor at a time is generally not a good strategy. For the moment, however, the emphasis is on simplification.

This experiment might be conducted by recording the process yield at the end of each day for a 2-week period. (Why 2 weeks? Why not 1 week or 1 day? This issue will be addressed shortly.) One temperature setting could be used for the first 2 weeks and the other setting for the following 2 weeks. The data might then appear as in Table 13.1. How might we analyze these data to determine if temperature has any effect on yield? It has often been stated that the first step in analyzing any set of data is to plot the data. This point was emphasized in Chapter 11, in which it was also emphasized that a particularly useful plot is to plot the data against time, assuming that the data have been collected over time and the time order has been preserved.

The time-sequence plot in Figure 13.4 is for the data in Table 13.1.

What can be gleaned from this plot? First, it is readily apparent that neither temperature setting is uniformly superior to the other over the entire test period since the lines cross several times. This coupled with the fact that the lines are fairly close together would suggest that increasing the temperature from 250 to

TABLE 13.1 Process Yield Data by Temperature (tons)

	250°F		300°F	
Day	Week 1	Week 2	Week 3	Week 4
Monday	2.4	2.5	2.6	2.7
Tuesday	2.7	2.8	2.4	2.3
Wednesday	2.2	2.9	2.8	3.1
Thursday	2.5	2.4	2.5	2.9
Friday	2.0	2.1	2.2	2.2

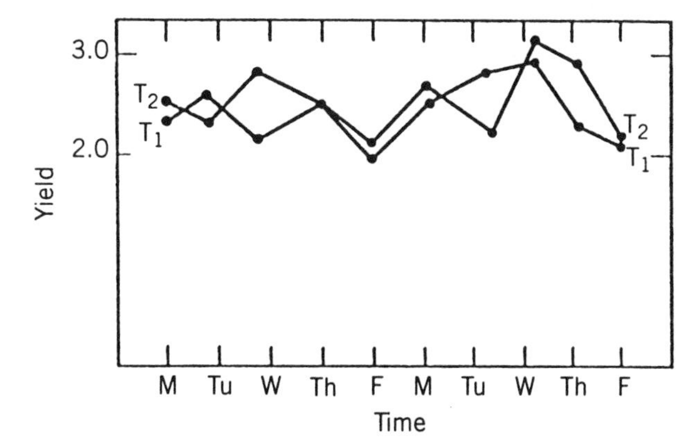

FIGURE 13.4 Time-sequence plot.

300°F might not have a perceptible effect on the process yield. Before this question is addressed statistically, rather than pictorially, it is important to recognize the other information that is apparent from Figure 13.4. In particular, it can be observed that the yield at each temperature setting is the lowest on Friday of each week. Although this discovery might be viewed as incidental relative to the main objective of determining whether or not the 50°F increase in temperature affects the process yield, it would certainly be viewed as important information. For example, some subsequent "detective" work might reveal that the workers were leaving an hour early on Friday afternoons! Or perhaps worker fatigue was a very real factor at the end of the week. In any event, this is an example of a discovery that might not have been made if a perfunctory statistical analysis had been performed without first plotting the data (although it is also apparent from Table 13.1).

We can also see from Figure 13.4 that there is considerable variability within each temperature setting, particularly relative to the magnitude of the difference between the temperature settings for each day. This coupled with the fact that the lines cross several times would tend to suggest that there is no real difference in the two settings. To elaborate on this somewhat, would this suggest that there would be a real (true) difference in the temperature settings if the lines never crossed? No, it would depend upon the variability within each setting. We have simply obtained a sample of 10 values from a (theoretically infinite) set of yield data for each setting. If there are a number of factors that are operating to cause random variability within each setting, we would logically expect some variability between the two settings, even when there is no difference in the effect of the two settings. On the other hand, if the two lines were both straight and parallel, we would conclude that there is indeed a real difference (although perhaps slight) even if the lines were almost touching! Such a difference would probably not be of any practical significance, however.

It is extremely important for the reader to understand the reasoning that has been presented in the preceding paragraph, as it essentially forms the foundation for the t-test that is presented in the next section as well as for analysis of variance, which is generally used in analyzing data from experimental designs.

13.4 t TESTS

Although the graphical analysis of the data in Table 13.1 is helpful for gaining insight, it is also desirable to analyze data "objectively" (i.e., with numbers). The t distribution was covered in Chapter 3, and it was stated that a t statistic is of the general form

$$t = \frac{\hat{\theta} - \theta}{s_{\hat{\theta}}}$$

where θ is the parameter of interest, $\hat{\theta}$ is the sample statistic that is the estimator of θ, and $s_{\hat{\theta}}$ is the estimator of the standard deviation of $\hat{\theta}$.

If we let μ_1 represent the true, but unknown, average yield that would be obtained using 250°F, and define μ_2 similarly for 300°F, our interest would then center upon the difference $\mu_1 - \mu_2$. (It should be noted that we are only approximating reality when we speak of "true" values of μ_1 and μ_2 or for any other parameter. Conditions change, so we might think of μ_1, for example, as representing what the process yield *would be* using 250°F if all other relevant factors could be held constant. Since the latter might hold true only for laboratory experiments, it should be noted that inferences made from experimental designs and extended to the future could be seriously undermined if conditions change greatly in the future. Since the future is obviously unknown, there is no statistical adjustment that can be made. It is simply a matter of combining common sense with statistical methods and recognizing the limitations of the latter.)

If we let θ represent this difference, then $\hat{\theta} = \bar{x}_1 - \bar{x}_2$, where $\bar{x}_1$ and $\bar{x}_2$ are the averages of the 250 and 300°F columns, respectively, in Table 13.1. It can be shown that

$$\sigma_{\bar{x}_1 - \bar{x}_2} = \sqrt{\frac{\sigma_1^2}{n_1} + \frac{\sigma_2^2}{n_2}} \tag{13.1}$$

(assuming the samples to be independent, which they are here), where σ_1^2 and σ_2^2 are the true (unknown) variances for 250 and 300°F, respectively, and n_1 and n_2 are the two sample sizes. (Here, of course, $n_1 = n_2 = 10$.) Of course, σ_1^2 and σ_2^2 are generally unknown and must be estimated. The logical estimators are s_1^2 and s_2^2, the two sample variances.

It would seem reasonable to substitute s_1^2 for σ_1^2 and s_2^2 for σ_2^2 in Eq. (13.1) and proceed from there. This would be logical if (1) n_1 and n_2 were both reasonably large (say, at least 30) or (2) n_1 and n_2 were small and differed considerably (e.g., 5 and 10), and it would be unreasonable to assume $\sigma_1^2 = \sigma_2^2$. The first condition is clearly not satisfied for this example since both sample sizes are equal to 10. The second condition would lead to the use of an approximate t test such as the one discussed in Section 13.4.2.

13.4.1 Exact *t* Test

The exact t test for testing the equality of two population means is of the form

$$t = \frac{(\bar{x}_1 - \bar{x}_2) - 0}{s_p\sqrt{(1/n_1) + (1/n_2)}} \tag{13.2}$$

where $\bar{x}_1, \bar{x}_2, n_1$, and n_2 are as previously defined and $s_p (= \sqrt{s_p^2})$ is the square root of the estimate of the (assumed) common variance, $\sigma_1^2 = \sigma_2^2 = \sigma^2$, where that estimate is obtained from

$$s_p^2 = \frac{(n_1 - 1)s_1^2 + (n_2 - 1)s_2^2}{n_1 + n_2 - 2}$$

The latter reduces to a simple average of s_1^2 and s_2^2 when $n_1 = n_2$, as in this example. The zero in Eq. (13.2) represents the fact that $\mu_1 - \mu_2 = 0$ if $\mu_1 = \mu_2$. Note that the denominator of Eq. (13.2) is in the general form of Eq. (13.1), with s_p^2 substituted for $\sigma_1^2 = \sigma_2^2 = \sigma^2$.

It is not necessary to always use zero in Eq. (13.2), however. In particular, if a new operating procedure is to be tested against the standard procedure, we might require that the new procedure be considerably better than the standard procedure, particularly if use of the new procedure would lead to an increase in costs (see Deming, 1975, p. 150). Thus, we might require $\mu_1 - \mu_2 < d$, where the value of d would be determined from monetary factors. This value would then be used in place of zero in Eq. (13.2).

A t statistic always has a certain number of "degrees of freedom" associated with it, specifically the degrees of freedom associated with the denominator of the statistic. For this test statistic μ_1 and μ_2 are estimated by $\bar{x}_1$ and $\bar{x}_2$, respectively, in the process of computing s_1^2 and s_2^2 for estimating σ^2 by s_p^2. Thus, two degrees of freedom (d.f.) are used in estimating μ_1 and μ_2, so there are $20 - 2 = 18$ d.f. left for estimating σ^2 and hence 18 d.f. for the t statistic.

Applying this t test to the data in Table 13.1 produces

$$\begin{aligned} t &= \frac{(2.45 - 2.57) - 0}{0.3005\sqrt{(1/10) + (1/10)}} \\ &= -0.893 \end{aligned}$$

It can be shown (by computer) that the probability of observing a value for t that is as small or smaller than -0.893 when $\mu_1 = \mu_2$ is .1919. Thus, there is almost a 20% chance of this occurring, so there is not sufficiently strong evidence to lead us to reject the hypothesis that $\mu_1 = \mu_2$ in favor of $\mu_1 < \mu_2$ (less than 5% would generally be viewed as constituting strong evidence, although this is subjective). Thus, the conclusion is that the difference between the sample averages is due to sampling variation and does not reflect a true difference (of any magnitude) in the effects of the two temperature settings. Specifically, we are not concluding that $\mu_1 = \mu_2$, as it is highly improbable that two process means would be equal. Rather, we simply conclude that the data do not provide sufficiently strong evidence to allow us to claim that $\mu_1 < \mu_2$. This can happen when μ_1 is approximately equal to μ_2 and/or the data in each sample are highly variable.

It should be noted that $\sigma_1^2 = \sigma_2^2$ was assumed but was not checked. In general, necessary assumptions should be checked, but this one is not crucial as long as $n_1 = n_2$. The (exact) t test is undermined when $\sigma_1^2 \neq \sigma_2^2$ and n_1 and n_2 differ considerably, however, particularly when at least one of the two sample sizes is relatively small. See Cressie and Whitford (1986) for a discussion of the robustness of the t test to violation of the equal-variance assumption for equal and unequal sample sizes and recommendations concerning the use of two-sample t tests. See also Posten, Yeh, and Owen (1982).

One problem in testing for equality of variances is that most of the standard tests are sensitive to nonnormality. Layard (1973) presents a test that is not sensitive, however. (The *t* test itself is not sensitive to slight-to-moderate departures from normality.) The test is also described in Miller and Wichern (1977, p. 167).

Before leaving this example, we should note the other requisite assumptions for this *t* test: (1) the two samples are independent, and (2) the observations are independent within each sample. The first requirement is clearly met since the data in each sample were obtained over different weeks, but the fact that the data were collected over time might suggest that the second requirement is not met. In particular, is the process yield on a given day related to the yield on the preceding day? The data in Table 13.1 are fictitious, and no trend is apparent in Figure 13.4 for either temperature setting, but in practice the issue of possible dependent observations should be addressed. If dependency is to be expected, this experiment could have been conducted by randomly assigning the temperature settings to the days (if frequent changes in the temperature could be tolerated).

We can also note that $\sigma_1^2 = \sigma_2^2$ is the requisite assumption that is usually stated for the exact *t* test, but Nelson (1984) points out that we would still have an exact *t* test if $\sigma_1^2 = c\sigma_2^2$, where c is a constant that would have to be known. Unfortunately, we are not likely to have such knowledge.

13.4.2 Approximate *t* Test

If n_1 and n_2 differ considerably and $c = \sigma_1^2/\sigma_2^2$ is unknown, an approximate *t* test should be used. The one to be discussed here (a number have been proposed) is due to Welch (1937) and is of the form

$$t^* = \frac{(\bar{x}_1 - \bar{x}_2) - 0}{\sqrt{s_1^2/n_1 + s_2^2/n_2}} \tag{13.3}$$

where the zero represents the fact that $\mu_1 - \mu_2 = 0$ *if* μ_1 and μ_2 are equal and the asterisk indicates that this is an approximate *t* test. It can be shown that t^* will have the same value as t for the exact *t* test when $n_1 = n_2$, but the d.f. will differ if $s_1^2 \neq s_2^2$. In general, the degrees of freedom are calculated as

$$\text{d.f.} = \frac{(s_1^2/n_1 + s_2^2/n_2)^2}{(s_1^2/n_1)^2/(n_1 - 1) + (s_2^2/n_2)^2/(n_2 - 1)}$$

Substituting the appropriate numbers produces d.f. = 17.98, so that 18 would be used. Thus, for this example the two *t* tests produce identical results, which is not surprising since $n_1 = n_2$ and s_1^2 and s_2^2 differ only slightly.

13.4.3 Confidence Intervals for Differences

We can also express the outcome of an experiment, such as the one given in Section 13.2, by constructing a confidence interval for the difference of two means. If we wanted to see if the experimental evidence suggested $\mu_1 < \mu_2$, we could construct a $100(1 - \alpha)\%$ confidence interval (see Chapter 3) of the form $(\overline{X}_1 - \overline{X}_2) + t_{\alpha,\nu} s_{\overline{X}_1 - \overline{X}_2}$, where ν is the degrees of freedom for the t statistic in Eq. (13.2). This would provide an upper bound on $\mu_1 - \mu_2$ that if less than zero would suggest that $\mu_1 < \mu_2$. Similarly, a two-sided interval could be constructed that would be of the form $(\overline{X}_1 - \overline{X}_2) \pm t_{\alpha/2,\nu}\, s_{\overline{X}_1 - \overline{X}_2}$, which could be used to determine if $\mu_1 \neq \mu_2$. Thus, a decision could be reached using a confidence interval, and the decision would be the same as that reached with the corresponding t test. An interval obviously provides an experimenter with more information than a "yes" or "no" test, however.

13.5 ANALYSIS OF VARIANCE FOR ONE FACTOR

The data in Table 13.1 could also have been analyzed by a technique known as analysis of variance (ANOVA). Historically, this technique has been used to analyze data obtained from the use of experimental designs. Such data can also be analyzed by regression methods that offer some comparative advantages; these will be discussed later in the chapter.

The expression *analysis of variance* is actually somewhat of a misnomer as it would tend to connote to a statistical neophyte that a "variance" is being analyzed. A variance in statistics is, of course, a single number, and it would be difficult to analyze a single number. What is actually being analyzed is *variation* — variation attributable to different sources.

The variation both within and between the temperature settings for the data in Table 13.1 was discussed in the graphical analysis of those data. This variation is handled more explicitly in ANOVA.

Before embarking upon an analysis of these data using ANOVA, it is necessary to introduce some new terminology. The variable *temperature* would generally be referred to as the experimental variable or *factor*, and the two temperature settings are *levels*. What is being measured is the process yield, and this would be labeled the response variable.

The use of graphical displays for ANOVA will be presented shortly when the current example is modified somewhat. At this point, however, we shall assume that the data have previously been studied with graphical methods and will proceed directly to the ANOVA calculations.

As was discussed previously, there are two types of variation in the data — variation between the temperature settings and variation within each temperature setting. Accordingly, these sources are often designated in ANOVA tables as "between" and "within."

A measure of the between variability would be the difference between the average value at each temperature setting and the overall average. Numerically we obtain $\overline{x}_{250°\text{F}} - \overline{x}_{\text{all}} = 2.45 - 2.51 = -0.06$ and $\overline{x}_{300°\text{F}} - \overline{x}_{\text{all}} = 2.57 - 2.51 = 0.06$. With two levels these two numbers must be the same, except for the signs that must be opposite, since the numbers must sum to zero. The extent to which these two numbers differ from zero is a measure of the temperature "effect" (another new term). Since the sum must always equal zero (even when there are more than two levels), the sum of the numbers will not provide any useful information. We can, however, add the squares of the numbers together, and this sum is then multiplied by n (= 10 here) and the product used in ANOVA tables. [The n results from summing the squares over the observations within each group. See Montgomery (1997) for the algebraic details.] Numerically, we obtain $0.0072 \times 10 = 0.072$, and we will let this be represented by SS_{temp}, which represents the *sum of squares* (SS) due to the temperature effect. A measure of variability for all of the data (20 numbers in this example) is obtained by squaring the difference between each number and $\bar{x}_{all}$ and then adding all of the squared differences. (The reader may recall from exercise 1 in Chapter 3 that this is also how the numerator of the sample variance is obtained and should recall that an equivalent formula was given for obtaining the numerator. That equivalent formula can be used in ANOVA.) For these data we obtain 1.698, and this will be represented by SS_{total}. The difference between SS_{total} and SS_{temp} can be represented by SS_{within}, so that $SS_{within} = SS_{total} - SS_{temp}$, and for these data we obtain $1.698 - 0.072 = 1.626$. It could be shown that SS_{within} is the same as the numerator of the expression that was given for s_p^2 in the preceding section for the exact t test. Thus, the (equivalent) relationship between the exact t test and ANOVA for a single factor with two levels will now begin to take shape.

An ANOVA table can be constructed using these sums of squares and other quantities. The ANOVA table is given in Table 13.2.

An explanation of the entries in this table is as follows. The degrees of freedom for "Total" will always be equal to the total number of data values minus 1. (This is true for *any* ANOVA table and for any type of design.) One is subtracted from the total number of values because there is one linear constraint on these values, namely, the sum of their deviations from the overall average must equal zero. The d.f. for the factor (temperature, in this case) will always be equal to the number of levels of the factor minus 1. The "1" results from the fact that the sum of the differences of each factor level average from the overall average must equal zero. The d.f. for within will always be equal to 1 less than the number of observations per level, multiplied by the number of levels. The entries in the MS column (where MS is an abbreviation for *mean square*) are obtained by taking the values in the SS column and dividing them by the corresponding values in the d.f. column (e.g., $0.090 = 1.626/18$). The F value is obtained by

TABLE 13.2 ANOVA Table for Temperature Data

Source of Variation	d.f.	SS	MS	F
Temperature	1	0.072	0.072	0.797
Within	18	1.626	0.090	
Total	19	1.698		

taking the top mean square and dividing it by the bottom one. The ratio of these mean squares is (before the numbers are plugged in) a random variable that has an F distribution (discussed in Chapter 3) with 1 d.f. for the numerator and 18 d.f. for the denominator. Thus, the analysis of these data could be carried out using the F table contained in Table D of the Appendix to the book. We might then ask what is the probability of obtaining a value of the F statistic that is as large or larger than 0.797 when there is no true difference in the effects of the temperature settings? The probability is .3838, which is exactly double the probability obtained using the exact t test. Is this a coincidence? No, this will always happen. This is due in part to the fact that the square of any value in the t table (Table C in the Appendix to the book) with "ν" d.f. and a tail area of a will equal the value in the F table (Table D in the Appendix to the book) with 1 d.f. for the numerator and ν d.f. for the denominator, and a tail area of $2a$. It can also be shown that the square of the t statistic given by Eq. (13.3) is equivalent to the F statistic. [It is easy to see this with numbers, such as the numbers in this example, as $(-0.893)^2 = 0.797$.] The reason that the tail area is doubled is because the F statistic cannot be negative (sums of *squares* and mean *squares* must obviously be nonnegative), whereas the t statistic could be either positive or negative. Thus, the tail areas for the negative and positive portions of the t distribution are added together to form the tail area for the F distribution.

Since the use of ANOVA for a single factor with two levels will produce results identical to the use of an exact t test, it would seem reasonable to assume that the ANOVA approach would be based upon the same assumptions as the exact t test. This is indeed the case. Thus, ANOVA is based upon the assumptions of normality of the populations and equality of the variances. As with the exact t test, it is relatively insensitive to slight-to-moderate departures from the first assumption and is also relatively insensitive to the assumption of equal variances, provided that the sample sizes are at least approximately equal. Unlike the t test, however, ANOVA can be used when a factor has more than two levels.

13.5.1 ANOVA for a Single Factor with More Than Two Levels

Let us assume now that the experiment to compare the effect of different temperature settings on process yield actually had three different settings, and the data for the third setting, 350°F, were temporarily mislaid. The full data set appears in Table 13.3.

For simplicity we will assume that these data have been collected over a period of 6 weeks, with 2 weeks at each temperature setting. This is the way the experiment would have to be designed if a company had only one plant that could be used in the experiment. If a company had three plants, however, one temperature setting could be used at each plant, and the experiment could then be conducted in 2 weeks rather than 6 weeks. That would certainly seem desirable, but one possible drawback is that the plants might not be identical (and probably would not be) in terms of productive capability. Differences in process yield for the three temperature settings might actually be due to plant differences,

TABLE 13.3 Process Yield Data for All Three Temperature Settings (in tons, 6-Week Period)

Day	250°F	300°F	350°F
	Week 1		
Monday	2.4	2.6	3.2
Tuesday	2.7	2.4	3.0
Wednesday	2.2	2.8	3.1
Thursday	2.5	2.5	2.8
Friday	2.0	2.2	2.5
	Week 2		
Monday	2.5	2.7	2.9
Tuesday	2.8	2.3	3.1
Wednesday	2.9	3.1	3.4
Thursday	2.4	2.9	3.2
Friday	2.1	2.2	2.6

and with this design it would be impossible to separate the temperature effect from the plant effect. This illustrates why the design of an experiment is so important — much more so than the analysis of the data. It has often been said that data from a well-designed experiment will practically analyze itself.

What about variation between the 2 weeks as well as day-to-day variation? It might seem as though we should be particularly concerned about the latter since it was previously determined in the analysis of the first two settings that the process yield drops off on Fridays.

Remember, however, that our objective is to determine whether or not these three temperature settings affect process yield, not to isolate every possible cause of variation in process yield. If the objective had been to identify factors that may influence process yield, then the experimental design should have been constructed with that objective in mind, and the design would likely have been different from the current design. Designs for the identification of such factors are called *screening designs*, and they will be discussed later in the chapter.

It would seem unlikely that there would be significant week-to-week variation, particularly if normal operating conditions existed in each of the 6 weeks. It can also be seen that the 2-week totals within each temperature setting do not differ greatly. Even if the totals did differ considerably, it would be a question of how the totals differed. For example, if the differences between the totals for weeks 1 and 2, 3 and 4, and 5 and 6 were virtually the same, the analysis to be presented would not be seriously affected unless the differences were large. (Large differences would cause the within mean square to be considerably inflated.) On the other hand, if weeks 1 and 6 differed considerably (and the difference was due to external factors) but the totals for the other weeks were virtually the same, the analysis would be of no value.

Day-to-day variation would not be a problem as long as the variation was essentially the same for each week and for each temperature setting. If not, this type of variation could also contaminate the analysis.

In general, much thought must go into the design of an experiment. The analysis of data from an experiment that has been designed properly is generally straightforward; the analysis of data from an experiment that is poorly designed is either difficult or impossible.

We will assume that the data in Table 13.3 are being analyzed for the first time, ignoring the fact that the data for the first two temperature settings were analyzed previously.

It has been stated that the first step in analyzing any set of data (regardless of the analysis that is to be performed) is to plot the data—usually in more than one way. The present use of graphical methods to analyze data from designed experiments is largely attributable to the teachings of Horace P. Andrews and Ellis R. Ott and the subsequent teachings of their students. The influence of the former is especially evident in *Experiments in Industry—Design, Analysis, and Interpretation of Results* (Snee, 1985), a collection of papers written by Andrews's former students and dedicated to his memory. Ott and Schilling (1990) also emphasize graphical procedures, particularly analysis of means. (The latter is discussed in detail in Chapter 16.)

A box-and-whisker plot was introduced in Chapter 11 as a simple way of providing a useful graphical summary of data. A first step in analyzing the data in Table 13.3 might well be to construct such a plot for each temperature setting.

Figure 13.5 shows the yield for the third temperature setting to be considerably higher than for the first two. The fact that the boxes appear to be of roughly the

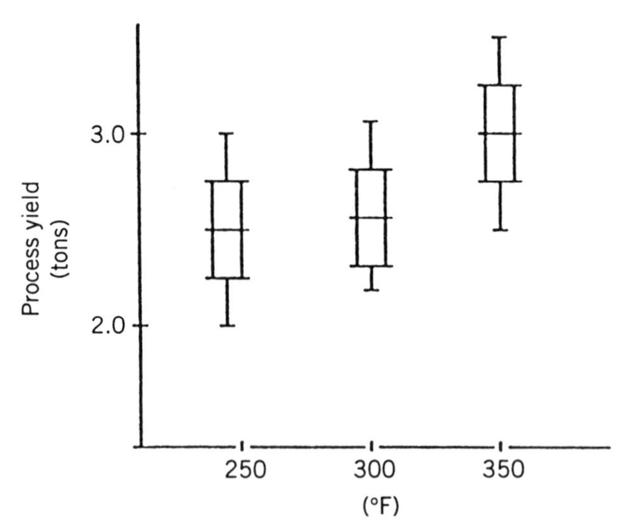

FIGURE 13.5 Box-and-whisker plot of data in Table 13.3.

same height coupled with the absence of any outlying values (none was labeled on the plot) would tend to suggest that the variability within each setting probably does not differ very much from setting to setting. (Of course, we have already seen that this was true for the first two settings.) This is important since, as noted in the preceding section, equality of variances is a prerequisite for the use of ANOVA, and the analysis is undermined somewhat when the requirement is not met, particularly when the sample sizes differ considerably.

How might we determine whether or not the requirement is met in a more precise manner than is provided by graphs? One approach would be to use the test given by Layard (1973), which was mentioned previously. Another test that does not require an assumption of normality is given by Conover (1980, p. 241), and there are various other tests that are sensitive to nonnormality.

There does not appear to be a need to use such a test for the present example, however, as the (sample) variances for each temperature setting appear to be roughly equal.

As with the data in Table 13.1, there is a need to compute the sum of squares for between (temperature), within, and total. The approach to obtaining the first sum of squares was motivated somewhat heuristically in the preceding section. The sum of squares for temperature can also be computed as

$$\mathrm{SS}_{\mathrm{temp}} = \sum_{i=1}^{3} \frac{T_i^2}{n_i} - \frac{\left(\sum_{i=1}^{3} T_i\right)^2}{N} \tag{13.4}$$

where T_i represents the total of the observations for the ith level of temperature and n_i represents the number of observations for the ith level. Similarly, $\sum_{i=1}^{3} T_i$ denotes the grand total of all of the observations, and N is the number of such observations. For ANOVA formulas in general, whenever a number is squared, it is always divided by the number of observations that were summed to produce that number.

For the current data we obtain

$$\begin{aligned} \mathrm{SS}_{\mathrm{temp}} &= \frac{(24.5)^2 + (25.7)^2 + (29.8)^2}{10} - \frac{(24.5 + 25.7 + 29.8)^2}{30} \\ &= 214.878 - 213.333 \\ &= 1.545 \end{aligned}$$

We should note, however, that such "sums of squares" formulas should not be routinely used for computations, as they can sometimes produce results that are very much in error due to roundoff. This can happen in particular when a sum of squares value is close to zero and the two terms in the formula are both large. [See Chan, Golub, and Leveque (1983) for further details.] Computer software with good numerical properties should generally be used for computations, rather than performing hand computations with the usual "cookbook" formulas. The latter are being used here simply for illustrative purposes, although their use with this data set is not going to cause any problems.

It was noted previously that the "total sum of squares" is always computed the same way. This can be expressed as

$$SS_{\text{total}} = \sum_{i=1}^{N} y_i^2 - \frac{\left(\sum_{i=1}^{N} y_i\right)^2}{N}$$

where y_i denotes the ith observation and, as before, N denotes the total number of such observations. For these data we obtain

$$\begin{aligned} SS_{\text{total}} &= 217.220 - \frac{(80)^2}{30} \\ &= 217.220 - 213.333 \\ &= 3.887 \end{aligned}$$

It then follows that the "within" sum of squares is $3.887 - 1.545 = 2.342$. These numbers can then be used to produce the ANOVA table that is given in Table 13.4. The probability of obtaining a value for F that is greater than or equal to 8.91 when there is no true difference in the effects of the temperature settings is .001. Since this is a very small probability, we would thus conclude that there is a difference in the effects of the temperature settings, and it is apparent from the boxplot given previously that the difference is due to the third setting. Thus, based on this and previous analyses, it would seem reasonable to conclude that the effect of the third setting differs significantly from the effect of the first two settings, but the latter do not differ significantly from each other.

13.5.2 Multiple Comparison Procedures

The formal assessment of such differences is generally made through the use of multiple comparison procedures. The problem in using such procedures is that a sizable number of them have been proposed, and there is no one procedure that can be deemed superior to all the others. For example, some procedures are conservative in that they produce few significant differences, some can only be used if the F test is significant, some can be used only to construct comparisons

TABLE 13.4 ANOVA Table for the Data in Table 13.3

Source of Variation	d.f.	SS	MS	F
Temperature	2	1.545	0.7725	8.91
Within	27	2.342	0.0867	
Total	29	3.887		

before the data are collected, and so on. As a result, one might successfully argue that graphical procedures, although approximate, might be a better choice for practitioners. The boxplot is one such approach. A slightly more refined graphical procedure is the reference distribution approach suggested by Box, et al. (1978, p. 190). Another graphical approach for detecting differences is analysis of means, which is discussed in Chapter 16. Readers interested in formal multiple comparison procedures are referred to Montgomery (1997), Neter et al. (1996), and Chapter 2 of Miller (1981).

13.5.3 Sample Size Determination

The question of how many days or weeks to run the temperature experiment was alluded to at the beginning of the discussion. There is no single correct answer to this question, so we might lead into a discussion by determining the shortest length of time that the experiment can be run. The answer is 6 days—2 at each setting. Why 6 days? Recall that we are actually pooling the variances for the different levels of the factor, and at least two numbers must be present before a variance can be calculated. So it would be possible to run the experiment using 2 days for each temperature setting, but would it be desirable? The answer is no. The variance of a sample variance is a function of both the population variance and the sample size, and for a sample of size 2 from a normal population that variance will be double the *square* of the population variance. Thus, judging whether or not the temperature effect is significant could depend on a poorly estimated common population variance.

In general, the determination of sample size in ANOVA can be done explicitly if the experimenter has in mind the minimum difference in the effects of (at least) two factor levels that he or she wants to detect with a given probability. The experimenter can then use tables for sample size determination such as those found in Nelson (1985).

Alternatively, it is useful to have simple "portable" formulas that can be used without having to resort to tables. Simple formulas of this type are given by Wheeler (1974), the most useful of which may be the following. If r denotes the number of levels of a factor, σ^2 is the variance of the observations, and Δ denotes the minimum absolute pairwise difference between the expected values of the means of the factor that one wishes to detect with probability .90 if the difference exists, and a significance level of $\alpha = 0.05$ is being used, the requisite sample size is given by

$$n = \left(\frac{4r\sigma}{\Delta}\right)^2$$

See Wheeler (1974) for details and for sample size expressions for other scenarios.

Software for sample size determination is also readily available. For example, O'Brien (1997) has provided a SAS macro for sample size determination. Graphs and charts for sample size determination have been provided by Brush (1988) and Odeh and Fox (1975).

13.5.4 Additional Terms and Concepts in One-Factor ANOVA

In this section we introduce some new terms and briefly discuss some common experimental designs.

An *experimental unit* is the unit (of material perhaps) to which a treatment is applied. The days are the experimental units in the experiment just analyzed.

If the temperature settings had been randomly assigned to the days, we would have had a *completely randomized design.* Would such a design have been advisable for the temperature experiment? Only if it was believed that production might be correlated over the days and weeks involved in the experiment, independent of the particular temperature setting. It has often been stated that one should *block* on factors that could be expected to influence the response variable (i.e., what is being measured) and randomize over factors that might be influential, but that could not be "blocked."

In the configuration for the tire experiment given in Figure 13.1, the cars were the blocks and the variation due to cars (blocks) would be isolated when the data were analyzed. If the tire brands had been randomly assigned to the wheel positions within each block (car), we would have had a *randomized block design.* As we saw at the time, however, this would be a poor design because it would (likely) result in an imbalance relative to wheel positions. The design settled upon and illustrated in Figure 13.2 is called a *Latin square design.*

It should not be inferred that the latter design is "good" and a randomized block design is "bad." Their appropriate use depends on how many extraneous factors are to be adjusted for—one in the case of a randomized block design (e.g., cars) and two in the case of a Latin square design (e.g., cars and wheel position). We may think of this relative to common-cause variation versus variation due to assignable causes when control charts are used. We do not want variation due to assignable causes to inflate control limits. Similarly, we do not want variation due to extraneous factors to inflate SS_{within}.

Factors are generally classified as *fixed* or *random.* When there is only one factor, the classification does not have any effect on how the data are analyzed, but it does make a difference when there are two or more factors. A factor is fixed if the levels of a factor are predetermined and the experimenter is interested only in those particular levels (e.g., 250, 300, or 350°F). A factor is classified as random if the levels are selected at random from a population of levels (e.g., 254, 287, and 326°F) and the inference is to apply to this population of levels rather than to the particular levels used in the experiment, as is the case with fixed factors.

A completely randomized design is sometimes (erroneously) equated with one-factor ANOVA, but the two are really not the same. The latter is simply a way to classify the general layout of an experiment (which is why it is sometimes termed a one-way layout), whereas the former refers to how the treatments are assigned to the experimental units.

Regression models were discussed in the preceding chapter. Similarly, ANOVA models can also be presented and discussed. The model for one-factor ANOVA can be written as

$$Y_{ij} = \mu + \tau_j + \varepsilon_{ij} \tag{13.5}$$

where j denotes the jth level of the single factor and i denotes the ith observation. Thus, Y_{ij} represents the ith observation for the jth level. Further, τ_j represents the effect of the jth level, μ is a constant that is the value that we would expect for each Y_{ij} if the effects of each level were zero, and ε_{ij} represents the error term (with the errors assumed to be random), analogous to the error term in a regression model. If the effects were all zero, there would be no need for a τ_j term in the model, so the model could then be written as

$$Y_{ij} = \mu + \varepsilon_{ij} \tag{13.6}$$

The F test illustrated previously determines whether the appropriate model is Eq. (13.5) or Eq. (13.6).

As stated previously, the data in one-factor ANOVA are analyzed in the same way regardless of whether the single factor is fixed or random, but the interpretation of the components of Eq. (13.5) does differ. Specifically, τ_j is a constant if the factor is fixed and a random variable if the factor is random. The error term, ε_{ij}, is NID $(0, \sigma^2)$ in both cases, and the Y_{ij} are assumed to be normally distributed in both cases. In the random-factor case, however, the Y_{ij} are not independent as the observations for the jth factor level are correlated since τ_j is a random variable and each observation is commonly influenced by the jth level. (It might seem that the same could be said of the fixed-factor case since each observation within the jth factor level shares the effect of the jth factor level, but these effects are constants, and variables are not correlated just because they have a constant in common.)

The data in the temperature experiment were "balanced" in that there was the same number of observations (10) for each level of the factor. When there is just a single factor, there is no difficulty in analyzing the data for an unequal number of observations. Specifically, the total sum of squares would be calculated in the same way, and Eq. (13.4) would be used to compute the "between" sum of squares, with the n_i not all equal.

13.6 REGRESSION ANALYSIS OF DATA FROM DESIGNED EXPERIMENTS

The material in the preceding sections of this chapter was intended to introduce the reader to ANOVA and, in general, to the analysis of data from designed experiments. For a single factor, ANOVA is intuitive and, if necessary, could be easily performed with a hand calculator. Regression could also be used as the method of analysis, however, and it has greater utility than ANOVA, particularly for complex designs. Regression provides the user with the tools for residual analysis, whereas the latter is not generally regarded as being part of ANOVA.

The value of residual analyses in regression was discussed in Chapter 12; residual analyses are also important in analyzing data obtained through the use of experimental designs.

The essential differences between regression and ANOVA will be noted. With the former, parameters are estimated, which is not always the case when the latter is used. The general form of an ANOVA table results from the implicit assumption of an underlying model, but the model is not always used explicitly. When the model is not used explicitly, diagnostic checking (through the use of residual analysis) cannot be performed. Analysis of variance is straightforward but mechanical and for fixed factors should either be supplemented or supplanted by regression analysis.

The least squares estimators in regression analysis resulted from minimizing the sum of squared errors. From Eq. (13.5) we obtain

$$\varepsilon_{ij} = Y_{ij} - \mu - \tau_j$$

so that

$$\sum_j \sum_i \varepsilon_{ij}^2 = \sum_j \sum_i (Y_{ij} - \mu - \tau_j)^2 \tag{13.7}$$

At this point we will make the assumption that the levels of the factor are fixed. (When this is the case, it is logical to estimate τ_j; when the factor is random, we would want to estimate σ_τ^2, the variance of the effects of the different possible levels, as there would be no point in estimating the effect when interest centers upon an entire range of possible levels.) We will also make the assumption that we are working with balanced data, as this results in some simplification in notation.

It is customary to think of the τ_j effect as a deviation from the overall mean μ. Specifically, $\tau_j = \mu_j - \mu$, where μ_j is the expected value of the response variable for the jth level of the factor. If we define μ to be the average of the μ_j components, it follows that $\sum_j (\mu_j - \mu) = 0$ so that $\sum_j \tau_j = 0$.

This restriction on the τ_j components, which is implied by the definitions of τ_j, μ_j, and μ, allows μ and each τ_j to be estimated using least squares. Specifically, the minimization of Eq. (13.7) produces

$$\hat{\mu} = \overline{Y}_{..}$$

and

$$\hat{\tau}_j = \overline{Y}_{.j} - \overline{Y}_{..}$$

where $\overline{Y}_{..}$ denotes the average of all of the observations and $\overline{Y}_{.j}$ denotes the average of the observations for the jth factor level. It then follows that

$$\begin{aligned} \hat{Y}_{ij} &= \hat{\mu} + \hat{\tau}_j \\ &= \overline{Y}_{..} + (\overline{Y}_{.j} - \overline{Y}_{..}) \\ &= \overline{Y}_{.j} \end{aligned}$$

This is quite intuitive, as we would logically estimate a particular Y_{ij} by using the average of the values for that particular j. It then follows that, analogous to the regression methodology presented in Chapter 12, the residuals are defined as

$$e_{ij} = Y_{ij} - \overline{Y}_{\cdot j}$$

The data in Table 13.3 were previously analyzed using ANOVA. We will now see how those data can be analyzed using regression.

Since we wish to emphasize residual analysis, as well as to provide additional insight, we will obtain the residual sum of squares from the residuals. The value obtained should, of course, be equal to 2.342, which is what was obtained using ANOVA. The residuals are given in Table 13.5.

These residuals uncover some facts that may not have been apparent when the data were analyzed using ANOVA. First, the production is higher for the second week at each temperature setting (since the sums are positive.) What could this mean? Does it take some time for the process to adjust to the new temperature setting? Second, not only is the production high during the second week, but it is especially high (relatively speaking) during Wednesday of that week since the residuals are considerably larger than any of the other residuals.

This illustrates one very important premise of data analysis—the more ways you look at data, the more you are apt to discover. It would take a person with a fairly sharp eye to detect this "peak day" by looking at the data in Table 13.3, whereas the comparatively large positive residuals stand out since about half of the residuals should be negative.

If the residuals in Table 13.5 are squared and the squares are summed, the sum is 2.342, which agrees with the result obtained previously using ANOVA.

The assumption of normality of the errors could be checked by constructing a *dot diagram* of the residuals (covered in Chapter 11). (Note that a histogram could also be constructed, but a histogram is not likely to be very informative when there is a small number of observations, as there will necessarily be a small number of classes using any of the methods for determining the number of classes

TABLE 13.5 Residuals for the Temperature Setting Data

	250°F		300°F		350°C	
Day	Week 1	Week 2	Week 3	Week 4	Week 5	Week 6
Monday	−0.05	0.05	0.03	0.13	0.22	−0.08
Tuesday	0.25	0.35	−0.17	−0.27	0.02	0.12
Wednesday	−0.25	0.45	0.23	0.53	0.12	0.42
Thursday	0.05	−0.05	−0.07	0.33	−0.18	0.22
Friday	−0.45	−0.35	−0.37	−0.37	−0.48	−0.38
Total	−0.45	0.45	−0.35	0.35	−0.30	0.30
$\overline{Y}_{ij}$	2.45		2.57		2.98	

given in Chapter 2.) The dot diagram for the residuals in Table 13.5 is given in Figure 13.6. There is no strong evidence from Figure 13.6 of serious departure from a normal distribution. The estimated standard deviation of the error term is, from Table 13.4, equal to $\sqrt{0.0867}$, which is 0.2944. With 32 residuals we would expect all of them to be within ± 3 standard deviations, which is the case. We would theoretically expect one or two to be outside of ± 2 standard deviations; there is actually none, although one, 0.53, is close to the upper boundary. We would also expect approximately 10 to be outside ± 1 standard deviation, and there are actually 11 that are outside. (These expectations were obtained by multiplying the probabilities in the standard normal table, Table B in the Appendix to the book, times 32.) Thus, there appears to be at least approximate normality.

The residuals for each temperature setting and for each week should also be examined. Since there are 2 weeks for each temperature setting, a diagram could be constructed for each temperature setting in which 'F' would denote a residual for the first week and 'S' would denote a residual for the second week.

The diagrams for each of the three temperature settings are given in Figure 13.7. It is clear from Figure 13.7 that the largest residuals occur in the second week and that the range of the residuals is roughly the same for each temperature setting. (The latter is reasonably important as it relates to the prerequisite assumption of equal variances.)

As was indicated in the chapter on linear regression, the residuals should also be plotted against time and against the predicted values of the response variable. The plot against time is given in Figure 13.8 and the plot against the predicted values in Figure 13.9. As can be seen from Figure 13.9, the spread of

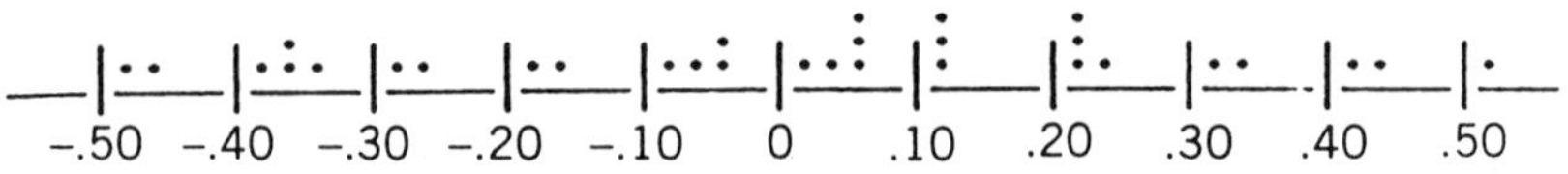

FIGURE 13.6 Dot diagram of the residuals in Table 13.5.

−.50 −.40 −.30 −.20 −.10 0 .10 .20 .30 .40 .50 250°F

−.50 −.40 −.30 −.20 −.10 0 .10 .20 .30 .40 .50 300°F

−.50 −.40 −.30 −.20 −.10 0 .10 .20 .30 .40 .50 350°F

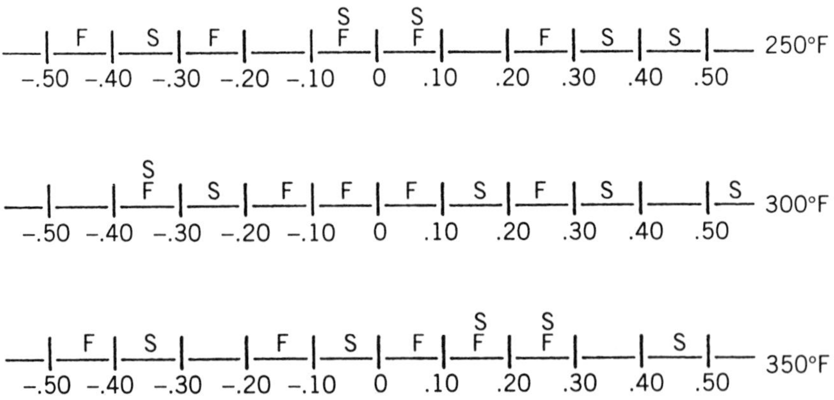

FIGURE 13.7 Residuals from Table 13.5 by week and by temperature.

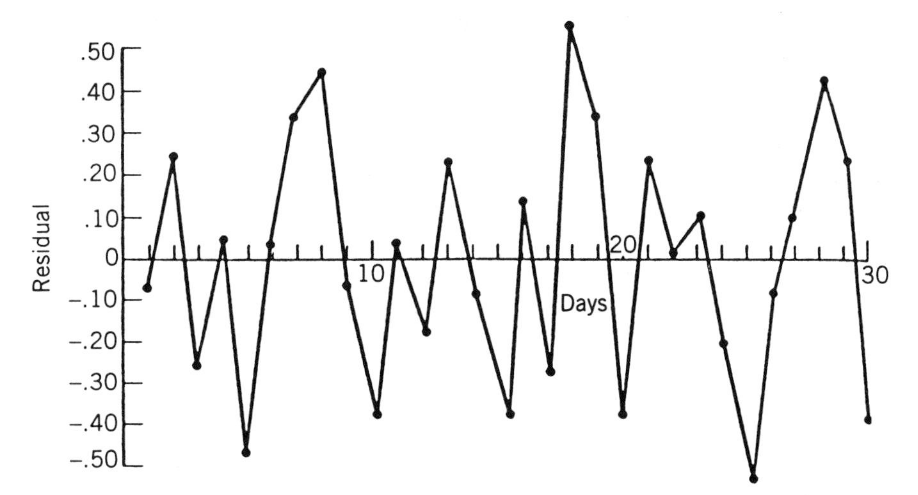

FIGURE 13.8 Plot of the residuals from Table 13.5 against time.

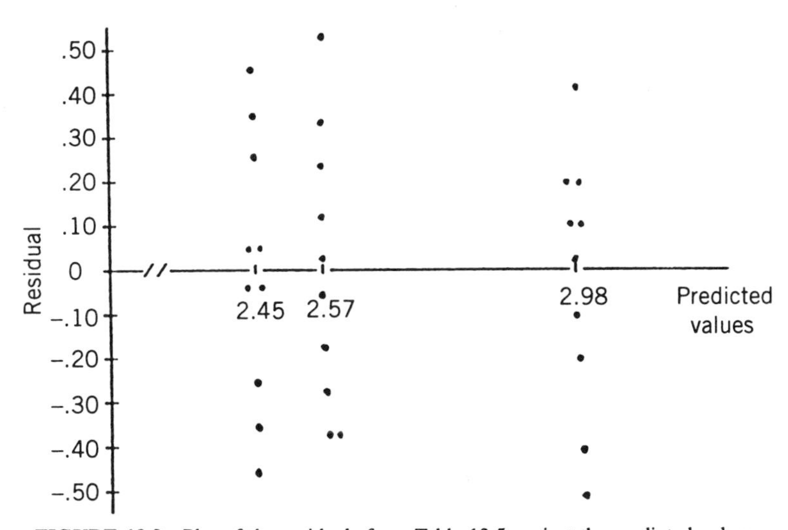

FIGURE 13.9 Plot of the residuals from Table 13.5 against the predicted values.

the residuals is virtually the same for each temperature setting, which is also part of the information provided by Figure 13.7. The points in residual plots are not generally connected, although it does make sense to connect points when "time" is the label of the horizontal axis (as with a control chart). Here, connecting the points provides fairly strong evidence of a cyclical pattern relative to the peaks. Subsequent analysis would then likely lead to the same discovery that was made when the data in Table 13.5 were analyzed, namely, an increase in production during the second week for each temperature setting.

It has been previously determined that there is a "week effect"; that is, the production clearly differs for the 2 weeks at each temperature setting. This effect may or may not be significant, however. Could an experimenter at this point reanalyze the data to determine if the week effect is significant? The answer is "yes," and this brings up an important point. Data from a designed experiment do not have to be analyzed in accordance with what the experimenter had in mind when the experiment was designed. Fewer factors can be examined than what the experimenter had originally intended to look at, and, occasionally, more factors can be examined. The latter will often be impossible, however, as it can be difficult to isolate factors if an experiment has not been designed in such a way as to permit this isolation.

13.7 ANOVA FOR TWO FACTORS

We will introduce this new topic by using "weeks" and "temperature" as the two factors. This might be called a *factorial design* or a *cross-classified design*, but some care needs to be exercised, as for the data in Table 13.3 it is really more complicated than that. Specifically, in a factorial design each level of every factor is "crossed" with each level of every other factor. Thus, if there are a levels of one factor and b levels of a second factor, there are then ab combinations of factor levels. Since there are actually 6 weeks involved in the temperature-setting experiment, there would have to be $6 \times 3 = 18$ combinations for it to be a cross-classified design. Since, there are obviously only six combinations of weeks and temperatures, it clearly cannot be a cross-classified design. (An actual cross-classified design will be presented later in this section.) What type of design is this then?

It is actually a *nested-factor design* as weeks are "nested" within temperature. The corresponding model is

$$Y_{ijk} = \mu + \tau_i + \beta_{j(i)} + \varepsilon_{k(ij)}$$

where $i = 1, 2, 3$; $j = 1, 2$; and $k = 1, 2, 3, 4, 5$. Here i designates the temperature, j the week, and k the replicate factor (days in this case). Further, $j(i)$ indicates that weeks are nested within temperature, and $k(ij)$ indicates that the replicate factor is nested within each i, j combination. The use of nested designs in process control and process variation studies has been discussed by Sinibaldi (1983), Snee (1983), and Bainbridge (1965), among others. These designs are also called *hierarchical designs* and are generally used for estimating *components of variance*.

The analysis of data obtained from a nested design is quite different from the analysis of data from a factorial design. Nested designs are covered in detail in Montgomery (1997). The intent here is to alert the reader that a design with two factors could be a nested-factor design or a crossed (classification) design, and it is desirable to be able to distinguish between them.

13.7.1 ANOVA with Two Factors: Factorial Designs

As we extend the discussion of designed experiments and consider more than one factor, it seems logical to pose the following question: Why not study each factor separately rather than simultaneously? Figure 13.10 provides an answer to this question. Assume that a company's engineering department is asked to investigate how to maximize process yield, where it is generally accepted that temperature and pressure have a profound effect upon yield. Three of the engineers are given this assignment, and these will be represented by A, B, and C, respectively. Each engineer conducts his own experiment. Assume that A and B each investigates only one factor at a time, whereas C decides to look at both factors simultaneously. Assume further that Figure 13.10 depicts what can be expected to result when both factors are studied together.

If A had used the low temperature (T_1) and varied the pressure from low to high, he would conclude that the best way to increase the yield is to increase the pressure, whereas he would have reached the opposite conclusion if he had used the high temperature. Similarly, if engineer B had set the pressure at the high level (P_2), she would have concluded that the best way to increase yield is to reduce the temperature, whereas she would have reached the opposite conclusion if she had used the low-pressure level.

Engineer C, on the other hand, would be in the proper position to conclude that interpreting the effects of the two factors would be somewhat difficult because of the *interaction effect* of the two factors. Interaction effects are depicted graphically by the lack of parallelism of lines as in Figure 13.10.

This type of feedback is not available when factors are studied separately rather than together. These "one-at-a-time" plans have, unfortunately, been used extensively in industry. They are considered to have very little value, in general, although Daniel (1973, 1976, p. 25) discusses their value when examining three factors.

The presence of interaction, particularly extreme interaction, can easily result in completely erroneous conclusions being drawn if an experimenter is not

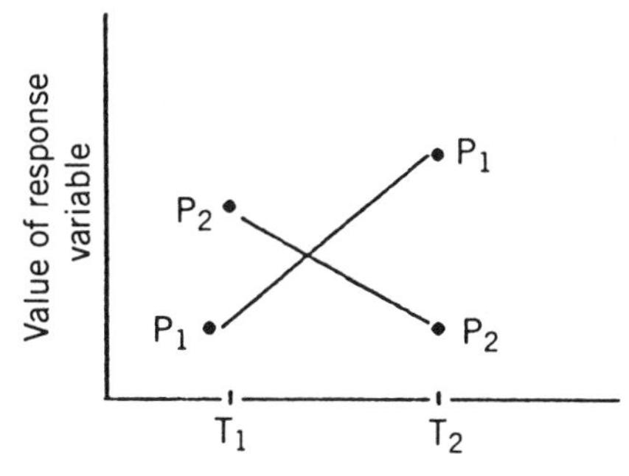

FIGURE 13.10 Interaction profile: P_1, low pressure; P_2, high pressure; T_1, low temperature; T_2, high temperature.

careful. Consider the configuration in Figure 13.11, which is a slight modification of Figure 13.10. It is clear that the value of the response variable varies by 10 units when either temperature or pressure is set at one of the two levels and the other factor is varied between its two levels. Yet when the data are analyzed, we would find that the sum of squares for each of the two factors is exactly zero. Thus, a "blind" analysis of computer output or a hand-constructed ANOVA table would lead an experimenter to conclude that there is neither a temperature nor a pressure effect, although each clearly has an effect on the response variable.

This falls in line with what has been stressed throughout this book, namely, there is much more to analyzing statistical data than just looking at numbers. Graphical displays are extremely important and should be used with virtually every statistical procedure.

Figures 13.10 and 13.11 are examples of *interaction profiles*, and it is important that these and other graphical displays be used in analyzing data from multifactor designs.

Although the configuration in Figure 13.11 is an extreme example, and not likely to occur exactly in practice, we should not be surprised to find significant interactions and nonsignificant factor effects. (Factor effects are generally called *main effects*.) Daniel (1976, p. 21) stresses that data from a designed experiment should not be reported in terms of main effects and interactions if an interaction is more than one-third of a main effect.

To this point in our discussion of multifactor experiments, the terms *main effects* and *factor effects* have been used rather generally, with implied reference to an ANOVA table. An effect of a factor can be defined more specifically as the change in the response variable that results from a change in the level of that factor.

13.7.2 Effect Estimates

Referring to Figure 13.11, if the pressure is set at P_1 and the temperature is increased from T_1 to T_2, the value of the response variable increases by 10

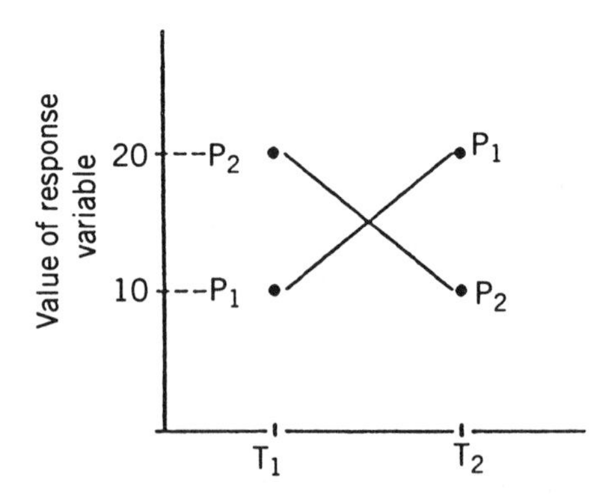

FIGURE 13.11 Extreme interaction. Abbreviations as in Figure 13.10.

units. On the other hand, if the pressure is set at P_2 and the temperature is increased from T_1 to T_2, the value of the response variable decreases by 10 units. It would seem logical to define the *temperature effect* as the average of these two changes (which of course is zero in this case). That is,

$$T = \tfrac{1}{2}(T_2P_1 - T_1P_1 + T_2P_2 - T_1P_2) \tag{13.8}$$

Similar reasoning for the *pressure effect* would lead us to define that effect, P, as

$$P = \tfrac{1}{2}(T_1P_2 - T_1P_1 + T_2P_2 - T_2P_1) \tag{13.9}$$

which is also zero for this example. The way in which the *interaction effect* would be defined might not be quite so intuitive, but another diagram may help. Although Figure 13.12 looks quite different from Figure 13.11, only one change has been made, namely, T_2P_2 was changed from 10 to 30. However, this now causes the lines to be parallel. It was stated previously that interaction results from nonparallelism of the two lines. It is clear from Figure 13.12 that the two lines are parallel because $T_2P_2 - T_2P_1 = T_1P_2 - T_1P_1$. Thus, the extent of the interaction will depend upon the difference of these two expressions, that is, $T_2P_2 - T_2P_1 - (T_1P_2 - T_1P_1)$. Accordingly, we could (and do) define the interaction effect, TP, as

$$\begin{aligned} TP &= \tfrac{1}{2}[T_2P_2 - T_2P_1 - (T_1P_2 - T_1P_1)] \\ &= \tfrac{1}{2}(T_2P_2 - T_2P_1 - T_1P_2 + T_1P_1) \end{aligned} \tag{13.10}$$

For Figure 13.12 this value is, of course, zero, but for the original interaction profile in Figure 13.11 the value is -10. Since we now know that for the data in Figure 13.11 $T = 0$, $P = 0$, and $TP = -10$, we could obtain the sum of squares for these three effects by squaring the effect estimates, where the effect estimates are given by Eqs. (13.8), (13.9), and (13.10). We thus have $\text{SS}_t = 0^2 = 0$, $\text{SS}_p =$

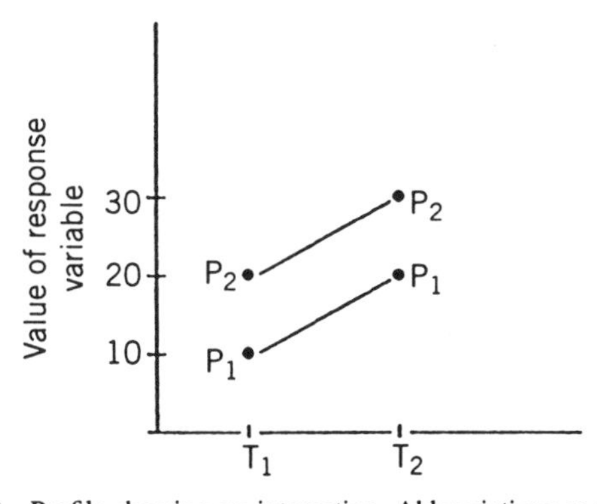

FIGURE 13.12 Profile showing no interaction. Abbreviations as in Figure 13.10.

$0^2 = 0$, and $SS_{tp} = (-10)^2 = 100$. In general, for a 2^k design with r observations per treatment combination, $SS_{effect} = r(2^{k-2})$ (effect estimate)^2. Here k, the number of factors, equals 2 and $r = 1$ so the result is simply (effect estimate)^2.

13.7.3 ANOVA Table for Unreplicated Two-Factor Design

Since we now have these sums of squares, we might attempt to construct an ANOVA table. Remembering that the degrees of freedom (d.f.) for "Total" is always the total number of observations minus 1 and the d.f. for a factor is always the number of factor levels minus 1, we thus have d.f.(Total) = 3, d.f.(T) = 1, and d.f.(P) = 1. The d.f. for any interaction effect is always obtained as the product of the separate d.f. of each factor that comprises the interaction. Thus, in this case we have d.f.(TP) = (1)(1) = 1.

If we add the d.f. for T, P, and TP, we recognize immediately that we have a problem. Specifically, there is no d.f. left for estimating σ^2. Thus, unless we have an estimate of σ^2 from a previous experiment (remember that experimentation should be thought of as being sequential), we have a case in which the interaction is said to be "confounded" (i.e., confused or entangled) with the "residual," where the latter might be used in estimating σ^2. We can summarize what we know to this point in the ANOVA table given in Table 13.6.

Notice that the F values are not filled in. It is "clear" that there is no temperature effect and no pressure effect since the sum of squares for each is zero. Remember, however, we recently saw that each does have an effect on the response variable; their effects are simply masked by the interaction effect. Under such conditions, it is necessary to look at *half effects*; that is, look at the effect of a factor at each level of the other factor. For these data the half effect of T is 10 at P_1 and -10 at P_2, and similarly for P.

It was stated in the section on one-factor ANOVA that the analysis is not influenced by whether the factor is fixed or random. This is not true when there is more than one factor, however. In general, when both factors are fixed, the main effects and the interaction (if separable from the residual) are tested against the residual. When both factors are random, the main effects are tested against the interaction effect, which, in turn, is tested against the residual. When one factor is fixed and the other one random, the fixed factor is tested against the interaction, the random factor is tested against the residual, and the interaction is

TABLE 13.6 ANOVA for the Data in Figure 13.11

Source of Variation	d.f.	SS	MS	F
T	1	0	0	
P	1	0	0	
TP (residual)	1	100	100	
Total	3	100		

tested against the residual. (By "tested against", we mean that the mean square for what follows these words is used in producing the F statistic.)

In this example the interaction is not separable from the residual because the experiment has not been "replicated"; that is, the entire experiment has not been repeated so as to produce more than one observation ($r > 1$) per treatment combination. This should be distinguished from *multiple readings* obtained within a *single* experiment which does *not* constitute a replicated experiment (i.e., the entire experiment is not being replicated). This may seem like a subtle difference, but it is an important distinction. For additional discussion on this topic the reader is referred to Box et al. (1978, p. 319).

If a prior estimate of σ^2 is available, possibly from a previous replicated experiment with perhaps slightly different factor levels, that estimate could be used in testing for significance of the main effects. [This idea of using an estimate of σ^2 from a previous experiment forms part of the foundation of evolutionary operation (EVOP), which is covered in Chapter 15.] If a prior estimate is not available, we might still be able to obtain an estimate of σ^2.

Tukey (1949) proposed a test for detecting interaction of a specific functional form for an unreplicated factorial. The test is described in detail in various sources, including Box et al. (1978, p. 222), Snedecor and Cochran (1980, p. 283), and Montgomery (1997, p. 253). The general idea is to decompose the residual into an interaction component and an experimental error component and perform an F test on the interaction. If the test is not significant, then σ^2 might be estimated using the residual. It should be recognized, however, that this test will only detect an interaction that can be expressed as a product of main effects times a constant.

It should be noted that there is a difference, conceptually, between "experimental error" and "residual," and the latter cannot be used, in general, as a substitute for the former. Experimental error should be thought of as the variability that results for a given combination of factor levels in a replicated experiment and is comprised of variability due to factors not included in the experiment, sampling variability, and perhaps variability due to measurement error. A residual (as in residual sum of squares) may consist of various interaction terms that are thought to be not significant, in addition to experimental error. Specifically, it would be logical to estimate σ^2 by 0.0867 from Table 13.4, but it would be totally illogical to estimate σ^2 by 100 from Table 13.6. With the latter σ^2 would be estimated using an interaction that from Figure 13.11 is obviously large.

It is interesting to note that Tukey's test would not detect this interaction. In fact, the test would indicate that the interaction is zero because the main effects are zero. We should remember that the test is not a general test for detecting the presence of interaction, nor can there be such a test for an unreplicated experiment.

This idea of experimental error versus residual is a very important one, and we will see later in the chapter how we can go wrong by using an interaction to estimate σ^2 for a set of real data.

Can the analysis begun in Table 13.6 be completed? The analysis was actually completed *before* the (attempted) construction of the ANOVA table, as the data are not appropriate for analysis by an ANOVA table. We have seen that there is indeed a temperature effect and a pressure effect, and the interaction profile in Figure 13.11 clearly shows the strong interaction.

In the absence of a prior estimate of σ^2, the only course of action that would allow completion of the ANOVA table would be to use the interaction as the residual and to test the two main effects against it. This, of course, would be sheer folly for these data as it would lead to the conclusion that nothing is significant, whereas in actuality all three effects are of significance.

This example was given for the purpose of illustrating how a routine analysis of data could easily lead to the wrong conclusions. This message can also be found in other sources, such as Box et al. (1978, p. 329) and Daniel (1976). In fact, the reader is referred to page 20 of Daniel for additional reading on the interpretation of data from a design of this type when a significant interaction effect exists.

It might appear that one solution to the problem of not being able to separate an interaction from a residual is simply to replicate the experiment. Although this is generally desirable, it is not always practical. One possible impediment is, of course, money, and as Ziegel (1984) points out, the data may be so expensive to collect as to preclude replication.

There are, however, some methods for assessing the possible significance of main effects and interactions in unreplicated experiments. One of these methods is due to Daniel (1959) and consists of plotting effect estimates on normal probability paper. This is illustrated in a later example.

13.7.4 Yates's Algorithm

For any 2^k design, where k is the number of factors and 2 is the number of levels of each factor, any treatment combination can be represented by the presence or absence of each of k lowercase letters, where "presence" would denote the high level and "absence" the low level. For example, if $k = 2$, ab would designate the treatment combination with factors A and B set at their high levels, a would represent A at its high level and B at its low level, b would represent just the opposite, and (1) would be used for each factor at its low level.

Yates (1937) presented a simple method for the analysis of 2^k designs, and the methodology has since been extended for use with other types of designs [see Daniel (1976, p. 38)]. We shall illustrate the procedure for a replicated 2^2 design with three observations per treatment combination. The data are given in Table 13.7.

The procedure is initiated by writing down the treatment combinations in *standard order*, by which we mean the following. The treatment combination (1) is always written first. The other combinations are listed relative to the natural alphabetic ordering, including combinations of letters. Thus, (1) would be followed by

TABLE 13.7 Data for Illustrating Yates's Algorithm

		A	
		Low	High
B	Low	10, 12, 16	8, 10, 13
	High	14, 12, 15	12, 15, 16

a, *b*, and *ab*, in that order, and if there were three factors, *ab* would precede *c* and *ac* would follow *c*.

The procedure can be employed using either the totals or averages for each treatment combination. Since totals are simpler to compute, we shall use those. (Of course, for an unreplicated experiment the two would be the same.) The computations are illustrated in Table 13.8.

The columns designated by (1) and (2) are columns in which addition and subtraction are performed for each ordered pair of numbers. (In general, there will be k such columns for k factors.) Specifically, the numbers in each pair are first added, and then the first number in each pair is subtracted from the second number. Thus, $38 + 31 = 69$, $41 + 43 = 84$, $31 - 38 = -7$, and $43 - 41 = 2$. This process is continued on each new column that is created until the number of such columns is equal to the number of factors. The last column that is created by these operations is used to compute the sum of squares for each effect. Specifically, each number except the first is squared and divided by the number of replicates times 2^k. When this operation is carried out, it produces the sum of squares of the effect for the corresponding treatment combination. For example, $(-5)^2/(3 \cdot 2^2) = 2.08$, which is the sum of squares for the main effect of factor A.

Notice that the first entry in the SS column is blank and that this corresponds to (1). The number could be filled in, but it is left blank here so as to emphasize the fact that the corresponding number in column (2) cannot be squared and divided by $3 \cdot 2^2$ to produce the sum of squares for any effect. Specifically, neither SS_{total} nor $SS_{residual}$ can be obtained directly by using Yates's algorithm. The squares of the individual observations are, of course, used in obtaining SS_{total}, but they are not used in Yates's algorithm. The first number in column (2) is used in calculating SS_{total}, however, since it is the sum of all of the observations.

TABLE 13.8 Illustration of Yates's Algorithm

Treatment Combination	Total	(1)	(2)	SS
(1)	38	69	153	
a	31	84	−5	2.08 (A)
b	41	−7	15	18.75 (B)
ab	43	2	9	6.75 (AB)

TABLE 13.9 ANOVA for Data in Table 13.7

Source	d.f.	SS	MS	F^a
A	1	2.08	2.08	< 1
B	1	18.75	18.75	3.36
AB	1	6.75	6.75	1.21
Residual	8	44.67	5.58	
Total	11	72.25		

[a] The F values are computed on the assumption that A and B are fixed factors.

Thus, $SS_{\text{total}} = 10^2 + 12^2 + 16^2 + \cdots + 15^2 + 16^2 - (153)^2/12 = 72.25$. The SS_{residual} would then be obtained by adding the sum of squares for A, B, and AB and subtracting the sum of those three from SS_{total}. This produces $SS_{\text{residual}} =$ 44.67. It can be shown that this is equivalent to computing the sum of the squared deviations from the average in each cell and adding those sums over the cells. An ANOVA table could then be constructed, and the results are displayed in Table 13.9. Since $F_{1,8,.95} = 5.32$, it appears as though neither A nor B nor their interaction has a significant effect upon the response variable. (Here a $1 - 0.95 =$ .05 "significance level" is being assumed.) Of course, the ANOVA table would generally be supplemented by other analyses, both graphical and nongraphical, but this will not be done with these data for two reasons: (1) the intent here was simply to introduce Yates's algorithm and (2) we do not wish to give undue emphasis to a 2^2 design, as it is not a commonly used design.

13.8 THE 2^3 DESIGN

Whereas the 2^2 design illustrated in Section 13.7.4 has only illustrative value, the 2^3 design does have practical value. Writing in an engineering journal, Box (1990) stated: "for there are hundreds of thousands of engineers in this country, and even if the 2^3 was the only kind of design they ever used, and even if the only method of analysis that was employed was to eyeball the data, this alone could have an enormous impact on the experimental efficiency, the rate of innovation, and the competitive position of this country" (pp. 367–368). Box's point is that simple experimental designs are so powerful that exclusive use of even the simplest designs will produce major gains.

It was mentioned previously that the analysis of an unreplicated factorial can be performed if there is an external estimate of σ^2 that is available, presumably from a prior experiment. (Another method of analyzing unreplicated factorials will be presented shortly that is not dependent upon such an external estimate.)

One approach that is often followed, but that is not easily defensible, is to estimate σ^2 using the mean square of pooled high-order interactions—either assuming that they are not significant or performing F tests on them and subsequently pooling those that are not significant. Daniel (1976, p. 72) takes a rather dim view of F tests resulting from the somewhat arbitrary pooling of high-order interactions (those involving three or more factors) when he states:

> This method, recommended in most textbooks, is frequently violated as soon as the data are in, first of all by the use of several levels of significance to indicate which effects are more and which less "significant." The lack of seriousness of the whole enterprise is revealed by the fact that no statistician has thought to investigate the operating characteristic (frequency of missing real effects) of the combined multilevel test. Examples will be given later of entirely jejune conclusions drawn in this way.

The point is that significance levels for F tests are somewhat arbitrary, and the practice of pooling interaction terms based upon the outcome of F tests with arbitrary significance levels creates a fair amount of arbitrariness. This is compounded somewhat by the fact that if enough F tests are constructed, even if they are independent, we are apt to find some that are significant due to chance alone.

Daniel (1976, p. 54) provides data from an actual experiment in which a 2^3 design was used. The factors were time of stirring (A), temperature (B), and pressure (C), and the objective was to investigate their effect on the thickening time of a certain type of cement. The data are given in Table 13.10.

The two levels of each factor were not given by Daniel, but the question of which levels to use had to be answered before the experiment was conducted. Box et al. (1978, p. 298) address this issue:

> The basic problem of experimental design is deciding what pattern of design points will best reveal aspects of the situation of interest.... The question of where the points should be placed is a circular one in the sense that, if we knew what the response function was like, we could decide where the points should be. But to find

TABLE 13.10 Data from an Example in Daniel (1976)

Treatment Combination	Response
(1)	297
a	300
b	106
ab	131
c	177
ac	178
bc	76
abc	109

out what the response function is like is precisely the object of the investigation. Fortunately, this circularity is not crippling, particularly when experiments may be conducted sequentially so that information gained in one set directly influences the choice of experiments in the next.

One point to keep in mind regarding the placement of the points is that if we use two levels that are considered to be equidistant from the center of the set of all possible levels for a factor, the levels are only $-\sigma$ and $+\sigma$ from the center if we assume that the set of all possible levels has a normal distribution. This follows from the fact that the transformation from the levels given in the actual units to $+1$ and -1 values that are implied is given by $Z = (\text{low}-\text{mid})/\sigma$ and $Z = (\text{high}-\text{mid})/\sigma$. Recall from Section 3.6.1 that a z-score is the number of standard deviations that an observation is from the mean. The numbers are worse if we consider, for example, a uniform distribution, as then the most extreme levels would be only $\pm 1.7\sigma$ from the mean, as can be seen using the results of Section 3.6.12. So if we assumed a uniform distribution and used two levels that are closer than the most extreme values, the difference between the levels might not be sufficient to cause a significant change in the response variable.

In general, the levels might not be far enough apart to indicate that a factor has a linear effect of practical significance.

We can eyeball the data in Table 13.10 and easily gain some insight into what effects are likely to be significant before beginning a formal analysis of the data.

Specifically, we can compare the value of the response variable at the high level of each factor with the value at the low level of that factor. With the way that the treatment combinations are ordered in Table 13.10, it is easy to compare high C with low C since the first four are at low C and the last four are at high C. (It should be understood, however, that the ordering in Table 13.10 is what would be used for implementation of Yates's algorithm. When the experiment is actually carried out, the treatment combinations should be run in some random order, if possible.) It should be apparent without even adding the two sets of numbers that low C differs considerably from high C. Thus, factor C appears to be important.

For factor B, the levels alternate in pairs (i.e., two lows, two highs, two lows, two highs), and it should be clear from a cursory inspection that the sums for low B and high B differ greatly, as all four responses at low B are considerably greater than any of the responses at high B. Thus, there appears to be a very strong B effect. Conversely, there appears to be virtually no A effect, as the levels alternate starting with low, and the responses in each pair obviously differ very little. (It should be noted that these conclusions are specifically for the factor levels used in the experiment. Other levels might produce different conclusions.)

The identification of interaction effects can be done somewhat similarly, but not quite as easily. Therefore, we shall not discuss how to identify interactions that are likely to be significant simply from eyeballing the data.

Where do we go from here in continuing the analysis? One possibility would be to use Yates's algorithm to obtain the appropriate sum of squares and then proceed to an ANOVA table. That approach will not be used for this example, however, because (1) it was illustrated previously, and (2) we do not wish to give undue emphasis to ANOVA tables. Although such tables have been used extensively in practice and are an integral part of computer output, they should be supplemented with other types of analyses, as was stated previously.

Yates's algorithm can be used for producing more than just sums of squares, however, so even if an ANOVA table is not to be the focal point of the analysis, the algorithm can still be used to good advantage.

It is left for the reader to verify that the last column of additions and subtractions, which would be denoted by (3) because there are three factors, would be as follows: 1374, 62, −530, 54, −294, 6, 190, 10.

These numbers by themselves have ready interpretations. As indicated previously, the first number will always be the total (regardless of the number of factors), so that number divided by 8 is the average response value for the eight treatment combinations. Similarly, the fifth number (−294), which when aligned vertically would be across from treatment combination *c*, represents the difference between the sum at high C and the sum at low C. Thus, when the number is divided by 4, it will be the difference between the two averages (at high C and low C) and will thus be the estimate of the "C effect." Dividing each of the other six numbers by 4 will produce estimates of the other main effects as well as the interaction effects.

Performing this division produces the effect estimates given in Table 13.11.

The figures in Table 13.11 reveal what we had discovered earlier from the simple calculations, namely that the main effects of B and C seem to be important, although the fact that the BC interaction is also of some magnitude may complicate the analysis somewhat. To determine the statistical significance of these effects, we need to know (or estimate) the standard deviation of each effect.

Recognizing that we have an unreplicated factorial, we know that we cannot obtain a clean estimate of σ. Remembering that interactions of third order (those

TABLE 13.11 Effect Estimates for the Data in Table 13.10

Effect	Estimate
Average	171.75
Main effects	
A	15.5
B	−132.5
C	−73.5
Interaction effects	
AB	13.5
AC	1.5
BC	47.5
ABC	2.5

containing three factors) and higher are generally not significant, we might "create" a residual term by using the ABC interaction as the residual. We could then use that to estimate σ and subsequently test for the significance of the various effects.

Of course, such an approach would be shaky at best, as *interaction* and *experimental error* are totally different concepts, as was previously discussed. For these data the estimate of σ using the ABC interaction would be $\sqrt{(10)^2/2^3} = 3.54$, where the 10 is from Yates's algorithm. Daniel (1976, p. 54) reports that σ was known to be about 12, however, so our estimate from this ad hoc approach is quite poor.

Earlier in this chapter the exact t test for two independent samples was illustrated. This can be used to test for the significance of the main effects and interactions since each effect estimate is simply the difference of two averages. Since the variance of the difference of two (independent) averages is equal to the variance of the first average plus the variance of the second, we have that the (estimated) $\text{Var}(\overline{y}_{\text{high}} - \overline{y}_{\text{low}}) = (s^2/n) + (s^2/n) = 2s^2/n$, where s^2 is the estimate of σ^2 and n is the number of observations from which each average is calculated. (Here the two numbers are assumed equal.) Using $(12)^2 = 144$ for s^2 and 4 for n, we obtain 72 as our estimate of the variance for each effect estimate, so that $\sqrt{72} = 8.49$ is our estimate of the standard deviation.

Dividing each effect estimate in Table 13.11 by 8.49 produces quotients that are less than 2 in absolute value for every effect except B, C, and BC. (Here 2 is used as a rough cutoff for the significance of a t statistic.)

Recalling Daniel's admonition that main effects should not be reported and interpreted separately when the interaction effect of those factors is more than one-third of the main effects, we need to look at the BC interaction profile to get a better handle on what is happening. The profiles are given in Figure 13.13.

These profiles illustrate why graphical displays should be used to supplement results such as those given in Table 13.11 and in ANOVA tables. The interaction

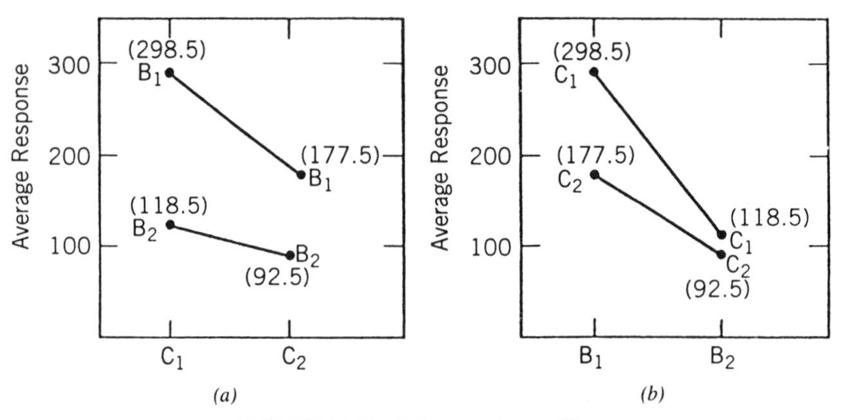

FIGURE 13.13 BC interaction profiles.

effect is apparent from either Figure 13.13*a* or *b*, and it is this BC interaction that makes the B effect considerably different for the two levels of C and the C effect considerably different for the two levels of B. Specifically, Table 13.11 indicates that the estimate of the B effect is −132.5. Using Figure 13.13*a*, the B effect is the average of 118.5 − 298.5 = −180 and 92.5 − 177.5 = −85. There is obviously a considerable difference between −180 and −85, and this difference results from the significant BC interaction. Thus, reporting that the estimate of the B effect is −132.5 is really not sufficient, since the effect is considerably greater than that (−180) at low C and considerably less than that (−85) at high C.

Although interaction profiles are generally constructed in such a way that the factor letter that comes second in the alphabet is on the horizontal axis, there is no compelling reason for always following such a procedure.

Figure 13.13*b* also illustrates the BC interaction and allows us to view the C effect for the different levels of B. Specifically, the estimate of the C effect must be equal to the average of 177.5 − 298.5 = −121 and 92.5 − 118.5 = −26. The average of these two numbers is, of course, equal to −73.5, which is the estimate of the C effect that is given in Table 13.11. Again, these two estimates differ greatly, so simply reporting the estimate to be −73.5 does not give the full story.

If the two lines in Figure 13.13*a* are not parallel, then the lines in Figure 13.13*b* are not going to be parallel either (and similarly when the lines are parallel), so the two interaction profiles simply provide a different view of the interaction.

13.9 ASSESSMENT OF EFFECTS WITHOUT A RESIDUAL TERM

In this example it was possible to assess the significance of the main effects and interaction effects because there was an estimate of the experimental error standard deviation that was available as prior information.

We cannot always expect prior information to be available, however, so we need a method of assessing the possible significance of main effects and interactions without having to use a residual term, or, for that matter, having to arbitrarily pool high-order interactions to create a residual term. Daniel (1959) developed a method for making this assessment that entails the plotting of the effect estimates on normal probability paper and seeing if they form roughly a straight line. If they do, none of the effects are likely to be significant. Conversely, points that lie some distance from a line that links the other points likely represent significant effects.

We would like to have a computer available for constructing such plots, but it is not difficult to construct them by hand. The use of normal probability paper for making such plots was mentioned briefly in Chapter 11, and the procedure has been presented in Nelson (1976) and other sources.

Software for constructing these graphs is readily available, however. The plot for these data that was produced using Minitab is given in Figure 13.14. Notice that the line goes through the point (0, 0), as it should since we are testing the hypothesis that all of the effects are normally distributed with a mean of zero,

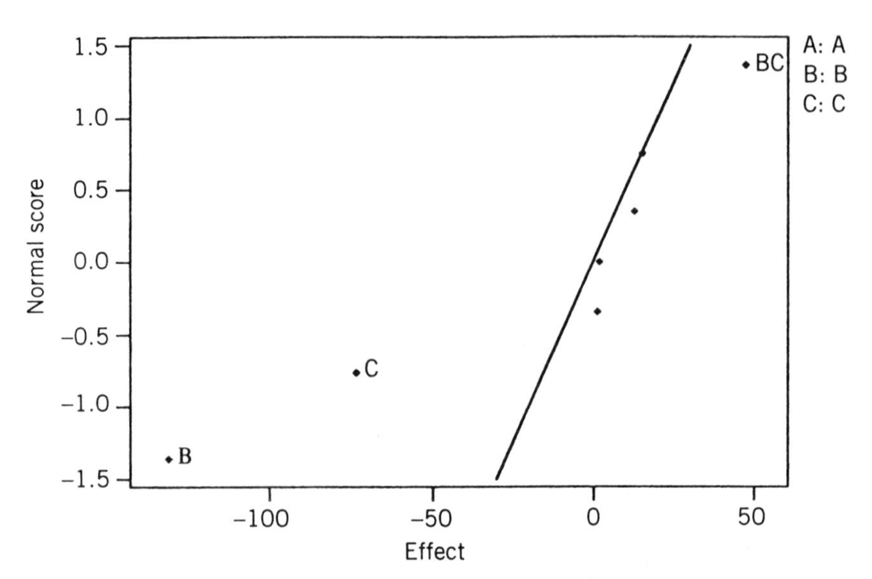

FIGURE 13.14 Normal probability plot for effects in 2^3 example ($\alpha = 0.15$).

and a z-score (i.e., a normal score) of zero results when the value of a normally distributed random variable is zero.

The plot gives the same message as was given by the other analyses that have been presented, namely, that the B, C, and BC effects appear to be significant. It should be noted, however, that the BC interaction effect is borderline since it is "significant" when a significance level of $\alpha = .15$ is used, whereas it is not significant when the Minitab default value of $\alpha = .10$ is used.

The alternative approach to interpreting normal plots, which is given in many books, is to look for effect estimates that are considerably off the line that will almost connect most of the effect estimates. When effects are not significant, their estimates should be reasonably close to zero, and a straight line will virtually connect the points. We should always have enough points to form such a line, as it would be somewhat rare to encounter a factorial design with three or more factors in which most of the effects are significant. [Daniel (1976, p. 75) roughly estimates that about four significant effects is average for a 2^4 design and seven is average for a 2^5.]

It should be noted that this plotting procedure does not incorporate an estimate of σ, and the magnitude of that standard deviation will essentially determine how far a point should be from the line before the effect that it represents could be judged a real (significant) effect. Therefore, this is not a precise method for determining real effects, but it is a very useful method.

We should not expect virtually all of the points to fall right on a line even when there are no real effects, however, since, assuming normality, we would expect the effect estimates to vary as would a normally distributed random variable with a mean of zero. Daniel (1976) provides a set of 40 normal probability plots in which each plot contains 16 points and each point is generated from a standard normal distribution. Not one of the 40 plots is a straight-line plot, and many are not even close, and this is due to the fact that 16 values generated from a standard normal distribution simply will not form an empirical standard normal distribution.

Box and Meyer (1986) presented a statistical procedure for unreplicated fractional factorials that can be used to supplement the graphical procedure of Daniel.

There are also various normal probability plot–type methods that have been proposed with "significant" effects identified on the plot. These methods all create a pseudo–error term for the purpose of identifying significant effects. Thus, they advocate doing what Daniel (1976) said should *not* be done. The comparative performance of these methods is quite similar, as has been demonstrated in studies by Berk and Picard (1991) and Hamada and Balakrishnan (1998). Perhaps the best-known method is due to Lenth (1989), whose approach has been implemented in Minitab.

One shortcoming of these approximate methods is that they all depend upon effect sparsity. Consequently, they could perform poorly if effect sparsity did not exist in a particular application. In discussing the available methods, Haaland (1998) stated, "there is no clear choice for large numbers of effects."

This was demonstrated by Loughin and Noble (1997), who provided a permutation test for determining significant effects. Although their method appeared to perform well in simulation studies, it is computationally intensive.

13.10 RESIDUAL PLOT

Another use of a normal probability plot in analyzing data from a designed experiment is in analyzing the residuals. For this type of plot we would expect the points (residuals) to come reasonably close to forming a straight line if the errors are normally distributed.

Residual plots and the calculation of residuals were discussed earlier in Section 13.5, in which a residual was presented as the difference between the observed value and the predicted value. Thus, in this section "residual" will not represent the residual sum of squares. Rather, it will represent $Y - \hat{Y}$, which when squared and summed over the design points equals the residual sum of squares.

Predicted values for experimental design data can be obtained by using the *reverse Yates algorithm* (see Daniel, 1976), in which the predicted values are obtained from a prediction equation that contains only those effects that are judged to be real.

The reverse Yates algorithm begins where the (forward) Yates algorithm leaves off. That is, the starting point is the last column of numbers obtained from the various sums and differences. The numbers are written in reverse order, however, with zeros substituted for those numbers that correspond to effects that are judged to be not significant. From that point on the additions and subtractions are performed in the same way, and when the numbers in column (3) are divided by the number of design points, the result is the column of predicted values for each of the treatment combinations. The residuals are then obtained by subtracting each of the predicted values from the corresponding observed values. The results are shown in Table 13.12.

The fact that the $\hat{Y}$ values repeat in pairs is due to the fact that every other number in the first column of numbers is a zero. The reader will observe that the numbers that repeat are the averages of the corresponding Y values. Again, this is due to the pattern of zeros in the first column and is not a general result.

The plot of the residuals, $Y - \hat{Y}$, is given in Figure 13.15. It can be observed that the points come reasonably close to forming a straight line and exhibit no gross abnormalities, although assessing normality is difficult with such a small number of points.

The predicted values could also be obtained from the prediction equation

$$\hat{Y} = 171.75 - 66.25X_1 - 36.75X_2 + 23.75X_1X_2$$

where the coefficients for X_1, X_2, and X_1X_2 are obtained by dividing by 2 the estimates of the B, C, and BC effects. The values for X_1 are -1 for the low level of B and $+1$ for the high level of B, and similarly for X_2 as it represents factor C. The values for X_1X_2 are simply the product of the individual values for X_1 and X_2. A measure of the adequacy of the prediction equation is R^2, which, the reader may recall, can be calculated as

$$R^2 = 1 - \frac{\Sigma(Y - \hat{Y})^2}{\Sigma(Y - \overline{Y})^2}$$

Using the appropriate values from Table 13.12, we obtain

$$R^2 = 1 - \frac{862}{51,291.5}$$
$$= 0.983$$

so that the use of B, C, and BC explains almost all of the variability in the data.

TABLE 13.12 Reverse Yates Algorithm for 2^3 Data

Effect	Column (3) from Forward Yates with Zeros Inserted	(1)	(2)	(3)	Divisor	$\hat{Y}$	Y	$Y - \hat{Y}$
ABC	0	190	−104	740	8	92.5	109	16.5
BC	190	−294	844	740	8	92.5	76	−16.5
AC	0	−530	−104	1420	8	177.5	178	0.5
C	−294	1374	844	1420	8	177.5	177	−0.5
AB	0	190	−484	948	8	118.5	131	12.5
B	−530	−294	1904	948	8	118.5	106	−12.5
A	0	−530	−484	2388	8	298.5	300	1.5
Average	1374	1374	1904	2388	8	298.5	297	−1.5

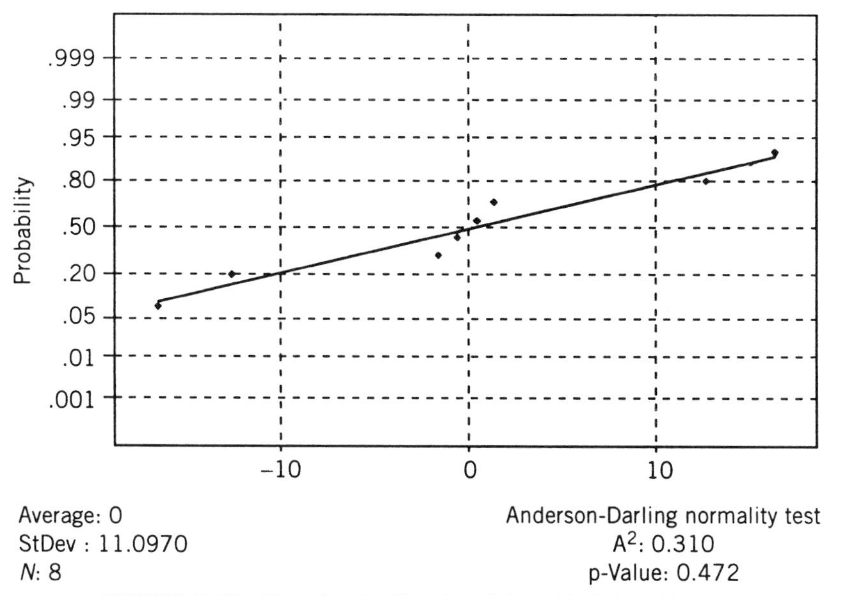

FIGURE 13.15 Normal probability plot of the residuals in Table 13.12.

Are we now finished with the analysis? It would appear so, but Daniel (1976, p. 59) reports that the experimenter believed strongly that factor A did indeed have some effect. Our analysis, however, revealed otherwise, so should we conclude that the experimenter must have been wrong? Not without taking a closer look at the data.

We saw previously in Table 13.11 that the estimate of the A effect is 15.5. This effect can be viewed as the average of four differences, in which each difference is obtained by subtracting the response with A at the low level from the response with A at the high level and the levels of B and C held constant in each pair. Thus, the differences are $a - (1)$, $ab - b$, $ac - c$, and $abc - bc$. The first and third differences are obviously at low B, and the second and fourth differences are at high B. It can be seen from Table 13.10 that the average of the first and third differences is 2.0, whereas the average of the second and fourth differences is 29.0. The estimated standard deviation of this second average is 12 [obtained from $\text{Var}(\overline{y}_{\text{high}} - \overline{y}_{\text{low}})$ with $n = 2$] so that the value of the t statistic for testing the significance of the A effect at high B would thus be 29/12, which is greater than the previously mentioned cutoff value of 2.0. Thus, there does appear to be a significant A effect at high B but obviously not at low B.

Again, the important point is that a routine analysis does not suffice, and even when it appears as though a thorough analysis has been conducted, it is quite possible that something may have been overlooked. In fact, it is not uncommon for a data set to be analyzed by a number of people (including prominent statisticians) and then someone would discover an important bit of information that had previously been undetected. [See Box et al. (1978, p. 496) for such an example.]

This concludes our discussion of the example for the 2^3 design. Although the design is favored by some, Daniel (1976, p. 53) indicates that 16-point designs are much more valuable and gives conditions under which a 2^3 might be used.

13.11 SEPARATE ANALYSES USING DESIGN UNITS AND UNCODED UNITS

In Section 13.9 the prediction equation was given in terms of the design units of the factors (i.e., +1 and −1). The analysis should always be performed in terms of the design units, with the prediction equation later converted to the form for the uncoded units, if desired.

The analysis should not be performed in terms of the uncoded (raw) units when the model includes terms other than main effect terms, as orthogonality will then be lost and the results could be misinterpreted. This can be illustrated with the following example.

Table 13.13 contains data on a response variable, X_1, X_2, and X_1X_2 in coded form, and the same variables in raw form. The values for Y were generated as $Y = 10 + 3X_1 + 4X_2 - 5X_1X_2 + \varepsilon$, with $\varepsilon \sim \text{NID}(0, 12.25)$ and the factors in coded form.

The analyses for the coded and raw-form data are given in Table 13.14. Notice that the basic summary information (R^2, the model F-statistic, etc.) are the same for the two analyses, but the p-values for X_1 and X_2 differ greatly for the two analyses. In particular, note that the p-values for the raw-form analysis are all rather small even though the value for R^2 is not very large. Note also that the p-values for the interaction term are identical. The latter can be explained as follows. The first factor would be coded as $[X_1(\text{raw}) - 150]/50$ and the second factor would be coded as $[X_2(\text{raw}) - 70]/20$. The coefficients will differ by the

TABLE 13.13 Simulated Data for Illustrative Example

Y	X_1	X_2	X_1X_2	X_1 (raw)	X_2 (raw)	X_1X_2 (raw)
1.7727	−1	−1	1	100	50	5,000
6.0556	1	−1	−1	200	50	10,000
18.2586	−1	1	−1	100	90	9,000
8.6289	1	1	1	200	90	18,000
0.3090	−1	−1	1	100	50	5,000
20.6989	1	−1	−1	200	50	10,000
17.0754	−1	1	−1	100	90	9,000
14.4546	1	1	1	200	90	18,000
0.7026	−1	−1	1	100	50	5,000
16.1993	1	−1	−1	200	50	10,000
10.9854	−1	1	−1	100	90	9,000
11.4810	1	1	1	200	90	18,000

TABLE 13.14 Analysis of Table 13.13 Data in Coded Form and Raw Form

a. Coded Form

The regression equation is

$\hat{Y} = 10.6 + 2.37X_1 + 2.93X_2 - 4.33X_1X_2$

Predictor	Coef	StDev	T	p
Constant	10.552	1.295	8.15	0.000
X_1	2.368	1.295	1.83	0.105
X_2	2.929	1.295	2.26	0.054
X_1X_2	−4.327	1.295	−3.34	0.010

$S = 4.488$ $R^2 = 71.0\%$ $R^2_{\text{adj}} = 60.2\%$

Analysis of Variance

Source	d.f.	SS	MS	F	p
Regression	3	394.90	131.63	6.54	0.015
Residual error	8	161.10	20.14		
Total	11	556.00			

Source	d.f.	SeqSS
X_1	1	67.28
X_2	1	102.94
X_1X_2	1	224.68

Unusual Observations

Obs	X_1	Y	Fit	StDev Fit	Residual	StResid
2	1.00	6.06	14.32	2.59	−8.26	−2.25R

R denotes an observation with a large standardized residual.

b. Raw Form

The regression equation is

$\hat{Y} = -52.2 + 0.350X_1(\text{raw}) + 0.795X_2(\text{raw}) - 0.00433X_1X_2(\text{raw})$

Predictor	Coef	StDev	T	p
Constant	−52.24	14.91	−3.50	0.008
Raw X_1	0.35025	0.09431	3.71	0.006
Raw X_2	0.7955	0.2048	3.88	0.005
Raw X_1X_2	−0.004327	0.001295	−3.34	0.010

$S = 4.488$ $R^2 = 71.0\%$ $R^2_{\text{adj}} = 60.2\%$

TABLE 13.14 ***(Continued)***

Analysis of Variance

Source	d.f.	SS	MS	F	p
Regression	3	394.90	131.63	6.54	0.015
Residual error	8	161.10	20.14		
Total	11	556.00			

Source	d.f.	SeqSS
Raw X_1	1	67.28
Raw X_2	1	102.94
Raw X_1X_2	1	224.68

Unusual Observations

Obs	raw X_1	Y	Fit	StDev Fit	Residual	StResid
2	200	6.06	14.32	2.59	−8.26	−2.25R

R denotes an observation with a large standardized residual.

Abbreviations: Coef, coefficient; StDev, standard deviation; SS, sum of squares; MS, mean square; d.f., degrees of freedom; SeqSS, sequential sum of squares; StResid, standardized residual.

product of the divisors, as can be seen by performing the appropriate algebra. But when the product of these two terms is simplified, there will be an "extra" term in X_1 and X_2 that will combine with the main effect terms. Consequently, the coefficients for X_1 and X_2 between the two model forms differ in a way that is not a simple linear transformation. Furthermore, transforming from coded form to raw form creates moderate correlations between X_1(raw) and X_1X_2(raw) and between X_2(raw) and X_1X_2(raw), 0.742 and 0.636, respectively.

Even though the values of Y were generated using all three terms, this does not mean that all three should be significant, as the noise term could be so large as to dwarf the effects. This almost happens here as evidenced by the value of R^2. Therefore, it would be quite inappropriate to conclude that all three effects are highly significant, as the raw-form analysis would lead us to believe.

13.12 TWO-LEVEL DESIGNS WITH MORE THAN THREE FACTORS

The analysis of designs such as the 2^4 and 2^5 can be carried out by following essentially the same steps that were illustrated for the 2^3. There is, of course, more computation that has to be performed (either by hand or by computer), and more "detective work" will generally be required as the number of factors increases. Extensive discussion and analysis of some actual experiments conducted using 2^4

and 2^5 designs (including some classic experiments) can be found in Chapters 6 and 7 of Daniel (1976), and the reader is urged to study that material carefully.

At this point the reader might wonder how one goes about determining the number of factors to study in an experiment. In some instances an experimenter may have a reasonably good idea as to the number of factors that seem to be worth investigating, and this number might be fairly small. In other applications, however, there might be a large number of potentially significant factors. If there were 10 such factors, a 2^{10} design would require 1024 treatment combinations. Not only would an experiment with this many design points be virtually impossible to carry out, but also the cost would likely be prohibitive. Accordingly, a fractional factorial design would be a logical alternative, in which a small fraction of the 1024 design points would actually be used. Once influential factors have been identified from the use of such a design, a two- or three-level design might then be used to study the k influential factors further, assuming that k is small. Fractional factorial designs are discussed in Section 13.15.

13.13 THREE-LEVEL FACTORIAL DESIGNS

Often there will be three levels of two or more factors that are logical to study, rather than just two levels. If a full factorial is to be used, however, the use of three levels imposes a practical limitation on the number of factors that can be investigated since the number of design points increases rapidly as k increases. Consequently, 3^2 and 3^3 designs are those that are of practical interest. They could be analyzed using the *extended Yates algorithm* (Davies, 1954), if desired.

When a factor has three levels, it will have two degrees of freedom, and the associated sum of squares can be decomposed into two components—one that represents the linear effect of the factor and the other that represents the quadratic effect. Assuming that the factor is continuous and has equally spaced levels, a linear effect is where the value of the response variable changes at (almost) a constant rate over the different levels. Conversely, a quadratic effect is where the value of the response variable changes along the lines of a quadratic relationship between the variable and the factor (i.e., a quadratic relationship would be of the general form $y = ax^2 + bx + c$, where y would be the response variable and x the factor).

If there was only one factor, the relationship between the response variable and the factor could be displayed as a two-dimensional graph. With two factors, however, a three-dimensional display is needed, and this is generally termed a surface, more specifically, a response surface. Thus, a 3^2 design enables the experimenter to gain insight into the response region (surface), although response surface designs (discussed briefly at the end of this chapter) are better suited for that purpose. (The same general conclusion can be drawn for a 3^3 design.)

Thus, although three-level factorial designs have been used to some extent, their usefulness is somewhat limited.

13.14 MIXED FACTORIALS

To this point there has been an implicit assumption that the number of levels of interest is the same for each of the factors to be studied. This will not always be the case, however. A mixed factorial (also called an asymmetrical factorial) in two factors is of the general form $a^{k_1}b^{k_2}$, where k_1 and k_2 are greater than or equal to 1 and $a \neq b$. Data from such designs could be analyzed using the extended Yates algorithm provided that $a = 2$ and $b = 3$ (Margolin, 1967). Mixed factorial designs have been discussed by, in particular, Addelman (1962) and in Chapter 18 of Kempthorne (1973), although the treatment in these sources is somewhat mathematical.

13.15 FRACTIONAL FACTORIALS

As was previously mentioned, the number of design points can be rather large when there are more than just a few factors of interest. As Steinberg and Hunter (1984) indicate, there might be as many as 50 or 100 potentially important factors in some applications. (A 2^{50} design might take a few lifetimes to run.)

We shall focus attention upon two-level fractional factorial designs, as this type of fractional factorial has been used extensively in practice. Fractional factorials were first presented by Finney (1945) and popularized in the landmark papers of Box and Hunter (1961a,b), and we shall adopt the notation of the latter throughout this section.

13.15.1 2^{k-1} Designs

A two-level fractional factorial can be written in the general form 2^{k-p}, where, as before, k denotes the number of factors and the fraction of the full 2^k factorial that is to be run is $1/2^p$. Thus, a 2^{3-1} design would be a $\frac{1}{2}$ fraction of a 2^3. Sixteen point designs (so that $k - p = 4$) are the ones that have been used the most often in industry.

For simplicity, however, we shall first illustrate a 2^{3-1}, which, although of limited usefulness, does have value for illustrative purposes. We should first recognize that with only four design points (here we are assuming that the design is not replicated), we will have only three degrees of freedom so we can estimate only three effects. Which three effects do we choose to estimate? Although in rare instances a two-factor interaction might be of more interest to an experimenter than a main effect, we would generally choose to estimate main effects over interactions, if we had to select one over the other. Thus, the logical choice would be to estimate the three main effects A, B, and C. Before we can undertake that task, however, we must determine what four design points to use. We cannot randomly select four treatment combinations from the eight that are available. For example, we obviously could not estimate the main effect of A if we happened to select four treatment combinations in which A was at the high level and none in which A was at the low level; similarly for B and C. Thus, we would clearly want to have two treatment combinations in which A is at the high level and two in which A is at the low level, and the same for B and C.

With a little trial and error we could obtain four treatment combinations that satisfy this property without too much difficulty, but it would obviously be preferable to use some systematic approach. Whenever a $\frac{1}{2}$ fraction is used, we have to select one effect to "confound" with the difference between the two fractions; that is, that particular effect would be estimated by the difference of the averages of the treatment combinations in each fraction (which of course is the way that we would logically estimate the difference between the two fractions). If we have to "give up" the estimate of one effect in this way (which is obviously what we are doing since we will run only one of the two fractions), it would be logical to select the highest order interaction to relinquish. For a 2^3, this is the ABC interaction.

A simple way to construct the two $\frac{1}{2}$ fractions (from which one would be randomly selected) would be to assign those treatment combinations with an even number of letters in common with ABC to one fraction and those with an odd number of letters in common with ABC to the other fraction. This produces the two fractions given in Figure 13.16.

For illustration, we shall use the data in Table 13.10, which were used previously to illustrate the full 2^3. It was suggested earlier that the reader use Yates's algorithm on that data. If that were done, the numbers in Table 13.15 should look familiar. It can be observed that the 10, which is the last number in column (3), is obtained by adding the response values of the treatment combinations in the first fraction in Figure 13.16 and subtracting off the response values of the treatment combinations in the second fraction.

Before proceeding any further, we should think about what we are giving up in addition to an estimate of the ABC interaction when we run one of the fractions in Figure 13.16. We assume that we are relinquishing information on the three two-factor interactions by using one of the fractions in Figure 13.16 (since we have only three degrees of freedom), but we also assume that we will be able to estimate the three main effects.

Let's verify this by looking at some of the other parenthetical expressions in Table 13.15. Assume that we have randomly selected the first fraction in Figure 13.16 to run. The sum and difference of the various treatment combinations beside the number −294 indicate that we would estimate the main effect of C by $(abc + c - a - b)/2 = (109 + 177 - 300 - 106)/2 = -60$, which does

Fraction 1	Fraction 2
a	(1)
b	ab
c	ac
abc	bc

FIGURE 13.16 Two $\frac{1}{2}$ fractions of 2^3.

TABLE 13.15 Yates's Algorithm Calculations Using the Data in Table 13.10

Treatment Combination	Response	(1)	(2)	(3)	Column (3) Representation
(1)	297	597	834	1374	(Sum of all responses)
a	300	237	540	62	$a + ab + ac + abc$ $-b - c - bc - (1)$
b	106	355	28	−530	$b + ab + bc + abc$ $-a - c - ac - (1)$
ab	131	185	34	54	$c + ab + abc + (1)$ $-a - b - ac - bc$
c	177	3	−360	−294	$c + ac + bc + abc$ $-a - b - ab - (1)$
ac	178	25	−170	6	$b + ac + abc + (1)$ $-a - c - ab - bc$
bc	76	1	22	190	$a + bc + abc + (1)$ $-b - c - ac - ab$
abc	109	33	32	10	$a + b + c + abc - ac$ $-ab - bc - (1)$

not differ greatly from the −73.5 that was obtained using all eight treatment combinations. Further, we can see from Table 13.15 that, using the first fraction, we would also estimate AB by $(abc + c - a - b)/2$. What this means is that the estimate of the C effect is *confounded* (i.e., confused) with the AB effect.

Similarly, it could be shown that A is confounded with BC and B is confounded with AC, as can be verified from Table 13.15. Fortunately, there are easier ways to determine what effects are confounded. One way is to write out how each effect is estimated using plus and minus signs and then identify those effects that have the same configuration of signs. The other method is even easier and is simply a matter of multiplying each effect times the effect that was confounded with the difference of the two fractions (ABC in this example), and removing any letter whose exponent is a 2. (This applies to any two-level fractional factorial.) Thus, $\mathrm{A(ABC) = A^2BC = BC}$, $\mathrm{B(ABC) = AB^2C = AC}$, and $\mathrm{C(ABC) = ABC^2 = AB}$. The effects that are confounded with each other are said to be *aliases* of each other, and the set of such aliases is said to be the *alias structure*.

Another way to view the alias structure is to use Yates's algorithm after filling in zeros for the treatment combinations that are in the fraction that is not used. Factors that are aliased will then have the same totals in column (3), as the reader is asked to demonstrate for these data in exercise 1.

This is not a recommended approach for determining the alias structure, however, as effect estimates [and, hence, numbers in column (3)] can be the same without the effects being confounded. It is also far more time consuming than the multiplication approach just illustrated, and of course we would generally prefer to use a computer anyway. It is simply another way of viewing the alias structure.

We saw from the analysis of the full factorial that the AB interaction was not significant, and we can also see from Table 13.11 that the estimate of C plus the estimate of AB equals the -60 that we just obtained using the four treatment combinations in the first fraction. Thus, we are actually estimating C + AB rather than just C, and the extent to which our estimate of C is contaminated depends upon the size of the AB interaction. Here there is no serious problem because the AB interaction is not large.

What if we had randomly selected the other fraction? A little arithmetic would reveal that our estimate of C would be -87 and that we would really be estimating C − AB.

We saw previously that the BC interaction was significant, and we know now that BC is aliased with A. Therefore, we can see to what extent the estimate of the A effect is contaminated by the presence of a strong BC interaction. Again assuming that we had used the first fraction, our estimate of the A effect would be obtained from $(abc + (1) - b - c)/2 = (109 + 300 - 106 - 177)/2 = 63$, which differs dramatically from our estimate of 15.5 obtained from the full factorial. As the reader might suspect, we are actually estimating A + BC (A − BC with the other fraction), so that, in this example, we would erroneously conclude that there is a strong A effect when in fact there was not. (We remember, however, that our detective work did reveal that A was somewhat influential when B was at its high level.)

The upshot of all of this is that when we run a fractional factorial we do take a risk, and the severity of the risk depends upon the order of the interactions that are lost. First-order interactions (i.e., involving two factors) and the second-order interaction were lost in the 2^{3-1} example, so that design should be considered only if there is a strong prior belief that none of the interactions will be significant. Even then, the advantage of the fractional factorial would be minimal, as four design points would be run instead of eight—not much of a saving.

The picture changes considerably, however, when there are more than three factors. What about a 2^{4-1}? The alias structure would certainly be more palatable in that the fractions could be constructed in such a way that the main effects would be aliased with second-order interactions, but, unfortunately, the first-order interactions are aliased in pairs. The ABCD interaction would be confounded with the difference between the two fractions, and we would then have A = BCD, B = ACD ... AB = CD, AC = BD, etc.

Snee (1985) considers the 2^{5-1}, 2^{6-2}, 2^{7-3}, and 2^{8-4} designs for the study of 5, 6, 7, and 8 factors, respectively, to be the most useful fractional factorial designs and provides an example of a 2^{5-1} and a 2^{7-3}. We will briefly discuss the former.

In the 2^{5-1} experiment the objective was to identify the process variables that affect the color of a product produced by a chemical process, so that a process control procedure could be implemented for controlling the variation in the color. Five process variables that were thought to be potentially important were (1) solvent/reactant ratio, (2) catalyst/reactant ratio, (3) temperature, (4) reactant purity, and (5) pH of reactant.

If we were to proceed as before, we would select ABCDE to confound with the difference between fractions. This is generally called the *defining contrast*. ABCDE is the best choice because it would produce the most palatable alias structure. Specifically, main effects will be aliased with four-factor interactions, and two-factor interactions will be aliased with three-factor interactions. Thus, we can estimate main effects and all two-factor interactions provided that second- and higher order interactions are negligible (which will usually be the case).

What is estimable for a particular fractional factorial design can be expressed compactly by indicating the *resolution* of the design. Coined by Box and Hunter (1961a), a design of Resolution III is one in which only the main effects are estimable, and they are estimable only if the two-factor interactions are negligible, since main effects and two-factor interactions are confounded. We recall that this was the alias structure for the 2^{3-1} design. Thus, this design is a Resolution III design. A Resolution IV design is one in which no main effects are confounded with two-factor interactions, but two-factor interactions are confounded with each other. An example was the 2^{4-1} design. The 2^{5-1} design is an example of a Resolution V design in which both main effects and two-factor interactions are estimable in the absence of second- and higher-order interactions.

It should be emphasized, however, that a 2^{5-1} design is not automatically a Resolution V design. It is such a design only when the best choice is made for the defining contrast (ABCDE), and it should be observed that the number of letters in the defining contrast determines the resolution of the design for any one-half fraction. (Determining the resolution for smaller fractions is slightly more involved and will be discussed later.)

To construct the two fractions of the 2^{5-1} design, we could proceed as before and allocate those treatment combinations with an even number of letters in common with ABCDE to one fraction, and those with an odd number of letters in common to the other fraction. This would require enumerating all $2^5 = 32$ treatment combinations, however, so that when there are more than four factors, such an approach is somewhat laborious.

An alternative procedure, following Box et al. (1978, p. 386), is to enumerate a full 2^4 design factorial using plus and minus signs (to designate the high and low levels, respectively) and then use the product of those signs for each treatment combination to form the signs for the fifth factor.

Specifically, for the full 2^4 factorial there are 16 treatment combinations, and these can be expressed by alternating plus and minus signs in a certain way, as is shown in Table 13.16. (I = ABCDE means that ABCDE is the defining contrast.) The pattern that is exhibited by the various columns should be apparent. The last column in the full factorial part (column D in this case) will always have 2^{k-1} minus signs followed by 2^{k-1} plus signs for a 2^k factorial, so that all of the columns begin with a consecutive number of minus signs that are each a power of 2 (i.e., 2^0, 2^1, 2^2, 2^3). (If we were constructing a 2^{6-1} fractional factorial, the last column in the full factorial part would have 16 consecutive minus signs followed by 16 consecutive plus signs.) Notice that the sign for the E factor in the first fraction is simply the product of the signs for the other four factors,

TABLE 13.16 The Two Half Fractions of a 2^5

Fraction 1 (I = ABCDE)						Fraction 2 (I = −ABCDE)					
Treatment Combination	A	B	C	D	E	Treatment Combination	A	B	C	D	E
e	−	−	−	−	+	(1)	−	−	−	−	−
a	+	−	−	−	−	*ae*	+	−	−	−	+
b	−	+	−	−	−	*be*	−	+	−	−	+
abe	+	+	−	−	+	*ab*	+	+	−	−	−
c	−	−	+	−	−	*ce*	−	−	+	−	+
ace	+	−	+	−	+	*ac*	+	−	+	−	−
bce	−	+	+	−	+	*bc*	−	+	+	−	−
abc	+	+	+	−	−	*abce*	+	+	+	−	+
d	−	−	−	+	−	*de*	−	−	−	+	+
ade	+	−	−	+	+	*ad*	+	−	−	+	−
bde	−	+	−	+	+	*bd*	−	+	−	+	−
abd	+	+	−	+	−	*abde*	+	+	−	+	+
cde	−	−	+	+	+	*cd*	−	−	+	+	−
acd	+	−	+	+	−	*acde*	+	−	+	+	+
bcd	−	+	+	+	−	*bcde*	−	+	+	+	+
abcde	+	+	+	+	+	*abcd*	+	+	+	+	−

and the negative of that product in the second fraction. Notice also that each of the treatment combinations in the first fraction has an even number of letters in common with ABCDE and an odd number of letters in common in the second fraction. Thus, this new approach gives us the same fractions that we would have obtained using the previously described approach. As Box et al. (1978, p. 386) point out, this approach will always produce a 2^{k-1} design with the highest possible resolution.

The first fraction is the one that was used in the experiment described by Snee (1985). The values of the response variable (color) were (in coded units) in order: −0.63, 2.51, −2.68, −1.66, 2.06, 1.22, −2.09, 1.93, 6.79, 6.47, 3.45, 5.68, 5.22, 9.38, 4.30, and 4.05.

How would we proceed to analyze these data? We *could* use the same approach as was illustrated for the 2^{3-1} example; that is, we could use Yates's algorithm to obtain the plus and minus signs of each treatment combination for the full 2^5 factorial and then use just the treatment combinations that actually occurred in the fraction. This would be quite laborious for five factors, however.

Another approach is to use a modification of Yates's algorithm for 2^{k-p} designs, in which the algorithm is applied to a full factorial in $k - p$ factors. This approach is discussed by Box et al. (1978, p. 407) and illustrated in Cochran and Cox (1957, p. 254). Berger (1972) shows how to generate the 2^{k-p} treatment combinations in the proper order so that Yates's algorithm can be applied, and McLean and Anderson (1984, p. 263) illustrate the approach.

A third approach would be to produce the column of plus and minus signs for each interaction that is estimable (the two-factor interactions in this case since this is a Resolution V design) and use those columns and the columns for A, B, C, D, and E in Table 13.16 in obtaining estimates of the two-factor interactions and main effects, respectively. The interaction columns would be obtained by multiplying together the columns of the factors that comprise each interaction. For example, the AB interaction column would be obtained by multiplying the A column times the B column, thus starting with the pattern $+--+$, with the pattern repeated three more times. The estimate of each effect would then be obtained by taking the average of the eight response values that are associated with a treatment combination that has a plus sign and subtracting off the average of the eight response values that are associated with a treatment combination that has a minus sign. The reader can use the A column to verify that the estimate of the A effect is 1.645 and the estimate of the AB effect is 0.11.

Since this is a Resolution V design, we know that the main effects and two-factor interactions are estimable, and since we have 15 of them, altogether, we have no d.f. for estimating σ with 16 design points. Therefore, there is a need to assess the possible significance of each effect by plotting the estimates of the effects on normal probability paper (or using some other method). Snee (1985) portrayed these estimates in a half-normal plot (recall that this differs from a normal plot only by dropping the sign on negative estimates and plotting the positive value). It was shown that D, B, E, and A have a significant effect, especially D.

It was stated previously that the procedure given by Box and Meyer (1986) can also be used to identify significant effects in unreplicated fractional factorials. A description of the procedure would be beyond the level of this text, however, as the procedure requires the use of posterior probabilities and numerical integration. Readers with a strong foundation in statistics are referred to their paper for details.

13.15.2 2^{k-2} Designs

It was observed that the 2^{5-1} design with 16 points allowed for the estimation of the main effects and two-factor interactions. What would be sacrificed if the experimenter had decided that 16 design points would be too expensive (or otherwise impractical) and chose to run a 2^{5-2} design instead?

In general, when the fraction is of the form $1/2^p$, there will be 2^{p-1} effects that must be confounded with the difference between the fractions, and their product(s) will also be confounded. Thus, for the 2^{5-2} we must select two effects to confound, and their product will also be confounded. This set of effects (i.e., defining contrasts) forms what is termed the *defining relation*. Since the resolution of a two-level fractional factorial is defined as the number of letters in the shortest defining contrast, it might seem as though we should select ABCDE and one of the four-factor interactions. This would be disastrous, however, as their product would contain only a single letter, and we would consequently lose a main effect. Similarly, a little reflection should indicate that any pair of four-factor interactions will have three factors in common and will thus produce a two-factor interaction when multiplied together. We can, however, choose a pair of three-factor interactions in such a way that the product is a four-factor interaction. For example, ABC and BDE would produce ACDE when multiplied together. (There is no advantage in selecting any particular pair, as only main effects are estimable, anyway.)

The alias structure is then obtained by multiplying each of the effects by the three defining contrasts. For example, A = BC = ABDE = CDE. The four $\frac{1}{4}$ fractions could be constructed by determining (1) the treatment combinations that have an even number of letters in common with both ABC and BDE, (2) those combinations with an even number in common with ABC and an odd number in common with BDE, (3) those combinations with an odd number in common with ABC and an even number in common with BDE [i.e., the reverse of (2)], and (4) those combinations with an odd number in common with both ABC and BDE.

Although theoretically desirable, it is not absolutely essential to write out all four fractions and then randomly select one. This is generally not done in practice. The four fractions are given in Table 13.17, however, for the sake of clarity.

The five main effects would then be estimated in the usual way — the average of the four values in which the factor is at the high level minus the average of the four values in which the factor is at the low level. For example, if the first fraction were used, the main effect of A would be estimated by $(ac + abe + abd + acde)/4 - [bcd + bce + de + (1)]/4$.

It should be noted that with this fractional factorial there will be two degrees of freedom for the residual, so the plotting of effect estimates on normal probability paper would not be absolutely essential, although probably desirable since the two degrees of freedom correspond to two-factor interactions that are not being estimated.

It was stated previously that obtaining the design configuration by enumerating the treatment combinations is somewhat laborious when there are more than just a few factors. For $\frac{1}{4}$ fractions, however, this is the most straightforward approach.

TABLE 13.17 Treatment Combinations of Four $\frac{1}{4}$ Fractions of a 2^5 (I = ABC = BDE = ACDE)

(1)	(2)	(3)	(4)
(1)	*ab*	*a*	*b*
ac	*acd*	*c*	*cd*
de	*ace*	*abcd*	*ce*
acde	*d*	*abce*	*bde*
abd	*e*	*bd*	*abc*
abe	*bc*	*be*	*ae*
bcd	*bcde*	*ade*	*ad*
bce	*abde*	*cde*	*abcde*

Another approach is described by Box et al. (1978, p. 397), who show how to obtain a 2^{5-2} from a 2^{7-4}.

For detailed information on the construction of two-level fractional factorials, the reader is referred to Chapter 12 of Box et al. (1978) and Chapters 11 and 12 of Daniel (1976), in particular, as well as Box and Hunter (1961a,b). Additional examples of these designs can be found in Daniel (1976) and Chapter 13 of Box et al. (1978).

13.15.3 More Highly Fractionated Two-Level Designs

With the aid of a computer, fractional factorial designs that are smaller than a $\frac{1}{4}$ fraction may be easily constructed. Such designs are necessary if very many factors are studied as, for example, a 2_{VI}^{9-2} design would require more design points (128) than can generally be afforded. (Here VI denotes the Resolution of the design.) A more practical alternative would be to use a 2_{IV}^{9-4} design, which would require only 32 points. For a 2^{k-4} design, four interactions must be selected, and the defining relation then consists of these four interactions plus all of the generalized interactions. Needless to say, the list of all aliases is then quite long.

Designs that are useful for screening purposes (i.e., identifying important variables) are those where the number of design points are one more than the number of factors. A 2_{III}^{7-4} design is one example of such a design. Such designs are always Resolution III and are referred to as *saturated designs* because the number of available degrees of freedom is equal to the number of factors (i.e., main effects to be estimated).

13.15.4 Fractions of Three-Level Factorials

In order to investigate nonlinear effects, such as possible quadrature, more than two levels must be used. Three-level factorials meet this requirement, although, in general, three-level factorials are not used as extensively as two-level factorials, and the same can be said for fractional factorials. The 3^{k-p} designs have the shortcoming that quite a few design points are required for such a design to be Resolution V. Response surface designs, such as central composite designs presented in Section 13.16.4, require fewer design points and are considered to be superior to the 3^{k-p} designs (as well as the 3^k designs). A similar view is expressed by Montgomery (1997, p. 461).

For readers who may have an interest in 3^{k-p} designs, we mention the catalog of such designs in Connor and Zelen (1959). Although this publication is now out of print, it has been reproduced as Appendix 2 in McLean and Anderson (1984).

13.15.5 Incomplete Mixed Factorials

The term *incomplete* is used here instead of *fractional*, as the number of design points is not generally a common fraction of the full mixed factorial. These designs are summarized in Chapter 13 of Daniel (1976), and the interested reader is referred to that material. More recent work includes Wang and Wu (1991, 1992).

13.15.6 Cautions

Fractional factorials have been successfully used for decades as screening designs, especially the 2^{k-p} series, where $k - p = 4$. Nevertheless, as Lucas (1994) states, the number of design points may not be enough to detect important effects. For example, using the sample size determination formula of Wheeler (1974) that was given in Section 13.5.3, we would need 16 runs to detect a factor effect that is equal to twice the experimental error and 64 runs to detect a factor effect that is equal to the experimental error. Thus, if there is a need to detect effects that are less than 2σ in magnitude, a 16-run screening design would be inadequate.

Another caveat, which should be apparent, is that the likelihood of interactions overwhelming main effects should be considered before any screening design is used.

13.16 OTHER TOPICS IN EXPERIMENTAL DESIGN AND THEIR APPLICATIONS

The following sections contain brief discussions of related material, including some recently developed methods. The reader is referred to the indicated sources for details.

13.16.1 Hard-to-Change Factors

It is not always possible to freely change the levels of factors. This induces restrictions on randomization and can also cause certain factor level combinations to be unattainable. This problem has received relatively little attention in the literature. Joiner and Campbell (1976) were apparently the first to discuss this problem in the statistics literature. Other work includes Wang and Jan (1995). See also Ganju and Lucas (1997).

13.16.2 Split-Lot Designs

Mee and Bates (1998) present a new class of designs and Mee, Bates, and Lynch (1998) illustrate their use in the semiconductor industry. These are called *split-lot designs;* they can be very useful when a product is manufactured in stages. These should not be confused with split-plot designs, which have their origin in agriculture but have also been used successfully in industrial applications, as illustrated by Box and Jones (1992).

13.16.3 Mixture Designs

An important application of experimental design techniques is in the area of mixture models. The general idea is to determine the best mixture of ingredients (factors) to optimize the response variable. A classic example is determining the best composition of gasoline to maximize miles per gallon. The proportions of the ingredients must add to 100%, which induces a type of restriction not generally found in other experimental design applications. Consequently, it requires the use of design procedures that were not discussed in this chapter. The development of such procedures dates from around 1970, although the initial impetus was provided by the pioneering paper of Scheffé (1958). For an introduction to mixture designs the reader is referred to Snee (1971, 1973, 1981) and Cornell (1979, 1983). A comprehensive treatment is given in the text by Cornell (1990). A lighthearted look at another approach to the gasoline mixture problem is given by Cornell (1994).

Piepel (1997) surveys software with mixture experiment capability.

13.16.4 Response Surface Designs

The emphasis in this chapter has been on designs for factors with two levels because those are the designs used most frequently in practice. There is a need to use more than two levels to detect curvature, however and, in general, to determine the shape of the "response surface." That is, how does the response vary over different combinations of values of the process variables? In what region(s) is the change approximately linear? Are there humps, and valleys, and saddle points, and, if so, where do they occur? These are the type of questions that response surface methodology (oftentimes abbreviated as RSM) attempts to answer.

The most frequently used response surface designs are *central composite designs (CCDs)*. These designs also permit the investigation of nonlinear effects and will usually be preferred over three-level fractional factorials. A CCD is constructed by starting with a two-level full factorial and then adding center points and axial (star) points that lie outside the square formed from connecting the factorial points. The design for two factors is shown in Figure 13.17. The value of α would be selected by the experimenter. Desirable properties of the design include orthogonality and rotatability. A design is rotatable if $\text{Var}(\hat{Y})$ is the same for all points equidistant from the center. For the CCD in Figure 13.17, rotatability is achieved if $\alpha = 1.414$. Both orthogonality and rotatability can be achieved by selecting the number of center points to achieve orthogonality, as the number of center points does not affect rotatability. For example, eight center points would be needed for the CCD in Figure 13.17 with $\alpha = 1.414$ in order to achieve both rotatability and orthogonality.

The work on RSM dates from the pioneering paper of Box and Wilson (1951). Other important papers include Box and Hunter (1957) and Box and Behnken (1960). A review of the early work on RSM is given by Hill and Hunter (1966). Cornell (1985) is one of the ASQC-published instructional booklets that are recommended for initial reading. An extensive treatment of RSM is given in the books by Box and Draper (1986), Khuri and Cornell (1996), and Myers and Montgomery (1995).

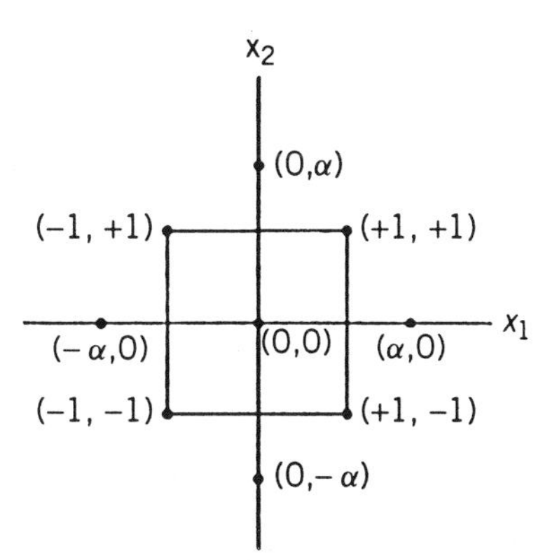

FIGURE 13.17 Central composite design for two factors.

13.16.5 Designs for Measurement System Evaluation

It was mentioned in Section 4.18 that experimental designs can be used to assess the effect of factors (such as operators and operating conditions) that can be expected to change during production. In the case of operators a random sample of operators might be used. It might be unwise to use a random sample of only two operators in an experiment, so something other than a two-level factorial design may need to be used.

Montgomery (1997, p. 473) describes a study for which the objective was to assess the variability due to operators and the variability due to a gauge. Three randomly selected operators measured each of 20 selected parts twice with a gauge. The objective was to assess variability due to operators, parts, and gauge. Letting σ_y^2 denote the total variability, σ_α^2 and σ_β^2 represent the variability due to operators and parts, respectively, and $\sigma_{\alpha\beta}^2$ represent the variability due to the interaction between operators and parts, it follows that

$$\sigma_y^2 = \sigma_\alpha^2 + \sigma_\beta^2 + \sigma_{\alpha\beta}^2 + \sigma^2$$

with σ^2 representing the variation that results when a part is repeatedly measured by the same operator using the same gauge. Thus, this is referred to as *gauge repeatability*, with $\sigma_\beta^2 + \sigma_{\alpha\beta}^2$ representing the *gauge reproducibility*.

The objective is to estimate these variance components and to determine significant effects. The estimates of effects that are not significant will often be negative when analysis of variance is used. This necessitates a different estimation approach since variances cannot, of course, be negative. See Montgomery (1997) for details.

In addition to point estimates, it is desirable to have confidence intervals for variance components in a reproducibility and repeatability (R&R) study. Burdick and Larsen (1997) compare the available methods for obtaining confidence intervals and indicate a preference for the ANOVA approach, as they found that the restricted maximum likelihood (REML) approach produced intervals for which the actual degree of confidence was much less than the stated level. [See, e.g., Searle, Casella, and McCulloch (1992) for a discussion of REML estimators in experimental design.]

Another measurement system experiment was described by Buckner, Chin, and Henri (1997). Three operators were used, with two levels used for each of the other five factors. The design had 16 points, so it was an irregular fraction of a mixed factorial. Designs of this type were given by Addelman (1962).

13.16.6 Computer-Aided Design and Expert Systems

The statistical design of experiments is not an easy task. There are literally hundreds of designs, and for a particular objective and physical setting there might be several designs that could be used. Consequently, the choice of a particular design can be difficult. Some type of guidance is necessary, not only for the experienced user who may be aware of some of his or her options, but especially for the experimenter whose knowledge of experimental designs is limited.

The assistance that is provided could be of several forms: (1) advice from a statistical consultant, (2) software for computer-aided design of experiments, and (3) software for expert systems.

Both 2 and 3 are relatively recent innovations. The difference between them is that with the latter there is minimal user input; the decision making is essentially performed by the software. The question naturally arises as to whether or not this is desirable. This and other issues are addressed by Hahn (1985) and the discussants of that paper.

Experimental design software that additionally guides users is a necessity, so it is really a matter of degree. A totally automated system would preclude the infusion of common sense and real-world experience, so some balance must be reached. Computer-aided design of fractional factorial experiments is discussed by Knight and Neuhardt (1983).

There are many statistical software packages that have excellent design construction and analysis capabilities, including general purpose software such as SAS Software and Minitab. Many users are attracted to software that is specifically for experimental design. The better known software in this category includes Design-Expert® and E-Chip®. GOSSETT, which has been used internally at Bell Labs, is perhaps the most comprehensive and powerful software for experimental design.

13.16.7 Sequential Experimentation

Experimentation does not end, of course, when a design has been selected and the data subsequently collected and analyzed. A physical system will generally change over time, and even if it did not, follow-up experiments should still be used to gain better insight into the nature of the system. Such experiments can often be used to resolve ambiguities from the previous experiment.

Consequently, the construction of subsequent designs should be performed using knowledge of how the previous design was constructed and the results of the analysis. This also requires, of course, that financial resources be allocated judiciously. Box et al. (1978) recommend that at most 25% of the resources be used for the first experiment, and Daniel (1976) recommends reserving 33–50% of the budget for subsequent experimentation. Meyer, Steinberg, and Box (1996) discuss augmenting 2^{k-p} designs with smaller follow-up experiments, with the objective of identifying significant main effects. Mee and Peralta (1999) gave follow-up designs, with the objective being to identify significant interactions. These designs result from semifolding 2^{k-p} designs. The term "semifolding" is due to Peter John and is illustrated, for example, in Barnett, Czitrom, John, and Leon (1997).

13.16.8 Supersaturated Designs and Analyses

A *supersaturated design* is a design for which there are fewer degrees of freedom than there are factors. They are useful for examining main effects when faced with a very large number of factors. Supersaturated designs are not orthogonal designs. Lin (1993) mentions the possibility of using stepwise regression to identify what effects seem to be important and then estimate only those effects. Variable selection techniques can perform poorly when variables are highly correlated, however, so it is important that the correlations between the columns in the design not be very large. Lin (1995) considered the maximum number of factors that can be used for a specified number of runs and when the degree of nonorthogonality is also specified.

13.16.9 Multiple Responses

Obviously analyzing data with multiple responses is more complicated than the analysis of data with a single response variable. That is, it is more difficult to analyze multivariate data than univariate data. Multivariate analysis of variance (MANOVA) is covered in many books, such as Morrison (1990).

A frequent objective when multiple responses are used is to try to determine settings for the factors that will produce values for the response variables that are optimal in some sense. Unfortunately, this cannot be done, in general, so multiple response optimization really involves compromises rather than optimization. That is, values for the response variables which are considered to be optimal generally cannot all be attained, due to correlations between the response variables. Accordingly, a "desirability function" approach (Derringer and Suich, 1980) is often used.

The complexity of the multiresponse optimization problem has resulted in some rather unrealistic assumptions being made in the literature. One such assumption is that all response variables are functions of the same independent variables and additionally that the model for each response variable has the same general form. Certain aspects of the multiresponse optimization problem remain unsolved.

We will illustrate the multiple response problem using the data in Czitrom, Sniegowski, and Haugh (1998). The data are given in Table 13.18.

TABLE 13.18 Data from a Silicon Wafer Experiment

Run	Bulk Gas Flow (sccm)	CF_4 Flow (sccm)	Power (Ws)	Selectivity	Etch Rate (Å/min)	Etch Rate Non uniformity
1	60	5	550	10.93	2710	11.7
2	180	5	550	19.61	2903	13.0
3	60	15	550	7.17	3021	9.0
4	180	15	550	12.46	3029	10.3
5	60	5	700	10.19	3233	10.8
6	180	5	700	17.5	3679	12.2
7	60	15	700	6.94	3638	8.1
8	180	15	700	11.77	3814	9.3
9	120	10	625	11.61	3378	10.3
10	120	10	625	11.17	3295	11.1

These data are from a silicon wafer experiment that was performed to determine the effect that three manufacturing factors (bulk gas flow, CF_4 flow, and power) have on three response variables—selectivity, etch rate, and etch rate nonuniformity. The design that was used was a 2^3 with two center points. The latter were used to check for changes over time.

It was also desirable to determine optimum levels of the manufacturing factors. The data were analyzed in detail by Czitrom et al. (1998). The reader is referred to their analysis, as a detailed analysis will not be given here. It is desirable to maximize *selectivity*, which is the ratio of the rate at which oxide is etched to the rate at which polysilicon is etched. Similarly, it is desirable to maximize the *etch rate*, which is the amount of oxide that is removed from the wafer per unit of time. *Etch rate nonuniformity* is the coefficient of variation (i.e., the standard deviation divided by the mean) of the oxide etch rate. This quantity should of course be minimized. Czitrom et al. (1998) found by using contour plots that selectivity is maximized at high bulk flow and low power, etch rate is maximized at low bulk flow and high power, and etch rate nonuniformity is minimized at low bulk flow and high power. (A contour plot shows constant response contours for different combinations of two factors.)

The authors' approach was to maximize selectivity and determine whether or not the settings that maximize selectivity produce acceptable values for etch rate and etch rate nonuniformity.

We will take a different approach and use these data to illustrate multiresponse optimization. The original settings of the factors were bulk gas flow $= 90$, $CF_4 =$ 5, and power $= 625$. These would produce selectivity $= 12.43$, etch rate $= 3068$, and etch rate nonuniformity $= 11.6$ from the fitted model that the authors used for each response. Czitrom et al. (1998) arrived at improved factor settings of bulk gas flow $= 180$, $CF_4 = 15$, and power $= 550$, which would produce selectivity $=$ 12.45, etch rate $= 3048$, and etch rate nonuniformity $= 10.3$. Recall that the authors were trying to maximize selectivity and were not trying to simultaneously maximize etch rate and minimize etch rate nonuniformity.

We can do better than these factor settings if we use a multiresponse optimization approach. The Multiple Response Optimizer in MINITAB allows the user to assign an importance weight to each response variable in accordance with the importance of each response being maximized, being minimized, or achieving a target value. If the emphasis is on maximizing selectivity, the weights for selectivity, etch rate, and etch rate nonuniformity could be 10, 0.1, and 0.1, respectively. This produces factor settings of bulk gas flow = 173.31, CF_4 = 13.74, and power = 561.04. The prediction equations then give predicted values of selectivity = 12.89, etch rate = 3087, and etch rate nonuniformity = 10.51. If an importance weight of 1.0 is used for each response, the settings are bulk gas flow = 169.07, CF_4 = 14.09, and power = 561.04. The predicted values using these settings are selectivity = 12.48, etch rate = 3091, and etch rate nonuniformity = 10.37. The experimenter would have to decide which set of values is preferable, but this illustrates the value of a multiresponse optimization approach.

The reader is referred to Section 6.6 of Myers and Montgomery (1995) for additional information on the subject.

13.17 SUMMARY

The statistical design of experiments is a very broad area, and dozens of books have been written on the subject. The focus for most of these, however, has been the analysis of data from standard designs. The first step must be the design of the experiment, and this is the hard part. Much thought needs to be given to the design, and the experimenter must also realize that a routine "cookbook" analysis of the data will generally be insufficient.

Since design of experiments is important, and much thought needs to be given to it, how can experimenters acquire the necessary expertise without having to spend a large amount of company funds and then possibly drawing erroneous conclusions?

The author has found the ideas and sample experiments presented in Hunter (1977) to be quite helpful in enabling students in a statistics course to "get their feet wet" in designing experiments and analyzing resultant data. Such ideas can be incorporated into corporate training programs to enable personnel to acquire the necessary expertise without committing the labor and materials that would be required for an actual company experiment.

REFERENCES

Addelman, S. (1962). Orthogonal main-effect plans for asymmetrical factorial experiments. *Technometrics 4*(1): 21–46.

Bainbridge, T. R. (1965). Staggered, nested designs for estimating variance components. *Industrial Quality Control 22*(1): 12–20.

Barnett, J., V. Czitrom, P. W. M. John, and R. V. Leon (1997). Using fewer wafers to resolve confounding in screening experiments. In V. Czitrom and P. D. Spagon, eds. *Statistical Case Studies for Industrial Process Improvement*, pp. 235–250. Philadelphia: SIAM.

Berger, P. D. (1972). On Yates' order in fractional factorial designs. *Technometrics 14*(4): 971–972.

Berk, K. N. and R. R. Picard (1991). Significance tests for saturated orthogonal arrays. *Journal of Quality Technology 23:* 79–89.

Box, G. E. P. (1990). George's column. *Quality Engineering 2*(3): 365–369.

Box, G. E. P. and D. Behnken (1960). Some new three level designs for the study of quantitative variables. *Technometrics 2*(4): 455–475.

Box, G. E. P. and N. R. Draper (1986). *Empirical Model Building and Response Surfaces.* New York: Wiley.

Box, G. E. P. and J. S. Hunter (1957). Multifactor designs for exploring response surfaces. *Annals of Mathematical Statistics 28:* 195–241.

Box, G. E. P. and J. S. Hunter (1961a). The 2^{k-p} fractional factorial designs, Part I. *Technometrics 3*(3): 311–351.

Box, G. E. P. and J. S. Hunter (1961b). The 2^{k-p} fractional factorial designs, Part II. *Technometrics 3*(4): 449–458.

Box, G. E. P. and S. Jones (1992). Split-plot designs for robust product experimentation. *Journal of Applied Statistics 19:* 3–26.

Box, G. E. P. and R. D. Meyer (1986). An analysis for unreplicated fractional factorials. *Technometrics 28*(1): 11–18.

Box, G. E. P. and K. B. Wilson (1951). On the experimental attainment of optimum conditions. *Journal of the Royal Statistical Society, Series B 13*(1): 1–45 (with discussion).

Box, G. E. P., W. G. Hunter, and J. S. Hunter (1978). *Statistics for Experimenters.* New York: Wiley.

Box, G., S. Bisgaard, and C. Fung (1990). *Designing Industrial Experiments.* Madison, WI: BBBF Books.

Brush, G. G. (1988). How to choose the proper sample size. In *Basic References in Quality Control: Statistical Techniques*, Vol. 12. Milwaukee, WI: American Society for Quality Control.

Buckner, J., B. L. Chin, and J. Henri (1997). Prometrix RS35e gauge study in five two-level factors and one three-level factor. In V. Czitrom and P. D. Spagon, eds. *Statistical Case Studies for Industrial Process Improvement*, pp. 9–17. Philadelphia: SIAM.

Burdick, R. K. and G. A. Larsen (1997). Confidence Intervals on Measures of Variability in R&R studies. *Journal of Quality Technology 29*(3): 261–273.

Chan, T. F., G. H. Golub, and R. J. Leveque (1983). Algorithms for computing the sample variance: Analysis and recommendations. *The American Statistician 37*(3): 242–247.

Cochran, W. G. and G. M. Cox (1957). *Experimental Designs*, 2nd ed. New York: Wiley.

Coleman, D. E. and D. C. Montgomery (1993). A systematic approach to planning for a designed industrial experiment. *Technometrics 35*(1): 1–12 (discussion: 13–27).

Connor, W. S. and M. Zelen (1959). *Fractional Factorial Experiment Designs for Factors at Three Levels.* Washington, DC: National Bureau of Standards, Applied Mathematics Series, No. 54.

Conover, W. J. (1980). *Practical Nonparametric Statistics*, 2nd ed. New York: Wiley.

Cornell, J. A. (1979). Experiments with mixtures: An update and bibliography. *Technometrics 21*(1): 95–106.

Cornell, J. A. (1983). How to run mixture experiments for product quality. *Basic References in Quality Control: Statistical Techniques*, Vol. 5. Milwaukee, WI: American Society for Quality Control.

Cornell, J. A. (1985). How to apply response surface methodology. *Basic References in Quality Control: Statistical Techniques*, Vol. 8. Milwaukee, WI: American Society for Quality Control.

Cornell, J. A. (1990). *Experiments with Mixtures: Designs, Models, and the Analysis of Mixture Data*, 2nd ed. New York: Wiley.

Cornell, J. A. (1994). Saving money with a mixture experiment. *ASQC Statistics Division Newsletter 14*(1): 11–12.

Cornell, J. A. and A. I. Khuri (1996). *Response Surfaces: Designs and Analyses.* New York: Dekker.

Cressie, N. A. C. and H. J. Whitford (1986). How to use the two sample T-test. *Biometrical Journal 28*(2): 131–148.

Czitrom, V., J. Sniegowski, and L. D. Haugh (1998). Improved integrated circuit manufacture using a designed experiment. In R. Peck, L. D. Haugh, and A. Goodman, eds. *Statistical Case Studies: A Collaboration Between Academe and Industry*. Society of Industrial and Applied Mathematics and the American Statistical Association.

Daniel, C. (1959). Use of half-normal plots in interpreting factorial two-level experiments. *Technometrics 1*(4): 311–341.

Daniel, C. (1973). One-at-a-time plans. *Journal of the American Statistical Association 68*(342): 353–360.

Daniel, C. (1976). *Applications of Statistics to Industrial Experimentation.* New York: Wiley.

Davies, O. L., ed. (1954). *Design and Analysis of Industrial Experiments.* New York: Hafner (Macmillan).

Deming, W. E. (1975). On probability as a basis for action. *American Statistician 29*(4): 146–152.

Derringer, G. and R. Suich (1980). Simultaneous optimization of several response variables. *Journal of Quality Technology 25:* 199–204.

Finney, D. J. (1945). Fractional replication of factorial arrangements. *Annals of Eugenics 12:* 291–301.

Ganju, J. and J. M. Lucas (1997). Bias in test statistics when restrictions in randomization are caused by factors. *Communications in Statistics—Theory and Methods 26*(1): 47–63.

Haaland, P. D. (1998). Comment. *Statistica Sinica 8*(1): 31–35.

Hahn, G. J. (1977). Some things engineers should know about experimental design. *Journal of Quality Technology 9*(1): 13–20.

Hahn, G. J. (1985). More intelligent statistical software and statistical expert systems: Future directions. *The American Statistician 39*(1): 1–8 (discussion: pp. 8–16).

Hamada, M and N. Balakrishnan (1998). Analyzing unreplicated factorial experiments: A review with some new proposals. *Statistica Sinica 8:* 1–41.

Hicks, C. R. (1982). *Fundamental Concepts in the Design of Experiments*, 3rd ed. New York: Holt, Rinehart, and Winston.

Hill, W. J. and W. G. Hunter (1966). A review of response surface methodology. *Technometrics 8*(4): 571–590.

Hunter, W. G. (1977). Some ideas about teaching design of experiments with 2^5 examples of experiments conducted by students. *The American Statistician 31*(1): 12–17.

Hunter, W. G. and M. E. Hoff (1967). Planning experiments to increase research efficiency. *Industrial and Engineering Chemistry 59*(3): 43–48.

Joiner, B. L. and C. Campbell (1976). Designing experiments when run order is important. *Technometrics 18:* 249–260.

Kempthorne, O. (1973). *Design and Analysis of Experiments.* New York: Krieger.

Khuri, A. and J. Cornell (1996). *Response Surfaces: Designs and Analyses*, 2nd ed. New York: Dekker.

Knight, J. W. and J. B. Neuhardt (1983). Computer-aided design of fractional factorial experiments given a list of feasible observations. *IIE Transactions 15:* 142–149.

Layard, M. W. J. (1973). Robust large-sample tests for homogeneity of variances. *Journal of the American Statistical Association 68*(341): 195–198.

Lenth, R. V. (1989). Quick and easy analysis of unreplicated factorials. *Technometrics 31:* 469–473.

Lin, D. K. J. (1993). A new class of supersaturated designs. *Technometrics 35*(1): 28–31.

Lin, D. K. J. (1995). Generating systematic supersaturated designs. *Technometrics 37*(2): 213–225.

Loughin, T. M. and W. Noble (1997). A permutation test for effects in an unreplicated factorial design. *Technometrics 39*(2): 180–190.

Lucas, J. M. (1994). How to achieve a robust process using response surface methodology. *Journal of Quality Technology 26*(3): 248–260.

Margolin, B. H. (1967). Systematic methods for analyzing $2^N 3^M$ factorial experiments. *Technometrics 9*(2): 245–260.

McLean, R. A. and V. L. Anderson (1984). *Applied Factorial and Fractional Designs.* New York: Dekker.

Mee, R. and R. Bates (1998) Constructing split lot designs. *Technometrics 40*(2): 127–140.

Mee, R. W. and M. Peralta (1999). Semifolding 2^{k-p} designs. *Technometrics* (to appear).

Mee, R., R. Bates, and R. Lynch (1998). Analysis of split lot experiments in the semiconductor industry. Manuscript.

Meyer, R. D., D. M. Steinberg, and G. Box (1996). Follow-up designs to resolve the confounding in multifactor experiments. *Technometrics 38:* 303–313.

Miller, R. G. Jr. (1981). *Simultaneous Statistical Inference*, 2nd ed. New York: Springer-Verlag.

Miller, R. B. and D. W. Wichern (1977). *Intermediate Business Statistics.* New York: Holt, Rinehart, & Winston.

Montgomery, D. C. (1996). Some practical guidelines for designing an industrial experiment. In J. B. Keats and D. C. Montgomery, eds. *Statistical Applications in Process Control.* New York: Dekker.

Montgomery, D. C. (1997). *Design and Analysis of Experiments*, 4th ed. New York: Wiley.

Morrison, D. F. (1990). *Multivariate Statistical Methods*, 3rd ed. New York: McGraw-Hill.

Myers, R. H. and D. C. Montgomery (1995). *Response Surface Methodology: Process and Product Optimization Using Designed Experiments.* New York: Wiley.

Nelson, L. S. (1976). Constructing normal probability paper. *Journal of Quality Technology 8*(1): 56–57.

Nelson, L. S. (1984). Some notes on Student's *t*. *Journal of Quality Technology 16*(1): 64–65.

Nelson, L. S. (1985). Sample size tables for analysis of variance. *Journal of Quality Technology 17*(3): 167–169.

Neter, J., M. H. Kutner, C. Nachtsheim, and W. Wasserman (1996). *Applied Linear Statistical Models*, 4th ed. New York: Irwin/McGraw-Hill.

O'Brien, R. G. (1997). UnifyPow: A SAS macro for sample-size analysis. In *Proceedings of the 22nd SAS Users Group International Conference (SUGI 22)*, pp. 1353–1358. Cary, NC: SAS Institute.

Odeh, R. E. and M. Fox (1975). *Sample Size Choice: Charts for Experiments with Linear Models.* New York: Dekker.

Ott, E. R. and E. G. Schilling (1990). *Process Quality Control: Troubleshooting and Interpretation of Data*, 2nd ed. New York: McGraw-Hill.

Piepel, G. F. (1997). Survey of software with mixture experiment capabilities. *Journal of Quality Technology 29*(1): 76–85.

Posten, H. O., H. C. Yeh, and D. B. Owen (1982). Robustness of the two sample *t*-test under violations of the homogeneity of variance assumption. *Communications in Statistics, Part A—Theory and Methods 11*(2): 109–126.

Scheffé, H. (1958). Experiments with mixtures. *Journal of the Royal Statistical Society, Series B 20*(2): 344–360.

Searle, S. R., G. Casella, and C. E. McCulloch (1992). *Variance Components.* New York: Wiley.

Sinibaldi, F. J. (1983). Nested designs in process variation studies. In *ASQC Annual Quality Congress Transactions*, pp. 503–508. Madison, WI: American Society for Quality Control.

Snedecor, G. W. and W. G. Cochran (1980). *Statistical Methods*, 7th ed. Ames, IA: Iowa State University Press.

Snee, R. D. (1971). Design and analysis of mixture experiments. *Journal of Quality Technology 3*(4): 159–169.

Snee, R. D. (1973). Techniques for the analysis of mixture data. *Technometrics 15*(3): 517–528.

Snee, R. D. (1981). Developing blending models for gasoline and other mixtures. *Technometrics 23*(2): 119–130.

Snee, R. D. (1983). Graphical analysis of process variation studies. *Journal of Quality Technology 15*(2): 76–88.

Snee, R. D. (1985). Experimenting with a large number of variables. In R. D. Snee, L. B. Hare, and J. R. Trout, eds. *Experiments in Industry—Design, Analysis and Interpretation of Results.* Milwaukee, WI: Quality Press.

Steinberg, D. M. and W. G. Hunter (1984). Experimental design: Review and comment. *Technometrics 26*(2): 71–97 (discussion: pp. 98–130).

Tukey, J. W. (1949). One degree of freedom for non-additivity. *Biometrics 5*(3): 232–242.

Wang, J. C. and C. F. J. Wu (1991). An approach to the construction of asymmetrical orthogonal arrays. *Journal of the American Statistical Association 86:* 450–456.

Wang, J. C. and C. F. J. Wu (1992). Nearly orthogonal arrays with mixed levels and small runs. *Technometrics 34:* 409–422.

Wang, P. C. and H. W. Jan (1995). Designing 2-level factorial experiments using orthogonal arrays when the run order is important. *Statistician 44*(3): 379–388.

Welch, B. L. (1937). The significance of the difference between two means when the population variances are equal. *Biometrika 29:* 350–362.

Wheeler, R. E. (1974). Portable power. *Technometrics 16*(2): 193–201.

Yates, F. (1937). *Design and Analysis of Factorial Experiments.* London: Imperial Bureau of Soil Sciences.

Ziegel, E. R. (1984). Discussion (of an invited paper by Steinberg and Hunter). *Technometrics 26*(2): 98–104.

EXERCISES

1. Assume that a 2^{3-1} design has been run using the treatment combinations (1), *ab*, *ac*, and *bc*, and the response values are 12, 16, 14, and 20, respectively. Use Yates's algorithm to show that the defining contrast is $\text{I} = -\text{ABC}$ by pairing the totals for column (3) that are the same.
2. It was stated at the beginning of the chapter that a plot of the values from two groups (as in Figure 13.3) in which the lines are straight and almost touching would indicate that $\mu_1 \neq \mu_2$. Assume that the values from the two groups are 13.00, 13.00, 13.00, 13.00, 13.001 from the first group and 13.01, 13.01, 13.01, 13.01, 13.011 from the second group. What would be the result if we use an exact or approximate t test to test whether $\mu_1 = \mu_2$? Do you think the difference is of any practical significance if this is yield data? What does this imply that we should keep in mind when using tests such as this?
3. Explain why the value of t^* is the same as the value for t using the data in exercise 2.
4. An experimenter decides to use a 2^{6-2} design and elects to confound ABD and CEF in constructing the fraction.

   (a) Determine what effect(s) would be aliased with the two-factor interaction AB.

   (b) Is there a better choice for the two defining contrasts?

5. A 2^2 design with two replicates is run and the following results are obtained. The response totals are 20 when both factors are at the low level, 30 when both factors are at the high level, 15 when A is high and B is low, and 20 when A is low and B is high. Assume both factors to be fixed and use $SS_{\text{total}} = 70$ to test for the significance of the A, B, and AB effects.

6. Explain why residual plots should be used with data from designed experiments.
7. How many degrees of freedom will be available for estimating σ^2 if a 2^3 design is run with three replicates?
8. Four factors, each at two levels, are studied with the design points given by the following treatment combinations: (1), *ab*, *bc*, *abd*, *acd*, *bcd*, *d*, and *ac*.

   (a) What is the defining contrast?

   (b) Could the design be improved upon using the same number of design points?

   (c) In particular, which three main effects are confounded with two-factor interactions?

9. Construct a sample of six observations from each of two populations in which the exact t-test for testing $\mu_1 = \mu_2$ produces a t value of zero.
10. Six similar (but not identical) machines are used in a production process. Each machine has a head that is not interchangeable between machines, although there are three different types of heads that can be used on a given machine. A study is to be performed to analyze process variability by studying the six machines and three heads. Could the data be analyzed as a 3×6 factorial design? Why or why not?
11. Assume that ANOVA is applied to three groups and the F test is significant. What graphical aid could be used to determine which groups differ in terms of their means?
12. Assume that data from a 2^3 design have been analyzed and one or more of the interactions are significant. What action should be taken in investigating main effects for factors that comprise those interactions?
13. Analyze the following data for two groups using both an exact t test and an F test. Show that $F = t^2$. Do the samples seem to be independent? Comment. Then construct a confidence interval for the difference between the two means.

1	2
14.3	14.8
16.2	16.6
13.5	13.6
14.6	14.9
14.9	15.3
15.4	15.8
16.0	16.4
15.7	16.2

14. Name one disadvantage in using a highly fractionated fractional factorial.

15. The following interaction profile shows the results of an unreplicated 2^2 design.

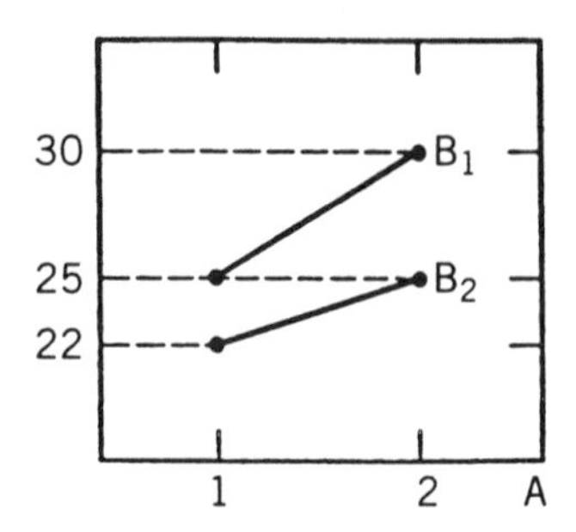

Estimate the A effect.

16. Consider the discussion in Section 13.2 regarding mean shifts. Assume that two production methods are to be compared but a 1-sigma increase in the process mean occurs after $n_2/2$ observations, with n_2 denoting the number of observations taken when the second production method is in effect. Assume that sigma is known and compute the expected value of the test statistic for testing the equality of the means for the two methods as a function of n_1 and n_2. Now assume that $n_1 = n_2 = 10$. What do you conclude?

17. Determine an appropriate model for the data given in Table 13.18 using selectivity as the response variable and considering main effects and interaction terms.

18. Using etch rate as the dependent variable for the data in Table 13.18, fit the model with the three main effects plus the interaction between the first two factors. Then fit the same model after first coding the data such that the coded values for each factor are -1, 0, and 1 and then forming the interaction term for the first two factors, using the coded values to form the interaction term. Explain why the p-values for the bulk gas term differ considerably.

CHAPTER 14

Contributions of Genichi Taguchi and Alternative Approaches

In this chapter we shall examine the methods that have been proposed by the Japanese engineer, Genichi Taguchi. His quality engineering ideas and statistical procedures have been used in Japan for decades, but it was the mid-1980s when the Western world became aware of his views toward process control and quality improvement in general and the set of tools that he advocates.

This eventual awareness led to the creation of the American Supplier Institute, Inc., in Romulus, Michigan. The objective has been to educate engineers, in particular, in statistical process control (SPC) and Taguchi methods. Presentations have also been made to engineers in the form of "executive briefings." Shin Taguchi, a nephew of Dr. Genichi Taguchi, and Professor Yuin Wu have taught Taguchi methods as well as standard control-charting procedures to American industrial personnel.

In this chapter we introduce the reader to Taguchi's quality engineering ideas, which for the most part have been enthusiastically received. We then discuss his statistical procedures, which have been highly controversial but which have sparked new research.

14.1 "TAGUCHI METHODS"

The pervasiveness of Taguchi's techniques cannot be denied, and the widespread use of these methods is undoubtedly due to a substantial entrepreneurial effort. Taguchi's methods have been used in a wide variety of applications; the following list is a sample of journal article titles since 1993:

"The application of Taguchi methods to a coil spring manufacturing process" (Caporaletti, Gillenwater, and Jaggers, 1993)

"Rating quality and selecting suppliers using Taguchi loss functions" (Snow, 1993)

"Taguchi analysis of heat treatment variables on the mechanical behavior of alumina/aluminum metal matrix composites" (Leisk and Saigal, 1995)

"Improving the quality of an optimal power flow solution by Taguchi method" (Bounou, Lefebvre, and Dai Do, 1995)

In many quarters design of experiments has become synonymous with "Taguchi design." This is unfortunate, as we will see later in the chapter.

14.2 QUALITY ENGINEERING

Specification limits (when they are, in fact, true tolerance limits) have generally been regarded as providing a range for the values of a process characteristic such that values within this range are acceptable. Implicit in this view is the idea that all values within this range are equally good.

Common sense should tell us, however, that if we have a complicated piece of machinery that consists of a large number of moving parts, the machinery is likely to perform better if the dimensions of the individual parts were made to conform to certain "optimal" values than if the dimensions were merely within tolerance limits.

This type of thinking is at the heart of the loss-function approach advocated by Taguchi, who contends that the "loss to society" increases as the value of a quality characteristic departs from its optimal value, regardless of whether or not a tolerance limit has been exceeded.

The loss to society idea of Taguchi has been replaced by "long-term loss to the firm" by those in the Western world, but the general idea is the same. That is, individual firms and society as a whole suffer a loss when products do not function as they could if they were made properly. Therefore, the objective should be to minimize variation about the optimal values. Accordingly, factors that influence variation should be identified.

14.3 LOSS FUNCTIONS

The simplest type of loss function is *squared error*, which is also referred to as *quadratic loss*. Specifically, if we let t denote a *target value* (i.e., optimal value) of a quality characteristic, Y the actual value of that characteristic, and L the loss incurred when $Y \neq t$, then

$$L = (Y - t)^2 \tag{14.1}$$

would be a simple (quadratic) loss function. To illustrate, if $t = 5$ and $Y = 8$, then $L = 9$ (dollars, say). If the tolerance limits on Y are (1, 9), there is thus a "loss" even though the value of Y is within the tolerance limits.

This should not be construed to mean that the loss will actually be 9 when $Y = 8$. Rather, a quadratic loss is simply being used as a model for the true loss function (which will generally be unknown). Quadratic loss will have the general shape shown in Figure 14.1 so that, for example, at Y_1 the (predicted) loss would be L_1.

If the loss were actually known for various values of Y, a more precise graph might be as in Figure 14.2. It would not be practical to try to determine a loss function that would generate these actual losses. Rather, quadratic loss would be a (useful in this case) approximation to the actual loss function.

A more general type of quadratic loss function, which has been advocated by Tagughi (1986) is

$$L = k(Y - t)^2 \tag{14.2}$$

where k is a constant that would have to be determined. A value for k can be easily determined if the actual loss is known for a particular value of Y. Specifically, if we solve Eq. (14.2) for k, we obtain

$$k = L/(Y - t)^2$$

so that if $L = 50$ when $Y = 10$ and $t = 5$, then k should be set equal to 2. Thus, the loss function would be

$$L = 2(Y - 5)^2 \tag{14.3}$$

Although the predicted loss at various values of Y is of interest [and could be obtained using Eq. (14.3)], a practitioner may be more interested in the average squared error, which is generally referred to as the *mean squared error* (MSE).

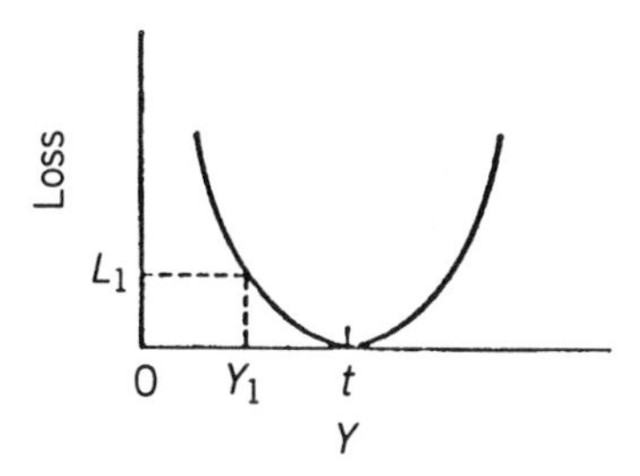

FIGURE 14.1 Quadratic loss.

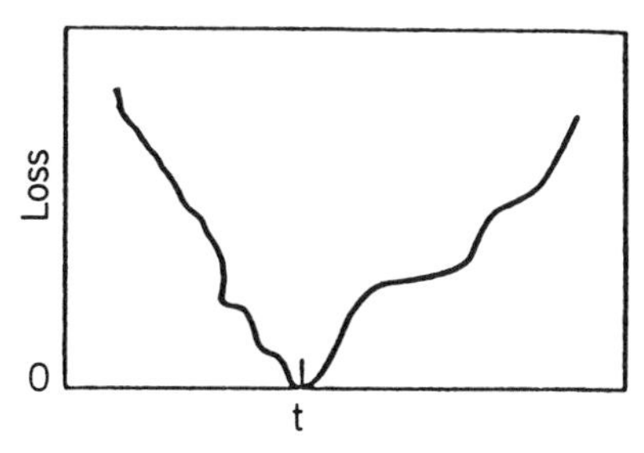

FIGURE 14.2 Actual loss.

For quadratic loss given by Eq. (14.1), MSE(Y) is given by

$$E(L) = E(Y - t)^2$$

where, as discussed in Chapter 3, E represents "expected value" (i.e., average). If we had $E(Y) = t$, then $E(L)$ would equal σ_y^2 and $E(L)$ could then be thought of as the (theoretical) average of the squared deviations, $(Y - t)^2$.

If $E(Y) \neq t$, then with μ substituted for $E(Y)$ we have

$$\begin{aligned} E(L) &= E(Y - t)^2 \\ &= E[(Y - \mu) + (\mu - t)]^2 \\ &= E[(Y - \mu)^2 + 2(Y - \mu)(\mu - t) + (\mu - t)^2] \\ &= E(Y - \mu)^2 + 0 + E(\mu - t)^2 \\ &= \sigma_y^2 + [E(Y) - t]^2 \end{aligned}$$

The last two lines follow from the fact that $E(Y - \mu) = E(Y) - \mu = 0$, $E(Y - \mu)^2$ is, by definition, σ_y^2, $E(\mu - t)^2 = (\mu - t)^2$, and $\mu = E(Y)$. The quantity $[E(Y) - t]^2$ is generally called the *squared bias*. It should be noted that if $L = k(Y - t)^2$, then $E(L) = k\{\sigma_y^2 + [E(Y) - t]^2\}$.

We shall use a few simple examples to illustrate the use of loss functions. Assume that a loss function is given by

$$L = (Y - t)^2$$

where $t = 5$ and $Y \sim N(\mu = 5, \sigma^2 = 1)$. Then $E(Y) = t$ so that $E(L) = \sigma_y^2 = 1$. How can this expected loss be reduced? Since the distribution of Y is already centered at the target value (as shown in Figure 14.3), the only way to reduce the expected loss is to reduce σ_y^2. If σ_y^2 is reduced to 0.5, then $E(L) = 0.5$, assuming that the distribution remains centered at t. This is, of course, intuitive because the loss depends on the distance that Y is from t, and by reducing σ_y^2, we increase

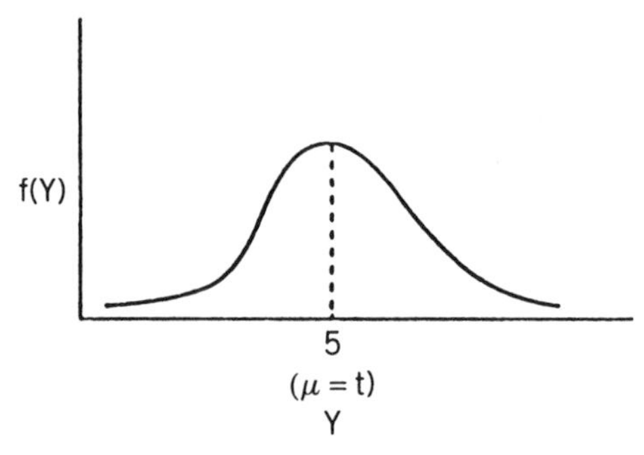

FIGURE 14.3 $Y \sim N(\mu = t = 5, \sigma^2 = 1)$.

the proportion of values of Y that are a given distance from t. *This reduction in expected loss resulting from a reduction of the variance is one reason why firms should continually strive to reduce the variance for all of their processes.*

The same point was made in an example given by Kacker and Ghosh (1998) in describing a comparison of the quality of television sets produced by a Sony factory in Japan and by a Sony factory in San Diego, with the comparison described in a prominent Japanese newspaper in 1979. One of the performance characteristics used in the comparison was color density. The distributions for Sony Japan and Sony USA were symmetric about the common target value, and the tolerance interval was $\mu \pm 5$. Sony USA did not ship any sets for which the color density was out of tolerance, whereas 0.27% of the sets shipped by Sony Japan were out of tolerance.

Thus, on the basis of nonconforming units alone, Sony USA would seem to have better quality sets. This is potentially very misleading, however, since the distribution for Sony USA was approximately uniform, whereas the distribution for Sony Japan was approximately normal. (Undoubtedly, the uniform distribution for Sony USA was caused at least in part by the fact that Sony USA did not ship any sets that were nonconforming in regard to color density, so the measurements for nonconforming sets were not included in the distribution.) As mentioned in Section 3.6.14, the variance for a uniform distribution defined on $\mu \pm 5$ is $\frac{25}{3}$. Since the distribution was centered at the target value, this is also $E(L)$. Since 0.27% is the percentage of nonconforming units that result when the tolerance limits are at $\mu \pm 3\sigma$, it follows that $\sigma = \frac{5}{3}$. Accordingly, for Sony Japan, $E(L) = \frac{25}{9}$, so the expected loss for Sony Japan is one-third of the expected loss for Sony USA. Thus, the quality was considerably better for Sony Japan.

14.4 DISTRIBUTION NOT CENTERED AT THE TARGET

If, on the other hand, the distribution of Y is not centered so that $E(Y) = t$, the expected loss can be reduced simply by shifting the center of the distribution closer to t, even if previous attempts at reducing the variance had failed. Centering the distribution will often cause a greater reduction in the expected loss than a reduction in the variance. For example, if $\sigma^2 = 1$, $\mu = 7$, and $t = 5$, there is no reduction in σ^2 that could cause a reduction in the expected loss equal to the reduction caused by shifting μ from 7 to 5.

As pointed out by Kacker (1985), in many engineering applications the sample average and sample variance are independent, so reducing one will not necessarily have any effect on the other. (Theoretically, this is true only for a normal distribution.)

14.5 LOSS FUNCTIONS AND SPECIFICATION LIMITS

To this point in the chapter there has been no mention of loss functions used in conjunction with specification limits. Under conventional thinking there has

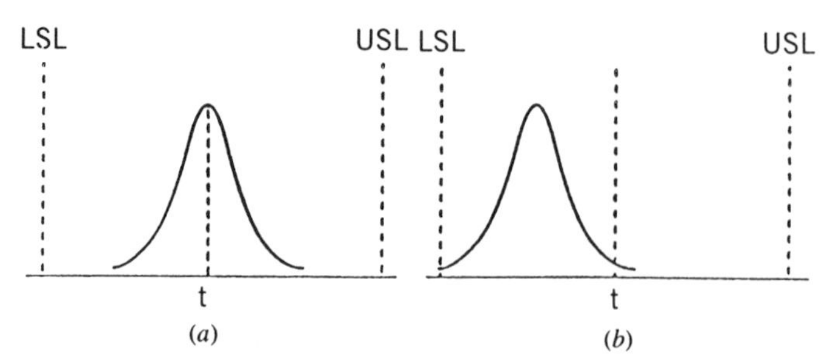

FIGURE 14.4 Normal distributions: (*a*) centered about the target and (*b*) not centered about the target. Both within specifications.

been the implicit assumption that the expected loss is zero as long as the value of a quality characteristic is within the specification limits. If, however, we accept the notion that a loss occurs whenever the target value is not met, the expected loss could be virtually independent of the specifications. For example, consider Figures 14.4*a* and 14.4*b*. The expected losses will differ considerably, but the differences will not depend to any extent upon the upper and lower specification limits, which are denoted by USL and LSL, respectively.

14.6 ASYMMETRIC LOSS FUNCTIONS

An asymmetric loss function would be appropriate if the loss differs for values of Y that are equidistant from the target. Although Taguchi-style experimentation assumes quadratic loss, Kros and Mastrangelo (1998) consider nonquadratic loss and point out, for example, that the losses due to underfilling would not likely equal the losses due to overfilling. Moorhead and Wu (1998) additionally give other scenarios for which the loss would be asymmetric. See also León and Wu (1992).

Consider that a value that exceeds the target may be more detrimental than a value that is below the target. Certainly overdosage of medicine could be much more harmful than underdosage. In that case, one could use a loss function of the form

$$L = \begin{cases} k_1(Y-t)^2 & \text{if } Y \le t \\ k_2(Y-t)^2 & \text{if } Y > t \end{cases}$$

where $k_2 > k_1$.

The appropriate values for k_1 and k_2 can be easily determined if the actual loss is known for one value of Y above t and for one value of Y below t. These might be at the specification limits or at some other values. (It would be impractical to use the specification limits, however, if the distribution of Y was well inside the limits as in Figure 14.4*a*.)

Consider the following example: $Y \sim N(\mu = 5, \sigma^2 = 4)$, $E(Y) = t$, and $L(7) = 8$ and $L(3) = 2$. We are thus assuming that the losses are known at points that are one standard deviation above and below the mean. These would likely have to be estimated, but, assuming that reasonably good estimates could be obtained, it would be better to use these estimates than to use losses at the specification limits, which would likely be easier to obtain. This is particularly true when the specification limits are well beyond the spread of the distribution of Y. This can be explained as follows. Assume that the USL is six standard deviations above the mean (and assume $\mu = t$). When the loss at the USL is used to determine k_2, it then provides an estimate of the loss for every other value of Y between t and the USL. Thus, the loss in a region that Y is not likely to reach is being used in estimating the loss for a region $(\mu,\ \mu + 3\sigma)$ that values of Y will fall in with probability of roughly 0.5. Granted, one value of k for the region (μ, USL) may not be completely satisfactory, but if a sizable error is to be made in estimating $L(Y)$ for various values of Y, such errors in the region ($\mu + 3\sigma$, USL) can be tolerated much more easily than errors in the region $(\mu,\ \mu + 3\sigma)$.

For the present example we would estimate k_2 as

$$\begin{aligned} k_2 &= \frac{L(Y)}{(Y-t)^2} \\ &= \frac{L(7)}{(7-5)^2} \\ &= \frac{\$8}{4} = \$2 \qquad Y > t \end{aligned}$$

Similarly, k_1 would be estimated as

$$\begin{aligned} k_1 &= \frac{L(Y)}{(Y-t)^2} \\ &= \frac{L(3)}{(3-5)^2} \\ &= \frac{\$2}{4} = \$0.50 \qquad Y \le t \end{aligned}$$

Thus, the loss function would be

$$L = \begin{cases} \$0.50(Y-5)^2 & \text{if } Y \le 5 \\ \$2(Y-5)^2 & \text{if } Y > 5 \end{cases}$$

How can the expected loss for this function be obtained? The formal approach to obtaining the expected loss utilizes the truncated normal distribution, which was covered in Chapter 3. Specifically,

$$\begin{aligned} E(L) &= \$0.50E[(Y-5)^2|Y \le 5]P(Y \le 5) \\ &\quad + \$2E[(Y-5)^2|Y > 5]P(Y > 5) \\ &= \$0.50\{\sigma^2_{y|y\le 5} + [E(Y|Y \le 5) - 5]^2\}(0.5) \\ &\quad + \$2\{\sigma^2_{y|y>5} + [E(Y|Y > 5) - 5]^2\}(0.5) \end{aligned}$$

where we would seemingly need to obtain values for the four components inside the two bracketed expressions—two expected values and two variances. (Each vertical line is read as "given.")

Actually, however, we only need one of each. Because the normal distribution is being split in half (since $\mu = t$), it follows that $\sigma^2_{y|y\leq 5}$ must equal $\sigma^2_{y|y>5}$. Furthermore, it also follows that the distance that $E(Y|Y \leq 5)$ is from μ must be the same as the distance that $E(Y|Y > 5)$ is from μ. Thus, the values of the two bracketed expressions are the same.

Accordingly, we need to utilize the general expression for either a left-truncated or right-truncated normal distribution and the single expression for the variance of a truncated normal. For a left-truncated normal distribution,

$$E(Y|Y > c) = \mu + \sigma f(u)[1 - \Phi(u)]^{-1}$$

where c is the truncation point, $u = (c - \mu)/\sigma$, $f(u)$ is the ordinate of the standard normal distribution, and $\Phi(u)$ is the cumulative area under the standard normal curve at point u.

Using the numbers in this example,

$$\begin{aligned} E(Y|Y > 5) &= 5 + 2f(0)[1 - \Phi(0)]^{-1} \\ &= 5 + 2(0.3989)[1 - 0.5]^{-1} \\ &= 5 + 1.5956 \\ &= 6.5956 \end{aligned}$$

where $u = (c - \mu)/\sigma = 0$ since $c = \mu$. It then follows that $E(Y|Y \leq 5) = 5 - 1.5956 = 3.4044$.

The variance of a left-truncated normal distribution is

$$\text{Var}(Y|Y > c) = \sigma^2\{1 + uf(u)[1 - \Phi(u)]^{-1} - [f(u)]^2[1 - \Phi(u)]^{-2}\}$$

so we thus have

$$\begin{aligned} \text{Var}(Y|Y > 5) &= 4\{1 + 0f(0)[1 - \Phi(0)]^{-1} - [f(0)]^2[1 - \Phi(0)]^{-2}\} \\ &= 4[1 - (0.3989)^2(0.5)^{-2}] \\ &= 1.454 \end{aligned}$$

Using these numbers for the expected value and the variance, we then obtain

$$\begin{aligned} E(L) &= \$0.50[1.454 + (3.4044 - 5)^2](0.5) \\ &\quad + \$2[1.454 + (6.5956 - 5)^2](0.5) \\ &= \$0.50(4)(0.5) + \$2(4)(0.5) \\ &= \$5 \end{aligned}$$

The reader may observe that when the general expressions for $\sigma^2_{y|y>5}$ and $[E(Y|Y > 5) - 5]^2$ are added together, the result is σ^2_y when $u = 0$, and similarly for $Y \leq 5$. Consequently, a shorter approach would have been to take the average of \$0.50 and \$2.00 and multiply that by σ^2_y.

If $u \neq t$, the formal approach would have to be used, and more work would also be involved. Specifically, both $E(Y|Y \leq t)$ and $E(Y|Y > t)$ would have to be evaluated. Also, both variances would likewise have to be calculated, but $\text{Var}(Y|Y \leq t)$ can be obtained using the expression for $\text{Var}(Y|Y > t)$ by using the fact that the normal distribution is symmetric and using a value for t, say t', such that the area under the curve to the right of t' is the same as the area under the curve to the left of t. Then $\text{Var}(Y|Y \leq t) = \text{Var}(Y|Y > t')$.

[It should be noted that most of the material presented to this point in the chapter was motivated by Jessup (1986).]

14.7 SIGNAL-TO-NOISE RATIOS AND ALTERNATIVES

The Taguchi approach to experimental design utilizes what are termed *signal-to-noise ratios.* As discussed by Kacker (1985), Taguchi has defined more than 60 signal-to-noise ratios (S/N) for various engineering applications. The general idea is to use an S/N that is appropriate for a particular situation. This S/N, which is to be maximized, is presumed to be a logical estimator of some performance measure.

For example, if a target value is to be used (as in the preceding section), squared error would be a logical loss function, and the expected loss will be equal to the variance plus $[E(Y) - t]^2$. If the sample average and sample variance are independent, factors that are thought to affect the value of a particular quality characteristic might be separated into two categories—those that affect the variance and those that affect the mean, but would have little, if any, effect on the variance.

The objective would then be to vary the levels of the first set of factors (those that are assumed to affect the variance) in an experimental design for the purpose of determining the levels of those factors that will minimize the variability of the values of a particular quality characteristic. In general, the objective is to use the results of such designed experiments to design products in which the product quality is relatively insensitive to "noise" (uncontrollable factors both during and after the time the product is manufactured).

Once the levels of the first set of factors have been determined, the second set of factors (called adjustment parameters by other authors) are used to drive the quality characteristic to its target value by using the appropriate levels of the factors.

Thus, it is a two-stage procedure. For the first stage it would seem logical to use some function of σ^2 as the performance measure or perhaps σ^2 itself. The latter could be used if it were not for the fact that performance measures are generally defined in such a way that their estimates are to be maximized. Accordingly, $-\log(\sigma^2)$ would be one such performance measure, which would logically be estimated by $-\log(s^2)$. The latter could not be called an S/N, however, as there is no "signal" since s^2 is a measure of the noise. Kacker (1985) calls this a performance statistic, and we shall also adopt this terminology for the remainder of the chapter.

Other performance statistics discussed by Kacker (1985) include $10\log(\overline{y}^2/s^2)$, which could be used when the average and variance are not independent. (In general, $\overline{y}$ and s^2 would be computed for each design point from replicates at each point. This is discussed further in the next section.) Hunter (1985) points out that a simple way to maximize the function $10\log(\overline{y}^2/s^2)$ would be to take the logarithm of all of the y values and then identify the factor-level settings (i.e., the design point) that produce the minimum value of s^2. Box (1986) demonstrates that this loss function can be rationalized if (and only if) a log transformation is needed to make the average and variance independent and to satisfy the other assumptions (normality, constant error variance). If we are using σ^2 as a performance measure, we certainly do not want σ^2 to be a function of μ, as μ would be "adjusted" in the second stage, after the attempted minimization of σ^2 in the first stage.

The use of signal-to-noise ratios has received considerable stated or implied criticism. Pignatiello and Ramberg (1985) point out that the use of the S/N discussed in the preceding paragraph implies the somewhat bold assumption that a unit increase in $\log(\overline{y}^2)$ is of equal importance as a unit decrease in $\log(s^2)$. They conclude that it would be better to simply study the variability by itself. A general criticism of performance statistics has been voiced by Lucas (1985).

León, Shoemaker, and Kacker (1987) proposed an alternative to the use of an S/N. They pointed out weaknesses in the latter and suggested that a performance measure independent of adjustment (PERMIA) be used instead. The general approach when a PERMIA is used is essentially the same as when an S/N is used. That is, a two-stage procedure is used where appropriate levels of the nonadjustment factors are determined using a PERMIA (rather than an S/N) as the response variable. The second stage entails using appropriate levels of the adjustment factors to drive the quality characteristic toward its target value, just as is done with the Taguchi approach.

The use of loss functions in quality improvement work was illustrated earlier in this chapter. León et al. (1987) show that for certain models and appropriate loss functions, the use of a suggested S/N will not minimize expected loss. Furthermore, León et al. gave an example of a measuring instrument problem in which the suggested S/N is not independent of an adjustment factor. This is obviously undesirable since the settings of the adjustment factors are considered to be arbitrary during the first stage when appropriate levels of the nonadjustment factors are determined.

The point is that some thought must be given to the selection of the response variable (e.g., PERMIA) that is to be used in determining the levels of the nonadjustment factors that are to be used for future production. It should be chosen after determining the model and an appropriate loss function, and its value should not depend on the values of any adjustment factors. The metric in which the analysis is to be performed is also of paramount importance.

Kros and Mastrangelo (1998) consider the relationship between nonquadratic loss functions and S/N ratios for the "smaller is better," "larger is better," and "nominal is best" scenarios and provide many references of related work.

In the next section we examine the experimental designs that Taguchi has indicated can be used in the first stage and compare them to the designs presented in Chapter 13, as well as examine recent developments.

14.8 EXPERIMENTAL DESIGNS FOR STAGE 1

For some of the models considered by León et al. (1987), the logical PERMIA was σ^2, where σ^2 was a function of the nonadjustment factors. Therefore, the experimental design(s) used in the first stage should first be used for determining the factors that affect process variability (if such information is not already available), and then appropriate levels of those factors should be selected. The Taguchi experimental design approach is not fashioned in this manner, however, nor is standard experimental design methodology. The former incorporates "noise factors" (uncontrollable factors) that are varied over levels likely to occur in practice, thus providing replicates for each design point. (Noise factors are factors that can be varied in a controlled experiment but are not controllable during normal production.) The variance, s^2, is computed at each design point and then used in estimating σ^2 if σ^2 is part of an S/N. (Alternatively, it could be used in estimating σ^2 in a PERMIA if a PERMIA is to be used.)

Using enough replicates to be able to obtain a reasonable estimate of σ^2 *at each design point* offsets, to some extent, one of the purported advantages of the designs advocated by Taguchi, namely, the ability to investigate the main effects of a moderate-to-large number of factors without having to use an inordinate number of design points.

Recognizing this, Box and Meyer (1986) provided a method for estimating the effects on variability of factors in an unreplicated fractional factorial design, in addition to the location effects of those factors. The intent here is somewhat different, however, from the intent of the Taguchi approach (or with a similar approach that utilizes a PERMIA). There the intent is to determine the levels of factors that have already been selected. It should be apparent, however, that the factors have to be selected first, and those selected should have a significant effect on variability (or, in general, upon the S/N or PERMIA used).

It should be noted that in the Taguchi terminology, the designs discussed in the next section would be used during the *parameter design* stage, during which the experimenter seeks to determine the optimum levels of the controllable factors. (Engineers use the term *parameter* in place of *factor*.) This is preceded by the *system design* stage in which attempts are made to reduce noise by using subject matter knowledge (only) and followed by the *allowance (tolerance) design* stage in which improvements are sought beyond what was realized during the parameter design stage. In the latter stage further quality improvement is sought by taking actions such as switching to higher quality (and more expensive) raw materials, and using narrower tolerances. The Japanese have used tolerance design only when the desired quality level has not been achieved through parameter design (Byrne and Taguchi, 1986).

14.9 TAGUCHI METHODS OF DESIGN

One of the striking developments in the field of experimental design during the 1980s and 1990s has been the gravitation of engineers and other industrial personnel in the United States and United Kingdom, in particular, toward what is generally referred to as "Taguchi methods of design." Taguchi (1987) contains the essentials of the Taguchi approach. This approach incorporates orthogonal arrays and linear graphs, as well as a number of other tools that are not a part of classical experimental design methodology but which have become widely adopted.

The term *orthogonal array* has not been used in any of the preceding sections although the common factorial and fractional factorial designs illustrated in Chapter 13 are indeed orthogonal arrays. A complete definition of an orthogonal array can be found in Raktoe, Hedayat, and Federer (1981, p. 169), which also contains rules for constructing such arrays, but a simpler definition will suffice for our purposes. For two-level factorial designs we have used the presence of a letter to indicate the high level of that factor and the absence of a letter to indicate the low level. If we now substitute "+1" and "−1", respectively, as in Section 13.11, we can think of an orthogonal array as a design in which all dot products of pairs of columns representing estimable effects are zero.

To illustrate, for a 2^3 design we would have the following:

Treatment Combination	Estimable Effects						
	A	B	C	AB	AC	BC	ABC
(1)	−1	−1	−1	1	1	1	−1
a	1	−1	−1	−1	−1	1	1
b	−1	1	−1	−1	1	−1	1
ab	1	1	−1	1	−1	−1	−1
c	−1	−1	1	1	−1	−1	1
ac	1	−1	1	−1	1	−1	−1
bc	−1	1	1	−1	−1	1	−1
abc	1	1	1	1	1	1	1

The reader who has the time and patience can easily verify that each of the 21 pairs of columns has a "dot product" of zero; each dot product is obtained by multiplying the corresponding numbers together and summing the eight products. For example, multiplying column A by column AB would produce $(-1)(1) + (1)(-1) + (-1)(-1) + (1)(1) + (-1)(1) + (1)(-1) + (-1)(-1) + (1)(1) = 0$.

Similarly, for a 2^{4-1} design we recall that it can be constructed by forming the full 2^3 factorial with the fourth column represented by ABC. Thus, verifying that the 2^3 is an orthogonal array also verifies that the 2^{4-1} design is an orthogonal array since the latter would consist of the columns A, B, C, and ABC (= D).

The reader should thus realize that the full and fractional factorials discussed in this chapter are orthogonal arrays. Not every fractional factorial can be viewed as an orthogonal array, however [see Raktoe et al. (1981, p. 174)].

14.9.1 Inner Arrays and Outer Arrays

An important contribution of Taguchi is the concept of a *noise array*, which is also called the *outer array*. As stated previously, the general idea is to use noise variables in a design that can be controlled in a laboratory setting but that are either impossible or impractical to control in actual production. We want to identify significant noise variables and ultimately design products and processes that are insensitive to variations in the uncontrollable noise variables. This is the idea behind *robust design*. A 2^k factorial design is usually advocated for the noise array, although there have been other proposals (e.g., Wang, Lin, and Fang, 1995). If a two-level factorial design is used, the problem of selecting the two values for each noise variable must be addressed. In general, the levels should be far enough apart to detect an effect due to a noise variable if an effect does exist, but not be so far apart as to be unrealistic since the values of the noise variables are not controllable during production. One approach would be to use values of the noise variables that are as far apart as possible and still be likely to occur during production. Steinberg and Bursztyn (1998) recommend that the levels of a noise factor N be set at $\pm 1.5\sigma_N$. Obviously this rule-of-thumb would require an estimate of the standard deviation of each noise factor, and such an estimate might be difficult to obtain. This would be well within reason for noise variables that have a normal distribution, however, and would also be reasonable for various other distributions, including a uniform distribution.

Much attention and discussion have been devoted to determining the form of the *inner array*, which contains the control factors. For the inner array we may choose between the orthogonal array approach and the fractional factorial approach, and we compare these approaches in subsequent sections. We also should consider whether we even want to use both an inner array and an outer array (and thus have a *product array*) or simply have one array, a *combined array*. We address this issue in Section 14.9.4.

14.9.2 Orthogonal Arrays as Fractional Factorials

We shall begin our discussion of the relationship between fractional factorial designs and the orthogonal arrays presented in Taguchi and Wu (1979) by considering their 8-point orthogonal array on page 66. (Their notation for such a design is L_8.) They use a "1" to indicate the low level and "2" the high level of a two-level orthogonal array with four factors, with the design configuration given in Table 14.1. There is no compelling reason for preferring one notational system over another. Other authors have used "0" and "1" to denote the two different levels. The advantage of using "+1" and "−1" is that it allows the orthogonality of the orthogonal array to be very apparent.

Accordingly, Table 14.1 becomes Table 14.2 when this substitution is made. The orthogonality could now be easily verified by taking products of columns, as was previously discussed.

Even with the "+1" and "−1" notation, however, we notice some new wrinkles in Table 14.2. In particular, the "e" as the heading for the last column simply indicates that not every degree of freedom will be used for estimating effects; one will be used for estimating σ. (We should question the worth of that estimate based on only one degree of freedom. We would likely be better off not even attempting to estimate σ and, instead, simply construct a normal probability plot of the effect estimates.) We also notice that the BC column is the *negative* of the product of the B and C columns. Therefore, we would actually be estimating −BC rather than BC (which is of no real consequence); similarly, we would be actually estimating −BD.

TABLE 14.1 L_8 Orthogonal Array Given in Taguchi and Wu

Treatment Combination	B	C	BC	D	BD	A	e
(1)	1	1	1	1	1	1	1
ad	1	1	1	2	2	2	2
ac	1	2	2	1	1	2	2
cd	1	2	2	2	2	1	1
b	2	1	2	1	2	1	2
abd	2	1	2	2	1	2	1
abc	2	2	1	1	2	2	1
bcd	2	2	1	2	1	1	2

Source: G. Taguchi and Y. Wu (1979).

TABLE 14.2 L_8 Orthogonal Array of Table 14.1 Using "+1" and "−1" for the Two Levels

Treatment Combination	B	C	(−)BC	D	(−)BD	A	e
(1)	−1	−1	−1	−1	−1	−1	−1
ad	−1	−1	−1	1	1	1	1
ac	−1	1	1	−1	−1	1	1
cd	−1	1	1	1	1	−1	−1
b	1	−1	1	−1	1	−1	1
abd	1	−1	1	1	−1	1	−1
abc	1	1	−1	−1	1	1	−1
bcd	1	1	−1	1	−1	−1	1

In our discussion of the resolution of a design in Chapter 13 we did not discuss the possibility that some (but not all) of the effects of a particular order could be estimable. With this particular orthogonal array we can do more than just estimate the main effects A, B, C, and D since we have seven degrees of freedom for estimating effects (provided that we are willing to forego an estimate of σ and settle for plotting the effect estimates).

What is not clear from Table 14.2, however, is the alias structure. We know that none of the six effects that are to be estimated could be aliased among themselves since the columns are obviously pairwise orthogonal. It would be helpful, however, to know the alias structure, and the structure is not readily apparent from the array.

This is a potential weakness of the orthogonal array approach, as the equivalent fractional factorial design (if one exists) needs to be identified for the alias structure to be clear. [Actually, the resolution of a fractional factorial that is an orthogonal array can be determined from the "strength" of the array in which strength 2 corresponds to Resolution III and strength 3 to Resolution IV. For details see Raktoe et al. (1981, p. 172).] It can be determined that this is actually a 2^{4-1} design with $-\text{ACD}$ as the defining contrast. (The reader can easily verify this by multiplying the C and D columns together and recognizing that the product is the negative of the A column. Thus, $\text{CD} = -\text{A}$, which is what one obtains when multiplying CD by $-\text{ACD}$ to find the alias of CD.) Therefore, the alias structure can now be determined, which is as follows:

$$
\begin{array}{ll}
\text{A} = -\text{CD} & \text{AB} = -\text{BCD} \\
\text{B} = -\text{ABCD} & -\text{BC} = \text{ABD} \\
\text{C} = -\text{AD} & -\text{BD} = \text{ABC} \\
\text{D} = -\text{AC} & \text{I} = -\text{ACD}
\end{array}
$$

We can see that the BC and BD interactions that are to be estimated are confounded with three-factor interactions. Thus, if the BC and BD interactions were deemed as likely to be significant before the experiment is carried out (as was true for this experiment), there is no problem in estimating them provided that the three-factor interactions are not significant. What could be quite risky, however, is the fact that three of the four main effects are aliased with two-factor interactions. The experimenter had better have a strong belief that those three interactions are not likely to be important. In other words, in using this design he or she would be assuming that two of the two-factor interactions are likely to be important (i.e., statistically significant) but not the others. That would be a rather bold assumption in the absence of data from a previous experiment.

Could this design be improved? Yes. There is no need to confound main effects with two-factor interactions. We can always construct a 2^{4-1} design in such a way that main effects are confounded with three-factor interactions (i.e., the design is Resolution IV). (This could be designated as 2_{IV}^{4-1}.) We simply use ABCD as the defining contrast so that the alias structure is as follows.

$$\begin{aligned} A &= BCD & BC &= AD \\ B &= ACD & BD &= AC \\ C &= ABD & AB &= CD \\ D &= ABC \end{aligned}$$

Of course, we now have the two-factor interactions that are anticipated as being important aliased with other two-factor interactions. This is a lesser evil, however, than having main effects aliased with two-factor interactions. If the estimates of the BC and BD effects were determined to be significant, additional design points could be used (if feasible) to disentangle BC from AD and BD from AC. [See Appendix 12B of Box et al. (1978) for general information concerning the selection of such additional points.]

The treatment combinations to be used in the 2_{IV}^{4-1} design are given in Table 14.3. The column headings in Table 14.3 are the same as those in Table 14.2, so the same effects would be estimable. The *e* in Table 14.3 represents AB − CD, whereas the *e* in Table 14.2 represents BCD + AB.

14.9.3 Other Orthogonal Arrays Versus Fractional Factorials

Taguchi and Wu (1979, p. 68) also present an L_{16} orthogonal array for the purpose of studying nine factors (each at two levels), with four two-factor interactions deemed to be important. The same objective could be met by constructing a 2_{III}^{9-5} design following the directions of Box et al. (1978, p. 410).

In the experiment for which the L_{16} was used, the objective was to estimate the nine main effects and four of the two-factor interactions: AC, AG, AH, and GH. In the preceding example we saw that the L_8 array was of lower resolution than the corresponding 2_{IV}^{4-1} design that allowed for estimation of the same effects. That will not be the case for the L_{16} versus the 2_{III}^{9-5}, however, as the L_{16} is also of Resolution III. The L_{16} array given by Taguchi and Wu is given in Table 14.4.

TABLE 14.3 2_{IV}^{4-1} Design with I = ABCD

Treatment Combination	A	B	C	D	BC	BD	e
(1)	−1	−1	−1	−1	1	1	1
ad	1	−1	−1	1	1	−1	−1
bd	−1	1	−1	1	−1	1	−1
ab	1	1	−1	−1	−1	−1	1
cd	−1	−1	1	1	−1	−1	1
ac	1	−1	1	−1	−1	1	−1
bc	−1	1	1	−1	1	−1	−1
abcd	1	1	1	1	1	1	1

TABLE 14.4 L_{16} Array from Taguchi and Wu[a]

Treatment Combination	Estimable Effects														
	A	G	(−)AG	H	(−)AH	(−)GH	B	D	E	F	I	e	e	(−)AC	C
(1)	−1	−1	−1	−1	−1	−1	−1	−1	−1	−1	−1	−1	−1	−1	−1
cdefi	−1	−1	−1	−1	−1	−1	−1	1	1	1	1	1	1	1	1
bch	−1	−1	−1	1	1	1	1	−1	−1	−1	−1	1	1	1	1
bdefhi	−1	−1	−1	1	1	1	1	1	1	1	1	−1	−1	−1	−1
bcfgi	−1	1	1	−1	−1	1	1	−1	−1	1	1	−1	−1	1	1
bdeg	−1	1	1	−1	−1	1	1	1	1	−1	−1	1	1	−1	−1
fhgi	−1	1	1	1	1	−1	−1	−1	−1	1	1	1	1	−1	−1
cdegh	−1	1	1	1	1	−1	−1	1	1	−1	−1	−1	−1	1	1
abcei	1	−1	1	−1	1	−1	1	−1	1	−1	1	−1	1	−1	1
abdf	1	−1	1	−1	1	−1	1	1	−1	1	−1	1	−1	1	−1
aehi	1	−1	1	1	−1	1	−1	−1	1	−1	1	1	−1	1	−1
acdfh	1	−1	1	1	−1	1	−1	1	−1	1	−1	−1	1	−1	1
aefg	1	1	−1	−1	1	1	−1	−1	1	1	−1	−1	1	1	−1
acdgi	1	1	−1	−1	1	1	−1	1	−1	−1	1	1	−1	−1	1
abcefgh	1	1	−1	1	−1	−1	1	−1	1	1	−1	1	−1	−1	1
abdghi	1	1	−1	1	−1	−1	1	1	−1	−1	1	−1	1	1	−1

[a]The fourth number in the (I) column is corrected from the error in Taguchi and Wu (1979, p. 68).

(Again, we use −1 and 1 in place of their 1 and 2, respectively. The fact that the design is Resolution III can be verified by simply observing that A = −DE.) The negative signs are placed in front of the two-factor interactions to indicate that the array actually estimates the negative of the interaction.

The reader can easily verify that each lowercase letter occurs exactly 8 times (as it must) over the 16 treatment combinations. As with the L_8 array, the "e" designates a column that will be used not for estimating any effect, but rather for estimating σ.

Tables of orthogonal arrays are needed to construct these designs, but a design configuration using the Box and Hunter (1961a,b) approach can be constructed that will provide for estimation of the same effects that are estimable in Table 14.4. The general idea is to construct a full 2^4 factorial ("4" because $9 - 5 = 4$) and then obtain the columns for estimating the other five main effects by taking all possible combinations of three of the four columns, and then the product of all four columns. Desired interactions can then be estimated by forming products of the appropriate columns. There is really no need to designate columns for estimating σ (i.e., the "e" columns) as d.f. that are not used for estimating effects are automatically used for this purpose anyway.

The 2^{9-5}_{III} design is given in Table 14.5. The effect in parentheses above each column level indicates the effect in the L_{16} array to which each column label corresponds. The representation of the e columns is also given for completeness, although, as indicated, such columns really are not necessary. The reader will observe that the treatment combinations in Table 14.5 are totally different from

TABLE 14.5 2_{III}^{9-5} **Design**

Treatment Combination	(D) A	(H) B	(G) C	(A) D	(−AC) E(= ABC)	(B) F(= BCD)	(I) G(= ACD)	(e − 13) H(= ABD)	(−C) I(= ABCD)	(AG) CD	(AH) BD	(GH) BC	(−AC) DI	(−DH) e	(e − 13) e
	Estimable Effects[a]														
i	−1	−1	−1	−1	−1	−1	−1	−1	1	1	1	1	−1	−1	−1
aegh	1	−1	−1	−1	1	−1	1	1	−1	1	1	1	1	1	1
befh	−1	1	−1	−1	1	1	−1	1	−1	1	−1	−1	1	1	1
abfgi	1	1	−1	−1	−1	1	1	−1	1	1	−1	−1	−1	−1	−1
cefg	−1	−1	1	−1	1	1	1	−1	−1	−1	1	−1	1	−1	−1
acfhi	1	−1	1	−1	−1	1	−1	1	1	−1	1	−1	−1	1	1
bcghi	−1	1	1	−1	−1	−1	1	1	1	−1	−1	1	−1	1	1
abce	1	1	1	−1	1	−1	−1	−1	−1	−1	−1	1	1	−1	−1
dfgh	−1	−1	−1	1	−1	1	1	1	−1	−1	−1	1	−1	−1	1
adefi	1	−1	−1	1	1	1	−1	−1	1	−1	−1	1	1	1	−1
bdegi	−1	1	−1	1	1	−1	1	−1	1	−1	1	−1	1	1	−1
abdh	1	1	−1	1	−1	−1	−1	1	−1	−1	1	−1	−1	−1	1
cdehi	−1	−1	1	1	1	−1	−1	1	1	1	−1	−1	1	−1	1
acdg	1	−1	1	1	−1	−1	1	−1	−1	1	−1	−1	−1	1	−1
bcdf	−1	1	1	1	−1	1	−1	−1	−1	1	1	1	−1	1	−1
abcdefghi	1	1	1	1	1	1	1	1	1	1	1	1	1	−1	1

[a]The column designations in Table 14.4 are given above each effect, with e − 13 designating column 13.

the treatment combinations in Table 14.4. That does not make any difference; what is important is that the columns be orthogonal (which they are), and that each factor occurs eight times at its high level and eight times at its low level (which they do).

We can observe that none of the estimable effects in Table 14.5 corresponds to either E or F in Table 14.4. This is because E = −AD and F = −DG in Table 14.4, with −AD in Table 14.4 = −AD in Table 14.5 and −DG in Table 14.4 = −AC in Table 14.5, and the columns for the five additional main effects in Table 14.5 are not formed by taking the product of any *pair* of columns (such as A and D).

We do observe one serious problem with the 2_{III}^{9-5} design, however. Specifically, the DI column is the same as the E column. Thus, the two are confounded. This is undesirable since DI in Table 14.5 corresponds to −AC in Table 14.4, and AC is one of the effects to be estimated. Therefore, by following the Box–Hunter approach directly we run into a problem, and this is due to the fact that this approach forces specific two-factor interactions (2 f.i.) to be nonestimable, and it happens that we need to estimate one of those interactions. Does this mean that we have to forsake the Box–Hunter approach for the orthogonal array approach? Not really. The Box–Hunter approach will guarantee that the resultant design will have maximum resolution, but the procedure can often be modified; the modified approach produces a design with the same resolution.

There are two obvious solutions to the problem. Since the problem results from the way in which E and I are defined in terms of the factors in the full factorial, we could solve the problem by redefining either of these. For example, we could let E = AC (as mentioned previously, E = −AD in the L_{16} array; that would also work here). We must simply avoid letting it equal one of the two-factor interactions that we are trying to estimate (CD, BD, and BC in Table 14.5 notation). (Notice that E = AC would not work if we wanted to estimate DG, e.g., since DG = AC.) Thus, we can still use the Box–Hunter approach, although sometimes a slight modification will be necessary. The "final" 2_{III}^{9-5} design with the appropriate modification is given in Table 14.6.

To this point we have seen one example in which the fractional factorial was superior to the orthogonal array and one in which the fractional factorial approach had to be modified slightly.

Taguchi and Wu (1979, p. 75) presented an L_{16} array for the purpose of investigating eight factors and three of the 2 f.i., AB, AD, and BD. (The column labels are in error, however, as two of them are labeled D when one of them should obviously be labeled C.) One of the columns created for a main effect (H) is formed by taking the negative of the product of two of the factors in the full factorial part (−DF). Thus, the design cannot be higher than Resolution III since H = −DF.

If we follow the Box and Hunter approach, however, we can construct a 2_{IV}^{8-4} design for the same purpose, with the desired interactions simply appended to the basic 2_{IV}^{8-4} as additional columns. Accordingly, the Box and Hunter approach produces a superior design since the user would not have to worry about the possible significance of any of the 2 f.i. involving any pair of factors D, F, and H.

TABLE 14.6 2^{9-5}_{III} Design Analogous to the L_{16} Array

Treatment	Estimable Effects[a]														
	(D)	(H)	(G)	(A)	(−F)	(B)	(I)	(e − 13)	(−C)	(AG)	(AH)	(GH)	(−AC)	(−DH)	(e − 13)
Combination	A	B	C	D	E(= AC)	F(= BCD)	G(= ACD)	H(= ABD)	I(= ABCD)	CD	BD	BC	DI	e	e
ei	−1	−1	−1	−1	1	−1	−1	−1	1	1	1	1	−1	−1	−1
agh	1	−1	−1	−1	−1	−1	1	1	−1	1	1	1	1	1	1
befh	−1	1	−1	−1	1	1	−1	1	−1	1	−1	−1	1	1	1
abfgi	1	1	−1	−1	−1	1	1	−1	1	1	−1	−1	−1	−1	−1
cfg	−1	−1	1	−1	−1	1	1	−1	−1	−1	1	−1	1	−1	−1
acefhi	1	−1	1	−1	1	1	−1	1	1	−1	1	−1	−1	1	1
bcghi	−1	1	1	−1	−1	−1	1	1	1	−1	−1	1	−1	1	1
abce	1	1	1	−1	1	−1	−1	−1	−1	−1	−1	1	1	−1	−1
defgh	−1	−1	−1	1	1	1	1	1	−1	−1	−1	1	−1	−1	1
adfi	1	−1	−1	1	−1	1	−1	−1	1	−1	−1	1	1	1	−1
bdegi	−1	1	−1	1	1	−1	1	−1	1	−1	1	−1	1	1	−1
abdh	1	1	−1	1	−1	−1	−1	1	−1	−1	1	−1	−1	−1	1
cdhi	−1	−1	1	1	−1	−1	−1	1	1	1	−1	−1	1	−1	1
acdeg	1	−1	1	1	1	−1	1	−1	−1	1	−1	−1	−1	1	−1
bcdf	−1	1	1	1	−1	1	−1	−1	−1	1	1	1	−1	1	−1
abcdefghi	1	1	1	1	1	1	1	1	1	1	1	1	1	−1	1

[a]The column designations in Table 14.4 are given above each effect, with e − 13 designating column 13.

As Taguchi and Wu (1979, Foreward) report, the L_8 and L_{16} were the most frequently used arrays for two-level factors employed by NEC (a Japanese company), as indicated in its 1959 report. This is not to suggest that all two-level orthogonal arrays should be discarded in favor of 2^{k-p} designs. Some of these orthogonal arrays do not correspond to an equivalent 2^{k-p} because the number of design points is not a power of 2. We should, however, strive to select the best design for a given number of design points.

Orthogonal arrays for three-level factors are also given by Taguchi and Wu as well as for mixed factorials. In particular, a nine-point array for examining four factors, each at three levels, is given by Taguchi and Wu (1979, p. 65) who indicate that this is a frequently used design. This design, which is obtained from a Graeco-Latin square, is also used for illustration by Kacker (1985). Such designs are criticized by Hunter (1985), however, on the grounds that all two-factor interactions must be assumed equal to zero.

Equivalencies can also be demonstrated between three-level orthogonal arrays and 3^{k-p} designs. Taguchi and Wu (1979, p. 109) provide an L_{27} array that could be used for examining the main effect of 13 factors, each at three levels. The same design could be obtained by constructing a 3^{13-10} design in a prescribed manner. Addelman (1962, p. 38) gives a 3^{13-10} design that can be seen to differ only slightly from the L_{27} array. The criticism of 3^{k-p} designs mentioned in Chapter 13 would thus also apply to these three-level arrays. Consequently, the potential user of such designs might wish to consider other (superior) designs (e.g., response surface designs).

14.9.4 Product Arrays Versus Combined Arrays

There has been considerable debate regarding whether a product array or a combined array should be used, just as there has been considerable debate regarding the statistical soundness of each of Taguchi's recommended statistical techniques. A *product array* consists of an inner array and an outer array, with the outer array repeated for each row in the inner array. In a *combined array*, there is only one array, which contains the columns for both the controllable and noise variables.

Shoemaker, Tsui, and Wu (1991) discuss the relative merits of each approach. One obvious disadvantage of the product array approach is that a very large number of experimental runs may be needed since the number of experimental runs is equal to the product of the number of runs in the inner and outer arrays. Thus, even if we had only three control factors that were examined in eight runs (e.g., a 2^3) and only two noise variables (e.g., a 2^2) in four runs, the product array would have 32 runs. We would generally expect to have more than three control factors and more than two noise variables, so the number of necessary runs will generally be much larger than this.

As discussed by Montgomery (1997, p. 641), many of the successful applications of the Taguchi design approach have been in industries characterized by high-volume, low-cost manufacturing. In other industries, and this would undoubtedly be the majority of industries, the application of the product array approach can result in an impractical number of design points being used.

What effects are estimable with a product array? Shoemaker et al. (1991) proved that the effects that are estimable consist of the effects that are estimable in the control and noise arrays plus all generalized interactions of those effects. Thus, if only main effects were estimable in each array, the total number of estimable effects would consist of the control main effects and the noise main effects plus all of the interactions between the control and noise factors.

Assume for the sake of illustration that we have only two control factors and two noise factors, with two levels to be used for each. Thus, the product design will consist of one 2^2 design crossed with another 2^2 design. That is, the replicates of each design point in the control array constitute a 2^2 design in the noise array. The design is given in Table 14.7. We would guess that this is simply a 2^4 design, written in an order that is not standard order. If the first column were switched with the third column and the second column were switched with the fourth column, the design would then be in standard order. Thus, with this design we could estimate the same 15 effects that we could estimate if we had simply constructed a combined array. Thus, in this instance the product array and the combined array are the same, and this will be true whenever each array consists of a complete factorial.

This will not be the case, in general, however, because the control array will almost certainly not be a full factorial. Since the noise array is frequently a full factorial, it is important to keep the number of points in the control array to a

TABLE 14.7 Product Array with Two Control Factors and Two Noise Factors

A	B	N_1	N_2
−1	−1	−1	−1
−1	−1	1	−1
−1	−1	−1	1
−1	−1	1	1
1	−1	−1	−1
1	−1	1	−1
1	−1	−1	1
1	−1	1	1
−1	1	−1	−1
−1	1	1	−1
−1	1	−1	1
−1	1	1	1
1	1	−1	−1
1	1	1	−1
1	1	−1	1
1	1	1	1

reasonable number so that the product array will not have an extremely large number of runs.

For another example, assume that the 2^{4-1}_{IV} design in Table 14.3 is to be used as the control array (except that "e" is replaced by the A × B interaction), and a 2^2 design is to be used for the noise array with noise factors E and F. The product array has 32 treatment combinations, so 31 effects are estimable. These are the 7 effects that are estimable from the control array, plus the 3 effects that are estimable from the error array, plus the $7 \times 3 = 21$ generalized interactions between the estimable effects in the two arrays. Specifically, the estimable effects are A, B, C, D, E, F, AB, BD, BC, AE, AF, BE, BF, CE, CF, DE, DF, EF, ABE, ABF, BDE, BDF, BCE, BCF, AEF, BEF, CEF, DEF, ABEF, BDEF, and BCEF.

We can compare the estimable effects for this product array with the estimable effects for a combined array. The logical comparison would be with a 2^{6-1}_{V} design, which would also have 32 runs and would thus allow 31 effects to be estimated. With I = ABCDEF, we could estimate all 15 two-factor interactions, rather than the specific 12 interactions that can be estimated with the product array. The difference is due to the fact that the product array is a Resolution IV design with the 6 two-factor interactions confounded in pairs, so three of the interactions are not estimable. The 20 three-factor interactions are confounded in pairs in the product array and in the combined array, so the designs are comparable in this respect. The product array specifically allows for the estimation of 3 four-factor interactions, which would not be estimable with the combined array since they are aliased with two-factor interactions, and it is the latter that we would certainly choose to estimate. Obviously we would rather estimate two-factor interactions than four-factor interactions, so the combined array is clearly preferable.

Although the two designs are somewhat similar, they are quite different in terms of alias structure. We may show that the product array is actually a 2^{6-1}_{IV} with I = ABCD. That is, the product array has inherited the defining contrast of the inner array. Thus, the product array is a suboptimal half-fraction in that it is a Resolution IV design, whereas the combined array is a Resolution V design. Note that this occurs despite the fact that the inner array was constructed in a manner so as to maximize the resolution of *that* array.

This result should not be a surprise because we have "tacked on" replicates of a 2^2 design to a Resolution IV design. Replicating each point in the control array is equivalent, as far as the alias structure for the control array is concerned, to replicating the control array. Since such replication will not affect the alias structure, the alias structure for the replicated control array will be the same as the alias structure for the unreplicated control array. Thus, if the control array has a certain Resolution, the product array cannot have a greater Resolution because the Resolution for the control factors will be the same in each array. It should also be obvious that the Resolution of the product array cannot exceed $\min(R_c, R_n)$, with R_c and R_n denoting the Resolution of the control array and the Resolution of the noise array, respectively.

This means that if we elect to model the control and noise variables, there will be many instances in which we should *not* use a product array, as the difference in Resolution for the combined and product arrays will increase as the number of noise variables increases. Furthermore, the difference will be even greater if we use an inner array that is equivalent to a suboptimal fractional factorial (e.g., the L_8 array given in Section 14.9.2). The following examples should make this point clear. Assume that a 2_{IV}^{4-1} design is used for the inner array, and a 2^3 design is used for the outer array, so that 64 points will be used. The product array will have 64 runs and will be Resolution IV (from the inner array). A combined array that is a 2_{VII}^{7-1} in 64 runs will be far superior, as even three-factor interactions are estimable. If the L_8 array had been used instead of the 2_{IV}^{4-1}, the difference in the Resolution of the combined and product arrays would have been four instead of three.

So why would anyone want to use a product array? Taguchi and others have recommended using a "pick-the-winner strategy" for robust design, selecting the combination of factor levels in the inner array that minimizes the deleterious effects of the noise variables. We should recognize, however, that it is difficult to compare variances when each variance is estimated from a small number of observations. In the examples given earlier in this section, either four or eight experimental runs were made at each point in the inner array. Is this sufficient? We need to remember that $\mathrm{Var}(s^2) = 2\sigma^4/(n-1)$, assuming normality, so if the noise array is small, the combination of factor levels at which the minimum value of s^2 occurs could easily be due to chance. Furthermore, when noise variables are included in an experiment, it is not possible to proceed as one could when replication is used and noise variables are not specified. Specifically, it is not possible to ignore the modeling aspect and test equality of variances over the different treatment combinations. This was discussed by Steinberg and Bursztyn (1998) in the context of a single control variable and a single noise variable, each at two levels. The problem is that the mean for each level of the noise variable causes the F-statistic for testing the equality of the two variances to have a doubly noncentral F distribution, not a central F distribution. Obviously this problem must also exist when there is more than a single noise variable and a single control variable.

So the usual tests for equality of variances cannot be applied, and a noise array of moderate size would be needed in order to make even an *informal* comparison of variances. But then the product array would almost certainly have an impractical number of design points. So it is difficult to justify the use of a product array under practically any circumstances (but see the discussion in Section 14.9.5).

The issue of a product array versus a combined array has also been considered by Rosenbaum (1994, 1996). Rosenbaum (1996) argues for the use of a (4, 3, 4) design, where a $(\gamma, \lambda, \alpha)$ design is one that denotes the design being of Resolution α for the control and error factors combined, of Resolution γ for the control factors only, and of Resolution λ for the error factors only. The advantage of the combined array for the designs considered by Rosenbaum (1996) is that by using a combined array rather than a product array, α will be 4 instead of 3. This is desirable because it is important to be able to estimate the control $\times$ noise interactions. Of course, with a Resolution IV design the two-factor interactions will be confounded among themselves, so having $\alpha = 4$ is helpful only if the control × noise interactions are confounded with noise × noise interactions, *and* the noise × noise interactions can be assumed to be quite small. One important point made by Rosenbaum (1996) is that control × noise interactions should not be aliased with any effect involving a control factor. Certainly that is true for Figure 14.5, as we want to be able to select the value of the control factor so as to reduce variability, but we will not necessarily be able to do that if that particular interaction is confounded with another two-factor interaction that is significant.

More generally, a practitioner could compute the value of some performance measure for each row of the inner array, using this as the "response" variable and performing an analysis of variance to determine significant effects and interactions. But as pointed out by Shoemaker et al. (1991), the model that would be appropriate for a performance measure may be more complex than the model for Y that the design permits.

The number of design points that will be needed will obviously depend on what effects need to be estimated. Are we interested in determining which noise variables, if any, are significant? Although such information would certainly be useful, we need to know what factor settings to use so as to create robust products. Obviously we cannot set the levels of the noise variables. Clearly top priority should be given to estimating the control × noise interactions. Ideally we would want these interactions to not be aliased with any second-order effect, but this would require that the design be Resolution V. Such a design might require more runs than would be considered practical.

As discussed by Shoemaker et al. (1991), there is considerable inflexibility inherent in the product array approach in regard to the effects that can be estimated. Specifically, a product array results in degrees of freedom being allocated to permit the estimation of all generalized interactions between the estimable effects in the inner array and the estimable effects in the outer array, at the expense of estimating control factor interactions that are believed to be important. Many of the control × noise interactions may not be important. Shoemaker et al. (1991) contend that it is wasteful to use a design that does not permit the selective estimation of interactions. They give an example of a product array for 8 control factors and 4 noise factors and state that a priori information suggests that 13 of 32 interactions are either insignificant or unimportant. Substituting a combined array for the product array permitted the estimation of 12 control × control interactions.

It is important to look at control × noise interactions, however, as these interactions hold the key to robust design. It is not the *magnitude* of the interactions that are important, however, but rather the *shape* of the interaction profiles. This is probably not well understood, but is illustrated, for example, by Shoemaker et al. (1991).

Consider Figure 14.5, which is similar to Figure 7 in Shoemaker et al. (1991). Here we have two levels of a noise factor and two levels of a control factor. Figure 14.5*a* shows that the noise factor has the same effect on the average response at each level of the control factor. Consequently, there is no opportunity

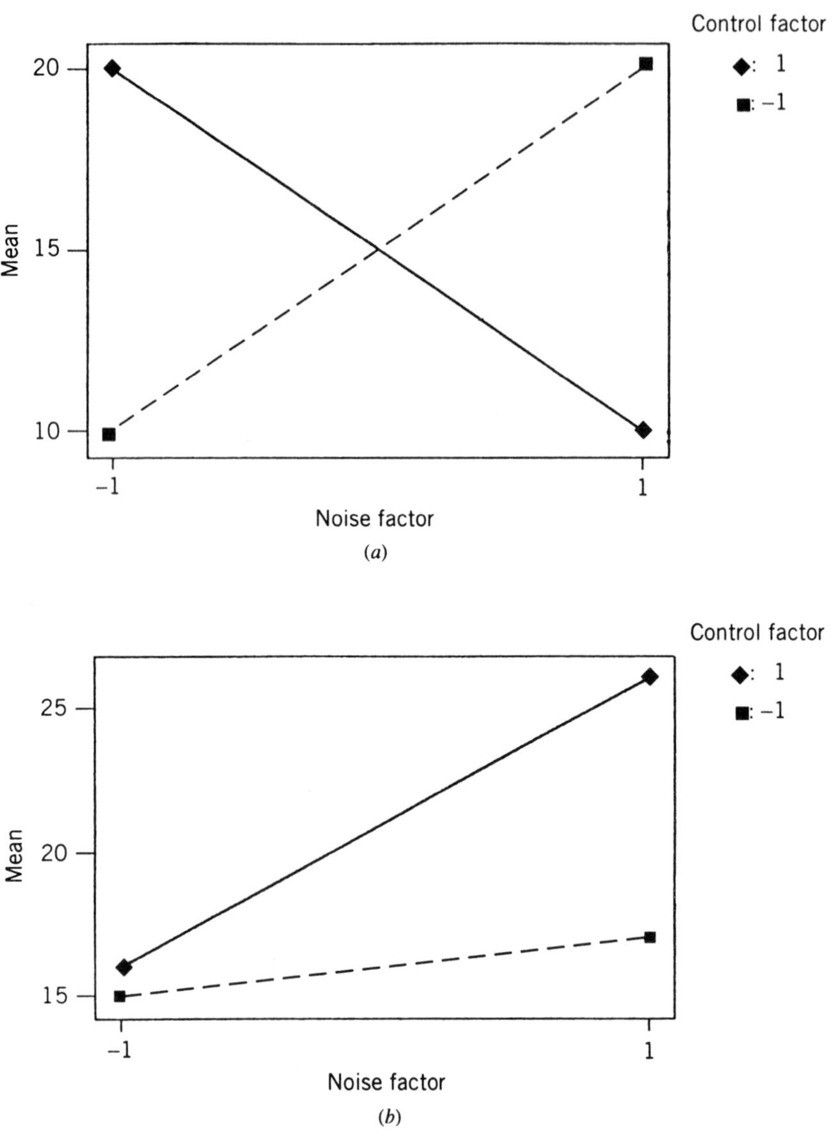

FIGURE 14.5 Interaction plots that illustrate (*a*) unfavorable conditions and (*b*) favorable conditions for selecting the level of a control factor so as to improve robustness.

to select a level of the control factor so as to minimize the effect of the noise variable. Figure 14.5*b* shows the opposite situation. Here the -1 level of the control factor is the obvious choice since the average response is approximately the same for each of the two levels of the noise variable.

The interaction effect is much larger for Figure 14.5*a* than for Figure 14.5*b*, however, so it is not sufficient to look at the magnitude of the interaction effect for a control factor and a noise variable. This suggests the need for a new algorithmic approach that incorporates the difference of the absolute values of the slopes of the lines in an interaction plot.

The need for such an approach is clearly illustrated by the following example. Assume that we have a 2^2 design for one control factor and one noise factor, with the treatment combinations (1), *a*, *b*, and *ab* given by 5, k, 10, and 5, respectively. If $k = 10$ we have the most extreme case of interaction, with the interaction profile having the same general configuration as was given in Figure 13.10 and Figure 14.5*a*. But such a configuration is of no help in creating a robust product, because regardless of whether A or B is the noise factor, it does not make any difference what setting is used for the other factor. Specifically, assume that A is the noise factor. If the high level of B is used, the response varies from 10 to 5 from the low level of the noise factor to the high level. Similarly, if the low level of B is used, the response varies from 5 to 10. Thus, the variation is the same.

At the other extreme (relative to robustification), if $k = 5$, then the obvious choice is to use the low level of B, as there is then no variation in the response. Unfortunately, when the interaction is less extreme, but more valuable from a robustification standpoint, the interaction can be declared not significant by statistical software. Considering the range $5 \leq k \leq 10$, only when $9.1 \leq k \leq 10$ is the interaction declared significant using the normal probability plot routine for effects in MINITAB (which is due to Lenth, 1989). Thus, we have the unpleasant result that the interaction has to be almost to the point where it is of no value from a robustification standpoint before it is declared significant!

Obviously the absolute values of the slopes of the lines in each profile should be compared, with a considerable difference in the absolute values showing the potential for robustification. The tie-in with the discussion in Section 13.7.3 is apparent, as this can be related to half effects. Specifically, still considering a 2^2 design, if the half effects of B differ considerably, we would say that there is a B effect, but if the absolute values of the half effects differ considerably, then there is the potential for robustification.

14.9.5 Application of Product Array

Since product arrays are undoubtedly used much more extensively than combined arrays, we will look at an application in which a product array has been used. Lewis, Hutchens, and Smith (1997) describe a designed experiment that was motivated by the dissatisfaction expressed by a customer of Electro-Scientific-Industries (ESI). Specifically, a customer in Japan stated it was dissatisfied with the mean-time-between-failure (MTBF) performance of one of the company's products. ESI had 6 months to find a solution to the problem; otherwise, the Japanese company would switch over to one of ESI's competitors.

The response variable in the study was "flat find" (FF) which was described as "on a scale of 0 to 1000, ... the degree of correlation between the image (provided by the pattern recognition system) of a corner of the primary flat and the learned (expected) image". The objective was to maximize FF. The data are given in Table 14.8.

TABLE 14.8 Design for Control Factors (A–H) and Noise Factors (J–L) with Response Values on Flat Find (FF)

Row	A	B	C	D	E	F	G	H	J	K	L	FF
1	−1	−1	−1	−1	−1	−1	−1	−1	−1	−1	1	916
2	−1	−1	−1	−1	−1	−1	−1	−1	1	−1	−1	934
3	−1	−1	−1	−1	−1	−1	−1	−1	−1	1	−1	912
4	−1	−1	−1	−1	−1	−1	−1	−1	1	1	1	922
5	1	−1	−1	−1	1	1	−1	1	−1	−1	1	967
6	1	−1	−1	−1	1	1	−1	1	1	−1	−1	964
7	1	−1	−1	−1	1	1	−1	1	−1	1	−1	959
8	1	−1	−1	−1	1	1	−1	1	1	1	1	972
9	−1	1	−1	−1	1	1	1	−1	−1	−1	1	906
10	−1	1	−1	−1	1	1	1	−1	1	−1	−1	934
11	−1	1	−1	−1	1	1	1	−1	−1	1	−1	880
12	−1	1	−1	−1	1	1	1	−1	1	1	1	926
13	1	1	−1	−1	−1	−1	1	1	−1	−1	1	856
14	1	1	−1	−1	−1	−1	1	1	1	−1	−1	967
15	1	1	−1	−1	−1	−1	1	1	−1	1	−1	822
16	1	1	−1	−1	−1	−1	1	1	1	1	1	959
17	−1	−1	1	−1	−1	1	1	1	−1	−1	1	939
18	−1	−1	1	−1	−1	1	1	1	1	−1	−1	952
19	−1	−1	1	−1	−1	1	1	1	−1	1	−1	926
20	−1	−1	1	−1	−1	1	1	1	1	1	1	943
21	1	−1	1	−1	1	−1	1	−1	−1	−1	1	957
22	1	−1	1	−1	1	−1	1	−1	1	−1	−1	954
23	1	−1	1	−1	1	−1	1	−1	−1	1	−1	964
24	1	−1	1	−1	1	−1	1	−1	1	1	1	961
25	−1	1	1	−1	1	−1	−1	1	−1	−1	1	879
26	−1	1	1	−1	1	−1	−1	1	1	−1	−1	941
27	−1	1	1	−1	1	−1	−1	1	−1	1	−1	847
28	−1	1	1	−1	1	−1	−1	1	1	1	1	935
29	1	1	1	−1	−1	1	−1	−1	−1	−1	1	738
30	1	1	1	−1	−1	1	−1	−1	1	−1	−1	973
31	1	1	1	−1	−1	1	−1	−1	−1	1	−1	776
32	1	1	1	−1	−1	1	−1	−1	1	1	1	972
33	−1	−1	−1	1	1	−1	1	1	−1	−1	1	944
34	−1	−1	−1	1	1	−1	1	1	1	−1	−1	948
35	−1	−1	−1	1	1	−1	1	1	−1	1	−1	948
36	−1	−1	−1	1	1	−1	1	1	1	1	1	948
37	1	−1	−1	1	−1	1	1	−1	−1	−1	1	956
38	1	−1	−1	1	−1	1	1	−1	1	−1	−1	963
39	1	−1	−1	1	−1	1	1	−1	−1	1	−1	957
40	1	−1	−1	1	−1	1	1	−1	1	1	1	958
41	−1	1	−1	1	−1	1	−1	1	−1	−1	1	957
42	−1	1	−1	1	−1	1	−1	1	1	−1	−1	954
43	−1	1	−1	1	−1	1	−1	1	−1	1	−1	964
44	−1	1	−1	1	−1	1	−1	1	1	1	1	961

TABLE 14.8 (*Continued*)

Row	A	B	C	D	E	F	G	H	J	K	L	FF
45	1	1	−1	1	1	−1	−1	−1	−1	−1	1	941
46	1	1	−1	1	1	−1	−1	−1	1	−1	−1	972
47	1	1	−1	1	1	−1	−1	−1	−1	1	−1	953
48	1	1	−1	1	1	−1	−1	−1	1	1	1	969
49	−1	−1	1	1	1	1	−1	−1	−1	−1	1	932
50	−1	−1	1	1	1	1	−1	−1	1	−1	−1	939
51	−1	−1	1	1	1	1	−1	−1	−1	1	−1	935
52	−1	−1	1	1	1	1	−1	−1	1	1	1	940
53	1	−1	1	1	−1	−1	−1	1	−1	−1	1	961
54	1	−1	1	1	−1	−1	−1	1	1	−1	−1	960
55	1	−1	1	1	−1	−1	−1	1	−1	1	−1	960
56	1	−1	1	1	−1	−1	−1	1	1	1	1	956
57	−1	1	1	1	−1	−1	1	−1	−1	−1	1	925
58	−1	1	1	1	−1	−1	1	−1	1	−1	−1	951
59	−1	1	1	1	−1	−1	1	−1	−1	1	−1	934
60	−1	1	1	1	−1	−1	1	−1	1	1	1	957
61	1	1	1	1	1	1	1	1	−1	−1	1	972
62	1	1	1	1	1	1	1	1	1	−1	−1	971
63	1	1	1	1	1	1	1	1	−1	1	−1	929
64	1	1	1	1	1	1	1	1	1	1	1	974

Obviously it is necessary to first identify the factors that influence FF. Eight factors were chosen to study (A–H), but there is no indication as to why those factors were selected. The control array was a 2^{8-4}_{IV} design and the noise array was a 2^{3-1}_{III} design. Thus, the product array was a (4,3,3) array, which, as argued by Rosenbaum (1996), is not as appealing as a (4,3,4) combined array. If we constructed a 2^{11-5}_{IV} design (i.e., a combined array), this would, of course, be a (4,4,4) array since in a fractional factorial design the array will be the same relative to all factors, both control and noise factors. Such designs were not considered by Rosenbaum (1996).

From the Lemma of Shoemaker et al. (1991), we know that the estimable effects for the product array are the effects that are estimable in each array plus all generalized interactions of these estimable effects, as stated previously. With A–H denoting the control factors and J, K, and L representing the noise factors, the estimable effects in the control array are the main effects A–H plus, if we elected to use all of the degrees of freedom, 7 of the two-factor interactions. Of course, trying to estimate any of these two-factor interactions could be quite risky since each of these interactions is aliased with 3 other two-factor interactions. The estimable effects in the noise array are the 3 main effects J, K, and L, so the total estimable effects would be the 15 estimable effects from the control array, the 3 estimable effects from the noise array, and the $15 \times 3 = 45$ generalized interactions. (Of course, the latter includes 21 three-factor interactions.) This accounts for the 63 degrees of freedom.

Lewis et al. (1997) stated "essentially, the experiment sought to identify interactions between design and environmental factors." Of course, this is what is generally done in seeking a robust process. Interestingly, the authors did not actually do this, however! Instead they summed over the points in the error array and analyzed only the 15 estimable effects in the control array. This accomplishes nothing in terms of seeking a robust process, and if the values of the noise variables are not likely to occur with a high probability, the analysis of the estimable effects in the control array may have little value.

Now assume that we want to do a proper analysis and look at control × noise (environmental) interactions. A key question then is "with what effects are these interactions aliased?" If they are aliased with two-factor interactions that are apt to be significant, the analysis could be undermined. We can determine the alias structure after combining the defining contrasts for the control array with the defining contrast(s) for the error array. For the control array we have I = ABDE = ABCF = BCDG = ACDH and for the error array I = JKL. When we merge these and take all generalized interactions, we have I = ABDE = ABCF = BCDG = ACDH = JKL = CDEF = ACEG = BCEH = ABDEJKL = ABDFG = BDFH = ABCFJKL = ABGH = BCDGJKL = ACDHJKL

We can "see," preferably with the help of computer software, that the control × noise (two-factor) interactions are aliased with three-factor interactions involving two of the noise factors, in addition to being aliased with higher-order interactions. For example, BJ is aliased with BKL. [Lewis et al. (1997) erroneously stated that the control × noise interactions are "confounded with five-factor and higher-order interactions involving the remaining factors."]

Shoemaker et al. (1991) contended that, in general, we can make better use of the available degrees of freedom by using a combined array. Before examining that issue for this problem, we will first look at the alias structure for the 2_{IV}^{11-5}. One possible set of defining contrasts (which is as good as any of the other possible sets in terms of the resolution of the design) is I = CDEG = ABCDH = ABFJ = BDEFK = ADEFL = ABEGH = ABCDEFGJ = BCFGK = ACFGL = CDFHJ = ACEFHK = BCEFHL = ADEJK = BDEJL = ABKL. From this set of contrasts (or, preferably, by using computer software), it can be determined that there are 34 two-factor interactions that are not confounded with other two-factor interactions. Remember that the product array forced the estimation of 21 three-factor interactions, which means that 24 two-factor interactions can be estimated, these being the interactions between the control and noise factors.

This does not mean that the 2_{IV}^{11-5} is necessarily superior, however, as we can easily see that there are four pairs of control × noise interactions that are confounded: AK = BL, AL = BK, FL = JK, and FJ = KL. (The first two pairs can be seen from the last of the 15 defining contrasts, the third pair can be seen by multiplying FL times ADEFL and JK times ADEJK and recognizing that both products equal ADE, so FL = JK = ADE, with the last pair obtained by considering ABFJ and ABKL, so that FJ = KL = AB.)

Clearly we would like to avoid having these control × noise interactions confounded and would prefer that these interactions be confounded with control ×

control interactions. Unfortunately, this appears to not be possible. See Section 14.9.6.1 in which this general issue is addressed.

Shoemaker et al. (1991) have contended that the automatic estimation of all control × noise interactions could be quite wasteful, stating that "in practice, engineering knowledge may allow us to go further in eliminating the need to estimate superfluous effects. Even quite limited knowledge of underlying physical mechanisms may imply that certain interactions can be ruled out."

This begs the question, however, "how can we tell when a control × noise interaction is apt to be small?" In discussing the a priori judgment of interaction effects in general, Montgomery, Borror, and Stanley (1997 p. 373) state:

> Our practical experience is that when factors are considered and there is little engineering or process knowledge that is applicable to the problem, experimenters have difficulty determining in advance which main effects are going to be important. Therefore, expecting them to make informed statements about the importance of interaction is usually asking a lot.

In general, it seems questionable how often certain control × noise interactions can be ruled out. Since the main objective has to be the examination of control × noise interactions, whether or not a combined array will be superior to a product array of the same size may hinge on what prior information may be available.

Since Lewis et al. (1997) did not perform an analysis of the control × error interactions, we will do so here and compare our analysis with their overall analysis. We will not assume any prior knowledge regarding control × noise interactions, nor did they express any prior knowledge. The data given in Table 14.8 are the data that were given in Lewis et al. (1997), but those data resulted from some averaging. Specifically, two wafers were cycled through the system five times and FF was determined (i.e., 10 values). Lewis et al. (1997) stated that both the wafer-to-wafer variation and within-wafer variation were generally small, relative to the variation in FF between treatment combinations, so each set of the 10 FF values was averaged. Obviously it would have been preferable to have all of the data, as the full data set might have provided some interesting clues.

As a starting point for the analysis, we will arbitrarily assume that the 7 two-factor interactions to be estimated from the control array are the first 7 in alphabetical order: AB, AC, AD, AE, AF, AG, and AH. (Of course, each of these is confounded with 3 other two-factor interactions, and later we will consider some of the confounding). So the effects that we will estimate will be the 11 main effects, these 7 two-factor interactions, the 24 control × noise two-factor interactions, and the 21 three-factor interactions.

We could use a normal probability plot (or perhaps a half-normal plot) in trying to identify significant effects, but some researchers prefer a Pareto chart of the absolute values of the effect estimates [see Haaland (1998) and Haaland and O'Connell (1995)]. The Pareto chart of effect estimates is given in Figure 14.6. Notice that the largest effect is the first noise factor, environmental light, and that most of the other relatively large effects are interaction effects involving this noise factor. We should bear in mind, however, that the environmental light

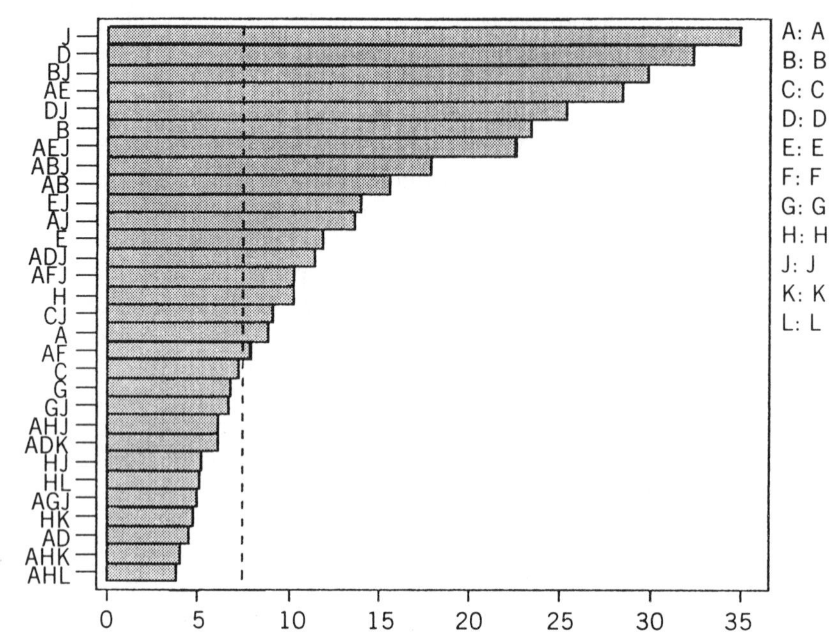

FIGURE 14.6 Pareto chart of effect estimates: response is FF; $\alpha = .10$; only 30 largest effects shown.

effect is aliased with the interaction effect of the other two environmental factors, so the presence of such an interaction would create havoc with the analysis.

Notice also that there are not just a few large effects, and the choice of a dividing line between significant and nonsignificant effects is certainly not obvious from the Pareto chart. The vertical line that is shown is intended to serve as a dividing line, however, with the line determined using the method for determining significant effects due to Lenth (1989). The application of that technique here identifies 18 significant effects, and fitting the model with these effects produces $R^2 = .932$.

When using all of the degrees of freedom for estimating effects, keep in mind that we cannot use diagnostics because we cannot estimate the error variance. It is generally unwise not to do a diagnostic analysis, however. That is especially true for this data set since there are some suspicious observations. When we fit the model with the aforementioned 18 terms, observations 13, 29, and 61 have standardized residuals of 3.26, −3.06, and 3.66, respectively. The first of these two occurred at treatment combinations where FF varied dramatically over the four noise readings. Specifically, observation 14 minus observation 13 = 111 and observation 16 minus observation 15 = 137. Then observation 30 minus observation 29 = 235 and observation 32 minus observation 31 = 196. [Lewis et al. (1997) do not indicate that the run order was different from the standard order for listing treatment combinations, so we assume that they were the same.]

The differences between consecutive observations for each of these two treatment combinations far exceed the differences for other treatment combinations, as all of the other differences are less than 100, and are generally far below 100.

Since the large differences occur for the high versus low levels of ambient light, the first noise factor, it would have been interesting to investigate why these large differences occurred. This would be important because these unusual observations have a substantial effect on the results. Lewis et al. (1997) state that "apparently, when ambient light is low and the far corner is the starting point of the flat find or when the illumination level is sub-par, then the system has difficulty recognizing the pattern" and discuss this relative to the equipment that was used, concluding that there is a straightforward explanation for this phenomenon. But their explanation is not entirely satisfactory, as the values for flat find (FF) range from 738 to 906 for this combination of factor levels for B, D, and J, with 906 not being small compared to the average of 937 for all 64 observations.

In the absence of any information to the contrary, we will assume that the unusual observations are valid data points and will proceed to look at relatively large interactions with factor J. All of the two-factor interactions are shown in Figure 14.7. Of course, we know that not all of these interactions are estimable. We can see that the largest interactions involving factor J are BJ and DJ, and fortunately each is of the type where the large interaction is due to the difference in the absolute values of the slopes. As emphasized previously, it is this difference

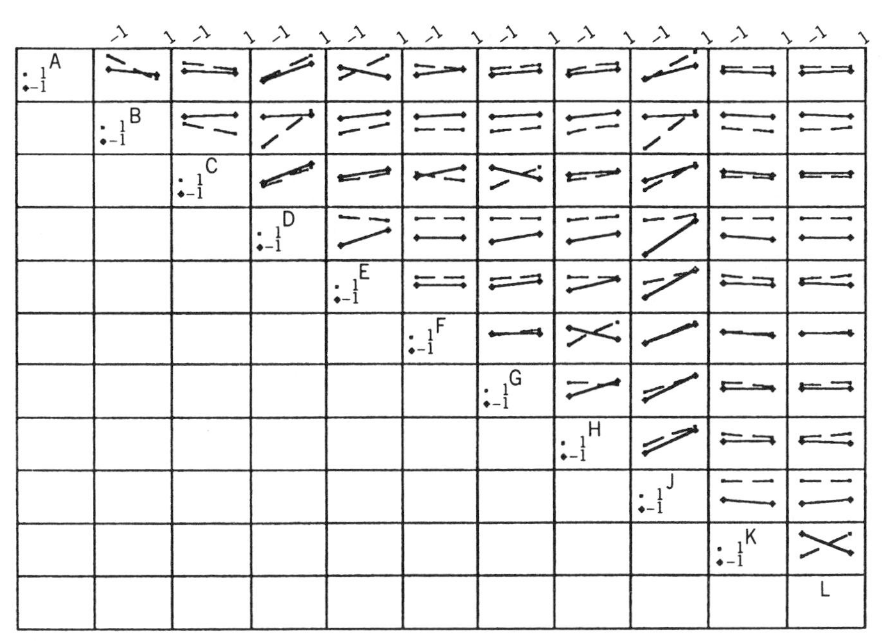

FIGURE 14.7 Set of all two-factor interactions for flat find data.

that is important, not simply the magnitude of the effect. Figure 14.7 suggests that the low level of B and the high level of D should be used. This will tend to not only maximize FF but also stabilize FF over J. Similarly, it would be desirable to use the high levels of A and E. The former may require some explanation since there is greater robustness at low A, but high A is better for producing a large FF value. Lewis et al. (1997) also recognized the desirability of using these factor settings for B and D, but did not consider settings for other control factors since they did not look at control × noise interactions.

What effect do the somewhat large three-factor interactions (AEJ, ABJ, ADJ, and AFJ) have on the analysis? Since we are concerned with control × noise interactions, we could view the AEJ interaction as the interaction between EJ and A. The AEJ interaction being relatively large means that the EJ interaction plots will differ noticeably for the two levels of A. These plots are given in Figure 14.8, in addition to the overall interaction plot. (Alternatively, we could have considered the AJ interaction plots for the two levels of E, but the EJ interaction is slightly larger than the AJ interaction.)

Each of the interaction plots in Figure 14.8 gives a different message. Figure 14.8*c* shows that it is considerably better to use the high level of factor E. This is also evident from Figure 14.8*b* (high A), which shows a much larger advantage from using the high level of E. Figure 14.8*a* (low A) suggests, however, that the low level of E should be used. Since we decided that high A was somewhat preferable to low A, Figure 14.8*b* shows how much better off we are by using high E rather than low E.

We could perform a similar analysis for each of the other significant three-factor interactions. For example, the ABJ interaction being relatively large means that the BJ interaction profile is probably misleading, and thus not representative of the interaction profile for each level of A. Remembering it was previously decided that it was best to use the high level of A, we should be particularly interested in the BJ interaction profile at high A. Although this profile has the same general shape as the profile at low A and the overall BJ interaction profile, the potential for improvement by using the low level of factor B is much greater when A is at the high level than when A is at the low level, as the reader is asked to demonstrate in exercise 8.

Since the DJ interaction was the fifth largest effect, and the ADJ interaction was also significant, it is similarly of interest to look at the DJ interaction profiles at each level of A. The three interaction plots are less dissimilar than the three plots in Figure 14.8 since the ADJ interaction is smaller than the ABJ interaction. Similarly, since the AFJ interaction is significant, we should look at the AJ interaction at each level of F. (There is no need to look at the FJ interaction at each level of A since the former is not significant.) These plots suggest that low A should be used, although there is very little gain if high F is used. The F effect was not significant, however, nor was the FJ interaction, so there is not a clear signal regarding the best choice of level for F.

The CJ interaction was also significant and the plot suggests that the low level of C should be used. Putting all of this together, the experimenter should use the high level of A, the low level of B, the low level of C, the high level of D, and the high level of E. The other control factors, F, G, and H, were not involved in any significant two-factor interactions with noise factors, and only the main effect of factor H was significant. Since the effect estimate was positive, we could use the high level of H, and use either level of F and G.

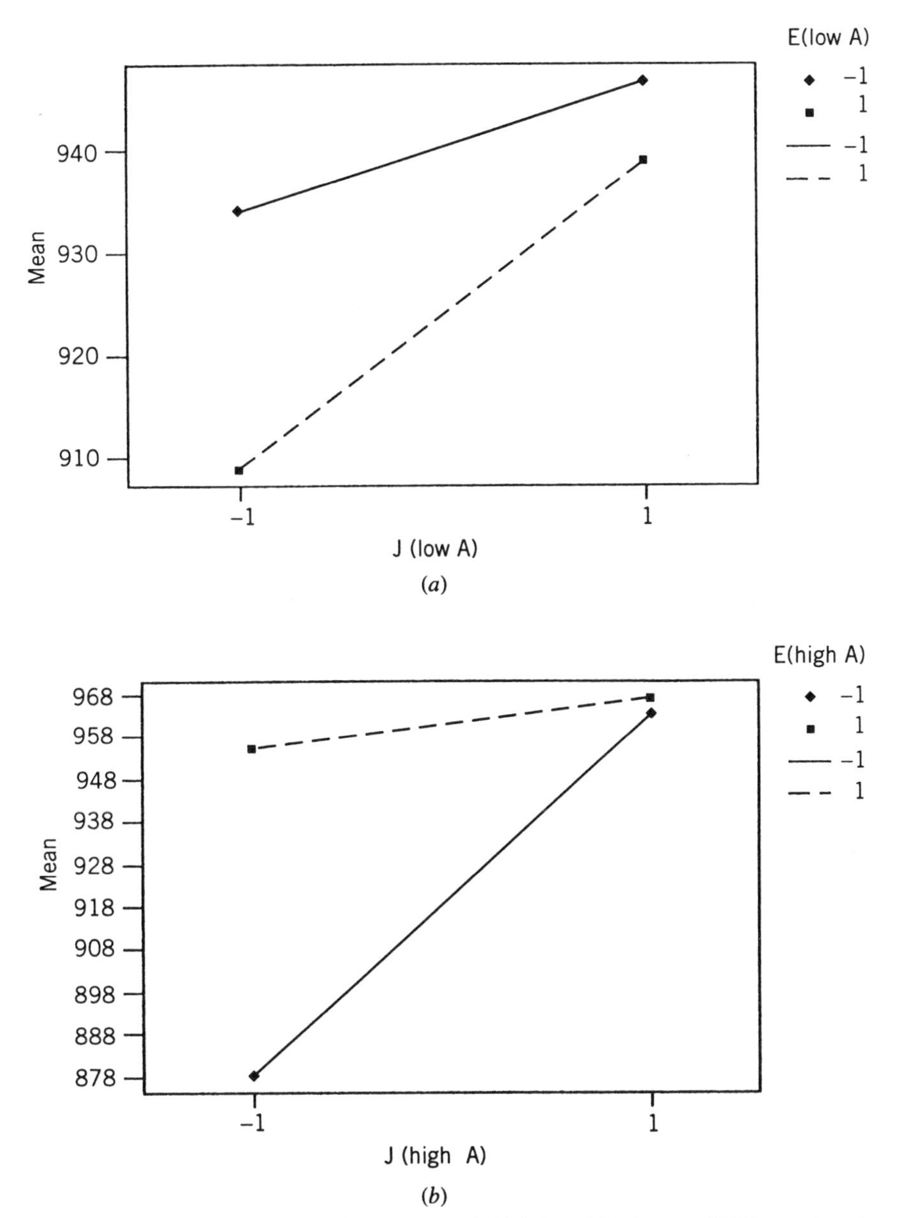

FIGURE 14.8 EJ interaction plots for (*a*) low A, (*b*) high A, and (*c*) the overall EJ interaction plot.

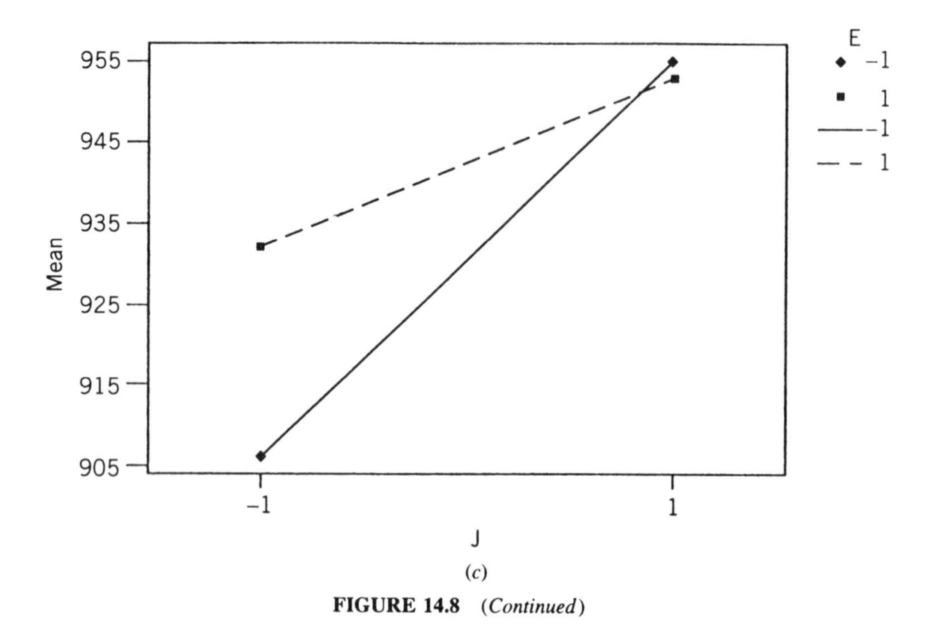

(*c*)

FIGURE 14.8 (*Continued*)

Notice that this combination of factor levels for A, B, C, D, E, and H did not occur in the design that was used. Treatment combinations 2 and 9 come the closest, and for the former the average FF value (966) was the highest of the treatment combinations, whereas the variability in FF was the smallest at combination 9. Therefore, the treatment combination suggested by our analysis might be a near-optimal combination in terms of both maximizing FF and minimizing the variability.

Accordingly, subsequent experimentation should probably start with the suggested levels for these six control factors. Specifically, these levels might serve as the center point in a central composite design (Section 13.16.4) that would be used to identify possible nonlinear effects.

Lewis et al. (1997) did provide an analysis for modeling the variability, using ln(s) as the dependent variable. Their analysis identified the B and D main effects as the only significant effects. Since the estimate of the B effect is positive and the estimate of the D effect is negative, this would suggest that the low level of B and the high level of D be used. Of course, the desirability of using these levels was identified in the analysis of location effects. There were four treatment combinations that have this combination of levels of B and D, and not surprisingly, these were the treatment combinations that had the smallest standard deviations (computed over the noise variables), and thus the smallest values of ln(s). The average value of FF at each of these combinations was 947, 958, 937, and 959, and these are not the best treatment combinations for maximizing FF.

What would have happened if we had used an S/N approach? We would need to use an S/N for the "larger the better" case since it is desirable to maximize FF.

We will not pursue this, however, because of the heavy criticism that analyses using S/N ratios have received.

14.9.5.1 Cautions

The preceding analysis depends very heavily on the levels of the noise factors that were used. Lewis et al. (1997) do not discuss the choice of levels; so it is natural to wonder how likely these levels are to occur in practice. It would be especially dangerous to select levels that are too extreme in each direction because in the control × noise interaction plots we would then be connecting points that have a very small chance of occurring, and then making robustness decisions based on the configuration of lines. It would be better to be somewhat conservative and use noise levels that are well within the expected range.

14.9.6 Desirable Robust Designs and Analyses

In Section 14.9.5 the validity of the analysis that was given was contingent on certain assumptions, such as interactions not existing between noise factors. Ideally, we would like to use a design for which we do not have to make assumptions that might be untenable. It would certainly be safer to use a Resolution V design, but for 11 factors that would require 128 points, and it will generally be too expensive to conduct experiments with so many runs.

14.9.6.1 Designs

If we had our choice, we would like to be able to estimate as many control × control interactions as possible, without having to sacrifice any control × noise interactions, unless there is prior information to suggest that some of the latter may be unimportant. In general, we would like to be able to choose which interactions to estimate and also have those interactions aliased with interactions that are not likely to be significant. The selection of a design to meet the first objective has been addressed by Wu and Chen (1992). Their method starts with the defining relations for a 2^{n-k} design determined using the criterion of minimum aberration and then constructs the set of all nonisomorphic graphs. [A *minimum aberration design* is one for which the defining relation with the smallest number of letters occurs the smallest possible number of times. The concept of minimum aberration is due to Fries and Hunter (1980). A *nonisomorphic graph* is a graph that cannot be created by simply relabeling the vertices (i.e., the graphs have different shapes).]

The objective would be to match a graph with the graph that results from the set of interactions that the practitioner wishes to estimate. It is possible that there will not be a match, in which case either the set of interactions to estimate must be altered or a larger design must be used. It is highly desirable that software be used to construct the set of nonisomorphic graphs.

14.9.6.2 Analyses

The analysis of the control × noise interaction plots in the example in Section 14.9.5 led to factor-level settings that seemed quite reasonable. We cannot necessarily expect to minimize the response variance with this somewhat ad hoc approach, however. More formal analysis methods have been given by Shoemaker and Tsui (1993), Tsui (1994), Tsui (1996), and McCaskey and Tsui (1997).

14.10 DETERMINING OPTIMUM CONDITIONS

For whichever type of design is used, it is desirable to use the resultant data to seek the best combination of levels of the process variables. This determination is not easily made from an initial analysis, however. Furthermore, the following quote from Myers and Montgomery (1995, pp. 480–481) should be kept in mind:

> With regard to modeling strategies, it should be emphasized that, in general, process optimization can be quite overrated. The more emphasis that is placed on learning about the process, the less important *absolute optimization* becomes. An engineer or scientist can always find (and often do) pragmatic reasons why the suggested optimum conditions cannot be adopted.

Nevertheless, Taguchi and Wu (1979, p. 37) and other writers have used graphs of marginal averages in attempting to arrive at the optimal levels of the process variables (factors). Marginal averages are obtained by computing the average of the response variable at each level of each process variable, while ignoring the other process variables. Unfortunately, this method will not identify optimum conditions, in general, but might identify conditions that are close to the optimum.

To illustrate, we shall assume that we have data from an unreplicated 3^2 design as in Table 14.9. The marginal averages for A and B are given in Figure 14.9. Following Taguchi and Wu (1979), if we were to use marginal averages in determining the levels of A and B so as to maximize the response variable, we would use the third level of A and the first level of B, as can be seen from Figure 14.9. It is apparent from Table 14.9, however, that the maximum occurs with the second level of B, not the first. Notice, however, that we do not miss the maximum by very much — for this example.

Of course, we can *see* in Table 14.9 the combination of A and B that produces the highest number; we do not need to construct marginal averages to determine that. All possible combinations of factor levels are not used in orthogonal arrays, however, so the user of marginal averages would be trying to infer what the best combination would be *if* all possible combinations had been used. This is what

TABLE 14.9 Data from an Unreplicated 3^2 Design — Small AB Interaction

		A		
		1	2	3
	1	8	8	10
B	2	6	8	11
	3	5	9	10

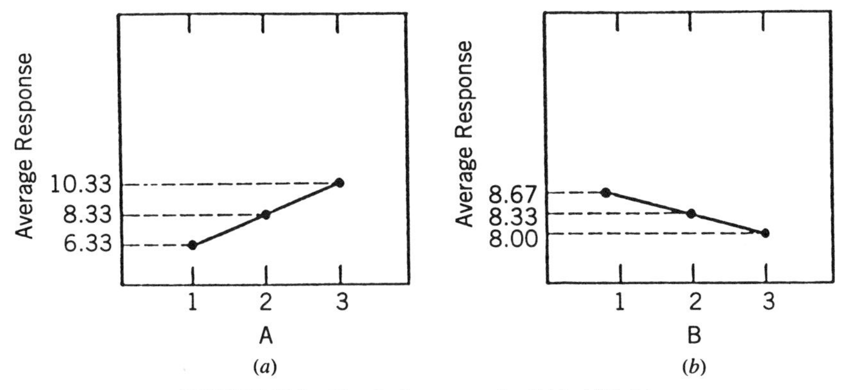

FIGURE 14.9 Marginal averages for Table 14.9 data.

is done in Taguchi and Wu (1979, p. 37) and in Barker (1986, p. 41). It should be clear, however, that since the marginal averages approach will frequently not work when we *do* have all combinations of factor levels, then obviously it will not work, in general, when we do not have all possible combinations.

Even when the marginal averages form straight lines, as in Figures 14.9*a* and 14.9*b*, the interactions will not necessarily be zero, so the optimum combination might not be selected.

For this example the interaction profile is given in Figure 14.10. We can see that the interaction is not extreme, and this is why the marginal averages approach closely approximates the maximum.

Consider, however, the data given in Table 14.10. For these data the marginal averages approach would lead to the selection of A_3B_3 as the best combination, but we can see from Table 14.10 that this is only the fifth best combination. The reason is that the AB interaction is more pronounced for these data than for the data in Table 14.9.

We can see how the marginal averages approach will be undermined by varying degrees of interaction by letting Y, the response variable that we are

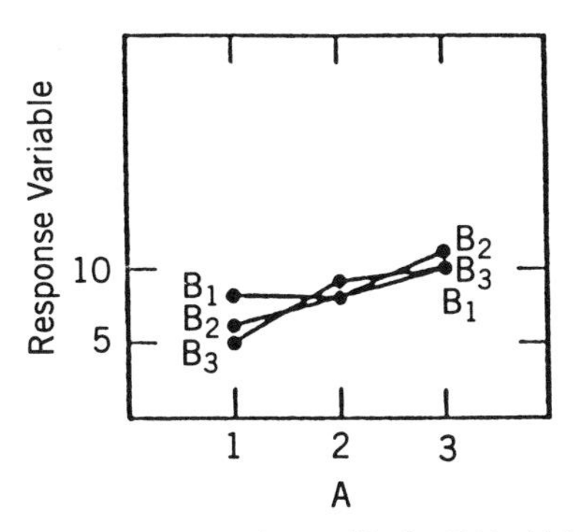

FIGURE 14.10 Interaction profile for Table 14.9 data.

TABLE 14.10 Data from an Unreplicated 3^2 Design—Moderate AB Interaction

		A		
		1	2	3
	1	10	17	24
B	2	17	20	23
	3	24	23	22

trying to maximize, have the functional form

$$Y = cX_1 + cX_2 - dX_1X_2$$

where X_1 (and X_2) $= -1$, 0, or 1, corresponding to the three levels of A and B. (This is analogous to the use of -1 and +1 for the two levels of each factor in two-level designs.) Notice that we are assuming a model without an error term for the purpose of simplification.

By using c as the coefficient for both X_1 and X_2, we are assuming that the main effects of A and B are equal, and the magnitude of d relative to c will determine the magnitude of the AB interaction relative to the two main effects. Table 14.11 contains the nine values of Y that result from using the nine combinations of X_1 and X_2. We can observe that the marginal totals (and thus the marginal averages) are not a function of d, and the use of the marginal averages approach would lead to the selection of A_3B_3 (equivalently, $X_1 = 1$ and $X_2 = 1$) as the best combination. That selection would be correct, however, only if $d < c$.

If $d = c$, there are five combinations that are "equally best," but the combination that is only the fifth best would be selected when $c < d < 2c$.

The interaction profiles for values of d that correspond to these three cases are given in Figure 14.11. We can see that large interactions will seriously undermine the marginal averages approach, but it might work moderately well, *as an approximation*, when the interactions are quite small.

In general, this use of marginal averages is analogous to the one-factor-at-a-time approach that was discussed in Chapter 13, and is thus fallible in the same way. In essence, the user of marginal averages is implicitly imputing values

TABLE 14.11 Data from a 3^2 Design in Terms of c and d

		X_1(A)			
		-1	0	1	Totals
	-1	$-2c - d$	$-c$	d	$-3c$
X_2(B)	0	$-c$	0	c	0
	1	d	c	$2c - d$	$3c$
Totals		$-3c$	0	$3c$	

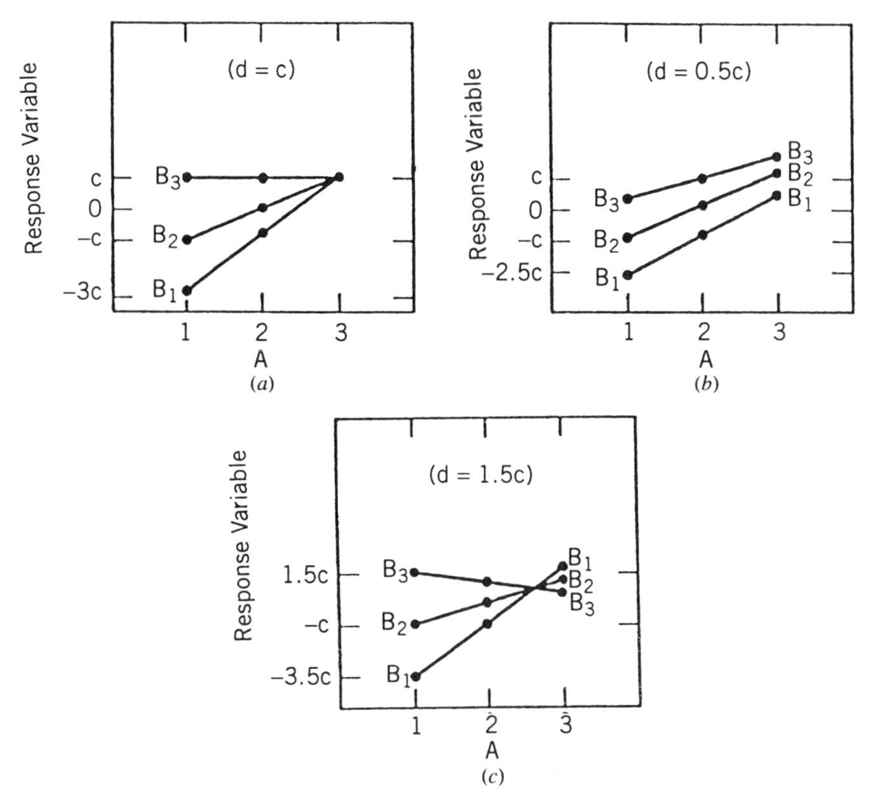

FIGURE 14.11 Interaction profiles for Table 14.11 data.

for the factor-level combinations that are not included in an orthogonal array, and the implied imputation forces the interactions to be (approximately) zero. It is difficult to estimate what the response variable would be for factor-level combinations that are missing without knowledge of the response surface.

We should also keep in mind that we should not expect to be able to identify the true optimum combination of factor levels from a single experiment. We looked at what was the best combination of A and B for *one set* of nine combinations of factor levels in *one* experiment.

In general, some optimization technique should be used in trying to determine optimum conditions, and the dual response approach introduced by Myers and Carter (1973) has been used by various researchers, including Vining and Myers (1990) and Kim and Lin (1998). The general idea is to seek to optimize the response subject to a constraint on the variance or to minimize the variance subject to a constraint on the response. The objective is to achieve a satisfactory balance between bias and variance, as one seeks to do with the loss function approach discussed in Section 14.3. Kim and Lin (1998) contend that their fuzzy modeling approach achieves a better balance between bias and variance than the previously proposed methods.

14.11 SUMMARY

Genichi Taguchi and his co-workers have taught us some valuable lessons concerning product quality. No one could dispute the idea of focusing attention on a target value rather than just operating within specification limits (which might be arbitrarily defined). Similarly, it certainly makes sense to try to design a product that will be relatively insensitive to manufacturing variations.

The extent to which it is important to operate at or very close to a target value will depend greatly upon the steepness of the actual loss function in the vicinity of the target value. A quadratic loss function will probably be a reasonable approximation in most cases, but losses within, say, two standard deviations of the target value would still have to be obtained or estimated to determine the steepness of the curve. The amount of money that is being saved could then be approximated as the variability is reduced and the mean of the process characteristic is brought closer to the target value.

Determining the values of design parameters is accomplished by Taguchi's methods through the use of orthogonal arrays. In using experimental designs we should not lose sight of the fact that we first need to determine the factors (variables) that affect the quality of the product. It would be unwise to initially select a moderate number of factors to use in a fractional factorial or orthogonal array design and then select the design point that minimizes or maximizes an S/N or PERMIA, acting as if these are the only factors that could be important.

It would be judicious to first use a screening design of some type with a large number of factors for the purpose of screening out factors that seem not to be important. A confirmatory experiment could then be conducted using a design that is similar to or identical to the design used in the first stage, for the purpose of corroborating the factor selection performed after the first experiment.

These selected factors could then be used in an experiment to determine the values of the factors that minimize (maximize) an S/N or PERMIA. The reader should recognize, however, that with this approach an S/N or PERMIA is not being minimized in a strict mathematical sense. Rather, the minimization is only over the design points used in the design. It would be only by sheer coincidence that the minimum for the design points would minimize the function over all plausible values of the factors that are being used.

Viewed with a broader perspective, the problem is actually one of trying to optimize a function subject to constraints on the possible values of the factors. A function such as $-\log(s^2)$ is clearly a nonlinear (objective) function, so the problem is logically a nonlinear programming problem.

Box and Fung (1986) discuss this approach in detail, using an example from Taguchi and Wu (1979) for illustration. The former show that the 36-point array used by the latter for studying five factors does not maximize the chosen S/N, even when only the three levels of each factor are considered.

To summarize what we have learned, we have seen that the orthogonal arrays illustrated by Taguchi and Wu (1979) are often inferior to the better known and understood fractional factorials, and if we wish to maximize or minimize an

S/N or PERMIA, it is logical to attempt to do so using known mathematical optimization procedures such as nonlinear programming, rather than trying to rely upon graphs of marginal averages.

We have also seen that the product array approach is generally inferior to the combined array approach. The identification of important control × noise interactions is highly desirable, but the key is not the magnitude of the interaction effects but rather the extent to which the absolute values of the slopes in an interaction plot differ. Software and new algorithms are needed for accomplishing this objective.

Taguchi's main contribution has been focusing our attention on new objectives in achieving quality improvement. During the past 10–12 years his suggested statistical methods have been rigorously debated, and the methods found wanting. Consequently, new methods have been developed that provide practitioners with alternatives.

One thing that is absent from the robust design literature is suggested strategies for the sequential use of robust designs. This should not be surprising, however, since there is very little in the literature on general strategies for the sequential use of experimental designs.

Readers seeking additional information are advised to read the following. For a recent, general discussion of robust design methods see Kacker and Ghosh (1998) who give references that show that Taguchi's parameter design approach is not new, although the widespread use of parameter design in manufacturing is, of course, new.

REFERENCES

Addelman, S. (1962). Orthogonal main-effect plans for asymmetrical factorial experiments *Technometrics 4*(1): 21–46.

Barker, T. R. (1986). Quality engineering by design: Taguchi's philosophy. *Quality Progress 19*(12): 32–42.

Bounou, M., S. Lefebvre, and X. Dai Do (1995). Improving the quality of an optimal power flow solution by Taguchi method. *International Journal of Electrical Power and Energy Systems 17*(2): 113–118.

Box, G. E. P. (1986). Studies in quality improvement: Signal to noise ratios, performance criteria and statistical analysis: Part I. Report No. 11, Center for Quality and Productivity Improvement, University of Wisconsin.

Box, G. E. P. and C. A. Fung (1986). Studies in quality improvement: Minimizing transmitted variation by parameter design. Report No. 8, Center for Quality and Productivity Improvement, University of Wisconsin.

Box, G. E. P. and J. S. Hunter (1961a). The 2^{k-p} fractional factorial designs Part I. *Technometrics 3*(3): 311–351.

Box, G. E. P. and J. S. Hunter (1961b). The 2^{k-p} fractional factorial designs Part II. *Technometrics 3*(4): 449–458.

Box, G. E. P. and R. D. Meyer (1986). Dispersion effects from fractional designs. *Technometrics 28*(1): 19–27.

Box, G. E. P., W. G. Hunter, and J. S. Hunter (1978). *Statistics for Experimenters*. New York: Wiley.

Byrne, D. M. and S. Taguchi (1986). The Taguchi approach to parameter design. In *Annual Quality Congress Transactions*, pp. 168–177. Milwaukee, WI: American Society for Quality Control.

Caporaletti, L., E. Gillenwater, and J. Jaggers (1993). The application of Taguchi methods to a coil spring manufacturing process. *Production and Inventory Management 34* (4): 22.

Fries, A. and W. G. Hunter (1980). Minimum aberration 2^{k-p} designs. *Technometrics 22*: 601–608.

Haaland, P. (1998). Comment. *Statistica Sinica 8* (1): 31–35.

Haaland, P. and M. A. O'Connell (1995). Inference for effect-saturated fractional factorials. *Technometrics 37* (1): 82–93.

Hunter, J. S. (1985). Statistical design applied to product design. *Journal of Quality Technology 17* (4): 210–221.

Jessup, P. (1986). The value of continuing improvement. *ASQC Automotive Division Newsletter*, pp. 5–10 (March).

Kacker, R. N. (1985). Off-line quality control, parameter design, and the Taguchi method. *Journal of Quality Technology 17* (4): 176–188 discussion: pp. 189–209.

Kacker, R. N. and S. Ghosh (1998). Robust design methods. In H. M. Wadsworth, ed. *Handbook of Statistical Methods for Engineers and Scientists*, 2nd ed. New York: McGraw-Hill

Kim, K-J. and D. K. J. Lin (1998). Dual response surface optimization: A fuzzy modeling approach. *Journal of Quality Technology 30* (1): 1–10.

Kros, J. F. and C. M. Mastrangelo (1998). Impact of nonquadratic loss in the Taguchi design methodology. *Quality Engineering 10* (3): 509–519.

Leisk, G. and A. Saigal (1995). Taguchi analysis of heat treatment variables on the mechanical behavior of alumina/aluminum metal matrix composites. *Composites Engineering 5* (2): 129–142.

Lenth, R. V. (1989). Quick and easy analysis of unreplicated factorials. *Technometrics 31*: 469–473.

León, R. V. and C. F. J. Wu (1992). A theory of performance measures in parameter design. *Statistica Sinica 2*: 335–358.

León, R. V., A. C. Shoemaker, and R. N. Kacker (1987). Performance measures independent of adjustment. *Technometrics 29* (3): 253–265; (discussion: pp. 266–285).

Lewis, D. K., C. Hutchens, and J. M. Smith (1997). Experimentation for equipment reliability improvement. In V. Czitrom and P. D. Spagon, eds. *Statistical Case Studies for Industrial Process Improvement*, Chapter 27. Jointly published by the Society of Industrial and Applied Mathematics (Philadelphia, PA) and the American Statistical Association (Alexandria, VA).

Lucas, J. M. (1985). Discussion (of Kacker, 1985). *Journal of Quality Technology 17* (4): 195–197.

McCaskey, S. D. and K.-L. Tsui (1997). Analysis of dynamic robust design experiments. *International Journal of Production Research 35* (6): 1561–1574.

Montgomery, D. C. (1997). *Design and Analysis of Experiments*, 4th ed. New York: Wiley.

Montgomery, D. C., C. M. Borror, and J. D. Stanley (1997). Some cautions in the use of Plackett-Burman designs. *Quality Engineering 10* (2): 371–381.

Morehead, P. R. and C. F. J. Wu (1998). Cost-driven parameter design. *Technometrics* *40*(2): 111–119.

Myers, R. H. and W. H. Carter (1973). Response surface techniques for dual response systems. *Technometrics 15*: 301–317.

Myers, R. H. and D. C. Montgomery (1995). *Response Surface Methodology: Process and Product Optimization Using Designed Experiments*. New York: Wiley.

Pignatiello, J. J. and J. S. Ramberg (1985). Discussion (of Kacker, 1985). *Journal of Quality Technology 17*(4): 198–206.

Raktoe, B. L., A. Hedayat, and W. T. Federer (1981). *Factorial Designs*. New York: Wiley.

Rosenbaum, P. R. (1994). Dispersion effects from fractional factorials in Taguchi's method of quality design. *Journal of the Royal Statistical Society, Series B 56*: 641–652.

Rosenbaum, P. R. (1996). Some useful compound dispersion experiments in quality design. *Technometrics 38*(4): 354–364.

Shoemaker, A. C. and K.-L. Tsui (1993). Response model analysis for robust design experiments. *Communications in Statistics—Simulation and Computation 22*(4): 1037–1064.

Shoemaker, A. C., K.-L. Tsui, and C. F. J. Wu (1991). Economical experimentation methods for robust design. *Technometrics 33*(4): 415–427.

Snow, J. (1993). Rating quality and selecting suppliers using Taguchi's loss functions. *Navy Engineer's Journal 105*(1): 51.

Steinberg, D. M. and D. Bursztyn (1998). Noise factors, dispersion effects, and robust design. *Statistica Sinica 8*(1): 67–85.

Taguchi, G. (1986). *Introduction to Quality Engineering*. Tokyo: Asian Productivity Organization.

Taguchi, G. (1987). *System of Experimental Design*, Vols. 1 and 2. White Plains, NY: UNIPUB.

Tsui, K.-L. (1994). Avoiding unnecessary bias in robust design analysis. *Computational Statistics and Data Analysis 18*(5): 535–546.

Tsui, K.-L. (1996). A multi-step analysis procedure for robust design. *Statistica Sinica 6*(3): 631–648.

Taguchi, G. and Y. Wu (1979). *Introduction to Off-Line Quality Control*. Central Japan Quality Control Association, Nagaya.

Vining, G. G. and R. H. Myers (1990). Combining Taguchi and response surface philosophies: A dual response approach. *Journal of Quality Technology 22*: 38–45.

Wang, Y., D. K. J. Lin, and K.-T. Fang (1995). Designing outer array points. *Journal of Quality Technology 27*(3): 226–241.

Wu, C. F. J. and Y. Chen (1992). A graph-aided method for planning experiments when certain interactions are important. *Technometrics 34*(2): 162–175.

EXERCISES

1. Consider the following orthogonal array (called a Plackett–Burman design) for 11 factors:

A	B	C	D	E	F	G	H	J	K	L
1	1	−1	1	1	1	−1	−1	−1	1	−1
−1	1	1	−1	1	1	1	−1	−1	−1	1
1	−1	1	1	−1	1	1	1	−1	−1	−1
−1	1	−1	1	1	−1	1	1	1	−1	−1
−1	−1	1	−1	1	1	−1	1	1	1	−1
−1	−1	−1	1	−1	1	1	−1	1	1	1
1	−1	−1	−1	1	−1	1	1	−1	1	1
1	1	−1	−1	−1	1	−1	1	1	−1	1
1	1	1	−1	−1	−1	1	−1	1	1	−1
−1	1	1	1	−1	−1	−1	1	−1	1	1
1	−1	1	1	1	−1	−1	−1	1	−1	1
−1	−1	−1	−1	−1	−1	−1	−1	−1	−1	−1

(a) Could this design be equivalent to a 2^{k-p} design?

(b) What effects are estimable with this design? (*Hint*: How many design points are there?)

2. For the loss function $L = (X - \mu)^2$, what is the expected loss for a process that is on target and has a standard deviation of 2.5?

3. Assume that an experimenter wishes to study the effects of four factors and recognizes that the cost for each design point will be sizable. For this reason he rules out a 2^4 design. Additionally, he believes that interactions above second order are not likely to be significant. He decides to use the first columns of the design given in exercise 1, as he believes that he can afford 12 design points.

(a) He would like to be able to estimate the two-factor interaction involving factor A. Consider, for example, the AB interaction. Is that estimable with this design? Specifically, would the AB column be orthogonal to the four columns for A, B, C, and D?

(b) What would this suggest about modifying Plackett–Burman designs to meet specific objectives?

(c) Could the experimenter obtain the AB interaction with a 2^{4-1} design, as well as the AC and AD interactions, if the other two-factor interactions are negligible?

4. Notice that the distribution of Y is not used explicitly in computing the expected loss unless an asymmetric loss function is used. Assume that a process is on target so that $E(L) = E(Y - t)^2 = \sigma^2$. To illustrate that $E(Y - t)^2$ is just the average of the $(Y - t)^2$ values, we do need to assume a distribution, however. Assume that $f(y) = 1/5$; $y = 1, 2, 3, 4, 5$. Compute $E(Y - 3)^2 = \sum_{i=1}^{5}(y_i - 3)^2/5$ and show that the result is 2, which is what

σ_y^2 can be shown to equal. [Note that an "unweighted" average is produced here because of the form of $f(y)$.]

5. Assume that a company has two production processes in which the first process is on target with a variance of 6, whereas the second process has a variance of 4 but is off target such that $E(Y) = 6$ but $t = 5$. With the loss function $L = \$2(Y - t)^2$, which process is costing the company the most money?

6. Explain how a company can have a "loss" if all of its processes are operating within specification limits.

7. Construct the interaction profile for the data in Table 14.10 and use this to explain why the marginal averages approach does not work for these data.

8. For the data in Table 14.8, construct the overall BJ interaction profile in addition to the profile at high A and at low A. Compare the three plots and comment.

9. Even though the construction of designs that allow specified interactions to be estimated was discussed in Section 14.9.6.1, explain why it is not a good idea to construct such designs *as a general rule*.

10. What is the Resolution of the design in Table 14.7?

CHAPTER 15

Evolutionary Operation

Evolutionary operation (EVOP) was introduced by G. E. P. Box (1957) as a technique to facilitate continuous process improvement. It is based upon experimental design concepts but is used much differently than the experimental designs presented in Chapter 13. With those designs the objective is to determine the relationship between a response variable and a number of process variables, and this determination is made by varying the process variables over a reasonable range. Doing so, however, will generally disrupt the normal production process. With EVOP only very small changes are made in the settings of the process variables so that the process is not disrupted, and, in particular, there is (hopefully) no increase in the percentage of nonconforming units.

Evolutionary operation has been successfully employed in the chemical industry, as well as in other industries. It is appropriate for continuous processes. As discussed by Box and Draper (1969), the evolution of an industrial process from the laboratory through a pilot plant and finally to a full-scale plant does not guarantee that the optimum operating conditions have been identified. Rather, the values of process variables that are used initially in production are probably only reasonable approximations, at best, to the optimum values. It should be noted that EVOP is not a true optimization procedure, as in mathematical optimization. Rather, the user of EVOP is attempting to find *improved* operating conditions.

The method is intended to be used by plant personnel and thus does not require the participation of someone with statistical expertise. The formation of an EVOP committee is generally recommended, however, and the committee is charged with periodically reviewing the results and determining what progress has been made.

Evolutionary operation also does not require the use of special equipment, although as Hunter and Chacko (1971) point out: "The results from the work sheet should be posted on an information board.... which is, in general, a large board prominently displayed near the process being studied." The work sheet that the authors refer to is used for performing the statistical calculations. Box and Draper (1969, p. 13) also discuss the use of an information board. Alternatively, computer software could be used for this purpose, although there is very little EVOP software that is commercially available.

The limited availability of software reflects the state of disuse into which EVOP has fallen. Indeed, Hahn (1984) asked: "Whatever happened to EVOP?" There is general agreement (among statisticians) that EVOP is being underutilized in industry and that this condition has existed for some time. A study conducted by Hahn and Dershowitz in 1972 revealed that many industrial personnel were well aware of the value of EVOP, however, and how it could be used to improve the production lines of their respective companies.

Why, then, is EVOP not being used? The primary reason seems to be a general reluctance to perturb a production process that is viewed as running smoothly. Other reasons cited by Hahn and Dershowitz (1974) include political reasons, lack of knowledge about EVOP, and lack of proper personnel. Only a small percentage of the respondents indicated that EVOP was not being used at their respective companies because of poor past experience.

The general admonition that is often heard—"If it isn't broke, don't fix it"—seems to have been adopted by many potential users of EVOP and thus serves as an impediment to its general usage. As Hahn (1984) points out: "One might even argue that the concept of EVOP runs contrary to the desire to maintain a process in statistical control." This could not be a valid argument against the use of EVOP, however, as the objective should be process *improvement*, not process control.

The real problem is perhaps captured in an important point made by Joiner (1985); namely that the proper managerial climate must exist before statistical procedures can be effectively applied on a wide scale. That climate might be changing somewhat, however, with the Six Sigma type training programs that have been initiated. (Six Sigma is discussed in Section 17.3.) In particular, Hoerl (1998) lists EVOP as being a component of a typical six-sigma black-belt training curriculum.

15.1 EVOP ILLUSTRATIONS

We shall illustrate the EVOP methodology developed by Box and a variation of that approach which has been used to some extent. Other variations will be mentioned briefly but not illustrated, as there is no evidence that they have ever been used.

The Box–EVOP procedure is generally performed using either a 2^2 or a 2^3 design when EVOP is used on a full-scale manufacturing process. The general idea is to keep the design as simple as possible since the data will be analyzed by plant personnel. When EVOP is used in a laboratory or pilot plant, more factors can be analyzed, and a fractional factorial might be used.

The form of the worksheets used in EVOP and the necessary calculations that are performed on them are illustrated in Box and Hunter (1959), and are reproduced in Box and Draper (1969).

We shall begin with a hypothetical example in which there are two process variables to be studied, temperature and pressure, and process yield is the response variable. These two variables might have been identified through the use of a screening design as apparently having a significant effect upon yield, or it simply might be the case that they were selected because they were believed to be important, with such a belief unsupported by data from prior experimentation.

In Chapter 13 an example was given in which the different levels of the factor "temperature" were 250, 300, and 350°F. Using such widely varying levels in an EVOP program could be disastrous, however, as the percentage of bad product might increase considerably. If the temperature used in the current plant operation is 300°, a safer strategy would be to use 290° and 310° in the first phase of the EVOP program. Similarly, if 30 pounds per square inch (psi) of pressure is currently being used, we might use 28 and 32 in the EVOP program.

A 2^2 design in an EVOP program is generally used with a center point, which denotes the *reference condition*. This is assumed to be the current best known operating conditions (i.e., values of the process variables). Assume that process yield is the response variable and that data have been collected for the previously mentioned levels of temperature and pressure, with the data given in Figure 15.1. The design points in Figure 15.1 are numbered using the same numbering system employed by Box and Draper (1969); namely zero denotes the current operating values, and 1, 2, 3, and 4 specify the order in which the four combinations are run. (The order is not "randomized" as is the usual case when experimental designs are used because several cycles are typically used, and changing the order each time a new cycle is run could easily confuse the plant personnel.)

With data as in Figure 15.1, it appears as though the current operating values are not the optimal values. How might we determine this statistically? In Chapter 13 we said that for an unreplicated factorial we need either a prior estimate of the error variance or to plot the effect estimates on normal

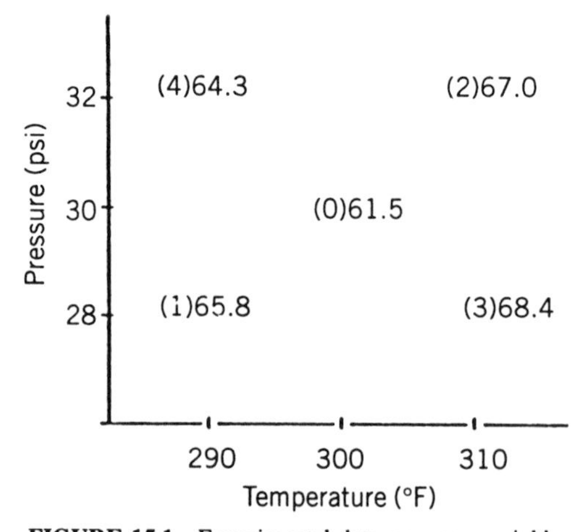

FIGURE 15.1 Experimental data on process yield.

probability paper, or use an approach suggested by Box and Meyer (1986). We need not do one of these in an EVOP program, however, as new data should be as easy to obtain as the initial data since there is no interference with the operation of the plant process.

Consequently, a new cycle should be run to obtain another set of five values. Of course, *if* a prior estimate of the error standard deviation was available from plant records, that estimate could be used in assessing the possible significance of the two main effects and the single interaction. Remembering how to estimate such effects from Chapter 13, it should be clear that the estimate of the temperature effect is 2.65 $\left[= \frac{1}{2}(68.4 - 65.8 + 67.0 - 64.3)\right]$, -1.45 is the estimate of the pressure effect, and the estimate of the temperature × pressure interaction is 0.05 $\left[= \frac{1}{2}(67.0 + 65.8 - 64.3 - 68.4)\right]$.

Then, if we had a prior estimate of the residual standard deviation, we could construct 2 standard error limits (i.e., approximately 95% confidence limits) for each effect, where *standard error* is the standard deviation of the estimator of the effect. Assume that we have such an estimate of the residual standard deviation (σ), and the estimate is 0.5. It can be shown that for a 2^2 design with one observation per design point, the standard deviation of the estimators of the two main effects and interaction effect is σ for each of the three effects.

For the main effects and interaction effect the limits are thus 2.65 ± 1.0 for T, -1.45 ± 1.0 for P, and 0.05 ± 1.0 for TP. Since the first two intervals do not cover zero, we would conclude there is likely a temperature effect and a pressure effect, assuming that the prior estimate of σ is reasonable.

It should be noted that 2 standard error limits in EVOP will not, in general, simplify to 2σ. The expression will be different, for example, when more than one cycle has been completed. This is discussed further later in the chapter.

The use of the center point (point zero) to represent the current operating values allows us to estimate another effect—the "change in mean" effect. Specifically, has the process suffered as a result of altering the current operating conditions or has there been an improvement? Using response surface terminology, Figure 15.1 provides evidence of a "valley" in the general region of the current operating values, with the yield increasing as movement is made in any one of the four directions. To determine whether or not this increase is statistically significant, we need to estimate the standard deviation of $\frac{1}{5}[(0) + (1) + (2) + (3) + (4)] - (0)$, which we will denote by the square root of

$$\begin{aligned}
\mathrm{Var}\left(\frac{y_0 + y_1 + y_2 + y_3 + y_4}{5} - y_0\right) &= \mathrm{Var}\left(\frac{y_1 + y_2 + y_3 + y_4 - 4y_0}{5}\right) \\
&= \mathrm{Var}\left[\frac{1}{5}(y_1 + y_2 + y_3 + y_4)\right] + \mathrm{Var}\left[\frac{4}{5}y_0\right] \\
&= \frac{1}{25}(4\sigma^2) + \frac{16}{25}\sigma^2 \\
&= \frac{20}{25}\sigma^2
\end{aligned}$$

The end result follows from the fact that the variance of the sum and/or difference of independent random variables is equal to the sum of the variances, and the variance of a constant times a random variable is equal to the constant squared times the variance of the random variable.

Our estimate of the standard deviation of the change-in-mean effect would then be $\sqrt{(20/25)\sigma^2} = 0.894\sigma = 0.447$ using our prior estimate of 0.50 for σ. The estimate of the change-in-mean effect is obviously $\frac{1}{5}(y_0 + y_1 + y_2 + y_3 + y_4) - y_0 = 65.4 - 61.5 = 3.9$, so the confidence limits are 3.9 ± 0.894, which clearly do not contain zero. Therefore, the change in mean is significant, which tells us that the current operating values should be changed.

If a prior estimate of σ is unobtainable, the standard approach is to repeat the same combination of factor levels and thus obtain a second set of five values. Assume that this new set of five values is as follows: (0) 62.3, (1) 66.0, (2) 66.8, (3) 68.2, and (4) 64.8.

At this point we could estimate σ by using one of the methods suggested in Chapter 13 (e.g., pooling the s^2 values at each design point or using Yates's algorithm). We should remember, however, that EVOP is meant to be used by plant personnel who generally should not be expected to produce variances or to use Yates's algorithm without appropriate computer software. Therefore, σ is estimated (in the literature on EVOP) the same way as it has historically been estimated by an $\overline{X}$ chart—by using the range.

Using the range method, the first step is to obtain the set of differences between the values in the second *cycle* (as it is called) and the corresponding values in the first cycle. The range of these differences is then divided by the appropriate value of d_2 (from Table E in the Appendix to the book with $n = 5$ since there are five design points) and then multiplied by $[(k-1)/k]^{1/2}$ where k is the number of cycles that has been carried out. Thus, for $k = 2$, our estimate of σ would be $(R/d_2)(\sqrt{0.5})$, and, in general, the estimate would be $(R/d_2)[(k-1)/k]^{1/2}$. [The derivation of the constant $(k-1)/k$ is given in the Appendix to this chapter, and can also be found in Appendix 1 of Box and Draper (1969).]

The necessary calculations for this example are given in Table 15.1. The estimate of sigma is thus 0.304. The effect estimates are obtained at the end of the second cycle analogous to the way they are obtained for the first cycle. When the calculations are performed, the effect estimates are as given in Table 15.2 along with the 2 standard error limits. It should be noted that the standard errors differ somewhat from those that were used at the end of the first cycle. This is due to the fact that the latter were based upon a prior estimate of sigma, but it is also true that the standard errors will generally differ from one cycle to the next, even when the estimate of sigma does not change. This is because the standard errors are a function of the number of cycles, k. Specifically, for a 2^2 design with one additional point (such as a center point), the standard error for the main effects and interaction effect is $\hat{\sigma}/\sqrt{k}$ where $\hat{\sigma}$ is the estimate of σ obtained from $(R/d_2)[(k-1)/k]^{1/2}$, and the standard error for the change in mean is $0.894\hat{\sigma}/\sqrt{k}$. The numerical values of these two different standard errors are then

TABLE 15.1 Estimating Sigma after the Second Cycle in an EVOP Program

	Operating Conditions				
	(0)	(1)	(2)	(3)	(4)
Response values					
Cycle I	61.5	65.8	67.0	68.4	64.3
Cycle II	62.3	66.0	66.8	68.2	64.8
Differences					
Cycle II minus cycle I	0.8	0.2	−0.2	−0.2	0.5

Range of differences $= R = 1.0$

Estimate of sigma $= (R/d_2)(\sqrt{0.5})$

$= (1.0/2.326)(\sqrt{0.5}) = 0.304$

$(0.304)/\sqrt{2} = 0.215$ and $0.894(0.304)/\sqrt{2} = 0.192$. If we used the prior estimate ($\sigma = 0.5$), the two standard errors would be 0.354 and 0.316, respectively.

Box and Hunter (1959, p. 85) and Box and Draper (1969, p. 108) suggest that the prior estimate of sigma (assuming that one is available) be used in calculating the standard errors for the first two cycles, with the data being first used to estimate σ upon completion of the third cycle. This is based upon the tacit assumption that a reliable estimate of σ will be obtainable from plant records, and the recognition that such an estimate should be more reliable than an estimate obtained from the small amount of data that will be available after only two cycles have been completed. Of course, if no prior estimate is available, or if such an estimate is not based upon factual information, it would be preferable to use the estimate obtained from the first two cycles.

The 2 standard error limits are given in Table 15.2 using both the prior estimate of sigma and the estimate obtained from the first two cycles.

The question naturally arises as to how many cycles should be performed. Critics of the Box–EVOP approach argue that this is one of the weaknesses of the method in that there is no obvious stopping point. Box and Draper (1969, p. 212) do provide a table that can serve as a guideline, however. Their table gives the number of cycles required to detect main effects, with a given probability, which increases the process standard deviation from σ to $k\sigma$. For example, if a 30% increase in σ could be tolerated without causing any serious problems, either four or five cycles would be appropriate when a 2^2 design is used (as in the present example). If only a 20% increase could be tolerated, then seven or eight cycles would be appropriate. Assume that two additional cycles are carried out (for a total of four) and the results are summarized in Tables 15.3 and 15.4.

It is worth noting that four cycles are sufficient to allow a sample variance, s^2, to be computed at each design point. These s^2 values could then be pooled to provide s^2_{pooled} (as discussed in Chapter 13), which would be used in estimating σ. This would produce a better estimate of σ than would be obtained using the

TABLE 15.2 Process Averages and Effect Estimates with Confidence Limits after Two Cycles

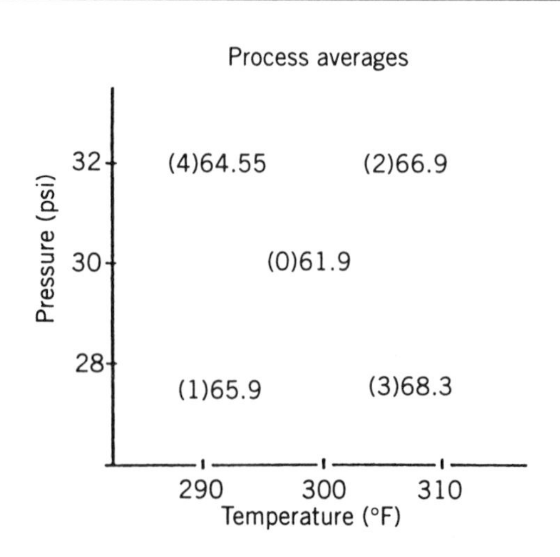

Effects with 2 standard error limits

	A. Using $\sigma = 0.50$ (prior estimate)	B. Using $\hat{\sigma} = 0.304$
Pressure	-1.375 ± 0.707	-1.375 ± 0.43
Temperature	2.375 ± 0.707	2.375 ± 0.43
Temperature × pressure	-0.025 ± 0.707	-0.025 ± 0.43
Change in mean	3.61 ± 0.632	3.61 ± 0.384

range of the differences (for reasons similar to those given in Chapter 5). For the data in Table 15.3 it could be shown that $\hat{\sigma} = \sqrt{s^2_{\text{pooled}}} = 0.244$, where s^2_{pooled} is the average of the s^2 values for the five design points. This estimate does not differ greatly from the estimate of 0.228 obtained using the range method, however, and would lead to the same conclusions regarding the significance of the four effects. (It should also be noted that the constant c_4, used in Chapter 5, would not be used here. For control charts s is divided by c_4 in estimating σ, but s is used in place of σ in other statistical procedures.)

The results shown in Table 15.4 suggest that there is a pressure effect and a temperature effect as well as a "change in mean" effect, since each of the three intervals does not cover zero. Furthermore, the fact that the temperature × pressure interval does cover zero makes the interpretation straightforward.

What if the interval for the interaction effect had *not* included zero? Recall the discussion in Section 13.7.3 regarding problems that can be caused by large interaction effects. That discussion also applies here. In particular, in this section we are considering essentially the same design as was illustrated in that section.

TABLE 15.3 Calculations for Four Cycles

Response Values	Operating Conditions (0)	(1)	(2)	(3)	(4)
Cycle I	61.5	65.8	67.0	68.4	64.3
Average (one cycle)	61.5	65.8	67.0	68.4	64.3
Cycle II	62.3	66.0	66.8	68.2	64.8
Cycle II minus average	0.8	0.2	−0.2	−0.2	0.5
Range of differences = 1.0 $(1.0/2.326)(\sqrt{0.5}) = 0.304 = \hat{\sigma}$					
Average (two cycles)	61.9	65.9	66.9	68.3	64.55
Cycle III	62.5	66.2	67.0	68.2	64.6
Cycle III minus average	0.6	0.3	0.1	−0.1	0.05
Range of differences = 0.7 $(0.7/2.326)(\sqrt{2/3}) = 0.246 = \hat{\sigma}$					
Average (three cycles)	62.1	66.0	66.93	68.27	64.57
Cycle IV	62.2	65.9	66.8	68.2	64.8
Cycle IV minus average	0.1	−0.1	−0.13	−0.07	0.23
Range of differences = 0.36 $(0.36/2.326)(\sqrt{3/4}) = 0.134 = \hat{\sigma}$					
Average (four cycles)	62.125	65.975	66.9	68.25	64.625
Combined estimate of sigma $= \dfrac{0.304 + 0.246 + 0.134}{3} = 0.228$					

Assume that at the end of the first phase when a 2^2 design has been used, the plotting of the averages is as in Figure 13.10, with the average at the current operating conditions corresponding to the point at which the lines intersect. In this instance the "change in mean" effect is zero, so this would tell the user that no changes should be made. Similarly, the effect estimates for the two process variables are also zero, which would *seem* to support the conclusion that no changes be made. Clearly a change *should* be made, however, as the average is 33% higher at two of the four experimental conditions relative to the current conditions.

Thus, we have an additional potential problem in EVOP: the change-in-mean effect could also be contaminated by a large interaction.

If we are unable to "transform away" the interaction, we should look at the averages that comprise the effect estimates and ignore the effect estimates. Thus, for the configuration in Figure 13.11, the best combinations are (low temperature, high pressure) and (high temperature, low pressure). If we were seeking to optimize the response, we could choose between these two but remember that EVOP is not a true optimization procedure. We are trying to move in a "better direction," but Figure 13.11 does not tell us how to change each variable. It does, however, give us some important information about *combinations* of levels, and this information might be used advantageously in the next phase. One approach would be to move away from the least desirable combination(s), similar to what is done in Simplex EVOP, which is discussed in Section 15.3. If we take the

TABLE 15.4 Process Averages and Effect Estimates with Confidence Limits after Four Cycles ($k = 4$)

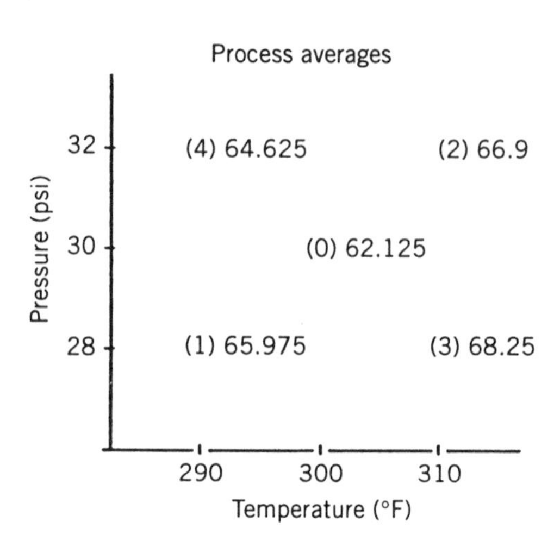

Effects with 2 standard error limits

Pressure	$\frac{1}{2}[(4)+(2)-(1)-(3)] \pm 2\frac{\hat{\sigma}}{\sqrt{k}} = -1.35 \pm 0.228$
Temperature	$\frac{1}{2}[(2)+(3)-(1)-(4)] \pm 2\frac{\hat{\sigma}}{\sqrt{k}} = 2.275 \pm 0.228$
Temperature × pressure	$\frac{1}{2}[(1)+(2)-(3)-(4)] \pm 2\frac{\hat{\sigma}}{\sqrt{k}} = 0 \pm 0.228$
Change in mean	$\frac{1}{5}[(0)+(1)+(2)+(3)+(4)] - (0) \pm 1.79\frac{\hat{\sigma}}{\sqrt{k}} =$ 3.45 ± 0.204

data that are represented by Figure 13.11 and graph the data in the form of Figure 15.1, we obtain Figure 15.2.

What factor levels should be used in the next phase? With only five points we do not have a very good idea of the response surface. We might make either point (3) or (4) the center point in the next phase, being mindful of how far we can stray from the current operating conditions without running the risk of significantly increasing the percentage of nonconforming units. If, at the end of the second phase, the interaction is not significant and neither are any of the effect estimates, we might construct the points about the point, (3) or (4), that was not the center for the second phase. Then the results for these two phases could be compared. If significant interactions were encountered in either phase, then, of course, the averages for each of the four new points would have to be considered. We should remember, though, that maintaining a rectangular experimental region is desirable for ease of computation by plant workers.

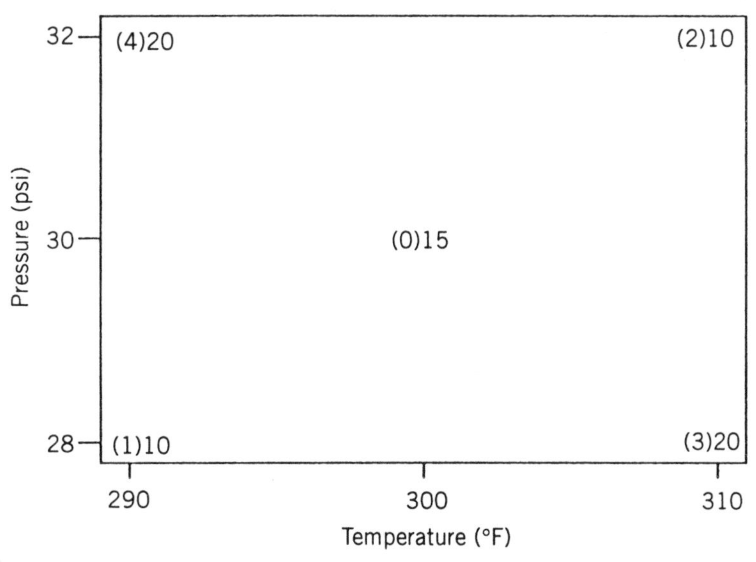

FIGURE 15.2 Data from Figure 13.11.

Returning to Table 15.4, the results suggest that temperature should be increased and pressure should be lowered. Following this suggestion would produce a new set of operating conditions for the next *phase*. Thus, we might use pressure set at 26, 28, and 30, and temperature set at 300, 310, and 320. Assume that four cycles are carried out in this phase, with the results given in Table 15.5, with sigma estimated at 0.225 from these four cycles. (*Note:* Whenever a new phase is entered, the estimate of sigma from the previous phase would generally be used for the first two cycles of the new phase, with the data from that phase being used to produce an estimate of sigma starting with the third cycle.) It is obvious from Table 15.5 that increasing the temperature to 320 and decreasing the pressure to 26 has a deleterious effect on the process yield. A three-dimensional display of these data would indicate a "peak" at operating condition (0), with the process yield dropping off rather sharply as changes are made in any of the indicated directions.

Thus, it would appear as though the optimum operating condition is in the general vicinity of 310°F and 28 psi for these two process variables. We should not stop here, however, because the "peak" might actually occur at a slightly different combination of values. Therefore, in the next phase the pressure might be set at 27.5, 28, and 28.5, and the temperature might be set at 308, 310, and 312.

As indicated previously, the use of new operating conditions should be almost a nonending process, with an EVOP program being continued as long as the cost of running the program is less than the savings being realized. [See Box and Draper (1969, p. 21) for a detailed explanation of cost vs. savings for an EVOP program.]

TABLE 15.5 Process Averages and Effect Estimates with Confidence Limits after Four Cycles of the Second Phase

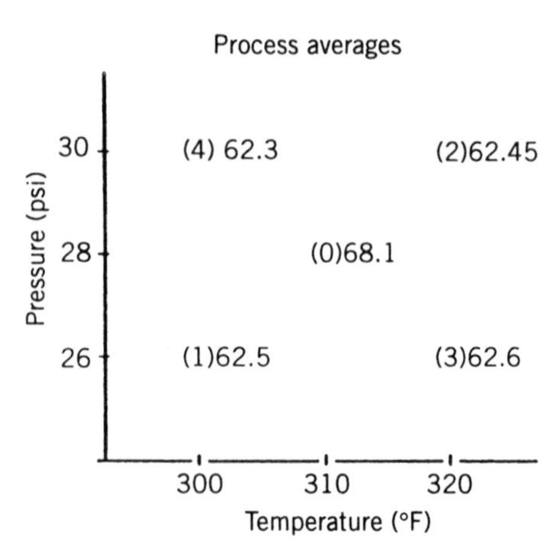

Effects with 2 standard error limits

Pressure	$\frac{1}{2}[(4)+(2)-(1)-(3)] \pm 2\frac{\hat{\sigma}}{\sqrt{k}} = -0.175 \pm 0.225$
Temperature	$\frac{1}{2}[(2)+(3)-(1)-(4)] \pm 2\frac{\hat{\sigma}}{\sqrt{k}} = 0.125 \pm 0.225$
Temperature × pressure	$\frac{1}{2}[(1)+(2)-(3)-(4)] \pm 2\frac{\hat{\sigma}}{\sqrt{k}} = 0.025 \pm 0.225$
Change in mean	$\frac{1}{5}[(0)+(1)+(2)+(3)+(4)] - (0) \pm 1.79\frac{\hat{\sigma}}{\sqrt{k}} = -4.51 \pm 0.201$

15.2 THREE VARIABLES

Box and Draper (1969, p. 99) indicate that no more than three variables could be considered practical for an EVOP program. Why? Remembering that an EVOP program is to be a permanent part of the normal operating procedure, it would be too much to expect plant operators to make changes in a half dozen or so process variables on a continuing basis. [See Box and Draper (1969, p. 176) for a similar view.] The necessary changes could be easily determined with appropriate software, but the operators would still have to make the changes. One thing to keep in mind is that by keeping EVOP programs as simple as possible, personnel who might be needed for assistance if an EVOP program were complicated can be used to provide assistance and expertise in ongoing off-line experimentation. In this way a company's human resources can be efficiently utilized.

A logical strategy would be to first use a fractional factorial design as a screening design to identify the process variables and two-factor interactions that seem to be important. If this initial study indicates that there are more than two or three process variables that seem to be important, and that some of the two-factor interactions are significant, two or more EVOP programs could be set up, with process variables involved in significant two-factor interactions being assigned to the same EVOP program.

Assume that the use of a screening design has led to the identification of six process variables that seem to be significantly related to the response variable, and that the few pairwise interactions that were significant have led to the assignment of the involved variables to the same EVOP program. How do we use three variables in an EVOP program? We could run all three variables simultaneously, with or without a reference condition. If a reference condition is used, a logical place for it would be the center of the cube, as shown in Figure 15.3.

Other possible approaches to the handling of three variables in an EVOP program include (1) using only two of the three variables in each phase, and (2) running the different operating conditions in two blocks.

The former is illustrated in Hunter and Chacko (1971), which describes how a company that manufactures a polymer latex used an EVOP program for the initial objective of minimizing the optical density of the latex. [This study is also described in Box et al. (1978, p. 365).] The three process variables were temperature, stirring rate, and addition time. A total of five phases were used. Stirring rate and addition time were studied in the first two phases, with temperature and stirring rate studied in the fourth phase, and temperature and addition time studied in the other two phases. (The decision to use only two variables in each phase was made for the sake of simplicity.) The analysis of the data resulting from an EVOP program conducted in this manner would be the same as that illustrated earlier in this chapter (assuming that two levels were used with a center point, which was the case). One of the most significant results of this particular study was that it led to the discovery that addition time could be reduced by 45 minutes,

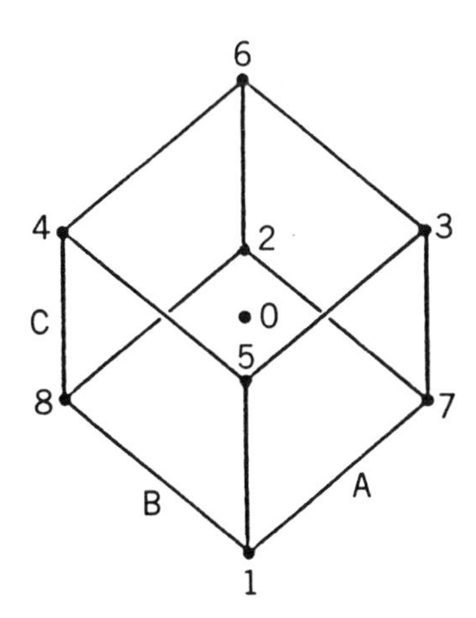

FIGURE 15.3 A 2^3 design with a center point, for use in a three-variable EVOP program.

which subsequently resulted in approximately a 25% increase in production, with beneficial results also being realized from changes in the other two variables.

The second alternative entails running a $\frac{1}{2}$ fraction of a 2^3 design in each block (with or without a reference condition), with the *ABC* interaction confounded with the difference between blocks. This approach is illustrated by Box and Draper (1969), and its advantage (over a regular 2^3 design) is also simplicity in that the plant worker would be working with the same number of observations within each block as he would have with an EVOP program that utilized a 2^2 design.

In each cycle (starting with the third), the standard deviation would be estimated by pooling the estimates obtained from each of the two blocks, and the data from the two blocks would be combined and analyzed using some method of analysis such as Yates's algorithm. [See Box and Draper (1969) for additional details.]

15.3 SIMPLEX EVOP

As mentioned at the beginning of this chapter, several alternatives to Box–EVOP have been proposed, but Simplex EVOP seems to be the only alternative procedure that has actually been used. The procedure is very easy to use, but it draws its name from the fact that it is based upon a geometrical figure termed a (regular) simplex. In two dimensions this is an equilateral triangle, as illustrated in Figure 15.4, which would be used for studying two variables. In general, the number of design points is always one more than the number of variables.

Simplex EVOP was originally proposed by Spendley, Hext, and Himsworth (1962), and it is discussed favorably by Lowe (1974), in particular, and Hahn (1976b). We shall illustrate the methodology using the same data that were utilized in introducing Box–EVOP, namely the data in Figure 15.1.

We shall now assume, however, that the current operating conditions are 290° for temperature and 28 psi for pressure, rather than 300° and 30 psi, respectively, which were originally assumed. Also, since we need only two more points, we will select those points that were labeled (0) and (3) in Figure 15.1 to form our simplex. In general, the other two points are chosen by determining what

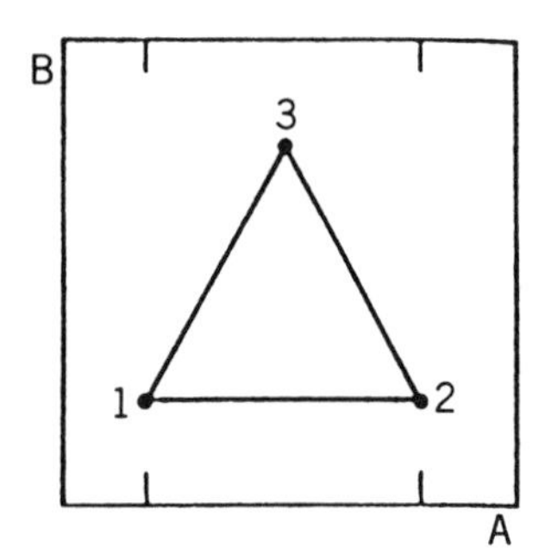

FIGURE 15.4 Simplex for studying two variables in an EVOP program.

changes in the two process variables might lead to improvement in the response variable that is being studied, without simultaneously running a substantial risk of causing a deterioration in the process (i.e., changes that are relatively small, but at the same time large enough to possibly cause some process improvement).

The initial setup is given in Figure 15.5. Point 2 obviously represents the least favorable operating condition. Since we are trying to maximize the response variable (yield), the next step is to create a new simplex by adding a new point that is as far from point 2 as possible. This requires that the new point be at 26 psi for pressure and 300° for temperature. The reader will recall that this combination was used in the second phase of the Box–EVOP illustration and that the average yield after four cycles was 62.5. With Simplex EVOP, however, there are no cycles; rather, only a single run is made at each operating condition. Assume that this run produces a value of 62.3. Now, if we create a new simplex by "reflecting" this second simplex about its least desirable operating condition, we would return to the original simplex. Therefore, we must introduce a new rule [termed rule 3 by Spendley et al. (1962)], which states that the second least favorable point is used in constructing the next simplex whenever the basic rule of using the worst point would return us to the same simplex that we just left.

Thus, the third simplex would be created by reflecting the second simplex about point 1. This new simplex and subsequent simplexes that would result from assumed observations are shown in Figure 15.6. The values displayed for points 4, 5, and 7 are in general agreement with the averages at those operating conditions given in Table 15.5, whereas point 6 was not part of the Box–EVOP illustration.

What can be said for Simplex EVOP versus Box–EVOP? First, the former is clearly much easier than the latter. With the simplex approach there are no statistical calculations to perform, and new operating conditions are obtained by following simple, well-defined rules. If the optimum value of the response variable drifts over time, the simplex approach could be expected to track such a

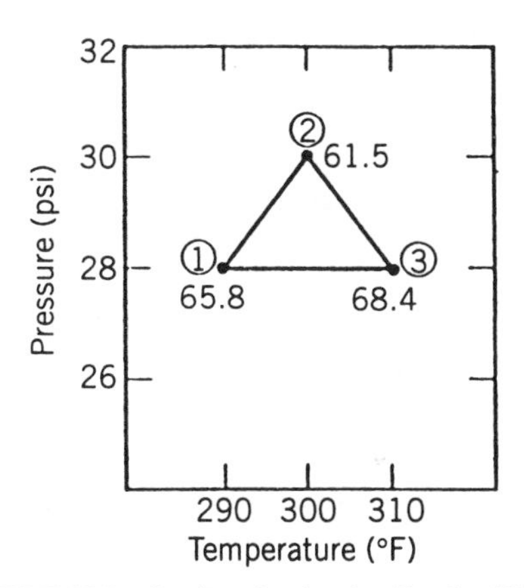

FIGURE 15.5 Starting simplex for Simplex EVOP.

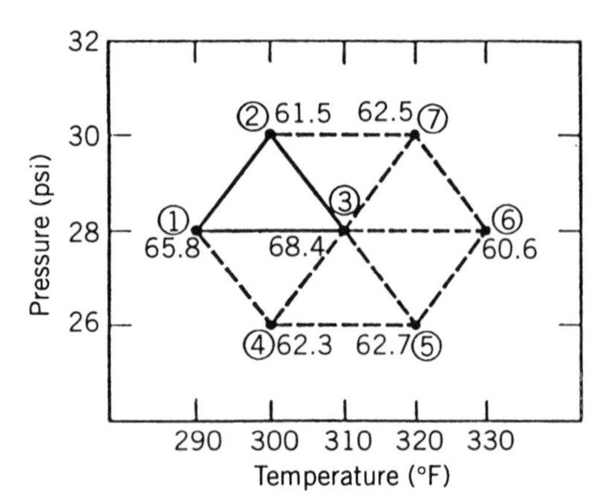

FIGURE 15.6 Starting simplex and subsequent simplexes.

drift much faster than the Box–EVOP approach. The simplex approach can also be applied when there are more than just a few process variables to be varied, whereas Box–EVOP should generally not be used with more than three process variables.

Spendley et al. (1962) presented Simplex EVOP as an approach that would be appropriate for "Automated EVOP"; that is, movement to new operating conditions would be performed automatically by computer. The desirability of using some automated or inflexible procedure is questionable, however. As G. E. P. Box has repeatedly emphasized, an experimenter learns about a process by perturbing it and seeing what happens, and, in general, communicating with the process. With Simplex EVOP, in its general form, there is really no explicit provision for flexibility, and the resultant "communication" is thus likely to be suboptimal.

Clearly, as the optimum value of the response variable is approached, smaller simplexes would need to be created [as Hahn (1976b) suggests]. Otherwise, we would be using simplexes that would just keep circling the optimum operating condition and producing values of the response variable that could differ considerably from the optimum value. This would put Simplex EVOP more in line with Box–EVOP, where changes in the process variables are generally not constant amounts, and the changes result from decisions coming from the EVOP committee, not from a computer.

A wholly automated, inflexible procedure is undesirable. For processes that drift over time and then temporarily stabilize, the use of Simplex EVOP followed by Box–EVOP would probably be the best approach.

Why not just modify the general simplex approach to allow for the construction of smaller simplexes rather than resorting to Box–EVOP? The answer is that it is generally easier to analyze data from a rectangular region than from a triangular region. Although the analysis of data from a simplex design is generally known, one could not expect the analysis to be performed by plant personnel, and the (statistical) analysis would be necessary in determining the sizes of the simplexes.

One of the main problems with Simplex EVOP is that changes in the operating conditions are based, at each stage, on a single number; specifically the number at the operating condition that is reflected to form the next simplex. This could easily result in a move in a bad direction, especially for a process that has considerable variability. Proponents of Simplex EVOP contend that a move into a bad area should be immediately followed by a move out of that area, but the fact remains that a sizable number of bad moves could result over a period of time. (Here a "bad area" is one that is somewhat removed from the optimum operating region, and where nonconforming units might be produced.)

Everything considered, Box–EVOP seems to be preferable to Simplex EVOP when a process is relatively stable. Readers seeking more details about Simplex EVOP are referred to the paper by Spendley et al. (1962), which contains information concerning how the coordinates of each process variable in each simplex would be obtained, for any number of variables, as well as additional information. Papers that describe industrial applications of Simplex EVOP include Kenworthy (1967). Lowe (1974) provides a comparison of Simplex EVOP versus Box–EVOP.

15.4 OTHER EVOP PROCEDURES

Other modifications of Box–EVOP that have been proposed include REVOP (random evolutionary operation) and ROVOP (rotating square evolutionary operation). Operating conditions in an EVOP program are generated randomly under REVOP, whereas under ROVOP successively larger (square) operating regions are generated that include the previous square, with the orientation of each square alternating between a regular square and a square oriented as a diamond. The reader is referred to Lowe (1964) for additional details of these two procedures, as well as to Box and Draper (1969).

Recommended reading on EVOP includes Box (1957) and the review paper by Hunter and Kittrell (1966). The latter contains some discussion regarding companies that have used EVOP and the amount of money that some companies have saved. The paper contains a considerable number of references [as does Box and Draper (1969)].

Other papers on EVOP include Draper and Box (1970) and Hahn (1976a), both of which are very general. The fundamental concepts of EVOP are given in Barnett (1974), and Himmelblau (1970) is a text that emphasizes the mathematics of some of the proposed variations of Box–EVOP.

15.5 MISCELLANEOUS USES OF EVOP

The EVOP method may be used for purposes other than trying to determine the optimum levels of process variables. For example, Box and Luceño (1995) discuss the possible use of EVOP for improving on the initial values of the two constants used in a discrete proportional-integral control scheme.

15.6 SUMMARY

Evolutionary operation (EVOP) is a program based on statistical concepts that is meant to be a permanent part of normal plant operation. It is intended for use by regular plant personnel, and the necessary calculations can be easily performed by hand computation or by computer. Box–EVOP has been used much more extensively than Simplex EVOP, although there are conditions under which Simplex EVOP might be used in tandem with Box–EVOP (e.g., process drift).

As indicated in the summary of Chapter 13, one learns about experimental design by designing experiments. Similar in spirit to Hunter (1977), Russell and Stephens (1970) present an EVOP teaching game that utilizes a simulated process. (It is obviously preferable to learn statistical principles of experimentation using something other than actual production processes that are critical to the quality of manufactured products.)

APPENDIX

Derivation of Formula for Estimating σ

The process standard deviation is estimated using differences between values at the same design point and then determining the range of those differences. The first time that the differences can be computed is after the second cycle. If we let y_{1i} and y_{2i} denote the value observed at the ith design point for the first and second cycle, respectively, then

$$\text{Var}(y_{2i} - y_{1i}) = 2\sigma^2$$

since the values are assumed to be independent between cycles. If we let $\text{Var}(y_{2i} - y_{1i})$ be denoted by σ^2_{diff}, it follows that we would estimate σ after the second cycle as

$$\hat{\sigma} = \hat{\sigma}_{\text{diff}}\left(\sqrt{\tfrac{1}{2}}\right)$$

where $\sqrt{\frac{1}{2}}$ follows from the form $\sqrt{(k-1)/k}$ for $k = 2$ cycles.

In general, after k cycles the differences would be obtained as

$$y_{ki} - \overline{y}_{(k-1)i}$$

where $\overline{y}_{(k-1)i}$ denotes the average at the ith design point for the first $k-1$ cycles. It follows that

$$\begin{aligned}\text{Var}(y_{ki} - \overline{y}_{(k-1)i}) &= \sigma^2_{\text{diff}} \\ &= \sigma^2 + \frac{\sigma^2}{k-1} = \frac{k\sigma^2}{k-1} = \frac{k}{k-1}\sigma^2\end{aligned}$$

Therefore, $\sigma^2_{\text{diff}} = [k/(k-1)]\sigma^2$ so that

$$\hat{\sigma} = \sqrt{\frac{k-1}{k}}\hat{\sigma}_{\text{diff}}$$

The standard deviation of the differences, σ_{diff}, is then estimated using the range of the differences.

REFERENCES

Barnett, E. H. (1974). Evolutionary operation. In J. M. Juran, F. M. Gryna, and R. S. Bingham, eds. *Quality Control Handbook*, 3rd ed., Section 27A. New York: McGraw-Hill.

Box, G. E. P. (1957). Evolutionary operation: A method for increasing industrial productivity. *Applied Statistics 6*(2): 81–101 .

Box, G. E. P. and N. R. Draper (1969). *Evolutionary Operation*. New York: Wiley.

Box, G. E. P. and J. S. Hunter (1959). Condensed calculations for evolutionary operation programs. *Technometrics 1*(1): 77–95.

Box, G. E. P. and A. Luceño (1995). Discrete proportional-integral control with constrained adjustment. *The Statistician 44*(4): 479–495.

Box, G. E. P. and R. D. Meyer (1986). An analysis for unreplicated fractional factorials. *Technometrics 28*(1): 11–18.

Box, G. E. P., W. G. Hunter, and J. S. Hunter (1978). *Statistics for Experimenters*. New York: Wiley.

Draper, N. R. and G. E. P. Box (1970). EVOP—makes a plant grow better. *Industrial Engineering 2:* 31–33.

Hahn, G. J. (1976a). Process improvement using evolutionary operation. *Chemtech 6:* 204–206.

Hahn, G. J. (1976b). Process improvement through Simplex EVOP. *Chemtech 6:* 343–345.

Hahn, G. J. (1984). Discussion (of an invited paper by Steinberg and Hunter). *Technometrics 26*(2): 110–115.

Hahn, G. J. and A. F. Dershowitz (1974). Evolutionary operation today–some survey results and observations. *Applied Statistics 23*(2): 214–218.

Himmelblau, D. M. (1970). *Process Analysis by Statistical Methods*. New York: Wiley.

Hoerl, R. W. (1998). Six Sigma and the future of the quality profession. *Quality Progress 31*(6): 35–42.

Hunter, W. G. (1977). Some ideas about teaching design of experiments with 2^5 examples of experiments conducted by students. *The American Statistician 31*(1): 12–17.

Hunter, W. G. and E. Chacko (1971). Increasing industrial productivity in developing countries. *International Development Review 13:* 311–316.

Hunter, W. G. and J. R. Kittrell (1966). Evolutionary operation: A review. *Technometrics 8*(3): 389–397.

Joiner, B. L. (1985). The key role of statisticians in the transformation of North American industry. *The American Statistician 39*(3): 224–227; (discussion: pp. 228–234).

Kenworthy, I. C. (1967). Some examples of simplex evolutionary operation in the paper industry. *Applied Statistics 16*(3): 211–224.

Lowe, C. W. (1964). Some techniques of evolutionary operation. *Transactions of the Institution of Chemical Engineers 42:* T332–344.

Lowe, C. W. (1974). Evolutionary operation in action. *Applied Statistics 23*(2): 218–226.

Russell, E. R. and K. S. Stephens (1970). An EVOP teaching game using a simulated process. *Journal of Quality Technology 2*(2): 61–66.

Spendley, W., G. R. Hext, and F. R. Himsworth (1962). Sequential applications of simplex designs in optimization and EVOP. *Technometrics 4*(4): 441–461.

EXERCISES

1. Explain the conditions under which an experimenter might decide to use (1) Box–EVOP instead of Simplex EVOP, (2) the latter instead of the former, and (3) a combination of the two.
2. Explain why an estimate of sigma is unnecessary when Simplex EVOP is used.
3. Assume that a Box–EVOP program is being initiated using two factors. How should sigma be estimated for the first couple of cycles of the first phase?
4. If an experimenter is ready to progress from the first phase to the second phase of a Box–EVOP program, how should he adjust the levels of his two factors if the 2 standard error limits are both below zero for his first factor, and also below zero for the second factor. (Assume that the limits for the interaction include zero.)
5. Referring to exercise 4, what would you recommend to the experimenter if the limits for the interaction effect did not include zero?
6. Consider the data given below, which represent averages after five cycles in the second phase of a Box–EVOP program.

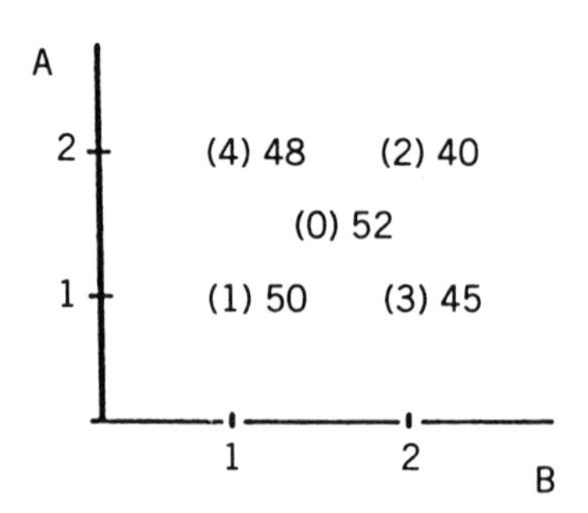

If $\hat{\sigma} = 5.14$, what are the two standard error limits for A, B, and AB?

7. An experimenter runs a Box–EVOP program using a 2^2 design with a center point. The average response values after two cycles are (0) 32.6, (1) 36.4, (2)

35.2, (3) 38.0, and (4) 34.2. Thinking of the two factors as A and B for the vertical and horizontal axes, respectively, and using 1.0 as a prior estimate of sigma,

(a) Determine what effects, if any, appear to be significant.

(b) What do these results suggest about how the values for A and B should be altered in the next phase?

8. Which of the two procedures, Box–EVOP or Simplex EVOP, would you choose if you believed that the optimum operating region could possibly be far removed from the initial conditions?

9. Using a 2^2 design with a center point, what is the estimate of sigma if the range of differences after the second, third, and fourth cycles of the first phase are 0.82, 0.64, and 0.29, respectively?

10. Critique the following statement: "We are interested in attaining optimum operating conditions, but we believe this can be better accomplished using Taguchi's methods than by using EVOP."

11. Consider an EVOP program for two variables so that the design is a 2^2 with a center point. Perform the following exercise. For the points in Figure 15.1 labeled (0), (1), . . ., (4), generate data from the following normal distributions: $N(10,1)$, $N(13,1)$, $N(14,1)$, $N(12,1)$, and $N(11,1)$, respectively. Do this four times so that there are four observations at each point. Use the fact that $\sigma = 1$ to construct the two standard error limits. What do your results suggest for the second phase? Since the data are generated from known distributions, we know what should happen theoretically. Determine the corresponding theoretical results and compare them with the empirical results.

CHAPTER 16

Analysis of Means

To some people the term *analysis of means* conjures up notions of analysis of variance (ANOVA), which is also concerned with the analysis of means, and which is much better known and more widely used than the statistical technique known as *analysis of means* (ANOM). (The relative obscurity of ANOM is indicated by the fact that it is not included among the approximately 2700 articles on statistical methods and other statistical topics included in the *Encyclopedia of Statistical Sciences*.) It is apt to have more appeal to engineers and other industrial personnel than does ANOVA, however, since ANOM is inherently a graphical procedure and is somewhat similar to a control chart.

ANOM was developed by E. R. Ott and presented in Ott (1958). It was introduced into the literature in Ott (1967), and Ott's text (1975) contains many illustrative examples. The original concept has been extended by Schilling (1973a,b) and L. S. Nelson (1983) provides values of the necessary mathematical constants.

Ott's 1967 paper appeared in the Walter Shewhart Memorial issue of *Industrial Quality Control*, and it is fitting that the January 1983 issue of the *Journal of Quality Technology* (which was previously named *Industrial Quality Control*) contained articles only on ANOM, with the purpose of serving as a tribute to Ellis Ott upon his passing.

Our first illustration of ANOM will be made using the temperature data in Table 13.3 that were analyzed in Chapter 13 using ANOVA. When the latter was used as the method of analysis, it was concluded that there was at least one mean that was different from the other means. It was mentioned that a multiple comparison procedure could be used to identify means that differ, although the box-and-whisker plot provided evidence that the mean at the highest temperature level was different from the means at the other two levels.

The reader will recall that with ANOVA the experimenter concludes either that all of the means are equal or that at least one of the means differs from the others. With ANOM, however, the user will see whether or not one or more means differs from the average of all of the means. Thus, what is being tested is different for the two procedures, so the results will not necessarily agree. In particular, when $k-1$ sample averages are bunched tightly together but the kth sample average (i.e., the other one) differs considerably from the $k - 1$ averages, the F value in ANOVA would likely be relatively small (thus indicating that the population means are equal), whereas the difference would probably be detected using ANOM. Conversely, if the differences between adjacent sample averages are both sizable and similar, the (likely) difference in the population means is more apt to be detected with ANOVA than with ANOM.

One procedure need not be used to the exclusion of the other, however. As Ott (1967) indicates, ANOM can be used either alone or as a supplement to ANOVA.

16.1 ANOM FOR ONE-WAY CLASSIFICATIONS

It was stated previously that with ANOM one compares $\overline{x}_i$ against the average of the $\overline{x}_i$, which will be denoted by $\overline{\overline{x}}$, analogous to the notation used for an $\overline{X}$ chart. The original ANOM methodology given by Ott (1967) was based upon the multiple significance test for a group of means given by Halperin, Greenhouse, Cornfield, and Zalokar (1955), which was based upon the studentized maximum absolute deviate. That approach provided an upper bound for the unknown critical value, but will not be discussed here since it is no longer used. The interested reader is referred to Schilling (1973a) for more details, including the theoretical development.

The current approach is based upon the exact critical value, h, and is described in L. S. Nelson (1983).

If we were testing for the significance of a single deviation, $\overline{x}_1 - \overline{\overline{x}}$, it would stand to reason that we would want to look at some test statistic of the form

$$\frac{\overline{x}_i - \overline{\overline{x}} - E(\overline{x}_i - \overline{\overline{x}})}{s_{\overline{x}_i - \overline{\overline{x}}}} \tag{16.1}$$

where E stands for expected value. If $\mu_i = (\mu_1 + \mu_2 + \cdots + \mu_k)/k$ then $E(\overline{x}_i - \overline{\overline{x}}) = 0$, and since the former is what would be tested, we take $E(\overline{x}_i - \overline{\overline{x}})$ to be zero. (We should note that some authors have indicated that the null hypothesis, which is tested with ANOM, is H_0: $\mu_1 = \mu_2 = \cdots = \mu_k$. Certainly if each μ_i is equal to the average of all of the means, then it follows that the μ_i must all be equal, since they are equal to the same quantity. But, stating the null hypothesis in this alternative way obscures the testing that is done.)

It can be observed that Eq. (16.1) becomes a t test when $k = 2$ since $\overline{x}_i - \overline{\overline{x}}$ is then $\overline{x}_1 - (\overline{x}_1 + \overline{x}_2)/2 = (\overline{x}_1 - \overline{x}_2)/2$ for $i = 1$ [and $(\overline{x}_2 - \overline{x}_1)/2$ for $i = 2$] so that

$$t = \frac{(\overline{x}_1 - \overline{x}_2)/2 - 0}{s_{(\overline{x}_1 - \overline{x}_2)/2}} = \frac{\overline{x}_1 - \overline{x}_2}{s_{\overline{x}_1 - \overline{x}_2}}$$

since the 2's cancel.

The two deviations $\bar{x}_1 - \bar{\bar{x}}$ and $\bar{x}_2 - \bar{\bar{x}}$ are thus equal; so we would conclude that the two means differ if

$$t = \frac{|\bar{x}_1 - \bar{x}_2|}{s_{\bar{x}_1 - \bar{x}_2}} > t_\alpha$$

for a selected value of α.

When $k > 2$, the t distribution cannot be used, however, so another procedure is needed. It can be shown that, assuming equal sample sizes, the deviations $\bar{x}_1 - \bar{\bar{x}}$ are equally correlated with correlation coefficient $\rho = -1/(k-1)$. If we let $T_i = (\bar{x}_i - \bar{\bar{x}})/s_{\bar{x}_i - \bar{\bar{x}}}$, the joint distribution of $T_1, T_2, \ldots, T_k$ is an equicorrelated multivariate noncentral t distribution, assuming that the sample averages are independent and normally distributed with a common variance (see P. R. Nelson 1982, p. 701).

Exact critical values for $k > 2$ were first generated by P. R. Nelson (1982), with a few tabular values subsequently corrected, and the corrected tables published in L. S. Nelson (1983). More complete and more accurate critical values given to two decimal places were given by P. R. Nelson (1993). These values differ by one in the second decimal place from some of the critical values given in L. S. Nelson (1983)

The general idea is to plot the averages against *decision lines* obtained from

$$\bar{\bar{x}} \pm h_{\alpha,k,\nu}\, s\sqrt{\frac{k-1}{kn}} \tag{16.2}$$

where n is the number of observations from which each average is computed, ν is the degrees of freedom associated with s, the estimate of σ, k is the number of averages, and $h_{\alpha,k,\nu}$ is obtained from the tables in P. R. Nelson (1993) for a selected value of α. It is demonstrated in the Appendix to this chapter that $s\sqrt{(k-1)/(kn)}$ is the estimate of $\sigma_{\bar{x}_i - \bar{\bar{x}}}$.

The value of α is the probability of (wrongly) rejecting the hypothesis that is being tested when, in fact, the hypothesis is true. (Here we are testing that each mean is equal to the average of all the k means, as indicated previously.)

The first step is to compute the overall average, $\bar{\bar{x}}$. For the data in Table 13.3 this value is 2.67. The next step is to compute the estimate of σ using s. This can be obtained from Table 13.4 as $\sqrt{0.0867}$. (Ott originally presented ANOM where σ was estimated from the range, but the current tables require the use of s, which might be obtained from an ANOVA table.)

The appropriate value for $h_{\alpha,k,\nu}$ is obtained from Table G in the Appendix to the book. If we use $\alpha = .05$, $h_{.05,3,27} = 2.485$, approximately, so the decision lines are obtained from

$$2.67 \pm 2.485\sqrt{0.0867}\sqrt{\frac{3-1}{(3)(10)}} = 2.67 \pm 0.19$$

(2.485 is obtained as the average of 2.47 and 2.50 since 27 d.f. is halfway between 24 and 30.) Thus, we have the two "limits" 2.48 and 2.86, which are termed decision lines rather than control limits, as they are used in reaching a decision about the k means instead of for controlling a process. Accordingly, with UDL representing the upper decision line and LDL representing the lower decision line, we have

$$\text{LDL} = 2.48 \qquad \text{UDL} = 2.86$$

The results can then be displayed graphically as in Figure 16.1. We would thus conclude that the true effects of 250° and 350° differ from the average process yield averaged over all three temperatures.

There is nothing sacrosanct about $\alpha = .05$, however. Ott (1967) suggests that .01 might be used in addition to or in place of .05, and in Ott (1975) one finds numerous displays in which both are used, and one display (p. 115) in which three sets of lines (for .01, .05, and .10) are shown.

For this example $h_{.01,3,27}$ is obtained as 3.18 (the average of 3.15 and 3.21), which would produce LDL(.01) = 2.43 and UDL(.01) = 2.91. If these lines had been displayed in Figure 16.1, only the average at 350° would have been outside these decision lines. Therefore, since the choice of α will often lead to different conclusions, the use of multiple decision lines is desirable.

The assumptions that need to be made when ANOM is applied to one-way classification data are the same as those needed for one-way ANOVA; namely, the k averages must be independent and normally distributed with a common variance. (We must make the additional assumption that the single classification factor is fixed, however, since ANOM is not for random effects.) Schilling (1973b) has extended ANOM to apply to nonnormal distributions.

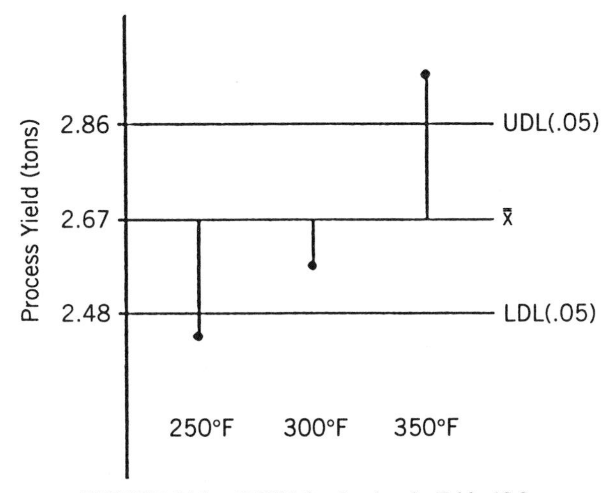

FIGURE 16.1 ANOM for the data in Table 13.3.

In the previous example the sample sizes for each level of the factor (temperature) were equal. This, of course, is not a requirement for ANOVA, nor is it a requirement for ANOM. L. S. Nelson (1983) gives approximate values of $h_{\alpha,k,\nu}$ to be used for unequal sample sizes.

16.2 ANOM FOR ATTRIBUTE DATA

16.2.1 Proportions

ANOM can be applied just as easily to attribute data as to measurement data. The following example is drawn from the author's consulting experience, and although the data are fictitious, it is the type of problem faced by many industrial managers.

A company in the agricultural industry employs a number of (human) harvesters and is interested in assessing the relative performance of each harvester. Specifically, the percentage of crops that do not meet specifications is to be recorded for each harvester, and the company is interested in identifying harvesters who perform significantly better than the overall average (for the purpose of meritorious recognition), as well as to identify those who are significantly below the overall average so that they might be retrained or reassigned.

Assume that the company has 20 harvesters and that the data for a particular month are as given in Table 16.1. The steps followed in producing the ANOM display for attribute data are similar to the steps followed for measurement data. Specifically, for proportions data the decision lines are obtained from

$$\overline{p} \pm h_{\alpha,k,\infty} s_p \sqrt{\frac{k-1}{k}} \tag{16.3}$$

where $\overline{p}$ is the average of all the proportions (20 in this example), $s_p = \sqrt{\overline{p}(1-\overline{p})/n}$ is the estimate of σ_p, and k is the number of proportions.

For this example we have

$$\overline{p} = 2.775\% = 0.02775$$

$$s_p = \sqrt{(0.02775)(0.97225)/1000} = 0.0052, \quad \text{and} \quad k = 20$$

The value of $h_{\alpha,k}$ is obtained from Table G in the Appendix to the book using infinity for the number of degrees of freedom. Thus, we have $h_{.05} = 3.01$ and $h_{.01} = 3.48$. Following Tomlinson and Lavigna (1983) we shall use $h_{.01}$ rather than $h_{.05}$ so as to pick the smaller of the two risks of telling a worker that he is worse than the average of the group when he really is not, and, in general, in recognition of the sensitivity of workers to being identified in this manner, even if they are, in fact, below par.

TABLE 16.1 Percentage of Nonconforming Crops for Each Harvester (March 1999)

Harvester Number	Nonconforming Crops (%) ($n = 1000$)
1	1.4
2	2.6
3	1.0
4	3.1
5	2.9
6	5.1
7	2.4
8	4.1
9	1.1
10	2.1
11	2.0
12	2.6
13	3.1
14	2.7
15	3.7
16	4.0
17	4.4
18	3.2
19	2.2
20	1.8

The decision lines are thus obtained from

$$0.02775 \pm 3.48(0.0052)\sqrt{\frac{19}{20}} = 0.02775 \pm 0.01764$$

so that the decision lines are LDL = 0.0101 and UDL = 0.0454. The display is given in Figure 16.2. It can be observed that 11 of the values are below the midline and 9 are above. Should the 9 that are above the midline be reprimanded? This is reminiscent of the story frequently told by W. Edwards Deming about the company that notified its workers who were below average, apparently not realizing that roughly half of them will always be below average regardless of how well or how poorly they are performing their job tasks.

In ANOM it is a question of how far each point is from the midline. In Figure 16.2 it is apparent that the proportion nonconforming for worker #6 is well above the UDL. Therefore, retraining or reassignment would seem to be indicated. At the other extreme, we can see that worker #3 is virtually at the LDL, and worker #9 is slightly above it. Thus, their respective performances seem to be especially meritorious.

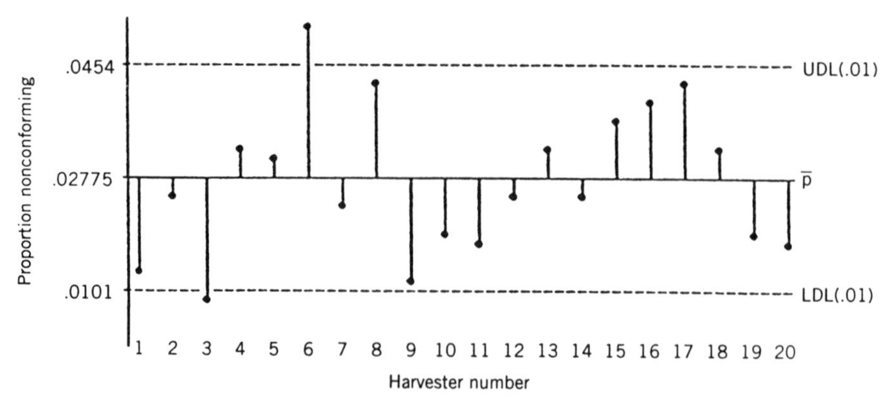

FIGURE 16.2 ANOM display for Table 16.1 data.

Students of statistics will recognize that this use of ANOM for data in the form of proportions serves as an alternative to a chi-square test, in which equality of the 20 proportions would be tested using the chi-square distribution. ANOM is more sensitive than a chi-square test in detecting a few extreme deviations from the average, however (vis-à-vis ANOM vs. ANOVA for measurement data), and the ability to detect extreme observations is what is really needed for this problem.

The requisite assumption for the application of ANOM to proportions data is the adequacy of the normal approximation to the binomial distribution. [Recall from Chapter 6 the rough rule-of-thumb that both np and $n(1-p)$ exceed 5 for the approximation to be of any value.] If the adequacy of the approximation is in doubt, then one of the transformations discussed in Chapter 6 for binomial (proportions) data should be used.

16.2.2 Count Data

ANOM can also be applied to count data (e.g., nonconformities) in which the Poisson distribution is an appropriate model for the data. The decision lines are obtained from

$$\bar{c} \pm h_{\alpha,k,\infty}\sqrt{\bar{c}}\sqrt{\frac{k-1}{k}}$$

where $\bar{c}$ is the average of k "counts" (e.g., nonconformities), and, for a given value of α, $h_{\alpha,k,\infty}$ is the same value as is used for proportions data. The use of ANOM with count data is based upon the assumed adequacy of the normal approximation to the Poisson distribution. (Recall from Chapter 6 that the approximation is generally assumed to be adequate when the mean of the Poisson distribution, estimated by $\bar{c}$, is at least 5.) When the adequacy is questionable, the counts should be transformed using one of the transformations for Poisson data given in Chapter 6, with ANOM then applied to the transformed data.

16.3 ANOM WHEN STANDARDS ARE GIVEN

16.3.1 Nonconforming Units

Ott (1958, 1975) also presented the methodology for ANOM when standards are given. In the example used to illustrate ANOM for proportions data, $\overline{p}$ was obtained as the overall proportion of nonconforming crops for the 20 harvesters. That value was 0.02775. How would the ANOM have been performed if the company had established a "standard" of 2% nonconforming? (Here the word *standard* is used analogously to *target value* in Chapter 14.)

If we denote the standard value by p, the decision lines are obtained from

$$p \pm h_{\alpha,k,\infty}\sqrt{\frac{p(1-p)}{n}} \qquad (16.4)$$

where, as before, n is the number of items inspected by each harvester, and $h_{\alpha,k,\infty}$ is the value obtained from Table G in the Appendix to the book for specified values of α and k, with infinity for the degrees of freedom.

If we compare Eq. (16.4) with Eq. (16.3), the main difference is that the former does not contain the factor $\sqrt{(k-1)/k}$ that is used in the latter. In general, this factor will be a part of the decision line calculations only when the midline is estimated from data. The other difference, of course, is that p is used in place of $\overline{p}$.

What would an ANOM display with decision lines obtained from Eq. (16.4) actually show? It would indicate whether or not any of the workers' true ability appears to differ from that of the standard. If the decision lines in Figure 16.2 had been obtained using Eq. (16.4) rather than Eq. (16.3), the values for UDL and LDL would have been 0.03541 and 0.00459, respectively. Comparing these values with the data in Table 16.1 leads to the identification of five workers as being worse than the standard and none being better than the standard.

The use of these decision lines should be given careful consideration, however, as the harvesters' performance might be affected by factors outside of their control (e.g., growing conditions), and the company standard thus might not be easily attainable. In general, this is comparable to determining whether or not a process is in a state of statistical control (with, say an $\overline{X}$ chart) with the control limits determined by the current process capability versus determined from a target value.

16.3.2 Nonconformities

When ANOM is applied to count data and a standard is to be used, the decision lines are obtained from

$$c \pm h_{\alpha,k,\infty}\sqrt{c}$$

where c is the standard acceptable count (e.g., the number of surface imperfections per 50 square yards of sheet metal), and $h_{\alpha,k,\infty}$ is as designated in the preceding section.

16.3.3 Measurement Data

The decision lines in Figure 16.1 were obtained under the assumption that μ and σ are unknown, so they were estimated by $\overline{\overline{x}}$ and s, respectively. If μ and σ were both known, the decision lines would be obtained from $\mu \pm h_{\alpha,k,\infty}(\sigma/\sqrt{n})$; if σ was known but μ unknown, the lines would be obtained from $\overline{\overline{x}} \pm h_{\alpha,k,\infty}(\sigma/\sqrt{n})\sqrt{(k-1)/k}$; and if μ was known but σ unknown, the lines would be obtained from $\mu \pm h_{\alpha,k,\nu}(s/\sqrt{n})$ where ν is the degrees of freedom upon which s is based. These special cases have been listed by Schilling (1973a).

16.4 ANOM FOR FACTORIAL DESIGNS

ANOM can also be used when there is more than one factor provided that at least one factor is fixed. (Remember that ANOM is used only for fixed factors.)

We begin with a simple example by using the data in Table 13.7 (for a 2^2 design), which were used in Chapter 13 to introduce the analysis of 2^k designs using Yates's algorithm. When that data set was analyzed using ANOVA, it was found that neither the two main effects nor the interaction effect were significant. The analysis of that data by ANOM could proceed as follows.

The decision lines for the main effects A and B could be computed from

$$\overline{\overline{x}} \pm h_{\alpha,k,\nu} \frac{s}{\sqrt{n}} \sqrt{\frac{k-1}{k}} \tag{16.5}$$

where, as previously defined, s is the estimate of σ based on ν degrees of freedom, k is the number of averages (i.e., factor levels) for each of the two factors, $\overline{\overline{x}}$ is the overall average, and $h_{\alpha,k,\nu}$ is obtained from Table G in the Appendix to the book for specified values of α, k, and ν.

Note that k is being used differently in this chapter from the way that it was used in the material on design of experiments (Chapter 13). In the latter it was used to denote the number of factors; in ANOM it denotes the number of plotted points (e.g., averages) that are compared per test. Thus, references to 2^k and 2^{k-p} designs in this section are for k = number of factors; all other uses of k in this section are for k = number of averages. This creates a slight inconsistency of notation, but the notation has been retained since it is the accepted notation in each of the two subject areas.

There are two ways in which the decision lines for the AB interaction can be generated, depending upon what is to be graphed in illustrating the interaction. The approach used by Ott (1975) will be discussed first.

It was shown in Chapter 13 that for factors T and P the TP interaction could be written as

$$TP = \tfrac{1}{2}(T_2P_2 - T_2P_1 - T_1P_2 + T_1P_1)$$

where the subscript 1 denotes the "low" level, and 2 the "high" level. This can obviously be rewritten as

$$TP = \tfrac{1}{2}(T_1P_1 + T_2P_2) - \tfrac{1}{2}(T_1P_2 + T_2P_1)$$

If we replace T and P by A and B, respectively, we thus have

$$AB = \tfrac{1}{2}(A_1B_1 + A_2B_2) - \tfrac{1}{2}(A_1B_2 + A_2B_1)$$

In Ott's notation this would be $AB = \overline{L} - \overline{U}$ where $\overline{L} = (1/2)(\overline{A_1B_1} + \overline{A_2B_2})$ and $\overline{U} = (1/2)(\overline{A_1B_2} + \overline{A_2B_1})$, with the bar above the treatment combination denoting the average response for that treatment combination. The choice of notation results from the fact that the first component is obtained from treatment combinations that have like (L) subscripts, and the second component is obtained from treatment combinations that have unlike (U) subscripts.

Therefore, since $AB = \overline{L} - \overline{U}$, one way to portray the magnitude of the interaction effect is to plot $\overline{L}$ and $\overline{U}$ as "averages," with the vertical distance between them equal to the interaction effect. The decision lines for $\overline{L}$ and $\overline{U}$ are also obtained from Eq. (16.5).

Whenever ANOM is applied to data from any 2^k or 2^{k-p} design, each main effect and each interaction effect will have two components (i.e., two "averages"), and, as Ott (1975, p. 215) has pointed out, ANOM is then a "graphical t-test," and the decision lines can be obtained from use of a t table. Specifically, for a 2^2 design, $h_{\alpha,k,\nu}$ becomes $t_{\alpha/2,\nu}$, $\sqrt{(k-1)/k}$ is then $\sqrt{2}/2$, and $n = 2r$ where r is the number of replicates. Thus, for a 2^2 design with r replicates the decision lines are obtained from

$$\overline{\overline{x}} \pm \frac{1}{2} t_{\alpha/2,\nu} \frac{s}{\sqrt{r}} \tag{16.6}$$

From the data in Table 13.7 we have $\overline{\overline{x}} = 153/12 = 12.75$, and from Table 13.9 we have that $s^2 = 5.58$ (based upon 8 degrees of freedom) so that $s = \sqrt{5.58} = 2.36$. For $\alpha = .01$, $t_{\alpha/2,8} = t_{.005,8} = 3.355$, and for $\alpha = .05$, $t_{\alpha/2,8} = t_{.025,8} = 2.306$. Therefore, with $r = 3$ the 0.01 decision lines are obtained from

$$12.75 \pm \frac{1}{2}(3.355)\frac{2.36}{\sqrt{3}} = 12.75 \pm 2.29$$

Thus, UDL(.01) = 15.04 and LDL(.01) = 10.46. Using $t_{.025,8} = 2.306$, the corresponding .05 decision lines are UDL(.05) = 14.32 and LDL(.05) = 11.18.

The interaction components that are to be plotted have been previously defined. The main effect components that are to be plotted are $\overline{A}_{\text{low}}$ and $\overline{A}_{\text{high}}$ for the A effect, and $\overline{B}_{\text{low}}$ and $\overline{B}_{\text{high}}$ for the B effect. Letting 1 represent "low" and 2 represent "high," we then find from Table 13.7 that $\overline{A}_1 = 13.17$, $\overline{A}_2 = 12.33$, $\overline{B}_1 = 11.50$, and $\overline{B}_2 = 14.00$. Plotting these together with $\overline{L} = 13.5$ and $\overline{U} = 12.0$ produces the ANOM display given in Figure 16.3.

This ANOM display shows what was seen numerically in the ANOVA table given in Table 13.9; namely that the B effect is the largest of the three effects, but none of the effects is significant since all of the points are inside both the .01 and .05 decision lines.

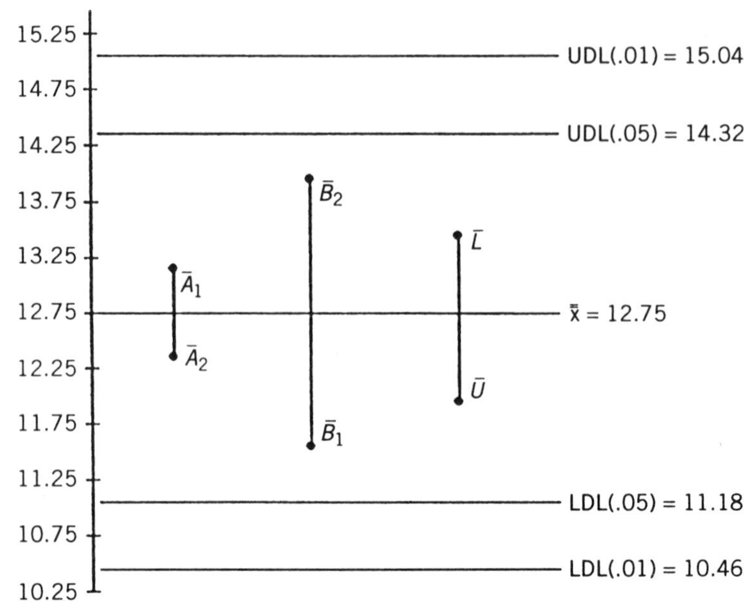

FIGURE 16.3 ANOM display for the data in Table 13.7.

It was stated at the beginning of this chapter that ANOM and ANOVA are similar but will not necessarily give the same results since what is being tested is slightly different for the two procedures. Although that is true in general, the two procedures will give identical results when each effect is represented by two means (as in the present example).

This can be demonstrated as follows. For a 2^2 design with r replicates, it can be shown, using ANOVA terminology, that the sum of squares due to the A effect, SS_A, can be represented by $r(\overline{A}_1 - \overline{A}_2)^2$. (The reader is asked to demonstrate this in exercise 1.) With $r = 3$ we then have

$$3(\overline{A}_1 - \overline{A}_2)^2 = SS_A$$

so that

$$\overline{A}_1 - \overline{A}_2 = \frac{\sqrt{SS_A}}{\sqrt{3}}$$

Since the F statistic for testing the significance of the A effect can be written as $F = SS_A/s^2$, we then have

$$\begin{aligned} \overline{A}_1 - \overline{A}_2 &= \frac{s\sqrt{F}}{\sqrt{3}} \\ &= \frac{st}{\sqrt{3}} \end{aligned}$$

since $t^2_{\alpha/2,\nu} = F_{\alpha,1,\nu}$. The length of the (equal) line segments above and below $\overline{\overline{x}}$ is $(\overline{A}_1 - \overline{A}_2)/2$ and

$$\frac{\overline{A}_1 - \overline{A}_2}{2} = \frac{1}{2} t \frac{s}{\sqrt{3}} \tag{16.7}$$

Notice that the right-hand side of Eq. (16.7) is of the same general form as the right-hand side of Eq. (16.6). Therefore, if the F test (or, equivalently a t test) for the A effect results in $F_{\text{calculated}} = F_{\alpha,1,\nu}$, then

$$\frac{\overline{A}_1 - \overline{A}_2}{2} = \frac{1}{2} t_{\alpha/2,\nu} \frac{s}{\sqrt{3}}$$

so that the line segment connecting $\overline{A}_1$ and $\overline{A}_2$ will extend from one of the α decision lines to the other with $\overline{A}_1$ and $\overline{A}_2$ lying directly on the lines. It follows that if $F_{\text{calculated}} < F_{\alpha,1,\nu}$, then

$$\frac{\overline{A}_1 - \overline{A}_2}{2} < \frac{1}{2} t_{\alpha/2,\nu} \frac{s}{\sqrt{3}}$$

and the line segment will lie entirely within the decision lines. Conversely, if $F_{\text{calculated}} > F_{\alpha,1,\nu}$, then

$$\frac{\overline{A}_1 - \overline{A}_2}{2} > \frac{1}{2} t_{\alpha/2,\nu} \frac{s}{\sqrt{3}}$$

and the ends of the line segment will fall outside the decision lines.

Companion results could be easily obtained for the B and AB effects that would show the same type of relationship.

16.4.1 Assumptions

The assumptions upon which ANOM for factorial designs is based are the same as for ANOVA for factorial designs. In particular, we must assume (approximate) normality (since a t table is being used) and equality of variances for every treatment combination. Nelson (1993) conjectures that ANOM is affected the same way as ANOVA when these two assumptions are violated.

Ott (1975) suggests using an R chart for verifying the second assumption. The four ranges are 6, 5, 3, and 4 so that $\overline{R} = 4.5$. For $r = 3$ we have $D_4\overline{R} = 2.57(4.5) = 11.565$. Since all of the ranges are below 11.565, we would thus conclude that there is no evidence of inequality of variances. [See Ott (1975, p. 261) for a ANOM-type approach to analyze variability, which is actually a rearrangement of the points plotted on an R chart.]

Ullman (1989) went a step further and proposed ANOM for ranges as a way of testing the assumption of equal variances. The control limits are based on a beta approximation.

Wludyka and Nelson (1997a) provided a ANOM procedure for variances, showing that pairs (N_i, N_j) are equicorrelated with correlation $\rho_{ij} = -1/(k-1)$, with $N_i = S_i^2 / \sum_{j=1}^{k} S_j^2$. Recall that this is the same equicorrelation structure stated in Section 16.1 as leading to the exact critical values discussed in that section. Tables of approximate values were given when $k \geq 3$. The authors also provide a large sample ANOM approach that utilizes the fact that a sample variance is asymptotically normally distributed. Since ANOM can be used to test for interactions (P. R. Nelson, 1988), the authors point out that ANOM can also be used to test for variance interactions. This idea is pursued further in Wludyka and Nelson (1997b), who also provide two ANOM-type tests for variances that are robust to nonnormality.

The assumption of normality is not crucial relative to the t values, however, as the latter are relatively insensitive to slight-to-moderate departures from normality. This is fortunate because it is difficult to check normality with only 12 numbers.

16.4.2 Alternative Way of Displaying Interaction Effects

Schilling (1973a) and Ramig (1983) illustrate ANOM for interaction effects by plotting interaction components that are different from those plotted in the preceding examples [although Schilling (1973b) does give an example of such a display.] For a two-factor design they display interaction components (say, AB_{ij}) where

$$AB_{ij} = \overline{AB}_{ij} - \overline{A}_i - \overline{B}_j + \overline{\overline{x}}$$

where $\overline{AB}_{ij}$, $\overline{A}_j$, and $\overline{B}_j$ are essentially the same as used in the preceding section with i and j denoting the factor levels. (Note that the symbols that these and other writers have used for factor-level combinations are slightly different from what was used earlier in this chapter and in Chapter 13. Specifically, they use AB_{ij} instead of A_iB_j.)

There are some problems inherent in this approach, however. First, the AB_{ij} values would be plotted on a display with a centerline of zero, which means that the interaction components would be plotted on a display different from the display used for the main effects. (This could be avoided, however, by adding $\overline{\overline{x}}$ to each AB_{ij} value.) More importantly, the interaction components do not have engineering significance when defined in this way, which is why some people believe that the interaction components should not be displayed. We should remember that ANOM was designed to show engineering significance, so we should determine what is to be plotted with that in mind.

Nothing would be gained by displaying these components for any two-level factorial design, anyway, as the four components will be the same except for the sign. There is a potential benefit when used with designs other than two-level designs, however, as they can be used to pinpoint the cell or cells that cause an interaction to be significant. When used for this purpose, there is hardly any need for decision lines, as it would already be known that the overall interaction is significant. This use of the interaction components will be illustrated later in the chapter.

16.5 ANOM WHEN AT LEAST ONE FACTOR HAS MORE THAN TWO LEVELS

16.5.1 Main Effects

The analysis is somewhat more involved when at least one of the factors has more than two levels. Ott (1975, p. 253) applied ANOM to data from a $2 \times 3 \times 4$ design with $r = 4$, and P. R. Nelson (1983) used these data as one of his examples for illustrating output from his ANOM computer program for which he listed the FORTRAN code. The data are given in Table 16.2. The output showed the main effects as deviations from $\bar{\bar{x}}$; alternatively, the means could be plotted with a center line of $\bar{\bar{x}}$, as in the previous examples in this chapter. Interactions are another matter, however, since only one of the factors has two levels.

When there is a sizable number of effects to be shown, it is preferable to use more than one ANOM display. Thus, we could show the main effects on one display, and the interaction effects on one or more additional displays.

We begin by computing the average value (length) at each level of each factor. For time we have $\overline{T}_1 = 3.78$, $\overline{T}_2 = 3.625$, and $\overline{T}_3 = 4.47$. For heat treatment we have $\overline{W} = 4.98$ and $\overline{L} = 2.94$. The averages for each machine are $\overline{A} = 3.42$, $\overline{B} = 5.875$, $\overline{C} = 0.875$, and $\overline{D} = 5.67$. The overall average is $\bar{\bar{x}} = 3.96$.

TABLE 16.2 Length of Steel Bars (Coded Data)[a]

	Heat Treatment							
	W				L			
	Machine				Machine			
Time	A	B	C	D	A	B	C	D
1	6	7	1	6	4	6	−1	4
	9	9	2	6	6	5	0	5
	1	5	0	7	0	3	0	5
	3	5	4	3	1	4	1	4
2	6	8	3	7	3	6	2	9
	3	7	2	9	1	4	0	4
	1	4	1	11	1	1	−1	6
	−1	8	0	6	−2	3	1	3
3	5	10	−1	10	6	8	0	4
	4	11	2	5	0	7	−2	3
	9	6	6	4	3	10	4	7
	6	4	1	8	7	0	−4	0

[a]These data originally appeared in Baten (1956).

Since heat treatment has only two levels, its decision lines can be obtained from Eq. (16.5) with $h_{\alpha,k,\nu}$ replaced by $t_{\alpha/2,\nu}$. From the computer output given by P. R. Nelson (1983, p. 48), we have $s = \sqrt{6.2153} = 2.493$. (The use of computer software is almost essential for obtaining s with designs that have several factors. If such software was not available, we could, for these data, compute s^2 for each of the 24 cells, and then take the average of the 24 values to obtain the 6.2153.) With 72 degrees of freedom for error we have $t_{.025,72} = 1.994$, so the decision lines are obtained from

$$\bar{\bar{x}} \pm t_{\alpha/2,\nu}\frac{s}{\sqrt{n}}\sqrt{\frac{1}{2}} = \bar{\bar{x}} \pm t_{\alpha/2,\nu}\frac{s}{\sqrt{2n}} \tag{16.8}$$

$$= 3.96 \pm 1.994\frac{2.493}{\sqrt{96}}$$

$$= 3.96 \pm 0.51$$

Thus, the .05 decision lines are UDL(.05) = 4.47 and LDL(.05) = 3.45. Since $\overline{W} = 4.98$ and $\overline{L} = 2.94$, these two averages obviously lie outside the .05 decision lines, so it would be desirable to compute the .01 decision lines and see whether or not the averages also lie outside these lines.

With $t_{.005,72} = 2.646$ we obtain UDL(.01) = 4.63 and LDL(.01) = 3.29, so the averages also lie outside the .01 decision lines. Thus, there is apparently a difference in the effects of the two heat treatments.

It should be noted that these decision lines differ slightly from those obtained by Ott (1975, p. 256). This is due in part to rounding, but primarily to the fact that Ott estimated σ by $\overline{R}/d_2^*$ whereas here we are using s (in accordance with the current methodology), with d_2^* slightly different from d_2. When the cell ranges are used, $(0.9)k(r-1)$ should be used for the degrees of freedom (before rounding to the nearest integer) in obtaining the value of $t_{\alpha/2,\nu}$, where r and k are as previously defined and $r > 1$.

It should also be noted that Eq. (16.8) is the general formula that should be used in producing the decision lines for the main effects and interaction effects for any two-level factorial design (full or fractional) as well as for other designs that have two-level factors (as in the current example).

The time factor has three levels so the decision lines will not be obtained using a t value. Instead, $h_{\alpha,k,\nu}$ can be obtained from the approximation formula given by L. S. Nelson (1983, p. 43), which generates the exact (almost all to two decimal places) tabular values given in P. R. Nelson (1993). Alternatively, since the latter does not provide, for a given α, the exact value for 72 degrees of freedom, this could be approximated using linear interpolation in the appropriate table.

The approximation formula is

$$h_{\alpha,k,\nu} \doteq B_1 + B_2K_1^{B_3} + (B_4 + B_5K_1)V_1 + (B_6 + B_7K_2 + B_8K_2^2)V_1^2 \tag{16.9}$$

where

$$k = \text{number of averages}$$

$$K_1 = \ln(k)$$

$$K_2 = \ln(k-2)$$

$$\nu = \text{degrees of freedom for error}$$

and

$$V_1 = 1/(\nu - 1)$$

The values for $B_1, B_2, \ldots, B_8$ are given in Table 16.3.

The use of Eq. (16.9) produces $h_{0.05,3,72} = 2.398$, where the third decimal place could be off by at most one digit. With Eq. (16.5) we then have the .05 decision lines determined from

$$3.96 \pm 2.398 \frac{2.493}{\sqrt{32}} \sqrt{\frac{2}{3}} = 3.96 \pm 0.86$$

so that UDL(.05) = 4.82 and LDL(.05) = 3.10. Since the three averages for the time factor are all between these numbers, they would also fall between the .01 decision lines, so the time factor appears not to be significant.

The decision lines for the machine factor would be obtained essentially the same way as for the time factor, the only difference being that the machine factor has four levels. The computations reveal that UDL(.05) = 5.07 and LDL(.05) = 2.85. Three of the four averages lie outside of these lines, so it is desirable to compute the .01 decision lines. Doing so produces UDL(.01) = 5.33 and LDL(.01) = 2.59. Since the three averages are also well beyond these decision lines, there is strong evidence of a machine effect.

TABLE 16.3 Constants for Eq. (16.9)

	α			
	0.1	0.05	0.01	0.001
B_1	1.2092	1.7011	2.3539	3.1981
B_2	0.7992	0.6047	0.5176	0.3619
B_3	0.6238	0.7102	0.7107	0.7886
B_4	0.4797	1.4605	4.3161	8.3489
B_5	1.6819	1.9102	2.3629	3.1003
B_6	−0.2155	0.2250	4.6400	27.7005
B_7	0.4529	0.6300	1.8640	5.1277
B_8	−0.6095	−0.2202	0.3204	0.7271

Source: L. S. Nelson (1983).

Although conclusions about each factor can obviously be reached without producing the graphical display, the magnitude of each main effect and the comparison of the effects for the different factors can be seen much better with the display, which is shown in Figure 16.4.

With this display we can clearly see how much larger the machine effect is when compared with the other two main effects, although some comparability is lost due to the fact that the three factors all have different decision lines. In particular, even though the machine effect is obviously greater than the heat treatments effect, the 0.01 decision lines for the machines are noticeably farther apart than those for the heat treatment.

Nevertheless, such an ANOM display can be used effectively.

The display given in Figure 16.4 is somewhat different from the ANOM displays given in other sources. Ott (1975, p. 256) uses lines that are slanted rather than vertical in providing a display of the main effects for these same data. Such a display requires more space than the one in Figure 16.4, however, and saving space can be important when there are several factors. As stated previously, other writers have plotted the distance of each average from $\overline{\overline{x}}$ rather than the averages themselves, using zero for the centerline. This tends to obscure the "engineering significance" of the averages, however, and ANOM was developed for the purpose of providing experimenters with a better "picture" of the results than can be obtained through the use of ANOVA.

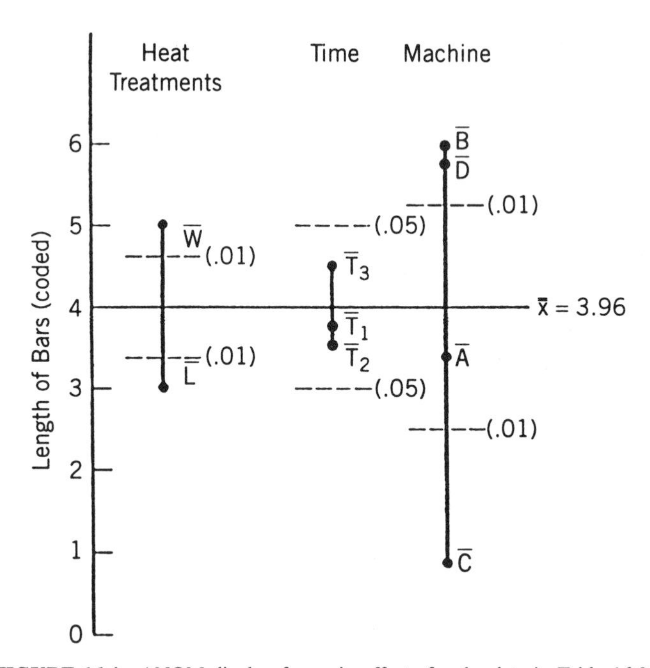

FIGURE 16.4 ANOM display for main effects for the data in Table 16.2.

16.5.2 Interaction Effects

Just as there are different ways in which the main effects can be displayed, there are also a couple of options for the interaction effects when one factor involved in the interaction has exactly two levels. Keeping in mind that we want to construct ANOM displays in such a way as to show the practical significance of the values that are charted, we recall that interactions were displayed in Chapter 13 by using interaction profiles.

We start with the heat treatments × machines interaction since the former has exactly two levels. The profile is given in Figure 16.5.

We can see that there is apparently no interaction effect due to the fact that the lines for $\overline{W}$ and $\overline{L}$ are almost parallel, meaning that the extra length of the steel bars for heat treatment W relative to the length for heat treatment L is virtually the same for each of the four machines.

What is needed in addition to Figure 16.5, however, is an objective way of assessing the interaction effect, preferably within the context of ANOM. Since the extent to which the differences $\overline{W} - \overline{L}$ vary (over the four machines) provides evidence as to the presence or absence of an interaction effect, it is logical to plot those differences against an appropriate midline and decision lines. This is the approach followed by Ott (1975, p. 259).

Since $\overline{W}$ and $\overline{L}$ are independent, it follows that $\text{Var}(\overline{W} - \overline{L}) = 2\sigma^2/n$, where n is the number of observations from which each average is computed. If we let $\overline{D}_i = \overline{W}_i - \overline{L}_i$ $(i = A, B, C, D)$, the midline will be $\overline{\overline{D}}$ and the decision lines computed from

$$\overline{\overline{D}} \pm h_{\alpha,k,\nu} \frac{s\sqrt{2}}{\sqrt{n}} \sqrt{\frac{k-1}{k}}$$

It can be shown that the four differences $\overline{W}_i - \overline{L}_i$ have pairwise correlations of $-\frac{1}{3}$, which is of the general form $-1/(k-1)$ for k means $(\overline{D}_i)$ to be plotted.

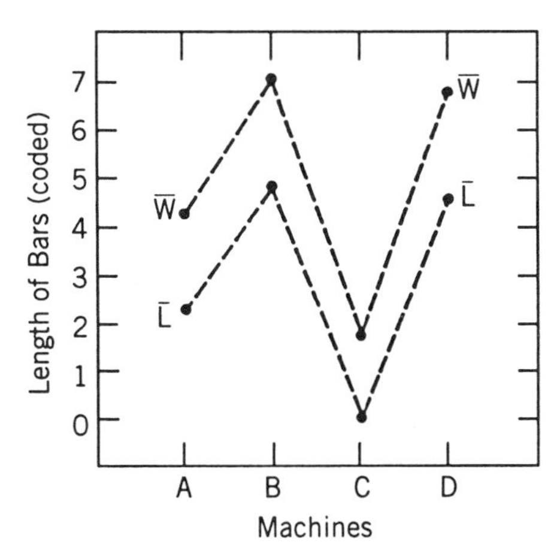

FIGURE 16.5 Heat treatments × machines interaction profile for the data in Table 16.2.

Consequently, the tables contained in L. S. Nelson (1983) can be used to obtain $h_{\alpha,k,\nu}$.

The values for $\overline{D}_i$ and $\overline{\overline{D}}$ are $\overline{D}_A = 1.83, \overline{D}_B = 2.25, \overline{D}_C = 1.75, \overline{D}_D = 2.33$, and $\overline{\overline{D}} = 2.04$. The 0.05 decision lines are then obtained from

$$2.04 \pm 2.53 \frac{2.493\sqrt{2}}{\sqrt{12}} \sqrt{\frac{3}{4}} = 2.04 \pm 2.23$$

so that UDL(.05) = 4.27 and LDL(.05) = −0.19. [These values differ slightly from those given by Ott (1975, p. 258) due primarily to the fact that Ott uses the range method for estimating sigma.] We can see that the $\overline{D}$ values are well within the values for the decision lines, which we would naturally expect since the $\overline{D}$ values differ very little. The ANOM display is given in Figure 16.6. See also P. R. Nelson (1988) for an alternative approach to constructing an ANOM display with an exact critical value for an interaction when at most one factor in the interaction has more than two levels (as in this case).

Another way to construct an ANOM display for this interaction is to compute

$$HM_{ij} = \overline{HM}_{ij} - \overline{H}_i - \overline{M}_j + \overline{\overline{x}}$$

for each of the eight combinations of i and j, where i denotes the ith heat treatment and j denotes the jth machine. These eight combinations would be plotted against decision lines obtained from

$$0 \pm h^*_{\alpha,k,\nu}\, s\sqrt{\frac{q}{N}}$$

where q is the number of degrees of freedom for the interaction effect, N is the total number of observations, s is as previously defined, and $h^*_{\alpha,k,\nu}$ is an upper bound on the true (unknown) value for $h_{\alpha,k,\nu}$. [An approximation is needed because the deviations $HM_{ij} - \overline{\overline{x}}$ are not equally correlated with $\rho = -1/(k-1)$

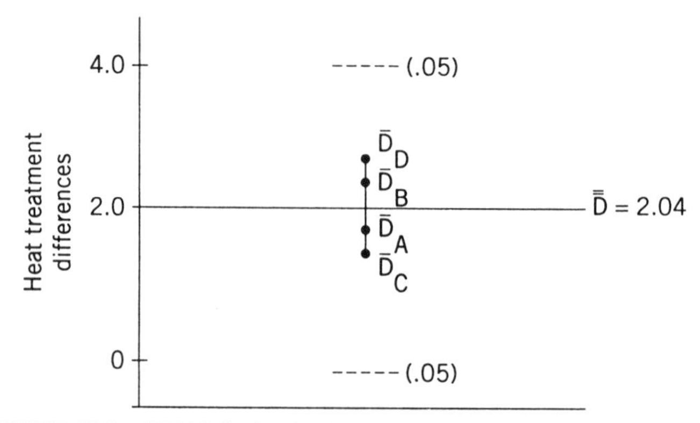

FIGURE 16.6 ANOM display for heat treatments × machines interaction.

for k deviations.] Using the approach for obtaining an upper bound suggested by L. S. Nelson (1983), we obtain

$$0 \pm 2.554(2.493)\sqrt{\frac{3}{96}} = 0 \pm 1.13$$

The eight values of HM_{ij} are ± 0.104 and ± 0.146, with each of these four values repeated once. The eight values are obviously well within the decision lines where UDL(.05) = 1.13 and LDL(.05) = −1.13.

This approach has two shortcomings—one major and one minor. The main problem is that the HM_{ij} values do not have engineering significance, and the other problem is that the exact value of $h_{\alpha,k,\nu}$ is unknown since the HM_{ij} do not have the requisite correlation structure. In particular, $HM_{ij} = -HM_{i'j}$ for each value of j, where i and i' denote the two heat treatments.

For fixed i, however, HM_{ij} and $HM_{ij'}$ are equally correlated with $\rho = -1/(k-1)$ for each j and j', so an exact value of $h_{\alpha,k,\nu}$ could be obtained if either of the four HM_{ij} or $HM_{ij'}$ values is plotted.

For the first heat treatment, W, we have $HM_{\text{WA}} = -0.104$, $HM_{\text{WB}} = 0.104$, $HM_{\text{WC}} = -0.146$, and $HM_{\text{WD}} = 0.146$. These four values could then be plotted against decision lines obtained from

$$0 \pm h_{\alpha,k,\nu}\, s\sqrt{\frac{q}{N}} = 0 \pm 2.53(2.493)\sqrt{\frac{3}{96}}$$
$$= 0 \pm 1.115$$

where, as before, $\alpha = .05$ and s and N are as previously defined. It should be noted that 1.115 is exactly half of the value (2.23) that was used to produce the decision lines in the approach suggested by Ott. This is due to the fact that, for example, $\overline{D}_A - \overline{\overline{D}}$ equals twice the absolute value of HM_{WA}, and similarly for the other three machines.

Accordingly, either approach could be used to identify the one or more cells that are causing an interaction to be significant. Daniel (1976, p. 41) asserts that, in his experience, a "one-cell" interaction is the type that occurs most frequently and that for factors with quantitative levels the cell is usually a corner cell.

A simple example should suffice. Consider the set of cell averages given in Table 16.4. If the 13 in the first cell were changed to a 5, the interaction profile

TABLE 16.4 Cell Averages for Illustrative Example

	B				Average
A	13	6	5	4	7
	3	4	3	2	3
Average	8	5	4	3	5

would exhibit two parallel lines, and there would be no interaction effect. Since only one cell is causing an interaction effect [either (1, 1) or (2, 1), where the first number designates the row and the second the column], the overall interaction might not be deemed statistically significant, but the AB_{ij} values can detect the discrepant cell (or cells), which will not always be as obvious as in this example. For the discrepant cell we have

$$\begin{aligned} AB_{11} &= \overline{AB}_{11} - \overline{A}_1 - \overline{B}_1 + \overline{\overline{x}} \\ &= 13 - 7 - 8 + 5 \\ &= 3 \end{aligned}$$

whereas $AB_{12} = AB_{13} = AB_{14} = -1$, and $AB_{2j} = AB_{1j}$ for $j = 1, 2, 3, 4$. Thus, a plot of the AB_{2j} or AB_{1j} values with a midline of zero would detect the corner cell (or the cell below it if the AB_{2j} values were plotted), and whether or not AB_{11} (or AB_{21}) is outside the decision lines would depend upon the values of s and the other components that determine the decision lines.

The same message would be received if the $\overline{D}$ values were calculated, where $\overline{D}_1 = 10, \overline{D}_2 = 2, \overline{D}_3 = 2$, and $\overline{D}_4 = 2$. The values for $(\overline{D}_i - \overline{\overline{D}})$ are 6, -2, -2, and -2, respectively, which, of course, are twice the AB_{1j} values.

The heat treatments $\times$ time interaction could also be analyzed using either of these two methods. For the three different times the $\overline{D}$ values are $\overline{D}_1 = 1.6875, \overline{D}_2 = 2.125$, and $\overline{D}_3 = 2.3125$. The .05 decision lines are UDL(.05) = 3.767 and LDL(.05) = 0.316, so the $\overline{D}$ values are well within the decision lines, and there is thus no evidence of an interaction effect.

The time $\times$ machines interaction presents a problem since neither of the factors has two levels. The approach using $h^*_{\alpha,k,\nu}$ as an approximation for $h_{\alpha,k,\nu}$ leads to the conclusion that the interaction is not significant for $\alpha = .05$. Ott (1975, p. 259) creates two sets of $\overline{D}$ values after pairing what seem to be the most similar and most dissimilar machines relative to the time factor, with this assessment made after viewing the overall time $\times$ machines interaction profile. This is an ad hoc procedure, but one that can provide important information. In particular, if the differences formed from the most similar machines are judged to be different, then the overall interaction wil! likely be significant. Conversely, if the differences formed from the least similar machines are not significant, then the overall interaction is probably not significant. On the other hand, when the results differ at these two extremes, then no conclusion can be drawn.

This approach suggested by Ott is essentially analogous to the use of a multiple comparison procedure for means; the difference is that there is no exact ANOM procedure for determining whether or not the overall interaction is significant before testing components of that interaction.

P. R. Nelson (1988) gave a Bonferroni approach for providing an upper bound on $h_{\alpha,k,\nu}$ for two-factor interactions where each factor has more than two levels and for three-factor interactions where at least two of the factors have exactly two levels. That approach could be applied to the time $\times$ machines interaction, but it could not be applied to the three-factor interaction since only one of the three factors has two levels.

16.6 USE OF ANOM WITH OTHER DESIGNS

We have shown how ANOM can be used with single-factor designs and factorial designs. P. R. Nelson (1993) extended the use of ANOM to other designs including Latin squares, Graeco-Latin squares, balanced incomplete block (BIB) designs, Youden squares, and axial mixture designs. [See Cornell (1990) for information on axial mixture designs, and Montgomery (1997) for information on the other designs.]

The key is recognizing that the correlation structure given in Section 16.1 [i.e., $\rho = -1/(k-1)$ for k means] also occurs with these designs. One slight difference occurs with BIB designs, Youden squares, and axial mixture designs, however, in that the center line is zero for these designs rather than the grand average.

Subramani (1992) proposed a step-by-step ANOM approach for analyzing data with missing values, illustrating the technique by applying it to data obtained from using a randomized block design.

16.7 NONPARAMETRIC ANOM

It will not always be reasonable to assume normality, and the assumption is very difficult to check with a small amount of data. Therefore, a nonparametric ANOM procedure will frequently be useful. Bakir (1989) gave an ANOM procedure that utilized the ranks (R_{ij}) of the data and was designated as ANOMR. Thus, it is applicable when there is a single factor and is therefore a nonparametric alternative to the parametric procedure given in Section 16.1. The statistic that is plotted on the ANOM display is the average of the ranks for each level rather than the average of the observations for each level. Thus, interest centers on the deviations $\overline{R}_i - \overline{\overline{R}}$, a large deviation suggesting that μ_i differs from the average of all of the means. The critical values, given in Tables II and III of Bakir (1989) for 3 and 4 levels and for equal and unequal sample sizes up to $n = 5$, were obtained by enumerating all configurations of rankings when the null hypothesis is true. Of course, exact significance levels are not possible due to the fact that the statistic $\overline{R}_i - \overline{\overline{R}}$ can assume only a finite number of possible values, with the number of possible values being greatly restricted when n and/or the number of factor levels is quite small.

For the case of equal sample sizes, Bakir (1989) provided an approximation based on Bonferroni's inequality and also gave a large sample approximation. The efficacy of the proposed procedure was demonstrated in a power study, but only symmetric distributions were used in the study, so the comparative value of the procedure for asymmetric distributions is not apparent.

16.8 SUMMARY

Analysis of means (ANOM) is a viable alternative or supplement to analysis of variance. It can be used with either measurement data or attribute data. When used with two-level designs, it becomes essentially a graphical t test. The use of ANOM with other designs was also illustrated. Two types of ANOM displays were discussed—one that has engineering significance and one that does not.

APPENDIX

We stated in Section 16.1 that $s\sqrt{(k-1)/(kn)}$ is the estimated standard deviation of $\overline{X}_i - \overline{\overline{X}}$. This can be demonstrated as follows:

$$\begin{aligned}
\operatorname{Var}(\overline{X}_i - \overline{\overline{X}}) &= \operatorname{Var}\left(\overline{X}_i - \frac{\overline{X}_1 + \cdots + \overline{X}_i + \cdots + \overline{X}_k}{k}\right) \\
&= \operatorname{Var}(\overline{X}_i) - 2\operatorname{Cov}\left(\overline{X}_i, \frac{\overline{X}_i}{k}\right) + \operatorname{Var}(\overline{\overline{X}}) \\
&= \frac{\sigma^2}{n} - \frac{2}{k}\left(\frac{\sigma^2}{n}\right) + \frac{\sigma^2}{kn} \\
&= \frac{\sigma^2(k-1)}{kn}
\end{aligned}$$

The result then follows after the square root of the last expression is taken, and s is substituted for σ.

REFERENCES

Bakir, S. T. (1989). Analysis of means using ranks. *Communications in Statistics—Simulation and Computation 18*(2): 757–776.

Baten, W. D. (1956). An analysis of variance applied to screw machines. *Industrial Quality Control 12*(10): 8–9.

Cornell, J. A. (1990). *Experiments with Mixtures: Designs, Models, and the Analysis of Mixture Data*, 2nd ed. New York: Wiley.

Daniel, C. (1976). *Applications of Statistics to Industrial Experimentation*. New York: Wiley.

Halperin, M., S. W. Greenhouse, J. Cornfield, and J. Zalokar (1955). Tables of percentage points for the studentized maximum absolute deviate in normal samples. *Journal of the American Statistical Association 50*(269): 185–195.

Montgomery, D. C. (1997). *Design and Analysis of Experiments*, 4th ed. New York: Wiley.

Nelson, L. S. (1983). Exact critical values for use with the analysis of means. *Journal of Quality Technology 15*(1): 40–44.

Nelson, P. R. (1982). Exact critical points for the analysis of means. *Communications in Statistics—Part A, Theory and Methods 11*(6): 699–709.

Nelson, P. R. (1983). The analysis of means for balanced experimental designs (computer program). *Journal of Quality Technology 15*(1): 45–54.

Nelson, P. R. (1988). Testing for interactions using the analysis of means. *Technometrics 30*(1): 53–61.

Nelson, P. R. (1993). Additional uses for the analysis of means and extended tables of critical values. *Technometrics 35*(1): 61–71.

Ott, E. R. (1958). Analysis of means. Technical Report No. 1, Rutgers University.

Ott, E. R. (1967). Analysis of means—a graphical procedure. *Industrial Quality Control 24*(2): 101–109.

Ott, E. R. (1975). *Process Quality Control.* New York: McGraw-Hill.

Ramig, P. F. (1983). Applications of the analysis of means. *Journal of Quality Technology 15*(1): 19–25.

Schilling, E. G. (1973a). A systematic approach to the analysis of means, Part I. Analysis of treatment effects. *Journal of Quality Technology 5*(3): 93–108.

Schilling, E. G. (1973b). A systematic approach to the analysis of means, Part II. Analysis of contrasts; Part III. Analysis of non-normal data. *Journal of Quality Technology 5*(4): 147–159.

Subramani, J. (1992). Analysis of means for experimental designs with missing observations. *Communications in Statistics—Theory and Methods 21*(7): 2045–2057.

Tomlinson, L. H. and R. J. Lavigna (1983). Silicon crystal termination—an application of ANOM for percent defective data. *Journal of Quality Technology 15*(1): 26–32.

Ullman, N. R. (1989). The analysis of means (ANOM) for signal and noise. *Journal of Quality Technology 21*(2): 111–127.

Wludyka, P. S. and P. R. Nelson (1997a). An analysis-of-means-type test for variances from normal populations. *Technometrics 39*(3): 274–285.

Wludyka, P. S. and P. R. Nelson (1997b). Analysis of means type tests for variances using subsampling and jackknifing. *American Journal of Mathematical and Management Sciences—Special Issue on Multivariate Statistical Analysis 17*: 31–60.

EXERCISES

1. Show that for a 2^2 design with r replicates SS_A can be written as $r(\overline{A}_1 - \overline{A}_2)^2$.

2. Consider the data in Table 13.10. Use 12 as an estimate of sigma and analyze the data using analysis of means, displaying both the .05 and .01 decision lines. Use the "column (3) representation" in Table 13.15 to compute the two averages (i.e., use the four plus signs to compute one average and the four minus signs to compute the other, for each effect). Compare your results with the results obtained in Chapter 13. Will the conclusions have to be the same? Why or why not?

3. When analysis of variance is applied to a one-way classification with k levels, what is being tested is $\mu_1 = \mu_2 = \cdots = \mu_k$. Write what is being tested, as a function of $\mu_1, \mu_2, \ldots, \mu_k$, when analysis of means is used.

4. Assume that ANOM is being used to analyze proportions data with 10 proportions. If $n = 900$ and $\overline{p} = .02$, what is the numerical value of UDL(.01)?
5. Consider the following data for a one-way classification:

1	2	3	4
11	10	10	13.6
12	14	13	12.6
9	11	8	15.6
13	12	12	11.6
7	8	8	16.6

Analyze the data using both ANOVA and ANOM (use $\alpha = .05$). Explain why the two procedures produce different results.

6. A company has five plants that produce automobile headlights. Given below are the number of surface scratches recorded at each plant for 1 month.

Plant Number	Number of Scratches
1	121
2	163
3	148
4	152
5	182

Should any plant(s) be singled out as being either particularly good or bad relative to the others?

7. Construct an ANOM display for the data in Table 13.7 and compare with the results given in Table 13.9.
8. How would you proceed, in general, if you had data from a 2^2 design with four replications? That is, would it be practical to check for normality with this number of observations or should a nonparametric ANOM approach be used?
9. For the data in Table 13.7, construct the ANOM display for the AB interaction using the approach described in Section 16.4.2, using the approximation due to L. S. Nelson (1983) described in Section 16.5.1. Compare the results with the ANOM display for *AB* using the approach advocated in the chapter.

CHAPTER 17

Using Combinations of Quality Improvement Tools

The statistical tools that have been presented in the preceding chapters should be used together to improve quality. Why is that? Assume that a process is out of control but an experiment is performed to compare the standard operating process with an experimental process that is conjectured to produce superior results. If the standard process is used for, say, 1 month and the experimental process is used for the same length of time, the results will not be comparable if the process is out of control in such a way as to affect the results of the experiment.

For example, if one or more process characteristics are out of control and this condition causes process yield to decrease when the experimental process is being used, a process engineer might be misled into perhaps falsely concluding that the experimental process is not superior. Similarly, we could erroneously conclude that a process change has been beneficial when we are actually seeing a data shift that is due to a parameter change. This was illustrated in Section 13.2 and is discussed further in Section 17.1.

Broadly stated, there is a need for a climate of "statistical thinking" to exist in order for combinations of statistical techniques to be used effectively. Britz, Emerling, Hare, Hoerl, and Shade (1997) give the following definition of the term (p. 67):

Statistical thinking is a philosophy of learning and action based on the following principles:

- All work occurs in a system of interconnected processes
- Variation exists in all processes
- Understanding and reducing variation are keys to success

See also Hare et al. (1995).

17.1 CONTROL CHARTS AND DESIGN OF EXPERIMENTS

In addition to the importance of using control charts to try to ensure that a process is in control before an experiment is conducted, it is also possible to use control charts to identify what variables to use in an experiment. This is illustrated by Hale-Bennett and Lin (1997), who showed how a sawtooth pattern on some control charts led to the identification of a position effect in a painting process, so that position was one of the variables that was used in a subsequent experiment.

This same general idea is discussed by Czitrom (1997), who states: "opportunties for process improvement identified during statistical process control can be addressed using a passive data collection or a designed experiment." [The term *passive data collection*, which is apparently due to Lynch (see Lynch and Markle, 1997, p. 85) refers to the collection of data when there has been no process intervention.]

It is also important that processes be in control when designed experimentation is performed, as was illustrated in Section 13.2.

As also stated in Section 13.2, Box et al. (1990, p. 190) recommended that a process be brought into the best state of control possible before an experiment is run, and that blocking and randomization be used. If we knew that the means of one or more process variables had changed and these changes would affect process yield, we would want to make sure that we carried out the experiment in such a way that the standard and experimental processes were compared when the means of these process variables were constant. Identifying periods when we have constant conditions (i.e., there are no lurking variables) may be difficult, however, and detecting process shifts of small amounts can also be difficult. Fortunately, small process shifts will usually not seriously undermine the analysis of data from designed experiments.

17.2 CONTROL CHARTS AND CALIBRATION EXPERIMENTS

Pankratz (1997) discusses a calibration study in which $\overline{X}$ and R charts were used in conjunction with the classical method of calibration (see Section 12.9). The control charts were used to try to help ensure that the measurements obtained would be representative of measurements that would be obtained on a daily basis.

17.3 SIX SIGMA PROGRAMS

One way to illustrate how various statistical tools can be used together to improve quality is to examine Motorola's Six Sigma program, which has also been adopted by other leading companies. Briefly,"Six-Sigma" refers (or should refer, more about this later) to products and processes whose range of values that are considered to be acceptable, for whatever is being measured, covers a range of 12 standard deviations (sigma)—6 on each side of the average or nominal value.

Various suppliers, especially suppliers to leading electronics firms, have been expected to adopt some type of Six Sigma program. (This is very similar to what occurred in the 1980s when automotive suppliers were expected to use control charts and, in general, demonstrate acceptable process capability.)

In 1988, Motorola was one of the recipients of the Malcolm Baldrige National Quality Award, an award that had been established in 1987 to improve the competitiveness of American industry with other countries, notably Japan. Motorola, which had become very serious about quality improvement during 1979–1981, attributed much of its success to a quality improvement program that it called Six Sigma. How successful has Motorola been? In an interview published in 1995 (Bemowski, 1995), Gary Tooker, CEO of Motorola, stated: "We've saved several *billion* [emphasis mine] dollars over the last year because of our focus on quality improvement and the six sigma initiative." Similarly, it was stated in early 1997 that GE hoped to save 7–10 billion dollars during the following decade (*Wall Street Journal*, January 13, 1997).

Motorola was later willing to share information on its Six Sigma program. In particular, IBM executives visited Motorola in 1989 to study the approach. A year later, IBM's Application Systems Division was a Baldrige Award winner. St. Clair (1995) describes IBM's implementation of a Six Sigma program.

McFadden (1993) discusses the assumptions upon which the 6-sigma approach is based. These include (1) normality, (2) known values of the process mean and standard deviation, (3) the process capability indices C_p and C_{pk} are assumed to be parameters, not sample statistics, (4) nonconformities are independent, and (5) that a change in the process mean of up to 1.5 sigma is likely to occur, but if the process were 6 sigma it would still be in a state of statistical control after such a change.

Let's examine these assumptions. The first assumption will never be met but will often be approximately met. The second assumption will rarely, if ever, be met, which implies that the third assumption will rarely be met. The fourth assumption will frequently not be met, and we will consider the apparent origin of the first part of the fifth assumption later in this section.

A normal distribution will never be strictly applicable for any measurement variable since the range of a normal random variable is from minus infinity to plus infinity, and negative measurements are obviously not possible. By definition, any measurement variable has to have some skewness, since values close to zero will be much more likely to occur than values close to plus infinity. So we can never have a normal distribution. The assumption of (approximate) normality should always be checked.

It is unrealistic to assume that a process mean and standard deviation are known, so it is equally unrealistic to fail to recognize that C_p and C_{pk} must be estimated and that the estimators used have sampling variability.

Although the assumption of independent occurrences of nonconformities and nonconforming units is generally made when control charts are used, this assumption will often be violated, as was discussed in Sections 6.1.7 and 6.2.6.

The apparent reason for the last assumption is discussed by Tadikamalla (1994), who questions how a 1.5-sigma shift can be "allowed." The author refers to an internal Motorola publication of M. J. Harry in which the latter claimed that a process mean could change from one *lot* to the next, with the average change being about 1.5-sigma. Harry claims that this assertion is supported by the literature (Bender, 1962; Evans, 1975; Gilson, 1951). As explained by Tadikamalla (1994), however, these latter sources discuss the desired value of the standard deviation of each component in a stack of components so that the half-width of a specified tolerance is equal to three times the standard deviation *for the stack assembly*. Bender (1962) suggests that 1.5 be used as the multiplier because of potential shifts in the means of the individual components.

If a stack assembly has, for example, five components, this is equivalent to assuming that the average change in the means for the five components is 0.3-sigma. This is obviously quite different from assuming that each mean changes by as much as 1.5-sigma. In general, it is reasonable to think that a process might shift by 0.3-sigma and still be considered "in control," since nothing is really static over time.

As pointed out in the Chapter 8 discussion of EWMA control charts, once an error appears in the literature (or in this case, "in practical use"), it is almost certainly going to be perpetuated to some extent. Evidence of this can be found in McFadden (1993), who cites Evans (1975) as a reference for the assumption that the mean of a single component can change by as much as 1.5-sigma. Similarly, Noguera and Nielsen (1992) state: "Motorola's choice of 1.5σ was based on both theoretical and practical grounds. Theoretically, an $\overline{X}$ control chart will not quickly detect a process shift until the magnitude of the shift is $\pm 1.5\sigma$ (based upon a sample size of $n = 4 \ldots$). This choice was also verified empirically by Motorola." Of course, an $\overline{X}$ chart will not quickly detect a process shift even when the shift is approximately 2σ, since this would be a $1\sigma_{\overline{x}}$ shift for a subgroup of size 4, and such a shift would be detected, on average, after 44 plotted points.

This confusion has apparently led some companies to believe that a 1.5-sigma change in the mean is to be expected. With such an expectation, it is easy to see why certain companies have given process control low priority. Since using control charts and/or other process control techniques to keep a process mean in a state of statistical control will generally be easier than reducing process variability, process control techniques should definitely be utilized as a company strives to attain 6-sigma capability.

Articles in which "Six-Sigma" is mentioned state that this corresponds to 3.4 nonconforming units per million. This is true only under some rather unrealistic conditions. First, this is based on the assumption of normality, which, as stated previously, is only an approximation to reality, and it is also based on the assumption that the process really has drifted *exactly* 1.5-sigma and stays there, with the specification limits assumed to be 6-sigma away from the process mean if there were no process drift. So this is actually 4.5-sigma capability. For true 6-sigma capability (i.e., the process mean was constant), there would be two nonconforming units per billion under the assumption of normality.

There is obviously much confusion about these issues in the literature. For example, McFadden (1993) states: "in a six-sigma process, $C_p = 2$." It is not possible to speak of 6-sigma in terms of C_p since the latter is not a function of the process mean. It is theoretically possible to have $C_p = 2$ and yet have virtually 100% of the units being nonconforming.

It should be pointed out that Motorola does not "allow" a 1.5-sigma change in the mean or a shift of any other specified amount. Rather, the company believes in keeping distributions centered.

How hard is it to achieve 6-sigma capability? St. Clair (1995) quotes one Motorola manager as saying: "You can reach five sigma with fairly traditional technology and approaches, but it requires hard work and smart decisions. You can't work hard enough to get six sigma. You have to work at the process level to predict the next defect and then head it off with some preemptive action. That's the way to get six sigma. You have to have the experience to practically smell a defect condition starting" (p. 153). For quality practitioners whose ability to sniff out problems is something less than perfect, this suggests that some type of process control approach be used, such as statistical process control. We should note that 5-sigma corresponds to 3.5-sigma with a fixed mean and 3.5-sigma capability would produce 2.33 nonconforming units per 10,000. Since the latter would be considered excessive by progressive companies that are seeking high-quality products, the need for process control techniques should be apparent. Five-sigma capability with a fixed mean would result in 5.7 nonconforming units per 10 million. Certainly the vast majority of companies would consider this to be acceptable.

Clearly, however, what should be considered acceptable depends on what is being measured. In trying to present a picture of 6 sigma in concrete terms, Boyett, Schwartz, Osterwise, and Bauer (1993) indicate that deaths from airplane disasters occur about 4.3 times per 10 million flights, which translates to 4.92-sigma quality if we assumed normality and used just one side of the distribution. (Of course, this is just for illustration, as a normal distribution probably would not be a suitable approximation to the actual distribution. The intent here is just to translate the "percent nonconforming" into a value of k in k-sigma.) For this application we would probably want to see more than 6-sigma capability, especially those of us who are frequent flyers.

Thus, a variation of the 6-sigma theme seems desirable. It also seems desirable to not seek 6-sigma capability for non-critical processes, as doing so may not be time well spent.

Another question is: Is it practical to expect that 6-sigma capability could be verified, even if it existed? This is analogous to estimating *very extreme* percentiles of a distribution. Clearly it is impractical to think that one could take a large enough sample to estimate far-tail percentiles with any degree of precision. If actual 6-sigma capability existed, a sample size of about 100 billion would probably be required to estimate the percentage of nonconforming units. One alternative would be to assume normality and estimate the percentage from the estimates of μ and σ and their relation to the specification limits, but this could produce a poor estimate.

It should be noted that Motorola's Six Sigma efforts have not been confined to manufactured products. Klaus (1997) describes Motorola's Total Customer Satisfaction competition, which involves almost half of Motorola's employees.

The objective is to provide motivation for continuous improvement. As explained by Gary Tooker in Bemowski (1995), the concept/goal is also applied to support nonmanufacturing and service activity. In particular, the finance department quickly adapted to the Six Sigma stratagem, with the books being closed about 3 days after the month, whereas it had previously taken a couple of weeks.

Not everyone has felt that Six Sigma can be easily applied to nonmanufacturing operations, however. Gary Wendt, former Chairman of GE Capital, was quoted as saying, "it's not quite as easy to say there are no defects in an automobile loan" (*Wall Street Journal*, January 13, 1997).

It seems safe to assume that the use of Six Sigma and related programs in nonmanufacturing operations will grow at a slower rate than in manufacturing operations.

17.3.1 Components of a Six Sigma Program

McFadden (1993) describes the components of a Six Sigma program, which include use of the statistical tools that were presented in the preceding chapters.

The program incorporates design of experiments, gauge repeatability and reproducibility, regression, correlation, hypothesis tests, failure mode and effects analysis (FMEA), statistical process control (SPC), and other statistical and non-statistical tools. Some companies that have instituted Six Sigma training have, to some degree, overlooked SPC as a necessary component of any Six Sigma program, or have mispositioned it in a sequence of steps to be performed. As discussed in Section 17.1, it is risky to conduct designed experiments without trying to maintain processes in a state of statistical control. Harry (1998) gives a sequence of steps to follow in a Six Sigma program, but process control is given as "phase 4", whereas design of experiments is used in phase 3. Furthermore, determining process capability was given as step 11, whereas process control was step 12. The importance of establishing reasonably good process control before conducting designed experiments was discussed in Section 17.1. Somewhat similarly, Hoerl (1998) lists SPC and advanced SPC as being covered in week 4 of a Six Sigma training curriculum, with design of experiments covered in week 3.

The objective should always be, in general, to improve quality by reducing variability. Clearly there can be no single Six Sigma training program as the tools that are used are dependent on the specific applications that are required. This will certainly vary among industries, among companies within industries, and of course will also vary considerably within companies for different applications (manufacturing versus nonmanufacturing, in particular).

Clearly there is the potential for Six Sigma programs to result in the same type of abuses and "number fudging" that were discussed in Section 7.2.3 regarding process capability indices.

17.3.2 Six Sigma Applications and Programs

Delott and Gupta (1990) describe a study to improve copperplating consistency within different and similar circuit boards so as to achieve Six Sigma quality.

Koons, Meyer, and Kurowski (1992) describe how design of experiments (DOE), regression, and simulation have been used together in Six Sigma work at Motorola Lighting, Inc.

Hoerl (1998) provides considerable insight into the use of Six Sigma programs and training at GE.

17.3.3 Six Sigma Concept for Customer Satisfaction

Fontenot, Behara, and Gresham (1994) explain the application of the Six Sigma concept in assessing the extent of customer satisfaction by Texas Instruments' Defense Systems and Electronics Group. The article is interesting, although the reader does encounter some misstatements, such as "six sigma is a way to measure the probability that companies can produce any given unit of a product (or service) with only 3.4 defects per million units or operations". Since the Six Sigma idea for measurement data involves specification limits, the way that the Six Sigma concept would be used with attribute data may not seem obvious. The authors describe the adaptation by matching the proportion of dissatisfied customers with the tail probability for measurement data. They apply the $\pm 1.5\sigma$ shift idea, however, which certainly seems inapplicable for attribute data. It would be more reasonable to use the percentage of dissatisfied customers to determine the value of k in k-sigma assuming no shift (and to do the same thing for measurement data).

Of course, companies could simply record the parts per million "defects" (i.e., dissatisfied customers) with conversion to k-sigma. Miller and Casavant (1994) describe this approach and the plotting of such values on a chart, which they term a "defects per unit chart."

17.3.4 Six Sigma Training

Before a company can implement an overall Six Sigma program, it obviously must decide how to proceed. In particular, how much should employees whose jobs are at different levels be taught, and what should be taught to each classification of employee? Borrowing from the martial arts, Motorola initiated the use of terms such as "green belt", "black belt", and "master black belt" to indicate different levels and amounts of training. [Note that these terms are now registered trademarks of Sigma Consultants, L.L.C., d/b/a Six Sigma Academy. See Harry (1998).]

Hoerl (1998) provides the typical black belt training curriculum at GE. It is interesting to note that this includes EVOP in week 3 (recall the discussion in Chapter 15 regarding the relative disuse of EVOP). GE's training has been performed using the statistical software package Minitab, as indicated by Hoerl (1998).

Although companies might use GE's curriculum as a rough guide, adaptations will obviously be necessary, and the software that is used should meet the specific company needs. Although Minitab is user friendly and popular, companies will probably have a need for additional general-purpose statistical software, in addition to special-purpose software.

17.3.5 Related Programs/Other Companies

We would expect companies to implement their own versions of Motorola's program. Craig (1993) describes the Six Sigma Quality Planning Process that was implemented by DuPont Connector Systems. The author makes two important points: (1) the process has no endpoint, and (2) customer expectations can change rapidly and necessitate a revised quality plan.

Rich (1997) mentions the Texas Instruments Defense System and Electronics Center "that is responsible for the development and deployment of statistical tools and techniques for the organization". The organization has over 100 "six sigma black belts." The "color of the belt" designates the level of training that an employee has received. Various other companies, including GE, have adopted the same practice.

Small companies have also been successful in implementing Six Sigma programs. Landes (1995) reports that Wainwright Industries, a Baldrige Award winner in 1994, sent one of its customers over 20 million parts during a 3-year period without a defect being found.

17.3.5.1 SEMATECH's Qual Plan

SEMATECH (Semiconductor Manufacturing Technology) is a nonprofit research and development consortium of U.S. semiconductor manufacturers located in Austin, Texas. Its members include Advanced Micro Devices, Digital Equipment Corporation (DEC), Hewlett-Packard Company, Intel Corporation, IBM, Lucent Technologies, Motorola, National Semiconductor Corporation, Rockwell International Corporation, Texas Instruments, and the U.S. Department of Defense.

Although not as well-known as Six Sigma programs, the SEMATECH Qualification Plan (Qual Plan) has been used successfully by companies in the semiconductor industry, with some companies, such as Intel Corporation, using a modification of Qual Plan. The plan actually evolved from the burn-in methodology developed by Intel, with efforts by SEMATECH engineers and statisticians to modify that methodology having begun in 1988.

The plan emphasizes engineering goals, with statistical methods used in seeking the attainment of the goals. A similar plan has been used in the chemical industry. Qual Plan is described in detail by Czitrom and Horrell (1997).

17.3.5.2 AlliedSignal's Operational Excellence Program

AlliedSignal launched their own Operational Excellence program in September 1994, although it is now referred to as a Six Sigma program. The program resulted in a saving of 200 million dollars in the first 2 years (Zinkgraf, 1997). Like Motorola and GE, the company has used various types of "belt" designations to reflect training and skill levels of its employees relative to Six Sigma. For example, in 1997 the company had 20 master black belts and 408 black belts.

Similar to the efforts of Motorola and GE, AlliedSignal has used combinations of non-statistical and statistical tools. AlliedSignal has included EVOP (Chapter 15), which unfortunately has not been used by very many companies in recent years but, as stated in Section 17.3.4, is included in GE's black belt training.

17.4 STATISTICAL PROCESS CONTROL AND ENGINEERING PROCESS CONTROL

Box and Luceño (1997) and Montgomery (1996), in particular, have advocated the combined use of statistical process control (SPC) and engineering process control (EPC). Why? Ideally, we want to identify and remove all assignable causes, as doing so permits the realization of a process in a state of statistical control. In the chapters on control charts there was no discussion of how a practitioner might proceed if all assignable causes cannot be identified and/or removed. Assume that a process is often out of control due to a single assignable cause that is either difficult to detect or detectable but impossible or impractical to eliminate. For this scenario we cannot maintain the process in statistical control by using any control chart because control charts will foster improvement only if assignable causes can be detected and removed.

Engineering process control can be effectively used for such a scenario, provided that one or more "adjustment variables" can be identified as well as an appropriate model that relates these variables to the output. If so, these variables would be manipulated to force the output variable closer to its target value, if one exists. For such a scenario, Box and Luceño (1997) advocate the use of *feedback control*. Specifically, process adjustments are made in accordance with where the process is predicted to be at the next time period.

One of several possible controllers could be used; these are discussed in detail by Box and Luceño (1997, Chapter 6).

We should remember that process adjustment is appropriate only for processes that cannot be maintained in a state of statistical control by eliminating assignable causes. Box and Luceño (1997, p. 142) show the numerical consequences of adjusting a process that is not out of control. For example, the variance of what is being monitored is doubled if the size of the adjustment that is made at time t is equal to $z_t = |X_t - \text{target}|$. This particular result was also obtained by Deming. See Deming (1986, p. 327), in which the author states: "if anyone adjusts a stable process to try to compensate for a result that is undesirable, or for a result that is extra good, the output that follows will be worse than if he had left the process alone." When adjustment is appropriate, the ideal would be to manipulate the adjustment variable at time t so that z_{t+1} is forced to be zero. Of course, this would be possible only if perfect prediction were possible. Since this ideal state cannot be reached and maintained, the adjustments must be made using a good model of X_t. Box and Luceño (1997) advocate the use of the EWMA for prediction and give some evidence of its robustness relative to various time series models. [*Note:* The use of an EWMA for prediction is quite different from the use of an EWMA control chart. The latter is used for controlling a *stationary* process, whereas the EWMA when used for prediction is for a *nonstationary* process. Since these uses are quite different, the weighting constants that are used for each purpose should be different (and they are).]

In general, SPC is inherently a hypothesis testing approach, whereas in EPC the focus is on model selection and parameter estimation.

REFERENCES

Bemowski, K. (1995). Motorola's fountain of youth. *Quality Progress 28*(10): 29–31.

Bender, A. (1962). Bendarizing tolerances—a simple practical probability method of handling tolerances for limit stack-ups. *Graphic Science*, p. 17.

Box, G. E. P. and A. Luceño (1997). *Statistical Control by Monitoring and Feedback Adjustment*. New York: Wiley.

Box, G., S. Bisgaard, and C. Fung (1990). *Designing Industrial Experiments*. Madison, WI: BBBF Books.

Boyett, J. H., S. Schwartz, L. Osterwise, and R. Bauer (1993). *The Quality Journey: How Winning the Baldrige Sparked the Remaking of IBM*. New York: Dutton.

Britz, G., D. Emerling, L. Hare, R. Hoerl, and J. Shade (1997). How to teach others to apply statistical thinking. *Quality Progress 30*(6): 67–79.

Craig, R. J. (1993). Six sigma quality, the key to customer satisfaction. In *Annual Quality Congress Transactions*, pp. 206–212. Milwaukee, WI: American Society for Quality Control.

Czitrom, V. (1997). Introduction to passive data collection. In V. Czitrom and P. D. Spagon, eds. *Statistical Case Studies for Industrial Process Improvement*, Chapter 6. Philadelphia: American Statistical Association and Society for Industrial and Applied Mathematics.

Czitrom, V. and K. Horrell (1997). SEMATECH qualification plan. In V. Czitrom and P. D. Spagon, eds. *Statistical Case Studies for Industrial Process Improvement*, pp. xxi–xxvii. Philadelphia: American Statistical Association and Society for Industrial and Applied Mathematics.

Delott, C. and P. Gupta (1990). Characterization of copperplating process for ceramic substrates. *Quality Engineering 2*(3): 269–284.

Deming, W. E. (1986). *Out of the Crisis*. Cambridge, MA: MIT Center for Advanced Engineering Study.

Evans, D. H. (1975). Statistical tolerancing: The state of the art, Part III: Shifts and drifts. *Journal of Quality Technology 7*: 72–76.

Fontenot, G., R. Behara, and A. Gresham (1994). Six sigma in customer satisfaction. *Quality Progress 27*(12): 73–76.

Gilson, J. (1951). *A New Approach to Engineering Tolerances*. London: Machinery Publishing Company.

Hale-Bennett, C. and D. K. J. Lin (1997). From SPC to DOE: A case study at Meco, Inc. *Quality Engineering 9*(3): 489–502.

Hare, L., R. W. Hoerl, J. D. Hromi, and R. D. Snee (1995). The role of statistical thinking in management. *Quality Progress 28*(2): 53–60.

Harry, M. J. (1998). Six sigma: A breakthrough strategy for profitability. *Quality Progress 31*(5): 60–64.

Hoerl, R. W. (1998). Six Sigma and the future of the quality profession. *Quality Progress 31*(6): 35–42.

Klaus, L. A. (1997). Motorola brings fairy tales to life. *Quality Progress 30*(6): 25–28.

Koons, G. S. Meyer and M. Kurowski (1992). Evaluating design sensitivity using DOE and simulation. In *ASQC Quality Congress Transactions*, pp. 711–717. Milwaukee, WI: American Society for Quality Control.

Landes, L. (1995). Leading the duck at mission control. *Quality Progress 28*(7): 43–48.

Lynch, R. O. and R. J. Markle (1997). Understanding the nature of variability in a dry etch process. In V. Czitrom and P. D. Spagon, ed. *Statistical Case Studies for Industrial Process Improvement*, Chapter 7. Philadelphia: American Statistical Association and Society for Industrial and Applied Mathematics.

McFadden, F. R. (1993). Six Sigma quality programs. *Quality Progress 26*(6): 37–42.

Miller, E. J. and C. Casavant (1994). Continuous improvement through defect management: A powerful technique for increasing customer satisfaction. In *ASQC Quality Congress Transactions*, pp. 210–217. Milwaukee, WI: American Society for Quality Control.

Montgomery, D. C. (1996). *Introduction to Statistical Quality Control*, 3rd ed. New York: Wiley.

Noguera, J. and T. Nielsen (1992). Implementing Six-Sigma for Interconnect Technology. In *ASQC Quality Congress Transactions*, pp. 538–544. Milwaukee, WI: American Society for Quality Control.

Pankratz, P. C. (1997). Calibration of an FTIR spectrometer for measuring carbon. In V. Czitrom and P. D. Spagon, eds. *Statistical Case Studies for Industrial Process Improvement*, pp. 19–37. Philadelphia: SIAM and ASA.

Rich, A. B. (1997). Continuous improvement: The key to future success. *Quality Progress 30*(6): 33–36.

St. Clair, J. F. (1995). Safety measurement: Towards continuous improvement. In *ASQC 49th Annual Quality Congress Proceedings*, pp. 150–158. Milwaukee, WI: American Society for Quality Control.

Tadikamalla, P. R. (1994). The confusion over six-sigma quality. *Quality Progress 27*(11): 83–85.

Zinkgraf, S. A. (1997). An overview of operational excellence at AlliedSignal. Presentation made at the Quality and Productivity Research Conference, Orlando, FL, May 11–13.

Answers to Selected Exercises

Chapter 3

1. $S^2 = 7.3$. Adding a constant to every number in a sample does not change the value of S^2.

5. (a) 0.95053; (d) 0.9545; (e) $z_0 = 1.96$.

7. $\frac{1}{12}$.

9. 0.06681. The probability is approximate since the particular probability distribution was not stated.

15. 2. We have 3% as the estimate of the proportion nonconforming, and conclude that the process needs improving.

17. 0.9727.

Chapter 4

5. The control limits could be adjusted or a control chart of residuals could be used.

Chapter 5

2. 13, 2.

3. The moving ranges and moving averages are correlated so interpreting the charts can be difficult.

5. An $\overline{X}$ chart is more sensitive in detecting a shift in the process mean (as the answers to exercise 3 indicate).

Chapter 6

1. 0.0462 and 0.2583.

3. Because there will not be a lower control limit.

6. The midline of the conventional u chart is 0.095 and all of the points are well within the (variable) limits.

Chapter 7

1. (a) $\frac{2}{3}$; (c) decrease, 0.02278 (before), 0.0027 (after).
3. 1.475, no.

Chapter 8

2. (a) 5.75 assuming a basic CUSUM scheme; (b) 15.
3. When doubt exists as to whether or not an assignable cause was removed.

Chapter 9

1. (a) 11.58.

Chapter 11

2. (a) No outliers are identified; (b) the estimate of sigma is 3.21.

Chapter 12

3. (c) $\hat{Y} = 4.0 - 0.214X$; (d) the plot is parabolic because the relationship between Y and X is parabolic, as can be seen from the scatter plot.
7. (b) $\hat{\beta}_0 = 12.69$, $\hat{\beta}_1 = 1.25$; (c) $R^2 = 0.945$; (d) (1.02, 1.48); (e) (40.39, 47.49).

Chapter 13

3. t will have the same value as t^* since $s_1^2 = s_2^2$.
5. The B effect and the AB interaction seem to be significant ($F = 10.59$).
7. 16.
11. Boxplot.

Chapter 14

1. (a) No, 12 is not a multiple of 2; (b) only main effects.
2. 6.25.
3. No, AB would not be independent of the C and D effects.
5. The first process.

Chapter 15

2. No standard error limits are computed with Simplex EVOP.
8. Simplex EVOP.

Chapter 16

3. k hypotheses are tested: $\mu_i = \dfrac{\mu_1 + \mu_2 + \cdots \mu_k}{k}$ $\quad i = 1, 2, \ldots, k$.
4. LDL = 0.0054, UDL = 0.0346.

APPENDIX

Statistical Tables

TABLE A Random Numbers[a]

1559	9068	9290	8303	8508	8954	1051	6677	6415	0342
5550	6245	7313	0117	7652	5069	6354	7668	1096	5780
4735	6214	8037	1385	1882	0828	2957	0530	9210	0177
5333	1313	3063	1134	8676	6241	9960	5304	1582	6198
8495	2956	1121	8484	2920	7934	0670	5263	0968	0069
1947	3353	1197	7363	9003	9313	3434	4261	0066	2714
4785	6325	1868	5020	9100	0823	7379	7391	1250	5501
9972	9163	5833	0100	5758	3696	6496	6297	5653	7782
0472	4629	2007	4464	3312	8728	1193	2497	4219	5339
4727	6994	1175	5622	2341	8562	5192	1471	7206	2027
3658	3226	5981	9025	1080	1437	6721	7331	0792	5383
6906	9758	0244	0259	4609	1269	5957	7556	1975	7898
3793	6916	0132	8873	8987	4975	4814	2098	6683	0901
3376	5966	1614	4025	0721	1537	6695	6090	8083	5450
6126	0224	7169	3596	1593	5097	7286	2686	1796	1150
0466	7566	1320	8777	8470	5448	9575	4669	1402	3905
9908	9832	8185	8835	0384	3699	1272	1181	8627	1968
7594	3636	1224	6808	1184	3404	6752	4391	2016	6167
5715	9301	5847	3524	0077	6674	8061	5438	6508	9673
7932	4739	4567	6797	4540	8488	3639	9777	1621	7244
6311	2025	5250	6099	6718	7539	9681	3204	9637	1091
0476	1624	3470	1600	0675	3261	7749	4195	2660	2150
5317	3903	6098	9438	3482	5505	5167	9993	8191	8488
7474	8876	1918	9828	2061	6664	0391	9170	2776	4025
7460	6800	1987	2758	0737	6880	1500	5763	2061	9373
1002	1494	9972	3877	6104	4006	0477	0669	8557	0513
5449	6891	9047	6297	1075	7762	8091	7153	8881	3367
9453	0809	7151	9982	0411	1120	6129	5090	2053	7570
0471	2725	7588	6573	0546	0110	6132	1224	3124	6563
5469	2668	1996	2249	3857	6637	8010	1701	3141	6147
2782	9603	1877	4159	9809	2570	4544	0544	2660	6737
3129	7217	5020	3788	0853	9465	2186	3945	1696	2286
7092	9885	3714	8557	7804	9524	6228	7774	6674	2775

TABLE A (*Continued*)

9566	0501	8352	1062	0634	2401	0379	1697	7153	6208
5863	7000	1714	9276	7218	6922	1032	4838	1954	1680
5881	9151	2321	3147	6755	2510	5759	6947	7102	0097
6416	9939	9569	0439	1705	4860	9881	7071	9596	8758
9568	3012	6316	9065	0710	2158	1639	9149	4848	8634
0452	9538	5730	1893	1186	9245	6558	9562	8534	9321
8762	5920	8989	4777	2169	7073	7082	9495	1594	8600
0194	0270	7601	0342	3897	4133	7650	9228	5558	3597
3306	5478	2797	1605	4996	0023	9780	9429	3937	7573
7198	3079	2171	6972	0928	6599	9328	0597	5948	5753
8350	4846	1309	0612	4584	4988	4642	4430	9481	9048
7449	4279	4224	1018	2496	2091	9750	6086	1955	9860
6126	5399	0852	5491	6557	4946	9918	1541	7894	1843
1851	7940	9908	3860	1536	8011	4314	7269	7047	0382
7698	4218	2726	5130	3132	1722	8592	9662	4795	7718
0810	0118	4979	0458	1059	5739	7919	4557	0245	4861
6647	7149	1409	6809	3313	0082	9024	7477	7320	5822
3867	7111	5549	9439	3427	9793	3071	6651	4267	8099
1172	7278	7527	2492	6211	9457	5120	4903	1023	5745
6701	1668	5067	0413	7961	7825	9261	8572	0634	1140
8244	0620	8736	2649	1429	6253	4181	8120	6500	8127
8009	4031	7884	2215	2382	1931	1252	8088	2490	9122
1947	8315	9755	7187	4074	4743	6669	6060	2319	0635
9562	4821	8050	0106	2782	4665	9436	4973	4879	8900
0729	9026	9631	8096	8906	5713	3212	8854	3435	4206
6904	2569	3251	0079	8838	8738	8503	6333	0952	1641

[a]Table A was produced using MINITAB, which is a registered trademark of Minitab, Inc., 3081 Enterprise Drive, State College, PA 16801, (814) 238–3280.

TABLE B Normal Distribution[a] $[P(0 \leq Z \leq z)$ where $Z \sim N(0, 1)]$

0 z

z	0.00	0.01	0.02	0.03	0.04	0.05	0.06	0.07	0.08	0.09
0.0	0.00000	0.00399	0.00798	0.01197	0.01595	0.01994	0.02392	0.02790	0.03188	0.03586
0.1	0.03983	0.04380	0.04776	0.05172	0.05567	0.05962	0.06356	0.06749	0.07142	0.07535
0.2	0.07926	0.08317	0.08706	0.09095	0.09483	0.09871	0.10257	0.10642	0.11026	0.11409
0.3	0.11791	0.12172	0.12552	0.12930	0.13307	0.13683	0.14058	0.14431	0.14803	0.15173
0.4	0.15542	0.15910	0.16276	0.16640	0.17003	0.17364	0.17724	0.18082	0.18439	0.18793
0.5	0.19146	0.19497	0.19847	0.20194	0.20540	0.20884	0.21226	0.21566	0.21904	0.22240
0.6	0.22575	0.22907	0.23237	0.23565	0.23891	0.24215	0.24537	0.24857	0.25175	0.25490
0.7	0.25804	0.26115	0.26424	0.26730	0.27035	0.27337	0.27637	0.27935	0.28230	0.28524
0.8	0.28814	0.29103	0.29389	0.29673	0.29955	0.30234	0.30511	0.30785	0.31057	0.31327
0.9	0.31594	0.31859	0.32121	0.32381	0.32639	0.32894	0.33147	0.33398	0.33646	0.33891
1.0	0.34134	0.34375	0.34614	0.34849	0.35083	0.35314	0.35543	0.35769	0.35993	0.36214
1.1	0.36433	0.36650	0.36864	0.37076	0.37286	0.37493	0.37698	0.37900	0.38100	0.38298
1.2	0.38493	0.38686	0.38877	0.39065	0.39251	0.39435	0.39617	0.39796	0.39973	0.40147
1.3	0.40320	0.40490	0.40658	0.40824	0.40988	0.41149	0.41308	0.41466	0.41621	0.41774
1.4	0.41924	0.42073	0.42220	0.42364	0.42507	0.42647	0.42785	0.42922	0.43056	0.43189
1.5	0.43319	0.43448	0.43574	0.43699	0.43822	0.43943	0.44062	0.44179	0.44295	0.44408
1.6	0.44520	0.44630	0.44738	0.44845	0.44950	0.45053	0.45154	0.45254	0.45352	0.45449
1.7	0.45543	0.45637	0.45728	0.45818	0.45907	0.45994	0.46080	0.46164	0.46246	0.46327
1.8	0.46407	0.46485	0.46562	0.46638	0.46712	0.46784	0.46856	0.46926	0.46995	0.47062
1.9	0.47128	0.47193	0.47257	0.47320	0.47381	0.47441	0.47500	0.47558	0.47615	0.47670

2.0	0.47725	0.47778	0.47831	0.47882	0.47932	0.47982	0.48030	0.48077	0.48124	0.48169
2.1	0.48214	0.48257	0.48300	0.48341	0.48382	0.48422	0.48461	0.48500	0.48537	0.48574
2.2	0.48610	0.48645	0.48679	0.48713	0.48745	0.48778	0.48809	0.48840	0.48870	0.48899
2.3	0.48928	0.48956	0.48983	0.49010	0.49036	0.49061	0.49086	0.49111	0.49134	0.49158
2.4	0.49180	0.49202	0.49224	0.49245	0.49266	0.49286	0.49305	0.49324	0.49343	0.49361
2.5	0.49379	0.49396	0.49413	0.49430	0.49446	0.49461	0.49477	0.49492	0.49506	0.49520
2.6	0.49534	0.49547	0.49560	0.49573	0.49585	0.49598	0.49609	0.49621	0.49632	0.49643
2.7	0.49653	0.49664	0.49674	0.49683	0.49693	0.49702	0.49711	0.49720	0.49728	0.49736
2.8	0.49744	0.49752	0.49760	0.49767	0.49774	0.49781	0.49788	0.49795	0.49801	0.49807
2.9	0.49813	0.49819	0.49825	0.49831	0.49836	0.49841	0.49846	0.49851	0.49856	0.49861
3.0	0.49865	0.49869	0.49874	0.49878	0.49882	0.49886	0.49889	0.49893	0.49896	0.49900
3.1	0.49903	0.49906	0.49910	0.49913	0.49916	0.49918	0.49921	0.49924	0.49926	0.49929
3.2	0.49931	0.49934	0.49936	0.49938	0.49940	0.49942	0.49944	0.49946	0.49948	0.49950
3.3	0.49952	0.49953	0.49955	0.49957	0.49958	0.49960	0.49961	0.49962	0.49964	0.49965
3.4	0.49966	0.49968	0.49969	0.49970	0.49971	0.49972	0.49973	0.49974	0.49975	0.49976
3.5	0.49977	0.49978	0.49978	0.49979	0.49980	0.49981	0.49981	0.49982	0.49983	0.49983
3.6	0.49984	0.49985	0.49985	0.49986	0.49986	0.49987	0.49987	0.49988	0.49988	0.49989
3.7	0.49989	0.49990	0.49990	0.49990	0.49991	0.49991	0.49992	0.49992	0.49992	0.49992
3.8	0.49993	0.49993	0.49993	0.49994	0.49994	0.49994	0.49994	0.49995	0.49995	0.49995
3.9	0.49995	0.49995	0.49996	0.49996	0.49996	0.49996	0.49996	0.49996	0.49997	0.49997

[a]These values were generated using MINITAB.

TABLE C *t* **Distribution**[a]

d.f. (ν)/α	0.40	0.25	0.10	0.05	0.025	0.01	0.005	0.0025	0.001	0.0005
1	0.325	1.000	3.078	6.314	12.706	31.820	63.655	127.315	318.275	636.438
2	0.289	0.816	1.886	2.920	4.303	6.965	9.925	14.089	22.327	31.596
3	0.277	0.765	1.638	2.353	3.182	4.541	5.841	7.453	10.214	12.923
4	0.271	0.741	1.533	2.132	2.776	3.747	4.604	5.597	7.173	8.610
5	0.267	0.727	1.476	2.015	2.571	3.365	4.032	4.773	5.893	6.869
6	0.265	0.718	1.440	1.943	2.447	3.143	3.707	4.317	5.208	5.959
7	0.263	0.711	1.415	1.895	2.365	2.998	3.499	4.029	4.785	5.408
8	0.262	0.706	1.397	1.860	2.306	2.896	3.355	3.833	4.501	5.041
9	0.261	0.703	1.383	1.833	2.262	2.821	3.250	3.690	4.297	4.781
10	0.260	0.700	1.372	1.812	2.228	2.764	3.169	3.581	4.144	4.587
11	0.260	0.697	1.363	1.796	2.201	2.718	3.106	3.497	4.025	4.437
12	0.259	0.695	1.356	1.782	2.179	2.681	3.055	3.428	3.930	4.318
13	0.259	0.694	1.350	1.771	2.160	2.650	3.012	3.372	3.852	4.221
14	0.258	0.692	1.345	1.761	2.145	2.624	2.977	3.326	3.787	4.140
15	0.258	0.691	1.341	1.753	2.131	2.602	2.947	3.286	3.733	4.073
16	0.258	0.690	1.337	1.746	2.120	2.583	2.921	3.252	3.686	4.015
17	0.257	0.689	1.333	1.740	2.110	2.567	2.898	3.222	3.646	3.965
18	0.257	0.688	1.330	1.734	2.101	2.552	2.878	3.197	3.610	3.922
19	0.257	0.688	1.328	1.729	2.093	2.539	2.861	3.174	3.579	3.883
20	0.257	0.687	1.325	1.725	2.086	2.528	2.845	3.153	3.552	3.849
21	0.257	0.686	1.323	1.721	2.080	2.518	2.831	3.135	3.527	3.819
22	0.256	0.686	1.321	1.717	2.074	2.508	2.819	3.119	3.505	3.792
23	0.256	0.685	1.319	1.714	2.069	2.500	2.807	3.104	3.485	3.768
24	0.256	0.685	1.318	1.711	2.064	2.492	2.797	3.091	3.467	3.745
25	0.256	0.684	1.316	1.708	2.060	2.485	2.787	3.078	3.450	3.725
26	0.256	0.684	1.315	1.706	2.056	2.479	2.779	3.067	3.435	3.707
27	0.256	0.684	1.314	1.703	2.052	2.473	2.771	3.057	3.421	3.690
28	0.256	0.683	1.313	1.701	2.048	2.467	2.763	3.047	3.408	3.674
29	0.256	0.683	1.311	1.699	2.045	2.462	2.756	3.038	3.396	3.659
30	0.256	0.683	1.310	1.697	2.042	2.457	2.750	3.030	3.385	3.646
40	0.255	0.681	1.303	1.684	2.021	2.423	2.704	2.971	3.307	3.551
60	0.254	0.679	1.296	1.671	2.000	2.390	2.660	2.915	3.232	3.460
100	0.254	0.677	1.290	1.660	1.984	2.364	2.626	2.871	3.174	3.391
Infinity	0.253	0.674	1.282	1.645	1.960	2.326	2.576	2.807	3.090	3.290

[a]These values were generated using MINITAB.

TABLE D F Distribution[a,b]

.05

0 $F_{\nu_1,\nu_2,.05}$

ν_1	1	2	3	4	5	6	7	8	9	10	11	12	13	14	15
							a. $F_{\nu_1,\nu_2,.05}$								
ν_2 1	161.44	199.50	215.69	224.57	230.16	233.98	236.78	238.89	240.55	241.89	242.97	243.91	244.67	245.35	245.97
2	18.51	19.00	19.16	19.25	19.30	19.33	19.35	19.37	19.39	19.40	19.40	19.41	19.42	19.42	19.43
3	10.13	9.55	9.28	9.12	9.01	8.94	8.89	8.85	8.81	8.79	8.76	8.74	8.73	8.71	8.70
4	7.71	6.94	6.59	6.39	6.26	6.16	6.09	6.04	6.00	5.96	5.94	5.91	5.89	5.87	5.86
5	6.61	5.79	5.41	5.19	5.05	4.95	4.88	4.82	4.77	4.74	4.70	4.68	4.66	4.64	4.62
6	5.99	5.14	4.76	4.53	4.39	4.28	4.21	4.15	4.10	4.06	4.03	4.00	3.98	3.96	3.94
7	5.59	4.74	4.35	4.12	3.97	3.87	3.79	3.73	3.68	3.64	3.60	3.57	3.55	3.53	3.51
8	5.32	4.46	4.07	3.84	3.69	3.58	3.50	3.44	3.39	3.35	3.31	3.28	3.26	3.24	3.22
9	5.12	4.26	3.86	3.63	3.48	3.37	3.29	3.23	3.18	3.14	3.10	3.07	3.05	3.03	3.01
10	4.96	4.10	3.71	3.48	3.33	3.22	3.14	3.07	3.02	2.98	2.94	2.91	2.89	2.86	2.85
11	4.84	3.98	3.59	3.36	3.20	3.09	3.01	2.95	2.90	2.85	2.82	2.79	2.76	2.74	2.72
12	4.75	3.89	3.49	3.26	3.11	3.00	2.91	2.85	2.80	2.75	2.72	2.69	2.66	2.64	2.62
13	4.67	3.81	3.41	3.18	3.03	2.92	2.83	2.77	2.71	2.67	2.63	2.60	2.58	2.55	2.53
14	4.60	3.74	3.34	3.11	2.96	2.85	2.76	2.70	2.65	2.60	2.57	2.53	2.51	2.48	2.46
15	4.54	3.68	3.29	3.06	2.90	2.79	2.71	2.64	2.59	2.54	2.51	2.48	2.45	2.42	2.40
16	4.49	3.63	3.24	3.01	2.85	2.74	2.66	2.59	2.54	2.49	2.46	2.42	2.40	2.37	2.35
17	4.45	3.59	3.20	2.96	2.81	2.70	2.61	2.55	2.49	2.45	2.41	2.38	2.35	2.33	2.31
18	4.41	3.55	3.16	2.93	2.77	2.66	2.58	2.51	2.46	2.41	2.37	2.34	2.31	2.29	2.27
19	4.38	3.52	3.13	2.90	2.74	2.63	2.54	2.48	2.42	2.38	2.34	2.31	2.28	2.26	2.23
20	4.35	3.49	3.10	2.87	2.71	2.60	2.51	2.45	2.39	2.35	2.31	2.28	2.25	2.22	2.20
21	4.32	3.47	3.07	2.84	2.68	2.57	2.49	2.42	2.37	2.32	2.28	2.25	2.22	2.20	2.18

TABLE D ***F*** **Distribution (*Continued*)**

ν_1	1	2	3	4	5	6	7	8	9	10	11	12	13	14	15
22	4.30	3.44	3.05	2.82	2.66	2.55	2.46	2.40	2.34	2.30	2.26	2.23	2.20	2.17	2.15
23	4.28	3.42	3.03	2.80	2.64	2.53	2.44	2.37	2.32	2.27	2.24	2.20	2.18	2.15	2.13
24	4.26	3.40	3.01	2.78	2.62	2.51	2.42	2.36	2.30	2.25	2.22	2.18	2.15	2.13	2.11
25	4.24	3.39	2.99	2.76	2.60	2.49	2.40	2.34	2.28	2.24	2.20	2.16	2.14	2.11	2.09
26	4.23	3.37	2.98	2.74	2.59	2.47	2.39	2.32	2.27	2.22	2.18	2.15	2.12	2.09	2.07
27	4.21	3.35	2.96	2.73	2.57	2.46	2.37	2.31	2.25	2.20	2.17	2.13	2.10	2.08	2.06
28	4.20	3.34	2.95	2.71	2.56	2.45	2.36	2.29	2.24	2.19	2.15	2.12	2.09	2.06	2.04
29	4.18	3.33	2.93	2.70	2.55	2.43	2.35	2.28	2.22	2.18	2.14	2.10	2.08	2.05	2.03
30	4.17	3.32	2.92	2.69	2.53	2.42	2.33	2.27	2.21	2.16	2.13	2.09	2.06	2.04	2.01
40	4.08	3.23	2.84	2.61	2.45	2.34	2.25	2.18	2.12	2.08	2.04	2.00	1.97	1.95	1.92

b. $F_{v_1,v_2,0.01}$

ν_2	1	2	3	4	5	6	7	8	9	10	11	12	13	14	15
1	4052.45	4999.42	5402.96	5624.03	5763.93	5858.82	5928.73	5981.06	6021.73	6055.29	6083.22	6106.00	6125.37	6142.48	6157.06
2	98.51	99.00	99.17	99.25	99.30	99.33	99.35	99.38	99.39	99.40	99.41	99.41	99.42	99.42	99.43
3	34.12	30.82	29.46	28.71	28.24	27.91	27.67	27.49	27.35	27.23	27.13	27.05	26.98	26.92	26.87
4	21.20	18.00	16.69	15.98	15.52	15.21	14.98	14.80	14.66	14.55	14.45	14.37	14.31	14.25	14.20
5	16.26	13.27	12.06	11.39	10.97	10.67	10.46	10.29	10.16	10.05	9.96	9.89	9.82	9.77	9.72
6	13.74	10.92	9.78	9.15	8.75	8.47	8.26	8.10	7.98	7.87	7.79	7.72	7.66	7.60	7.56
7	12.25	9.55	8.45	7.85	7.46	7.19	6.99	6.84	6.72	6.62	6.54	6.47	6.41	6.36	6.31
8	11.26	8.65	7.59	7.01	6.63	6.37	6.18	6.03	5.91	5.81	5.73	5.67	5.61	5.56	5.52
9	10.56	8.02	6.99	6.42	6.06	5.80	5.61	5.47	5.35	5.26	5.18	5.11	5.05	5.01	4.96
10	10.04	7.56	6.55	5.99	5.64	5.39	5.20	5.06	4.94	4.85	4.77	4.71	4.65	4.60	4.56
11	9.65	7.21	6.22	5.67	5.32	5.07	4.89	4.74	4.63	4.54	4.46	4.40	4.34	4.29	4.25
12	9.33	6.93	5.95	5.41	5.06	4.82	4.64	4.50	4.39	4.30	4.22	4.16	4.10	4.05	4.01
13	9.07	6.70	5.74	5.21	4.86	4.62	4.44	4.30	4.19	4.10	4.02	3.96	3.91	3.86	3.82
14	8.86	6.51	5.56	5.04	4.69	4.46	4.28	4.14	4.03	3.94	3.86	3.80	3.75	3.70	3.66

15	8.68	6.36	5.42	4.89	4.56	4.32	4.14	4.00	3.89	3.80	3.73	3.67	3.61	3.56	3.52
16	8.53	6.23	5.29	4.77	4.44	4.20	4.03	3.89	3.78	3.69	3.62	3.55	3.50	3.45	3.41
17	8.40	6.11	5.18	4.67	4.34	4.10	3.93	3.79	3.68	3.59	3.52	3.46	3.40	3.35	3.31
18	8.29	6.01	5.09	4.58	4.25	4.01	3.84	3.71	3.60	3.51	3.43	3.37	3.32	3.27	3.23
19	8.18	5.93	5.01	4.50	4.17	3.94	3.77	3.63	3.52	3.43	3.36	3.30	3.24	3.19	3.15
20	8.10	5.85	4.94	4.43	4.10	3.87	3.70	3.56	3.46	3.37	3.29	3.23	3.18	3.13	3.09
21	8.02	5.78	4.87	4.37	4.04	3.81	3.64	3.51	3.40	3.31	3.24	3.17	3.12	3.07	3.03
22	7.95	5.72	4.82	4.31	3.99	3.76	3.59	3.45	3.35	3.26	3.18	3.12	3.07	3.02	2.98
23	7.88	5.66	4.76	4.26	3.94	3.71	3.54	3.41	3.30	3.21	3.14	3.07	3.02	2.97	2.93
24	7.82	5.61	4.72	4.22	3.90	3.67	3.50	3.36	3.26	3.17	3.09	3.03	2.98	2.93	2.89
25	7.77	5.57	4.68	4.18	3.85	3.63	3.46	3.32	3.22	3.13	3.06	2.99	2.94	2.89	2.85
26	7.72	5.53	4.64	4.14	3.82	3.59	3.42	3.29	3.18	3.09	3.02	2.96	2.90	2.86	2.81
27	7.68	5.49	4.60	4.11	3.78	3.56	3.39	3.26	3.15	3.06	2.99	2.93	2.87	2.82	2.78
28	7.64	5.45	4.57	4.07	3.75	3.53	3.36	3.23	3.12	3.03	2.96	2.90	2.84	2.79	2.75
29	7.60	5.42	4.54	4.04	3.73	3.50	3.33	3.20	3.09	3.00	2.93	2.87	2.81	2.77	2.73
30	7.56	5.39	4.51	4.02	3.70	3.47	3.30	3.17	3.07	2.98	2.91	2.84	2.79	2.74	2.70
40	7.31	5.18	4.31	3.83	3.51	3.29	3.12	2.99	2.89	2.80	2.73	2.66	2.61	2.56	2.52

[a]These values were generated using MINITAB.
[b]ν_2 = degrees of freedom for the denominator; ν_1 = degrees of freedom for the numerator.

TABLE E Control Chart Constants

	For Estimating Sigma		For $\overline{X}$ Chart		For $\overline{X}$ Chart (Standard Given)	For R Chart		For R Chart (Standard Given)		For s chart (Standard Given)			
n	c_4*	d_2	A_2	A_3*	A	D_3	D_4	D_1	D_2	B_3	B_4	B_5*	B_6*
2	0.7979	1.128	1.880	2.659	2.121	0	3.267	0	3.686	0	3.267	0	2.606
3	0.8862	1.693	1.023	1.954	1.732	0	2.575	0	4.358	0	2.568	0	2.276
4	0.9213	2.059	0.729	1.628	1.500	0	2.282	0	4.698	0	2.266	0	2.088
5	0.9400	2.326	0.577	1.427	1.342	0	2.115	0	4.918	0	2.089	0	1.964
6	0.9515	2.534	0.483	1.287	1.225	0	2.004	0	5.078	0.030	1.970	0.029	1.874
7	0.9594	2.704	0.419	1.182	1.134	0.076	1.924	0.205	5.203	0.118	1.882	0.113	1.806
8	0.9650	2.847	0.373	1.099	1.061	0.136	1.864	0.387	5.307	0.185	1.815	0.179	1.751
9	0.9693	2.970	0.337	1.032	1.000	0.184	1.816	0.546	5.394	0.239	1.761	0.232	1.707
10	0.9727	3.078	0.308	0.975	0.949	0.223	1.777	0.687	5.469	0.284	1.716	0.276	1.669
15	0.9823	3.472	0.223	0.789	0.775	0.348	1.652	1.207	5.737	0.428	1.572	0.421	1.544
20	0.9869	3.735	0.180	0.680	0.671	0.414	1.586	1.548	5.922	0.510	1.490	0.504	1.470
25	0.9896	3.931	0.153	0.606	0.600	0.459	1.541	1.804	6.058	0.565	1.435	0.559	1.420

*Columns marked with an asterisk are from The American Society for Quality Control Standard A1, Table 1, 1987. Reprinted with permission of the American Society for Quality Control. The balance of the table is from Table B2 of the *A.S.T.M. Manual of Quality Control of Materials.* Copyright American Society for Testing Materials. Reprinted with permission.

TABLE F Percentage Points of the Sample Range for Producing Probability Limits for R Charts[a]

D/n	4	5	6	7	8	9	10
0.0005	0.158	0.308	0.464	0.613	0.751	0.878	0.995
0.0010	0.199	0.367	0.535	0.691	0.835	0.966	1.085
0.0050	0.343	0.555	0.749	0.922	1.075	1.212	1.335
0.0100	0.434	0.665	0.870	1.048	1.205	1.343	1.467
0.9900	4.403	4.603	4.757	4.882	4.987	5.078	5.157
0.9950	4.694	4.886	5.033	5.154	5.255	5.341	5.418
0.9990	5.309	5.484	5.619	5.730	5.823	5.903	5.973
0.9995	5.553	5.722	5.853	5.960	6.050	6.127	6.196

[a]These values have been adapted from Table 1 of H. L. Harter, Tables of range and studentized range. *The Annals of Mathematical Statistics 31*(4), December 1960. Reprinted with permission of the Institute of Mathematical Statistics.

TABLE G Analysis of Means Constants[a]

	Number of Means, k																	
d.f.[b] (v)	3	4	5	6	7	8	9	10	11	12	13	14	15	16	17	18	19	20
								a. $h_{0.05}$										
3	4.18																	
4	3.56	3.89																
5	3.25	3.53	3.72															
6	3.07	3.31	3.49	3.62														
7	2.94	3.17	3.33	3.45	3.56													
8	2.86	3.07	3.21	3.33	3.43	3.51												
9	2.79	2.99	3.13	3.24	3.33	3.41	3.48											
10	2.74	2.93	3.07	3.17	3.26	3.33	3.40	3.45										
11	2.70	2.88	3.01	3.12	3.20	3.27	3.33	3.39	3.44									
12	2.67	2.85	2.97	3.07	3.15	3.22	3.28	3.33	3.38	3.42								
13	2.64	2.81	2.94	3.03	3.11	3.18	3.24	3.29	3.34	3.38	3.42							
14	2.62	2.79	2.91	3.00	3.08	3.14	3.20	3.25	3.30	3.34	3.37	3.41						
15	2.60	2.76	2.88	2.97	3.05	3.11	3.17	3.22	3.26	3.30	3.34	3.37	3.40					
16	2.58	2.74	2.86	2.95	3.02	3.09	3.14	3.19	3.23	3.27	3.31	3.34	3.37	3.40				
17	2.57	2.73	2.84	2.93	3.00	3.06	3.12	3.16	3.21	3.25	3.28	3.31	3.34	3.37	3.40			
18	2.55	2.71	2.82	2.91	2.98	3.04	3.10	3.14	3.18	3.22	3.26	3.29	3.32	3.35	3.37	3.40		
19	2.54	2.70	2.81	2.89	2.96	3.02	3.08	3.12	3.16	3.20	3.24	3.27	3.30	3.32	3.35	3.37	3.40	
20	2.53	2.68	2.79	2.88	2.95	3.01	3.06	3.11	3.15	3.18	3.22	3.25	3.28	3.30	3.33	3.35	3.37	3.40
24	2.50	2.65	2.75	2.83	2.90	2.96	3.01	3.05	3.09	3.13	3.16	3.19	3.22	3.24	3.27	3.29	3.31	3.33
30	2.47	2.61	2.71	2.79	2.85	2.91	2.96	3.00	3.04	3.07	3.10	3.13	3.16	3.18	3.20	3.22	3.25	3.27
40	2.43	2.57	2.67	2.75	2.81	2.86	2.91	2.95	2.98	3.01	3.04	3.07	3.10	3.12	3.14	3.16	3.18	3.20
60	2.40	2.54	2.63	2.70	2.76	2.81	2.86	2.90	2.93	2.96	2.99	3.02	3.04	3.06	3.08	3.10	3.12	3.14
120	2.37	2.50	2.59	2.66	2.72	2.77	2.81	2.84	2.88	2.91	2.93	2.96	2.98	3.00	3.02	3.04	3.06	3.08
Infinity	2.34	2.47	2.56	2.62	2.68	2.72	2.76	2.80	2.83	2.86	2.88	2.90	2.93	2.95	2.97	2.98	3.00	3.02

b. $h_{0.01}$

3	7.51																	
4	5.74	6.21																
5	4.93	5.29	5.55															
6	4.48	4.77	4.98	5.16														
7	4.18	4.44	4.63	4.78	4.90													
8	3.98	4.21	4.38	4.52	4.63	4.72												
9	3.84	4.05	4.20	4.33	4.43	4.51	4.59											
10	3.73	3.92	4.07	4.18	4.28	4.36	4.43	4.49										
11	3.64	3.82	3.96	4.07	4.16	4.23	4.30	4.36	4.41									
12	3.57	3.74	3.87	3.98	4.06	4.13	4.20	4.25	4.31	4.35								
13	3.51	3.68	3.80	3.90	3.98	4.05	4.11	4.17	4.22	4.26	4.30							
14	3.46	3.63	3.74	3.84	3.92	3.98	4.04	4.09	4.14	4.18	4.22	4.26						
15	3.42	3.58	3.69	3.79	3.86	3.92	3.98	4.03	4.08	4.12	4.16	4.19	4.22					
16	3.38	3.54	3.65	3.74	3.81	3.87	3.93	3.98	4.02	4.06	4.10	4.14	4.17	4.20				
17	3.35	3.50	3.61	3.70	3.77	3.83	3.89	3.93	3.98	4.02	4.05	4.09	4.12	4.14	4.17			
18	3.33	3.47	3.58	3.66	3.73	3.79	3.85	3.89	3.94	3.97	4.01	4.04	4.07	4.10	4.12	4.15		
19	3.30	3.45	3.55	3.63	3.70	3.76	3.81	3.86	3.90	3.94	3.97	4.00	4.03	4.06	4.08	4.11	4.13	
20	3.28	3.42	3.53	3.61	3.67	3.73	3.78	3.83	3.87	3.90	3.94	3.97	4.00	4.02	4.05	4.07	4.09	4.12
24	3.21	3.35	3.45	3.52	3.58	3.64	3.69	3.73	3.77	3.80	3.83	3.86	3.89	3.91	3.94	3.96	3.98	4.00
30	3.15	3.28	3.37	3.44	3.50	3.55	3.59	3.63	3.67	3.70	3.73	3.76	3.78	3.81	3.83	3.85	3.87	3.89
40	3.09	3.21	3.29	3.36	3.42	3.46	3.50	3.54	3.58	3.60	3.63	3.66	3.68	3.70	3.72	3.74	3.76	3.78
60	3.03	3.14	3.22	3.29	3.34	3.38	3.42	3.46	3.49	3.51	3.54	3.56	3.59	3.61	3.63	3.64	3.66	3.68
120	2.97	3.07	3.15	3.21	3.26	3.30	3.34	3.37	3.40	3.42	3.45	3.47	3.49	3.51	3.55	3.55	3.56	3.58
Infinity	2.91	3.01	3.08	3.14	3.18	3.22	3.26	3.29	3.32	3.34	3.36	3.38	3.40	3.42	3.44	3.45	3.47	3.48

[a]From Tables 2 and 3 of L. S. Nelson, Exact critical values for use with the analysis of means. *Journal of Qiality Technology 15(1)*, January 1983. Reprinted with permission of the American Society for Quality control.

[b] Degrees of freedom for s.

Author Index

Subject Index

WILEY SERIES IN PROBABILITY AND STATISTICS
ESTABLISHED BY WALTER A. SHEWHART AND SAMUEL S. WILKS

Probability and Statistics Section

*ANDERSON · The Statistical Analysis of Time Series
ARNOLD, BALAKRISHNAN, and NAGARAJA · A First Course in Order Statistics
ARNOLD, BALAKRISHNAN, and NAGARAJA · Records
BACCELLI, COHEN, OLSDER, and QUADRAT · Synchronization and Linearity: An Algebra for Discrete Event Systems
BASILEVSKY · Statistical Factor Analysis and Related Methods: Theory and Applications
BERNARDO and SMITH · Bayesian Statistical Concepts and Theory
BILLINGSLEY · Convergence of Probability Measures, *Second Edition*
BOROVKOV · Asymptotic Methods in Queuing Theory
BOROVKOV · Ergodicity and Stability of Stochastic Processes
BRANDT, FRANKEN, and LISEK · Stationary Stochastic Models
CAINES · Linear Stochastic Systems
CAIROLI and DALANG · Sequential Stochastic Optimization
CONSTANTINE · Combinatorial Theory and Statistical Design
COOK · Regression Graphics
COVER and THOMAS · Elements of Information Theory
CSÖRGŐ and HORVÁTH · Weighted Approximations in Probability Statistics
CSÖRGŐ and HORVÁTH · Limit Theorems in Change Point Analysis
DETTE and STUDDEN · The Theory of Canonical Moments with Applications in Statistics, Probability, and Analysis
DEY and MUKERJEE · Fractional Factorial Plans
*DOOB · Stochastic Processes
DRYDEN and MARDIA · Statistical Analysis of Shape
DUPUIS and ELLIS · A Weak Convergence Approach to the Theory of Large Deviations
ETHIER and KURTZ · Markov Processes: Characterization and Convergence
FELLER · An Introduction to Probability Theory and Its Applications, Volume I, *Third Edition,* Revised; Volume II, *Second Edition*
FULLER · Introduction to Statistical Time Series, *Second Edition*
FULLER · Measurement Error Models
GHOSH, MUKHOPADHYAY, and SEN · Sequential Estimation
GIFI · Nonlinear Multivariate Analysis
GUTTORP · Statistical Inference for Branching Processes
HALL · Introduction to the Theory of Coverage Processes
HAMPEL · Robust Statistics: The Approach Based on Influence Functions
HANNAN and DEISTLER · The Statistical Theory of Linear Systems
HUBER · Robust Statistics
HUSKOVA, BERAN, and DUPAC · Collected Works of Jaroslav Hajek—with Commentary
IMAN and CONOVER · A Modern Approach to Statistics

*Now available in a lower priced paperback edition in the Wiley Classics Library.

Probability and Statistics (Continued)

JUREK and MASON · Operator-Limit Distributions in Probability Theory
KASS and VOS · Geometrical Foundations of Asymptotic Inference
KAUFMAN and ROUSSEEUW · Finding Groups in Data: An Introduction to Cluster Analysis
KELLY · Probability, Statistics, and Optimization
KENDALL, BARDEN, CARNE, and LE · Shape and Shape Theory
LINDVALL · Lectures on the Coupling Method
McFADDEN · Management of Data in Clinical Trials
MANTON, WOODBURY, and TOLLEY · Statistical Applications Using Fuzzy Sets
MORGENTHALER and TUKEY · Configural Polysampling: A Route to Practical Robustness
MUIRHEAD · Aspects of Multivariate Statistical Theory
OLIVER and SMITH · Influence Diagrams, Belief Nets and Decision Analysis
*PARZEN · Modern Probability Theory and Its Applications
PRESS · Bayesian Statistics: Principles, Models, and Applications
PUKELSHEIM · Optimal Experimental Design
RAO · Asymptotic Theory of Statistical Inference
RAO · Linear Statistical Inference and Its Applications, *Second Edition*
RAO and SHANBHAG · Choquet-Deny Type Functional Equations with Applications to Stochastic Models
ROBERTSON, WRIGHT, and DYKSTRA · Order Restricted Statistical Inference
ROGERS and WILLIAMS · Diffusions, Markov Processes, and Martingales, Volume I: Foundations, *Second Edition;* Volume II: Îto Calculus
RUBINSTEIN and SHAPIRO · Discrete Event Systems: Sensitivity Analysis and Stochastic Optimization by the Score Function Method
RUZSA and SZEKELY · Algebraic Probability Theory
SCHEFFE · The Analysis of Variance
SEBER · Linear Regression Analysis
SEBER · Multivariate Observations
SEBER and WILD · Nonlinear Regression
SERFLING · Approximation Theorems of Mathematical Statistics
SHORACK and WELLNER · Empirical Processes with Applications to Statistics
SMALL and McLEISH · Hilbert Space Methods in Probability and Statistical Inference
STAPLETON · Linear Statistical Models
STAUDTE and SHEATHER · Robust Estimation and Testing
STOYANOV · Counterexamples in Probability
TANAKA · Time Series Analysis: Nonstationary and Noninvertible Distribution Theory
THOMPSON and SEBER · Adaptive Sampling
WELSH · Aspects of Statistical Inference
WHITTAKER · Graphical Models in Applied Multivariate Statistics
YANG · The Construction Theory of Denumerable Markov Processes

Applied Probability and Statistics Section

ABRAHAM and LEDOLTER · Statistical Methods for Forecasting
AGRESTI · Analysis of Ordinal Categorical Data
AGRESTI · Categorical Data Analysis
ANDERSON, AUQUIER, HAUCK, OAKES, VANDAELE, and WEISBERG · Statistical Methods for Comparative Studies
ARMITAGE and DAVID (editors) · Advances in Biometry
*ARTHANARI and DODGE · Mathematical Programming in Statistics

*Now available in a lower priced paperback edition in the Wiley Classics Library.

Applied Probability and Statistics (Continued)

ASMUSSEN · Applied Probability and Queues
*BAILEY · The Elements of Stochastic Processes with Applications to the Natural Sciences
BARNETT · Comparative Statistical Inference, *Third Edition*
BARNETT and LEWIS · Outliers in Statistical Data, *Third Edition*
BARTHOLOMEW, FORBES, and McLEAN · Statistical Techniques for Manpower Planning, *Second Edition*
BATES and WATTS · Nonlinear Regression Analysis and Its Applications
BECHHOFER, SANTNER, and GOLDSMAN · Design and Analysis of Experiments for Statistical Selection, Screening, and Multiple Comparisons
BELSLEY · Conditioning Diagnostics: Collinearity and Weak Data in Regression
BELSLEY, KUH, and WELSCH · Regression Diagnostics: Identifying Influential Data and Sources of Collinearity
BHAT · Elements of Applied Stochastic Processes, *Second Edition*
BHATTACHARYA and WAYMIRE · Stochastic Processes with Applications
BIRKES and DODGE · Alternative Methods of Regression
BLOOMFIELD · Fourier Analysis of Time Series: An Introduction, *Second Edition*
BOLLEN · Structural Equations with Latent Variables
BOULEAU · Numerical Methods for Stochastic Processes
BOX · Bayesian Inference in Statistical Analysis
BOX and DRAPER · Empirical Model-Building and Response Surfaces
BOX and DRAPER · Evolutionary Operation: A Statistical Method for Process Improvement
BUCKLEW · Large Deviation Techniques in Decision, Simulation, and Estimation
BUNKE and BUNKE · Nonlinear Regression, Functional Relations and Robust Methods: Statistical Methods of Model Building
CHATTERJEE and HADI · Sensitivity Analysis in Linear Regression
CHERNICK · Bootstrap Methods: A Practitioner's Guide
CHILÈS and DELFINER · Geostatistics: Modeling Spatial Uncertainty
CHOW and LIU · Design and Analysis of Clinical Trials: Concepts and Methodologies
CLARKE and DISNEY · Probability and Random Processes: A First Course with Applications, *Second Edition*
*COCHRAN and COX · Experimental Designs, *Second Edition*
CONOVER · Practical Nonparametric Statistics, *Second Edition*
CORNELL · Experiments with Mixtures, Designs, Models, and the Analysis of Mixture Data, *Second Edition*
*COX · Planning of Experiments
CRESSIE · Statistics for Spatial Data, *Revised Edition*
DANIEL · Applications of Statistics to Industrial Experimentation
DANIEL · Biostatistics: A Foundation for Analysis in the Health Sciences, *Sixth Edition*
DAVID · Order Statistics, *Second Edition*
*DEGROOT, FIENBERG, and KADANE · Statistics and the Law
DODGE · Alternative Methods of Regression
DOWDY and WEARDEN · Statistics for Research, *Second Edition*
DUNN and CLARK · Applied Statistics: Analysis of Variance and Regression, *Second Edition*
ELANDT-JOHNSON and JOHNSON · Survival Models and Data Analysis
EVANS, PEACOCK, and HASTINGS · Statistical Distributions, *Second Edition*
FLEISS · The Design and Analysis of Clinical Experiments
FLEISS · Statistical Methods for Rates and Proportions, *Second Edition*
FLEMING and HARRINGTON · Counting Processes and Survival Analysis
GALLANT · Nonlinear Statistical Models
GLASSERMAN and YAO · Monotone Structure in Discrete-Event Systems

*Now available in a lower priced paperback edition in the Wiley Classics Library.

Applied Probability and Statistics (Continued)

GNANADESIKAN · Methods for Statistical Data Analysis of Multivariate Observations, *Second Edition*
GOLDSTEIN and LEWIS · Assessment: Problems, Development, and Statistical Issues
GREENWOOD and NIKULIN · A Guide to Chi-Squared Testing
*HAHN · Statistical Models in Engineering
HAHN and MEEKER · Statistical Intervals: A Guide for Practitioners
HAND · Construction and Assessment of Classification Rules
HAND · Discrimination and Classification
HEIBERGER · Computation for the Analysis of Designed Experiments
HEDAYAT and SINHA · Design and Inference in Finite Population Sampling
HINKELMAN and KEMPTHORNE: · Design and Analysis of Experiments, Volume 1: Introduction to Experimental Design
HOAGLIN, MOSTELLER, and TUKEY · Exploratory Approach to Analysis of Variance
HOAGLIN, MOSTELLER, and TUKEY · Exploring Data Tables, Trends and Shapes
HOAGLIN, MOSTELLER, and TUKEY · Understanding Robust and Exploratory Data Analysis
HOCHBERG and TAMHANE · Multiple Comparison Procedures
HOCKING · Methods and Applications of Linear Models: Regression and the Analysis of Variables
HOGG and KLUGMAN · Loss Distributions
HOSMER and LEMESHOW · Applied Logistic Regression
HØYLAND and RAUSAND · System Reliability Theory: Models and Statistical Methods
HUBERTY · Applied Discriminant Analysis
JACKSON · A User's Guide to Principle Components
JOHN · Statistical Methods in Engineering and Quality Assurance
JOHNSON · Multivariate Statistical Simulation
JOHNSON and KOTZ · Distributions in Statistics
Continuous Multivariate Distributions
JOHNSON, KOTZ, and BALAKRISHNAN · Continuous Univariate Distributions, Volume 1, *Second Edition*
JOHNSON, KOTZ, and BALAKRISHNAN · Continuous Univariate Distributions, Volume 2, *Second Edition*
JOHNSON, KOTZ, and BALAKRISHNAN · Discrete Multivariate Distributions
JOHNSON, KOTZ, and KEMP · Univariate Discrete Distributions, *Second Edition*
JUREČKOVÁ and SEN · Robust Statistical Procedures: Aymptotics and Interrelations
KADANE · Bayesian Methods and Ethics in a Clinical Trial Design
KADANE AND SCHUM · A Probabilistic Analysis of the Sacco and Vanzetti Evidence
KALBFLEISCH and PRENTICE · The Statistical Analysis of Failure Time Data
KELLY · Reversability and Stochastic Networks
KHURI, MATHEW, and SINHA · Statistical Tests for Mixed Linear Models
KLUGMAN, PANJER, and WILLMOT · Loss Models: From Data to Decisions
KLUGMAN, PANJER, and WILLMOT · Solutions Manual to Accompany Loss Models: From Data to Decisions
KOVALENKO, KUZNETZOV, and PEGG · Mathematical Theory of Reliability of Time-Dependent Systems with Practical Applications
LACHIN · Biostatistics Methods: The Assessment of Relative Risks
LAD · Operational Subjective Statistical Methods: A Mathematical, Philosophical, and Historical Introduction
LANGE, RYAN, BILLARD, BRILLINGER, CONQUEST, and GREENHOUSE · Case Studies in Biometry
LAWLESS · Statistical Models and Methods for Lifetime Data
LEE · Statistical Methods for Survival Data Analysis, *Second Edition*
LEPAGE and BILLARD · Exploring the Limits of Bootstrap

*Now available in a lower priced paperback edition in the Wiley Classics Library.

Applied Probability and Statistics (Continued)

LINHART and ZUCCHINI · Model Selection
LITTLE and RUBIN · Statistical Analysis with Missing Data
LLOYD · The Statistical Analysis of Categorical Data
MAGNUS and NEUDECKER · Matrix Differential Calculus with Applications in Statistics and Econometrics, *Revised Edition*
MALLER and ZHOU · Survival Analysis with Long Term Survivors
MANN, SCHAFER, and SINGPURWALLA · Methods for Statistical Analysis of Reliability and Life Data
McLACHLAN and KRISHNAN · The EM Algorithm and Extensions
McLACHLAN · Discriminant Analysis and Statistical Pattern Recognition
McNEIL · Epidemiological Research Methods
MEEKER and ESCOBAR · Statistical Methods for Reliability Data
MILLER · Survival Analysis
MONTGOMERY and PECK · Introduction to Linear Regression Analysis, *Second Edition*
MYERS and MONTGOMERY · Response Surface Methodology: Process and Product in Optimization Using Designed Experiments
NELSON · Accelerated Testing, Statistical Models, Test Plans, and Data Analyses
NELSON · Applied Life Data Analysis
OCHI · Applied Probability and Stochastic Processes in Engineering and Physical Sciences
OKABE, BOOTS, and SUGIHARA · Spatial Tesselations: Concepts and Applications of Voronoi Diagrams
PANKRATZ · Forecasting with Dynamic Regression Models
PANKRATZ · Forecasting with Univariate Box-Jenkins Models: Concepts and Cases
PIANTADOSI · Clinical Trials: A Methodologic Perspective
PORT · Theoretical Probability for Applications
PUTERMAN · Markov Decision Processes: Discrete Stochastic Dynamic Programming
RACHEV · Probability Metrics and the Stability of Stochastic Models
RÉNYI · A Diary on Information Theory
RIPLEY · Spatial Statistics
RIPLEY · Stochastic Simulation
ROLSKI, SCHMIDLI, SCHMIDT, and TEUGELS · Stochastic Processes for Insurance and Finance
ROUSSEEUW and LEROY · Robust Regression and Outlier Detection
RUBIN · Multiple Imputation for Nonresponse in Surveys
RUBINSTEIN · Simulation and the Monte Carlo Method
RUBINSTEIN and MELAMED · Modern Simulation and Modeling
RYAN · Statistical Methods for Quality Improvement, *Second Edition*
SCHUSS · Theory and Applications of Stochastic Differential Equations
SCOTT · Multivariate Density Estimation: Theory, Practice, and Visualization
*SEARLE · Linear Models
SEARLE · Linear Models for Unbalanced Data
SEARLE, CASELLA, and McCULLOCH · Variance Components
SENNOTT · Stochastic Dynamic Programming and the Control of Queueing Systems
STOYAN, KENDALL, and MECKE · Stochastic Geometry and Its Applications, *Second Edition*
STOYAN and STOYAN · Fractals, Random Shapes and Point Fields: Methods of Geometrical Statistics
THOMPSON · Empirical Model Building
THOMPSON · Sampling
THOMPSON · Simulation: A Modeler's Approach
TIJMS · Stochastic Modeling and Analysis: A Computational Approach
TIJMS · Stochastic Models: An Algorithmic Approach

*Now available in a lower priced paperback edition in the Wiley Classics Library.

TITTERINGTON, SMITH, and MAKOV · Statistical Analysis of Finite Mixture Distributions
UPTON and FINGLETON · Spatial Data Analysis by Example, Volume 1: Point Pattern and Quantitative Data
UPTON and FINGLETON · Spatial Data Analysis by Example, Volume II: Categorical and Directional Data
VAN RIJCKEVORSEL and DE LEEUW · Component and Correspondence Analysis
VIDAKOVIC · Statistical Modeling by Wavelets
WEISBERG · Applied Linear Regression, *Second Edition*
WESTFALL and YOUNG · Resampling-Based Multiple Testing: Examples and Methods for p-Value Adjustment
WHITTLE · Systems in Stochastic Equilibrium
WOODING · Planning Pharmaceutical Clinical Trials: Basic Statistical Principles
WOOLSON · Statistical Methods for the Analysis of Biomedical Data
*ZELLNER · An Introduction to Bayesian Inference in Econometrics

Texts and References Section

AGRESTI · An Introduction to Categorical Data Analysis
ANDERSON · An Introduction to Multivariate Statistical Analysis, *Second Edition*
ANDERSON and LOYNES · The Teaching of Practical Statistics
ARMITAGE and COLTON · Encyclopedia of Biostatistics: Volumes 1 to 6 with Index
BARTOSZYNSKI and NIEWIADOMSKA-BUGAJ · Probability and Statistical Inference
BENDAT and PIERSOL · Random Data: Analysis and Measurement Procedures, *Third Edition*
BERRY, CHALONER, and GEWEKE · Bayesian Analysis in Statistics and Econometrics: Essays in Honor of Arnold Zellner
BHATTACHARYA and JOHNSON · Statistical Concepts and Methods
BILLINGSLEY · Probability and Measure, *Second Edition*
BOX · R. A. Fisher, the Life of a Scientist
BOX, HUNTER, and HUNTER · Statistics for Experimenters: An Introduction to Design, Data Analysis, and Model Building
BOX and LUCEÑO · Statistical Control by Monitoring and Feedback Adjustment
BROWN and HOLLANDER · Statistics: A Biomedical Introduction
CHATTERJEE and PRICE · Regression Analysis by Example, *Third Edition*
COOK and WEISBERG · Applied Regression Including Computing and Graphics
COOK and WEISBERG · An Introduction to Regression Graphics
COX · A Handbook of Introductory Statistical Methods
DILLON and GOLDSTEIN · Multivariate Analysis: Methods and Applications
DODGE and ROMIG · Sampling Inspection Tables, *Second Edition*
DRAPER and SMITH · Applied Regression Analysis, *Third Edition*
DUDEWICZ and MISHRA · Modern Mathematical Statistics
DUNN · Basic Statistics: A Primer for the Biomedical Sciences, *Second Edition*
FISHER and VAN BELLE · Biostatistics: A Methodology for the Health Sciences
FREEMAN and SMITH · Aspects of Uncertainty: A Tribute to D. V. Lindley
GROSS and HARRIS · Fundamentals of Queueing Theory, *Third Edition*
HALD · A History of Probability and Statistics and their Applications Before 1750
HALD · A History of Mathematical Statistics from 1750 to 1930
HELLER · MACSYMA for Statisticians
HOEL · Introduction to Mathematical Statistics, *Fifth Edition*
HOLLANDER and WOLFE · Nonparametric Statistical Methods, *Second Edition*
HOSMER and LEMESHOW · Applied Survival Analysis: Regression Modeling of Time to Event Data

*Now available in a lower priced paperback edition in the Wiley Classics Library.

Texts and References (Continued)

JOHNSON and BALAKRISHNAN · Advances in the Theory and Practice of Statistics: A Volume in Honor of Samuel Kotz
JOHNSON and KOTZ (editors) · Leading Personalities in Statistical Sciences: From the Seventeenth Century to the Present
JUDGE, GRIFFITHS, HILL, LÜTKEPOHL, and LEE · The Theory and Practice of Econometrics, *Second Edition*
KHURI · Advanced Calculus with Applications in Statistics
KOTZ and JOHNSON (editors) · Encyclopedia of Statistical Sciences: Volumes 1 to 9 wtih Index
KOTZ and JOHNSON (editors) · Encyclopedia of Statistical Sciences: Supplement Volume
KOTZ, REED, and BANKS (editors) · Encyclopedia of Statistical Sciences: Update Volume 1
KOTZ, REED, and BANKS (editors) · Encyclopedia of Statistical Sciences: Update Volume 2
LAMPERTI · Probability: A Survey of the Mathematical Theory, *Second Edition*
LARSON · Introduction to Probability Theory and Statistical Inference, *Third Edition*
LE · Applied Categorical Data Analysis
LE · Applied Survival Analysis
MALLOWS · Design, Data, and Analysis by Some Friends of Cuthbert Daniel
MARDIA · The Art of Statistical Science: A Tribute to G. S. Watson
MASON, GUNST, and HESS · Statistical Design and Analysis of Experiments with Applications to Engineering and Science
MURRAY · X-STAT 2.0 Statistical Experimentation, Design Data Analysis, and Nonlinear Optimization
PURI, VILAPLANA, and WERTZ · New Perspectives in Theoretical and Applied Statistics
RENCHER · Linear Models in Statistics
RENCHER · Methods of Multivariate Analysis
RENCHER · Multivariate Statistical Inference with Applications
ROSS · Introduction to Probability and Statistics for Engineers and Scientists
ROHATGI · An Introduction to Probability Theory and Mathematical Statistics
RYAN · Modern Regression Methods
SCHOTT · Matrix Analysis for Statistics
SEARLE · Matrix Algebra Useful for Statistics
STYAN · The Collected Papers of T. W. Anderson: 1943–1985
TIERNEY · LISP-STAT: An Object-Oriented Environment for Statistical Computing and Dynamic Graphics
WONNACOTT and WONNACOTT · Econometrics, *Second Edition*

WILEY SERIES IN PROBABILITY AND STATISTICS

ESTABLISHED BY WALTER A. SHEWHART AND SAMUEL S. WILKS

Editors
Robert M. Groves, Graham Kalton, J. N. K. Rao, Norbert Schwarz, Christopher Skinner

Survey Methodology Section

BIEMER, GROVES, LYBERG, MATHIOWETZ, and SUDMAN · Measurement Errors in Surveys

*Now available in a lower priced paperback edition in the Wiley Classics Library.

Survey Methodology (Continued)

COCHRAN · Sampling Techniques, *Third Edition*
COUPER, BAKER, BETHLEHEM, CLARK, MARTIN, NICHOLLS, and O'REILLY (editors) · Computer Assisted Survey Information Collection
COX, BINDER, CHINNAPPA, CHRISTIANSON, COLLEDGE, and KOTT (editors) · Business Survey Methods
*DEMING · Sample Design in Business Research
DILLMAN · Mail and Telephone Surveys: The Total Design Method
GROVES and COUPER · Nonresponse in Household Interview Surveys
GROVES · Survey Errors and Survey Costs
GROVES, BIEMER, LYBERG, MASSEY, NICHOLLS, and WAKSBERG · Telephone Survey Methodology
*HANSEN, HURWITZ, and MADOW · Sample Survey Methods and Theory, Volume 1: Methods and Applications
*HANSEN, HURWITZ, and MADOW · Sample Survey Methods and Theory, Volume II: Theory
KISH · Statistical Design for Research
*KISH · Survey Sampling
KORN and GRAUBARD · Analysis of Health Surveys
LESSLER and KALSBEEK · Nonsampling Error in Surveys
LEVY and LEMESHOW · Sampling of Populations: Methods and Applications, *Third Edition*
LYBERG, BIEMER, COLLINS, de LEEUW, DIPPO, SCHWARZ, TREWIN (editors) · Survey Measurement and Process Quality
SIRKEN, HERRMANN, SCHECHTER, SCHWARZ, TANUR, and TOURANGEAU (editors) · Cognition and Survey Research

*Now available in a lower priced paperback edition in the Wiley Classics Library.